PHYSICAL SYSTEMS MODELLING and MANAGEMENT APPLICATIONS

SUSHIL
Abdulaziz Alsagar Chair Professor
Department of Management Studies
Indian Institute of Technology, Delhi

Publishing Globally

NEW AGE INTERNATIONAL (P) LIMITED, PUBLISHERS
New Delhi • Bangalore • Chennai • Cochin • Guwahati
Hyderabad • Kolkata • Lucknow • Mumbai
Visit us at **www.newagepublishers.com**

Published by New Age International (P) Ltd., Publishers
First Edition: 2014

BRANCHES

- **Bangalore** 37/10, 8th Cross (Near Hanuman Temple), Azad Nagar, Chamarajpet, Bangalore-560 018
  Tel.: (080) 26756823, Telefax: 26756820, **E-mail: bangalore@newagepublishers.com**
- **Chennai** 26, Damodaran Street, T. Nagar, Chennai-600 017
  Tel.: (044) 24353401, Telefax: 24351463, **E-mail: chennai@newagepublishers.com**
- **Cochin** CC-39/1016, Carrier Station Road, Ernakulam South, Cochin-682 016
  Tel.: (0484) 2377004, Telefax: 4051303, **E-mail: cochin@newagepublishers.com**
- **Guwahati** Hemsen Complex, Mohd. Shah Road, Paltan Bazar, Near Starline Hotel, Guwahati-781 008
  Tel.: (0361) 2513881, Telefax: 2543669, **E-mail: guwahati@newagepublishers.com**
- **Hyderabad** 105, 1st Floor, Madhiray Kaveri Tower, 3-2-19, Azam Jahi Road, Nimboliadda, Hyderabad-500 027
  Tel.: (040) 24652456, Telefax: 24652457, **E-mail: hyderabad@newagepublishers.com**
- **Kolkata** RDB Chambers (Formerly Lotus Cinema) 106A, 1st Floor, S.N. Banerjee Road, Kolkata-700 014
  Tel.: (033) 22273773, Telefax: 22275247, **E-mail: kolkata@newagepublishers.com**
- **Lucknow** 16-A, Jopling Road, Lucknow-226 001
  Tel.: (0522) 2209578, 4045297, Telefax: 2204098, **E-mail: lucknow@newagepublishers.com**
- **Mumbai** 142C, Victor House, Ground Floor, N.M. Joshi Marg, Lower Parel, Mumbai-400 013
  Tel.: (022) 24927869, Telefax: 24915415, **E-mail: mumbai@newagepublishers.com**
- **New Delhi** 22, Golden House, Daryaganj, New Delhi-110 002
  Tel.: (011) 23262368, 23262370, Telefax: 43551305, **E-mail: sales@newagepublishers.com**

ISBN: 978-81-224- 3628-0

₹ 599.00

C-13-08-7080

Printed in India at Glorious Printers, Delhi.
Typeset at In House, Delhi.

PUBLISHING FOR ONE WORLD
NEW AGE INTERNATIONAL (P) LIMITED, PUBLISHERS
7/30 A, Daryaganj, New Delhi-110002
Visit us at **www.newagepublishers.com**

PREFACE

Physical systems provide a link with the reality in a tangible manner. The functioning of any practical enterprise cannot take place without managing physical systems of various kinds. The physical systems consist of tangible and quantifiable variables. These variables could be real or conceptual and accordingly the physical systems could be real or conceptual physical systems. In the context of business organizations and socio-economic systems, most of the tangible systems of interest are conceptual physical systems consisting of flow of material, energy, money, man power, orders, and information.

This book provides a theory of physical systems which is a general systems theory to act as a base for modelling a vast variety of physical systems. Though the modelling methodology is theoretically based on a rigid scientific framework, rooted in the concept of a linear graph, in practical implementation it can be adapted to suit the requirement of the system under consideration. This theory, which is termed as physical systems theory (PST) can be suitably amended and integrated with other modelling methodologies to cover a range of physical systems from operational to tactical to strategic.

The book is aimed to provide a comprehensive flow modelling framework to suit a variety of physical systems in managerial context, both at micro and macro levels. Though a large number of applications of 'Physical Systems Theory' are reported in book form in the context of engineering systems, its applications in managerial context are available only in research papers.

The objective of the book is to provide the fundamentals and developments on methodology front with flexibility in modelling by suitable integration with other methodologies, to be illustrated by a variety of applications at operational, tactical and macro levels. This will amply illustrate the potential of the proposed flexible modelling methodology that can be taken up by management researchers, students, and practitioners.

This book is based on methodology, which is illustrated by applications at operational, tactical and macro levels. The audience for such a book would include any professional, interested to address the managerial problems at these levels through modelling and analysis. This would include management researchers carrying out research about operational, tactical and macro-level systems, MBA and MS students taking up projects in modelling, analysis and management support, and practitioners who want to develop a model base for management support systems. All of these may address a variety of problems linked with physical systems, such as manufacturing systems, organizational systems, information systems, and socio-economic systems. The book will also be of interest to methodological researchers who are working on developing novel methodologies for addressing a variety of managerial issues in an integrative manner.

All the scholarly content of the book is original and can be a unique contribution to information science and management literature giving a comprehensive presentation of 'Physical Systems Theory' in managerial context. This also propounds application of 'Flexible Systems Methodology' to suitably integrate a variety of modelling techniques for addressing the managerial problems in a flexible manner. It advocates to adapt the modelling methodology to suit the requirements of the problem situation rather than being purist in application.

The two major contributions of the proposed book are:

(i) It highlights the significance of management of physical systems and use of a flexible modelling methodology for the same.

(ii) It presents a novel flow modelling methodology with applications so that modelling of managerial systems can be addressed in a fresh and comprehensive manner.

The book is organized in six parts covering the basics of methodology, variety of applications, and future perspectives.

Part I outlines the physical view of management systems and the relevance of a theory of physical systems. It justifies the need for flexibility in management of physical systems and discusses about flexibility in modelling methodology at the level of conceptualization, modelling, and analysis.

Part II provides details of modelling methodology in terms of its foundations, philosophy, fundamental postulates, and basic processes in socio-economic physical systems. It outlines the application areas and models of basic processes. The steps of modelling and analysis are provided along with possible simplification and generalization to suit to a wide variety of socio-economic systems.

Part III covers applications in the area of manufacturing management. It provides a flexible systems approach for integrative operations management. It gives applications in the context of multistage manufacturing systems with and without rework. Specific applications for a small system of manufacturing compression springs and a large system of manufacturing turbogenerators are provided. It also provides an illustration for a closed chain production system with storage and gives an optimization model for technology selection in multistage manufacturing systems. Other applications include analysis of Cost and Quality, Inspection Policy, and Integrated Material Handling Systems.

Part IV deals with applications in organizational and tactical management. These are non-traditional applications of PST to the areas such as Integrated Financial Decision-making, Organizational Analysis, Project Management, Management Information Systems, and Decision Support System for Maintenance Management. The methodology used is flexible by integrating a variety of techniques, such as Goal Programming, Fuzzy Set Theory, Analytic Hierarchy Process, and System Dynamics with PST.

Part V provides applications at the macro level socio-economic systems. These are comparatively traditional applications of PST but provided with a flexible approach by integrating with other techniques. A variety of applications are provided, such as Intersectoral Modelling of National Economy, National Budget, Energy Planning, Steel Planning, Transport Systems, and Waste Management Systems.

Part VI looks beyond the applications made in other parts. In order to make the modelling of physical systems more organic and realistic, it brings out the issues of non-linearity, dynamics, uncertainty, and ambiguity. It projects new areas of physical system applications such as value chain analysis, supply chain management, and enterprise modelling. Finally, it provides the application in a gaming and learning environment using a synthetic and creative framework of modelling, which is flexible enough to cater to the diversity of physical systems and associated managerial issues.

The book is an outcome of teaching the course on 'Systems Analysis and Methodology' at Indian Institute of Technology (IIT) Delhi for over a period of nearly 15 years. The applications reported in the book came out of the course assignments, and master and doctoral level projects supervised by the author.

I am extremely thankful to my ex-students and co-authors of a series of research papers on the subject, whose efforts have been adapted and synthesized in this book to give a total picture to the reader. Specifically, I would like to thank Dr Anjana Kak, Prof. Nanua Singh, Mr DK Choudhury, Dr Pankaj, Dr Rajiv Vijay, Mr RG Kulkarni, Prof. PK Jain, Mr B Raghunathan, Mr A Bajaj, Prof. MP Gupta, Prof. N Kumar, Prof. Prem Vrat, Mr M Gupta, Dr Anil Vij, Prof. Abid Haleem, (Late) Prof. Arun Kanda, Dr VK Sibal, Dr RK Jalan, and Dr VK Srivastava among others for their inputs of various kinds.

I would like to thank Mr Raj Kumar and Mrs. Rejani Raghu who have, over the time, contributed by word processing various drafts of the manuscript.

I owe a great deal to my family members with whose patience and cooperation over the years the project has seen the light of the day. Special personal thanks are due to my wife Mukta, and inspiring daughters, Vertika and Arushee.

Sushil

CONTENTS

PART I

OPENING THE DOOR

1 PHYSICAL MANAGEMENT SYSTEMS AND FLEXIBILITY

1.1 INTRODUCTION

Any enterprise is composed of both physical and non-physical (or conceptual) management systems. Even the conceptual systems which have a real physical interface can be treated as physical systems, e.g. organizational systems, information systems, project systems, monetary systems, and so on. Only the systems which are highly qualitative and ill-structured, such as ideas, vision, strategic thinking, culture, political linkages and so on are highly abstract and should not be treated as physical systems. Thus, the systems which are strategic and ill-structured are conceptual and operational and tactical systems with transparent structure are physical in nature. The ultimate delivery of the conceptual vision and ideas manifests through the physical operations systems of finance, manufacturing, distribution and logistics, quality control, and so on.

Thus, the significance of proper management of physical systems in any enterprise can hardly be over emphasized. There is a symbiotic relationship between operational and tactical physical systems and conceptual strategic systems. For an enterprise to be successful, it must have a proper management approach that can integrate the two types of systems effectively. The ideas at policy level systems can be converted into guidelines and structures and a suitable model of the resultant physical system can be developed.

This book is primarily concerned with a modelling methodology, which is flexible to cater to a variety of aspects of the physical management systems with a scientific rigour. The central hub of the proposed flexible modelling methodology is physical system theory rooted into the concept of a linear graph. This is very useful in depicting the input-output structure of the system in a graphical manner. The methodology has been simplified for effective use in managerial applications and illustrated with a variety of applications both at the micro and macro levels.

At the micro level, applications in manufacturing management and other organizational and tactical management issues are presented with generalized models and illustrative examples/case studies.

At the macro level, intersectional modelling of national economy, energy planning, steel planning, transportation systems, and waste management systems are illustrated.

This chapter presents a preamble to the main text in terms of nature of management, physical view of management systems, relevance of a theory of physical systems for management, concept of flexibility in management, flexible systems framework for application of modelling methodology of physical systems, and an overview of the book.

1.2 NATURE OF MANAGEMENT

Management is an elusive and abstract concept, which is increasingly being realized to be complex, ambiguous, and paradoxical. From a traditional normative view point, management is understood as the process of optimal utilization of resources for effective achievement of the intended goals. Such traditional management is governed by the functions of planning, organizing, and controlling.

On the other hand, management is also seen as a learning process of understanding the socio-political reality and knowing the mechanistic-organic systems in operation. It provides a renewal capability to continuously innovate and change to adapt to the changing requirements of the customer, competition and business environment.

Ultimately, the management is a synthesis and dynamic interplay of "optimization" and "learning" paradigms treating them to be complementary rather than competitive to each other. This requires a flexible systems approach to management.

Though an enterprise can have an overall management approach, the requirements at different levels of management, viz. operational, tactical and strategic, are different. The structure of the problem at the operational level is transparent, which becomes ill-defined as we go up in the hierarchy. Thus, we need a flexible methodology to deal with a vast variety of problems ranging from highly structured to semi-structured to almost unstructured problems.

Keeping in view the variety of problem situations and systems to be handled, the techniques of systems modelling and analysis are of three types, viz. descriptive, explanatory, and normative. These techniques attempt to answer the three basic questions in management, i.e. "what", "why", and "how" respectively. The descriptive techniques try to describe the system under consideration and attempt to answer the basic question "what is happening". The explanatory techniques attempt to explain the behaviour of the system exploring the answer of the fundamental question "why the system behaves that way". Finally, the normative techniques contribute towards design and prescription in terms of "what should be done". For example, in a manufacturing system we want to know what is the system and how is it working. Then, at the second place, we have to inquire why the manufacturing system is having low/high productivity or capacity utilization. Finally, we would like to get a solution in terms of how to improve or redesign it. Thus, we need a flexible modelling methodology to capture the three kinds of researches in the management systems.

In terms of modelling, an enterprise can be treated as a mass-energy flow system linked with information flow for decision-making. Thus, a part of the enterprise dealing with physical or tangible flows is governed by the principle of conservation, whereas, the other part having intangible and information flows is non-conserved. Thus, the enterprise modelling requires a methodology that can define the laws of conservation on the one hand, and is able to model non-conserved flow at the other end. Since the central concern in this book is on the physical management systems, a methodology with principles of conservation would be the central methodology supported by other techniques.

Thus, the management of physical systems in an enterprise requires requisite flexibility in approach to deal with a variety of situations in an effective manner. The next section gives a synoptic account of physical view of management systems in an enterprise.

1.3 PHYSICAL VIEW OF MANAGEMENT SYSTEMS

The world of management is both physical and non-physical in nature. A significant part of management literature deals with the non-physical view of management. In reality, the manager has to take a physical view of management systems and manage the physical systems interactively.

The physical quantities are the ones that can be measured. The three fundamental physical quantities used in engineering systems are length, mass, and time. Based on these, a host of derived quantities are obtained, e.g. area, volume, speed, etc. In the context of management systems, the tangible or physical quantities are quantity of production, manufacturing capacity, inventories, etc. and the systems dealing with physical quantities are treated as physical management systems.

A physical management system could be real physical system such as a material flow or manufacturing systems or a conceptual physical system such as money flow or financial system or a material requirements planning system. Some of the flows in an organization are more abstract such as authority and responsibility, but if a quantitative measure of the same is generated, they can be treated to form a physical organizational system.

In simple terms, a physical management system is a system that deals with physical quantities, has a tangible structure and the flows in the systems are governed by some principle of conservation; for example, the systems dealing with flows of materials, money, energy, manpower, and orders. Such a description of the physical management systems deals with mechanistic metaphor of the organization.

However, if a system is not having a transparent structure, certain assumptions are made and part of the system is treated as a physical system. For example, quality in a manufacturing system may be varying over time and in the absence of complete data, an estimated deterministic value of quality is assumed to model the system. Further, the system may not hold strict conservation in flow of some variables like quality. In that case, quality transfer functions can be estimated and the system can be modelled as a physical system as discussed in chapter five. This kind of extended framework of physical systems deals with the organismic metaphor of the organization in which information flows are also incorporated extensively.

The important characteristics of the physical systems can be summarized as follows:

i. The system has a tangible structure or can be conceptualized into some form of structure, which is transparent in terms of components and interactions.

ii. The system has a flow structure, in which flows and stocks are conserved. If certain flows are not conserved, generalized transfer functions are to be established.

iii. The system deals with tangible variables such as material flow, money flow, etc. If the intangible variables can be quantified, then they can be treated to be composing a physical system.

iv. The system behaviour can be deduced from the model of the system structure to a great extent.

Though, a variety of problems in the management systems do not strictly adhere to the theoretical definition of physical systems, with acceptable assumptions, majority of operational and tactical systems can be modelled as physical systems with reasonable degree of accuracy. For example, the systems dealing with manufacturing, finance, distribution, manpower planning, quality control, projects, inventory, costing, budgeting, technology, resource and waste management, organizational structure, monitoring and control, transportation and handling, and energy flow can be treated as physical systems for all practical purposes among many others.

However, despite the fact that physical systems are conserved, they need to be modified, upgraded, and improved continuously to adapt to the changing situation. Thus, the physicalness of the system should not be misinterpreted as rigidness. A physical system can be flexible whereas a conceptual system can be rigid, e.g. fixed mind sets. In fact, flexibility in the working of organizations came first at the level of physical systems in terms of "flexible manufacturing systems" and then taken up at the level of conceptual systems such as flexi-time and empowerment.

1.4 RELEVANCE OF A THEORY OF PHYSICAL SYSTEMS

Management is dealing with both the real physical and conceptual physical systems to meet its objectives. In this physical world, any action entails the use of physical systems and resources. Nothing can be physically achieved without interfacing with the physical systems. Thus, in order to give shape to its ideas, management requires physical systems and resources. At the operational level, the contact of management with the physical systems is strongest which reduces as one goes up in the hierarchy.

In order to manage physical systems, management is supposed to understand their functioning and then plan, organize and control their utilization. At the same time, in order to perform the functions of planning, organizing, implementing and controlling, management requires to design some new physical real as well as physical conceptual systems.

A sound theoretical basis is needed to manage a host of physical systems effectively and efficiently. A theory of physical systems provides a sound basis to analyze, design and effectively utilize physical systems interfacing with management.

Physical systems, unlike human and social systems, by their very nature demand more precision in their handling. Thus, it requires a scientific basis to manage physical systems, which can be generated by using a physical system theory for their analysis and design. Managerial decision-making at the operational and tactical levels needs a rational and scientific support, which can be effectively provided by physical system theory as discussed in this book.

Management of physical systems with reference to their analysis, planning, design, operation and control requires interaction and communication at various levels. A theory of physical systems provides a language that can facilitate such a communication at various levels of management, with reference to policies, operational plans, schedules, budgets, costing, authority and responsibility, monitoring and control information, etc. This provides the necessary framework to practice the science of management, which can be amalgamated with the art of management by the flexibility of managers. Physical system theory provides a rational and scientific basis to the manager (actor) for developing the appropriate 'process' of managing the physical systems in a creative and interactive manner.

The role of such a rational and technical support becomes far more eminent at the operational and tactical levels, where the contact with the physical systems is direct. The managers at these levels require higher technical skills, which can be effectively created by using a systematic modelling and analysis methodology. An interactive and participative use of physical system theory in managing physical systems as well as a learning exercise will help generating the requisite technical skills in the manager, which are needed to cope with the emerging dynamics of the environment.

A theory of physical systems can help in management of a wide variety of functions in the areas of manufacturing management, technology management, financial management, organization management, and sectoral management. The specific functions, for which application of physical system theory is presented in this book, include costing, production planning, materials requirement planning, technology selection, materials handling, financial decision-making, budgeting, project planning, organizational analysis, intersectoral analysis, resource and waste flow, energy planning, productivity analysis, and analysis and design of systems.

Since physical systems in modern day turbulent environment are subject to continuous pressures from all sides and are expected to be adaptive, responsive and agile, any theory to deal with physical systems should have requisite flexibility.

1.5 CONCEPT OF FLEXIBILITY IN MANAGEMENT

In the previous sections, we have highlighted the need for flexibility in management of physical systems and the associated modelling methodology. It would be important to first understand the concept of flexibility at a systemic level and then discuss the flexibility in the modelling methodology.

Flexibility is a multi-dimensional concept, it might mean different things to different people in different contexts. It might connote adaptiveness to the changes in the environment, adjustment to situation, agility in action, amiability in relationships, autonomy in functioning, balance in competing opposites, broadening of mind, compromise for betterment, contingency in planning, customize solutions, elasticity under tension, freedom of thought and expression, informal attitude, liberalization from controls, localness in organization, openness in thinking, resilience in implementation, responsiveness to customers' requirements, variability in parameters and specifications, and versatility in solutions and operations. This is only a representative list and many more connotations of flexibility can be identified.

Flexible Systems Management: An Overview

Since the connotations of flexibility are so varied, a cohesive framework of flexibility in management is lacking. Flexible systems management[1] provides a framework of systemic flexibility, which is based on a simple concept of a paradox between the thesis and antithesis, i.e. above-below, light-dark, day-night, and so on. It emphasizes that moving from the thesis (rigidity) to antithesis in management will not necessarily generate flexibility. For example, moving from centralization to decentralization in organizations is not creating flexibility in the true sense. Rather, if we decide to attach the modern management with the antithesis, on the one hand, we again become rigid leaving the whole range of options available from the thesis to antithesis, and on the other hand, lose the very identity of the system under consideration. Flexibility does not mean shifting to extremes, rather dynamically balancing them. Flexible systems management is evolving in the form of a third wave of change, i.e. the synthesis of the thesis and antithesis. This would mean carrying the seemingly opposite concepts in a dynamic interplay such as continuity and change, stability and dynamism, and uniformity and diversity in the working of organizations.

The concept of *"systemic flexibility"*[2] dwells on three central issues, viz. options, change, and freedom of choice. In order to create flexibility in any system, firstly, we must have a range of options on the continuum from the thesis to antithesis; secondly, we must have change mechanisms to be able to shift from one option to the other or able to integrate a few of the options; and thirdly, we must have more freedom of choice with the actors to be able to carry the change by conscious choice rather than changing in a random manner.

In order to effectively implement flexible systems management in organizations, a simple yet powerful *"SAP-LAP framework"*[3] (Figure 1.1) is generated, which is well-tested in learning and empirical settings. The three basic components that define the dynamic interplay of the reality in flexible systems management paradigm are 'situation', 'actor' and 'process' (SAP). In any managerial context, a situation is to be managed, by one or a group of actors, who use some processes, which recreate the situation. In this interplay, the freedom of choice lies with the actor, which creates the flexible and adaptive processes.

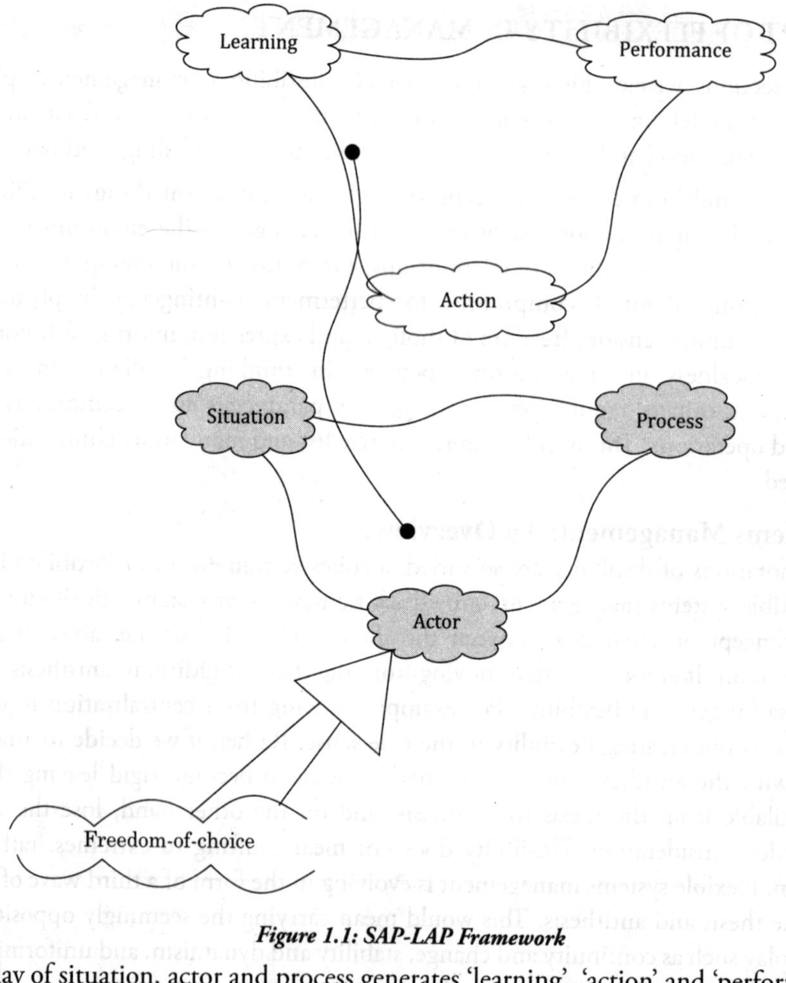

Figure 1.1: SAP-LAP Framework

The interplay of situation, actor and process generates 'learning', 'action' and 'performance' (LAP). Interplay of action and performance is much highlighted in the traditional management, whereas interplay of learning and performance is more emphasized in the modern management. The holistic flexible management takes into consideration the interplay of all the three, i.e. learning, action and performance. We learn about the situation, actor and process, take actions on all the three, get a feedback of performance at all the three levels, and the cycle in repeated.

The SAP-LAP framework can be effectively used for case analysis, and can be applied to create SAP-LAP models for managerial inquiry, problem solving, strategy formulation, and decision-making.

1.6 FLEXIBILITY IN MODELLING METHODOLOGY

Flexible System Methodology

Flexible system methodology[4] is based on the idea that the design of study should match with the problem attributes. The approaches that use one technique to solve all types of problems can be termed as 'isolationist'. The 'opportunistic' approaches are those, in which a particular methodology is used to solve a given problem. In both the approaches, the assumption is that one methodology is sufficient to solve a given problem. No innovation, combination, merger or superimposition is attempted to suit the problem attributes. Flexible system methodology is based on the principle which takes into consideration the problem attributes and view points of actors, and then makes use of one or more techniques, thereby using the integrated models. The steps involved in research study-design using flexible system methodology are, conceptualization, fuzzy clustering, matching problem attributes, selection of techniques, integration and innovation, implementation, and dynamic shift.

Problem Conceptualization

The problem situation is to be conceptualized in terms of its attributes, such as the nature of system/systems, people involved, structure of the problem, nature of the outcome desired, level in the organization, clarity, uncertainty, data availability, functional area, situation specific characteristics, etc.

Fuzzy Clustering

In terms of the nature of systems, people involved and the attributes of the problem situation, the research methods need to be clustered in a fuzzy manner, that is, a problem situation as well as a technique will have a membership function in different possible clusters. Different alternative clustering approaches can be used. It can be done in a creative manner by the research team or by using an expert system.

Matching Problem Attributes

The characteristics of research methods/techniques which can handle the problem attributes such as structure, clarity, uncertainty, etc. can be matched. The idea is that the problem characteristics should not be changed to suit to a research method or technique, but a group of techniques/methods can be used to handle the problem attributes for developing an effective solution.

Selection

Depending upon the nature of attributes, either one or a set of methods or techniques can be selected which are best suited for a specific problem situation. The selected methods/techniques can be further verified for their applicability in context to the given problem. The selected techniques or methods are then used either in succession or in combination to take care of different components of the problem situation.

Integration and Innovation

Some of the possible schemes for integration of techniques are as follows:

- integration by results,
- using the techniques in succession,
- using different techniques for different components in the problem situation,
- using a both way integration of techniques, and
- using an amalgamation of techniques leading to a new technique.

Implementation

Once a suitable integration approach is designed, an implementation scheme has to be worked out as to which part of the problem has to be handled by which technique(s). Implementation plan will determine the effectiveness of the research design.

Dynamic Shift

As we decide to handle the problem and devise a scheme of integrated techniques, there may take place a dynamic shift in the problem attributes. The problem attributes may change dimensions, become favourable or unfavourable, or may change the format itself. The approach of the evolved methodology should be flexible enough to take care of dynamic shift on a continuum from more unfavourable to most favourable.

The above mentioned steps are only broad guidelines to design an appropriate approach to a research problem. The process of selection and integration of techniques is to be adopted flexibly, keeping the practical situation in mind.

Physical System Theory in Flexible Systems Framework

By applying flexible system methodology, an appropriate mix of the modelling methods can be identified to design the study of physical system as discussed in chapter three. In this study design, physical system theory (PST) can provide a basic definition of the structure, which can either be optimized or simulated.

Keeping in view the variety of problems and their diverse requirements, PST should be applied in a flexible manner. It is proposed to have flexibility at the levels of conceptualization of the problem, modelling process, and analysis.

Flexibility in Conceptualization

Conceptualization of the problem is the first attempt to understand the relevant physical system. The development of the model would depend to a great extent, on the manner in which the problem is conceptualized. There can be flexibility in conceptualization on various planes such as aggregation level, system, view, structure, and boundary as shown in Figure 1.2.

i. **Aggregation level:** The problem can be treated with few components and variables at an aggregate level or more details can be dealt with at disaggregate level depending upon the purpose of modelling and analysis. For example, the national economy can be conceptualized at an aggregate level to be composed of three sectors, viz. primary, secondary and tertiary or it can be conceptualized to be composed of major aggregated sectors such as agriculture, industry, power, steel, transport, and so on. The economy can also be conceptualized at a disaggregated level in terms of a large number of sectors (80 to 100) based on the input-output table used for national planning.

ii. **System:** The system under consideration can be conceptualized either as a physical real system or as a physical conceptual system. For example, if we are dealing with a material flow system we can conceptualize it as a physical real system of material transportation and handling, storage and manufacturing machines on which material is processed. Alternatively it can be conceptualized as a materials requirement planning (MRP) system, which is a physical conceptual system.

iii. **View:** The system can be viewed in a mechanistic or organic manner. If we take strong assumptions about strict continuity and compatibility, it leads to a mechanistic description of the system. Alternatively, the constraints can be generalized and quality and other soft variables can be incorporated to view it as an organic system. In this book, both the views are presented in different applications. Further, the processes in the system can be viewed to be static or dynamic with time response. The simple transformation process based structure is static, which is illustrated in a number of applications. The inclusion of storage processes in the model makes it dynamic. Depending upon the purpose of analysis, either of the views could be conceptualized about the same system.

iv. **Structure:** The process of conceptualization leads to the definition of the structure, which could be well-structured and transparent or semi-structured or unstructured with ambiguities. However, in the context of PST, modellers are required to deal with ambiguous structures, which to some extent is made possible by extending the framework (chapter five) and use of fuzzy set theory[5].

v. **Boundary:** The boundary of the system is important for the purpose of modelling. The boundary could be conceptualized as either physical or conceptual. For example, in case of a manufacturing system the physical boundary would be limited to the manufacturing plant, whereas conceptually it may get extended to include vendors on the one side and distributors on the other.

Figure 1.2: Flexibility at the Level of Conceptualization

Flexibility in Modelling

There exists flexibility in modelling any physical system. In the first place, the process of modelling should start with the problem conceptualization followed by the selection and design of the modelling methodology. This provides openness about the features of the model that best suit the requirements of the problem under consideration. Rather than fitting the problem situation to the modelling method, the modelling method should be adapted to suit the problem situation and thus a spectral and integrative paradigm of modelling would provide a flexible system methodology of modelling. The flexibility in modelling can be at various planes including number of models, paradigm, methodology, integration, technique, basis, nature, variables, constraints, relationships, database, and responsibility as shown in Figure 1.3.

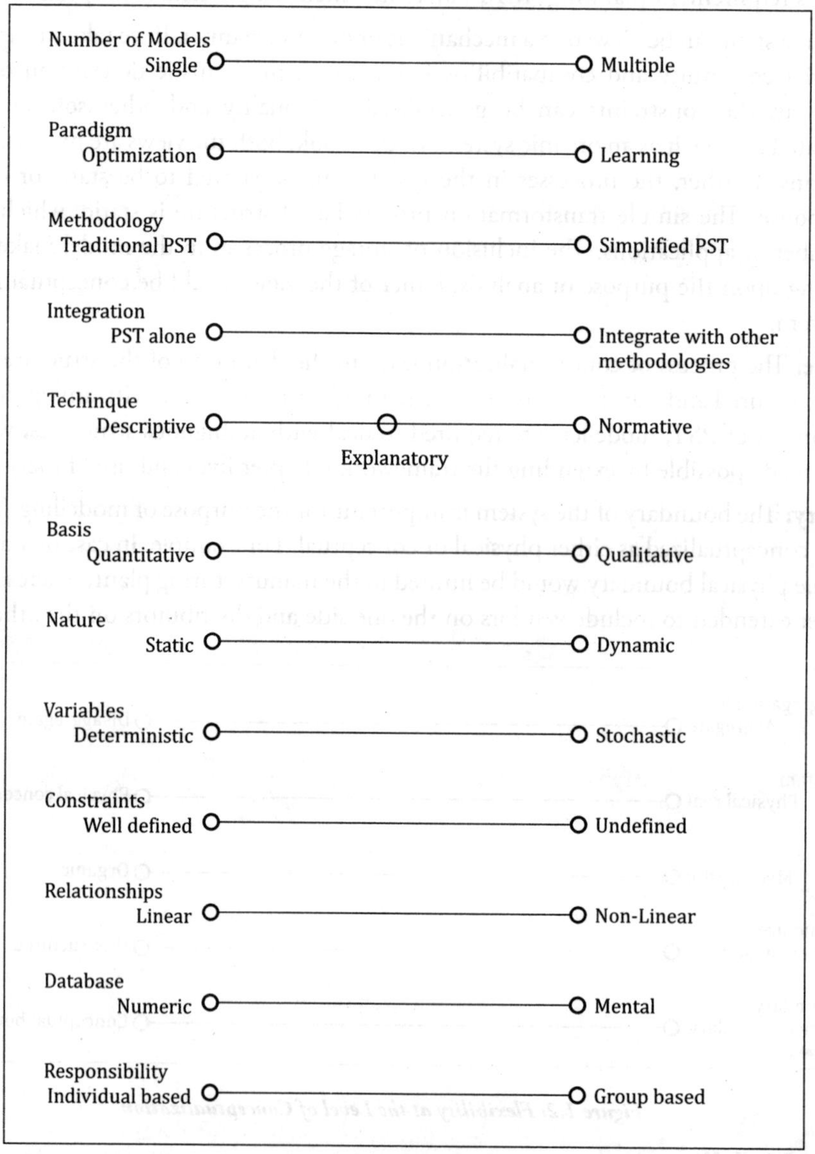

Figure 1.3: Flexibility at the Level of Modelling

i. **Number of Models:** While modelling a system, the number of models could be single or multiple. Either the whole problem is comprehensively modelled in one model or different views of the problem could be obtained by multiple models.

ii. **Paradigm:** The paradigm of modelling could be optimization or system design on the one hand or the modelling could be treated as a learning process to gain insight about the working of the system on the other. The paradigm can also be analytic or synthetic[6], keeping in view whether we want to look into the system and answer "how it is functioning" or we want to look out of the system and relate it with the suprasystem to answer "why the system exists the way it is". Largely, an analytic paradigm is used in the models presented in this book. However, PST can also be applied in a synthetic manner as given in the chapter on "looking beyond".

iii. **Methodology:** Depending upon the nature of system, we may apply either classical PST methodology or the simplified PST methodology (chapter four). Usually, in case of pure hierarchical systems the simplified methodology would be easily applicable. Further, the methodology can also be selected based on the extended framework, if all the variables are not following the constraints of continuity and compatibility (chapter five). In this book, most of the applications are modelled by using the simplified methodology and a few have used the extended framework (e.g. chapters 11, 16 and 17).

iv. **Integration:** PST can be used as the only modelling method if it captures all the required features or can be integrated with other modelling methods. In this book, some applications are made with PST modelling only and in some others various modelling methods such as system dynamics, simulation, heuristics, mathematical programming, fuzzy set theory and AHP are integrated.

v. **Technique:** The techniques of the modelling used would be either descriptive, explanatory or normative as discussed earlier. If the learning paradigm is used, the descriptive and explanatory techniques are useful, whereas for the optimization paradigm, normative techniques would be required.

vi. **Basis:** The basis of modelling could be quantitative, qualitative or mixed in character. The graphical model is qualitative and can be supported by qualitative variables quantified by fuzzy set theory. Most of the PST applications are quantitative in nature.

vii. **Nature:** The nature of the model would be static or dynamic depending upon whether the system is conceptualized to be static or dynamic. The dynamic models presented, in this book, are few (chapters 9, 18 and 24).

viii. **Variables:** The variables/parameters of the model could be either deterministic or stochastic in nature. The stochastic models are usually developed through simulation modelling (e.g. chapters 11 and 12). Similarly, the variables could be precise and crisp or ambiguous and fuzzy. Usually, the abstract variables like quality are treated as fuzzy variables (chapter 17).

ix. **Constraints:** The constraints in the model could be explicit and well defined or implicit or undefined depending upon the problem situation at hand. Normally, a few of the constraints can be well defined, whereas the remaining are left implicit and explored through scenario building.

x. **Relationships:** The relationships of variables in the model could be linear or non-linear in nature. In most of the applications, linear relationships are assumed. However, in the simplified framework when only one equation is to be solved at a time, the non-linearity can be easily handled as is done in system dynamics modelling.

xi. Database: The database to be used for modelling would be either numeric and mathematical or mental and conceptual. For quantitative modelling, numeric database is used, whereas for qualitative modelling conceptual database will be useful.

xii. Responsibility: The responsibility of modelling may be either on individual modeller or on the group. In a unitary mode, individual basis is possible, whereas in case of plurality of opinion group based modelling is to be preferred.

Flexibility in Analysis

Flexibility in analysis lies in terms of options for analyzing the physical systems models and change from one option of analysis to another, as per the requirements. The flexibility in analysis can be on many fronts, such as purpose, mode, validation, software, and responsibility, as shown in Figure 1.4.

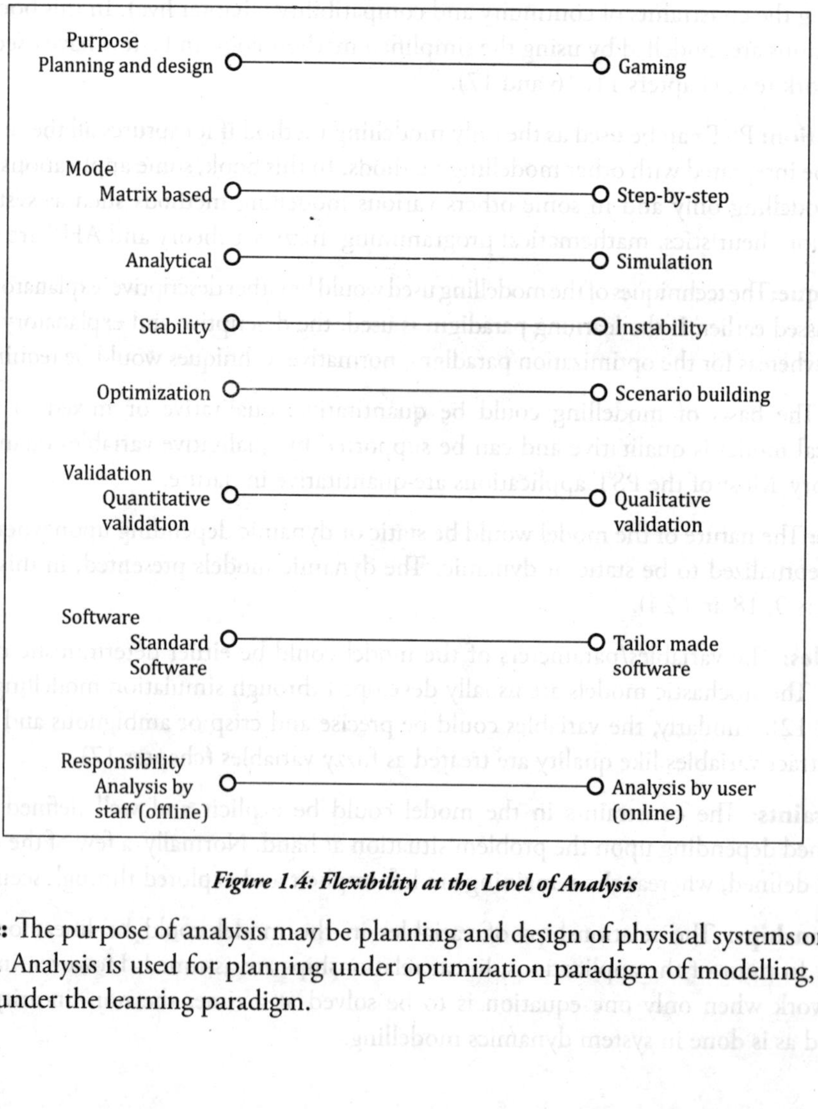

Figure 1.4: Flexibility at the Level of Analysis

i. Purpose: The purpose of analysis may be planning and design of physical systems or gaming and learning. Analysis is used for planning under optimization paradigm of modelling, and used for gaming under the learning paradigm.

ii. Mode: The analysis of PST models can be done using matrix based analytical methods or could be done in a step-by-step and simulation framework. The mode of analysis is to test the stability or instability of the model, which can be performed by sensitivity analysis. Finally, the analysis can be done in optimization mode or to build scenarios.

iii. Validation: The validation of models may be carried out quantitatively or qualitatively or both. The qualitative validation can be performed on the structure and purpose, whereas quantitative validation is to be done using sensitivity analysis as discussed in chapter three.

iv. Software: The software to be used for analysis of PST models could be standard such as spreadsheets and system dynamics software (DYNAMO, i-think). Alternatively, tailor made software could be developed for a class of models such as manufacturing systems models, financial decision-making, organizational analysis, and so on as discussed in chapter three.

v. Responsibility: The responsibility of the analysis may be with the staff to be done offline, or with the line managers and users to be done online depending upon the problem context. Usually, analysis for planning is done off-line, whereas analysis for learning is done on-line by the users.

More discussion about flexibility in PST modelling and analysis is given in chapter two.

1.7 AN OVERVIEW OF THE BOOK

The book deals with applications of physical system theory in a flexible modelling methodology for scientific management of physical systems. The book is organized in six parts, which contains, in all, twenty-five chapters. A schematic representation of various parts of the book is shown in Figure 1.5.

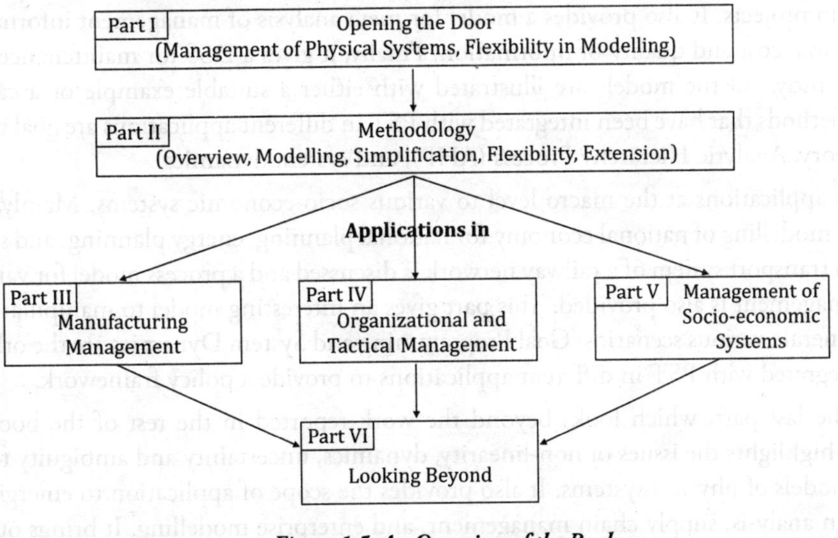

Figure 1.5: An Overview of the Book

Part I is the preamble of the book that opens the door to the sky of methodology of the book. This part has one chapter that outlines the nature of management, physical view of management systems, and the relevance of a theory of physical systems. It brings out the need for flexibility in management of physical systems, defines the systemic flexibility and provides a framework of flexibility in the modelling methodology. Finally, it gives an overview of the structure of the book.

Part II deals with details of the methodology used in the subsequent applications through four chapters. It provides an overview of physical system theory in terms of its foundations, philosophy, fundamental postulates, modelling of management and socio-economic systems, relationships with system dynamics, and flexibility in methodology. Using this methodology, the steps of modelling and analysis have been provided with the proposed simplification in application to managerial and socio-economic systems. Finally, the framework of PST has been extended and generalized to deal with organic variables in the systems such as quality.

Part III provides eight applications in the context of manufacturing management. It provides an integrated framework for operations management of multi-stage manufacturing systems using PST as the central methodology. It provides three case studies: first, in the manufacturing of turbogenerator, second, in manufacturing of compression springs, and third, in a spinning unit with extended framework for quality planning and analysis. Some specific areas include mass-energy flow in a closed chain production system, technology selection, inspection policy, and integrated material handling system, which are illustrated with suitable examples. The other systems techniques integrated with PST in different applications include mathematical programming, simulation, and heuristics.

Part IV covers a variety of applications in organizational and tactical management in five chapters. It provides an integrated model of financial decision-marking to deal with capital structure, capital budgeting and dividend decisions in a single framework. The model of an organizational system is provided to analyze authority-responsibility relationships in a Decision Support Systems (DSS) framework. Another application is in the area of project management to aid in the analysis of flow, cost and quality in projects. It also provides a model for meta analysis of management information systems to analyze flows, cost and quality of information. Finally, it gives a DSS for maintenance management with a case study. All the models are illustrated with either a suitable example or a case study. The modelling methods that have been integrated with PST in different applications are goal programming, fuzzy set theory, Analytic Hierarchy Process (AHP), and system dynamics.

Part V gives applications at the macro level to various socio-economic systems. Mainly, it deals with intersectoral modelling of national economy for national planning, energy planning, and steel planning. A model of a transport system of a railway network is discussed and a process model for various methods of waste management is also provided. This part gives an interesting model to manipulate the national budget to generate various scenarios. Goal Programming and System Dynamics are the other modelling methods integrated with PST in different applications to provide a policy framework.

Part VI is the last part, which looks beyond the work reported in the rest of the book. It has one chapter that highlights the issues of non-linearity, dynamics, uncertainty and ambiguity to be captured in realistic models of physical systems. It also provides the scope of application to emerging areas such as value chain analysis, supply chain management, and enterprise modelling. It brings out the organic view of physical systems and utility of actor oriented models. The issues regarding gaming, learning and virtual reality are highlighted and finally, it is summed up so as to bring out the emergence of a flexible system theory to manage physical systems flexibly.

REFLECTIONS

Physical systems play a significant role in the working of any organization and thus need to be managed carefully. In order to manage physical systems objectively, there is a need to evolve a flexible theory of physical systems. Flexibility in physical systems as well as in modelling methodology is required to adapt to the changing requirements and diversity of problems to be managed.

Notes

1. Sushil (1999) *Flexibility in Management*, Global Institute of Flexible Systems Management, Vikas Publishing House, New Delhi.
2. Sushil (2000) Systemic Flexibility, *Global Journal of Flexible Systems Management*, (1), 77-80; Sharma M.K., Sushil and Jain P.K. (2010) Revisiting Flexibility in Organizations: Exploring its impact on Performance, *Global Journal of Flexible Systems Management*, 11(3), 51-68.
3. Sushil (2001) SAP-LAP Framework, *Global Journal of Flexible Systems Management*, 2(1), 51-55; Sushil (2009) SAP-LAP Linkages – A Generic Interpretive Framework for Analyzing Managerial Contexts, *Global Journal of Flexible Systems Management*, 10(2), 11-20.
4. Sushil (1994) Flexible System Methodology, *Systems Practice*, 7(6).
5. Zadeh L.A. (1965) Fuzzy Sets, *Information and Control*, 8, 338-353.
6. Sushil (1999) *Op.cit.*

PART II

METHODOLOGY

2 FUNDAMENTALS OF METHODOLOGY

2.1 INTRODUCTION

A large class of real world problems can be easily comprehended by taking a physical view of systems associated with them. These systems could be real physical systems such as electrical, mechanical, hydraulic, or other engineering systems; or conceptual physical systems such as managerial or socio-economic systems. The concrete and real physical systems exist in space and time, made of matter and energy, and organized by information. The relationships among things or units in these systems are measurable. However, in case of conceptual physical systems, which are usually the living systems, the relationships such as love, hate, prestige, richness etc. are of the kinds which involve meaning as interpreted by an observer[1]. Thus, in case of the conceptual physical systems or the managerial or socio-economic systems the mechanistic metaphor gives only one viewpoint through which we can conceptualize the real world. Depending upon the problem situation, the system needs to be conceptualized using either the mechanistic, organismic, cultural or political metahpors[2] or a combination of these. Thus, a theory of physical systems can help us developing a mechanistic view of the systems to begin with, which can then be developed further by using other approaches along with the physical system theory.

According to Etizioni[3], a physical system theory is easy to understand. The world is usually seen as a collection of concrete objects in space-time, through relationships that are less obvious. Conceptualization of biological and social aspects of life and behaviour in terms of concrete system in space-time enables us to benefit from a lifetime of experience in thinking that way.

Though a physical view of the real world has its own limitations of not being able to capture many complexities in terms of growth, reproduction, multiple perspectives and conflict, it has its own merits of a tangible assessment and clearer conception. In a variety of systems, such a physical view may be either sufficient or provides a sound basis, which can then be refined and enriched by subtler understanding of the real world. Thus, a theory of physical systems can help a long way in assisting the analysis and design of various systems of interest. In this chapter, fundamentals and developments of the linear graph theoretic physical system theory (PST) are discussed, which is then compared and interfaced with a more organismic methodology, i.e. system dynamics (SD) to generate flexibility in modelling. Using this kind of integration as a base, finally an overview of flexible systems methodology is provided to design a flexible modelling methodology using PST as the central pillar.

2.2 HISTORICAL DEVELOPMENT

The historical development of PST can be traced into three phases, viz. developments prior to the formulation of the theory; formulation of PST principles and concepts; and development towards its generalization to the application to managerial and socio-economic systems.

The principle of conservation of energy offers a sound basis for the development of PST. The theory of electrical networks developed by Kirchhoff in 1847 laid the foundation of PST. The Kirchhoff's nodal law states that the charge can neither be stored or lost at a node converted into two complementary laws, viz. current law, and voltage law. This law is analogous to Newton's law that the sum of all the forces acting on a body at rest is equal to zero[4].

Another set of important developments took place in the area of developing electrical analogs of non-electrical systems, for example, oscillographic solutions of electro-mechanical systems[5], dynamic analogies[6], and engineering analogies[7]. While developing the analogies, Firestone[8] introduced the distinction between two types of physical variables; across variables, and through variables. The across variables are measurable across two spatially different points, such as voltage, velocity, pressure, temperature, etc. The through variables are measurable at a single point such as current, force, fluid flow, heat flow, etc.

On the other hand, developments in the area of linear/control system theory developed a sound basis for the modelling of physical systems.

The third set of developments took place in mathematical graph theory to provide isomorphism between oriented linear graphs and lumped physical systems[9]. On the basis of this development, many more developments took place, which culminated into the development of a new field of physical system engineering through the electro-mechanical system theory[10]. Here the concepts of graph theory were applied to develop free body models and system linear graph was used to write interconnection constraints. This has shifted the emphasis from developing analogies to linear graphs. The work was further developed by Koenig et al.[11] to develop a discipline of analyzing discrete physical systems in the form of PST, which was applied to electrical, mechanical, and hydraulic systems. Contemporarily another development was made by Roe[12] to use graph theoretic techniques for analysis of physical systems.

Further developments in generalization based on the use of graph theoretic concepts resulted into the development of PST as a theory of general systems by Satsangi and Ellis.[13] This has provided general principles of modelling to include conceptual physical systems (socio-economic) along with the real physical system (engineering). Koenig et al.[14] have then applied PST to engineering–economic and ecological systems and discussed the models of basic processes in an ecosystem, which is then used as a basis for its further application in managerial and socio-economic systems. In the context of managerial and socio-economic systems, it is simplified and the paradigm is enlarged to generalize the variables by Sushil[15].

PST is interfaced with many other modelling methodologies to suit the requirements of the problem situation and systems in a flexible systems methodological framework[16]. In this chapter, it has been critically compared with SD and the schemes of its integration with SD are outlined.

2.3 FOUNDATIONS

In order to appreciate PST methodology and modelling of systems using this methodology, it would be desirable to develop understanding about its roots. It initially was developed as a general theory of physical-real systems, which is later on generalized to model and analyze physical-conceptual systems also. Thus, the foundations were more led with the principles of physical-real systems, with inputs from mathematics to facilitate abstraction as a means of making it a general system theory. This fact is amply reflected in the nature of PST and its roots can be traced to linear graph theory, electrical network theory, and linear system theory that facilitate PST modelling process. The mathematical tools such as matrix algebra and differential equations are then used for analyzing and studying the system behaviour. These foundations and their provision to PST are shown in Figure 2.1.

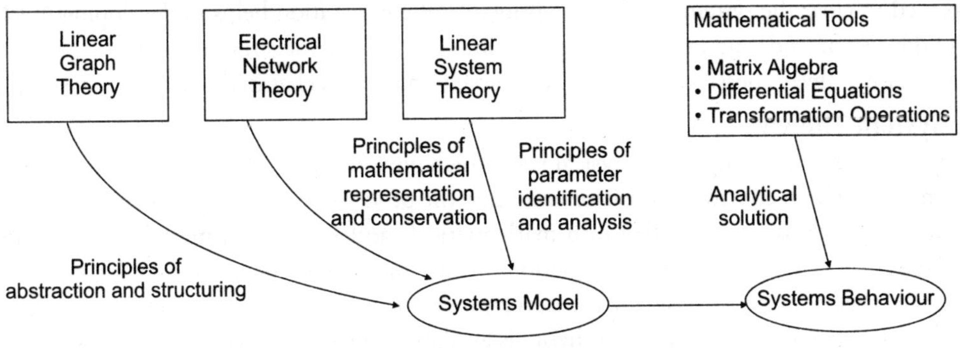

Figure 2.1: Foundations of Physical System Theory

Linear Graph Theory

PST is primarily developed to model and analyze discrete physical systems. In case of the physical real discrete systems, the structure is transparent, components are discrete, and interfaces are finite. Thus, these systems can be conveniently represented in the form of abstract graphs having nodes or vertices and directed edges connecting these nodes. The graphic representation abstracts the real system in terms of its structure, which is the focus of study in system theory. Thus, it can act as a means through which systems can be generalized transcending the disciplinary boundaries. The PST develops on the assumption whether it is an electrical, mechanical, hydraulic, managerial or socio-economic system, that its structure can be represented in the form of a linear graph[17]. This requires a well-defined system structure, which is true for the physical real systems and also for a class of physical-conceptual systems (managerial and socio-economic) such as material flow systems, energy flow systems, money flow systems, etc. As PST is primarily interested in flow modelling, the directed graphs are used to represent the structure.

The beginning point is, thus, qualitative definition of the system/process to be modelled. In case of managerial and socio-economic systems, the graphic representation acts as a means of system conceptualization and communication of system structure. Though theoretically it demands a transparent structure, its practice helps in clearing the ambiguity and making systems more transparent to the users.

The use of graph theoretic consideration in PST not only helps in abstraction of the system, it facilitates system comprehension, and at the same time provides a framework through which the power of graph theoretic tools and techniques can be effectively used for systems modelling and analysis.

Electrical Network Theory

The graphic representation acts as a basis of system conceptualization, but for the purposes of analytical treatment it needs to be converted into a mathematical model. The graphic representation facilitates the development of mathematical equations representing the internal behaviour of the components of the systems as well as their interconnections.

The mathematical model makes use of Electrical Network Theory[18] to portray the mathematical representation and principles of conservation in the system in representing the components and their interconnections. The principle of conservation holds well in most of the physical real systems, but in case of physical conceptual systems it may not be true in all cases and for all parts of the system. Thus, while extending the application of PST to physical conceptual systems, certain adaptations/modifications are to be performed to close the system model. The principle of conservation helps in developing mechanistic models of the system, and thus, delimits the applications of PST in managerial and socio-economic systems where it is neither always possible nor always desirable to have a mechanistic view of the systems.

Linear System Theory

According to Glass[19], "Although the physical laws which govern the dynamic behaviour of ocean waves differ from those which control a transistor amplifier, the processes by which we analyze them are quite similar. The physical laws are distilled into mathematical models......". A mathematical model for a dynamic system is a set of differential equations which predicts its behaviour under typical inputs. Linear system theory provides basis for the development of mathematically tractable models, whereas in case of nonlinear systems little general theory is available. The state-space approach has come to play a central role in the theories of optimal control of constrained as well as unconstrained systems[20]. This provides a strong basis for mathematical analysis.

The mathematical model so developed can be manipulated analytically to study the system behaviour by using mathematical tools such as matrix algebra, and differential equations. The use of graph theory in PST facilitates the model building process as well as it unbolts the power of these analytical mathematical tools for analysis purposes. Thus, it can be seen that PST is developed on strong theoretical background and uses powerful tools and techniques which increase its utility.

2.4 PHILOSOPHY

PST is clearly governed by the underlying systems philosophy that the system behaviour is dependent upon the structure of the components as well as their interrelationships. There are two basic philosophies that help in developing such structures viz.,

- **i.** Causal philosophy; and
- **ii.** Chance philosophy.

According to the causal philosophy, there is always a cause behind an effect seen. The Indian philosophy is essentially causal, which is deterministic in nature. The philosophy of '*karma*' is causal in nature, according to which the fruits or effects obtained are an outcome of the '*karma*' caused by an individual. Thus, every happening can be traced back in the causal chain and, in principle, some cause can be assigned for the happening of any event in the world.

On the other hand, the chance philosophy treats the events to be as chance happenings. The outcome of an event is not certain or deterministic rather probabilistic. Same action might result into many possible happenings according to some probability distribution. Thus, any event in real life can be assessed by the probability of its happening.

PST is based on a philosophy which is essentially causal in nature in an implicit sense. The PST models, thus, in a way represent the causal relationships in the system. The fundamental premises of PST are discussed as follows:

Fundamental Premise I

"Systems behaviour is primarily determined by the structure of its components and their interrelationships."

This premise directs to view the structure of the system in terms of its components and interactions to study the system behaviour. A corollary of this dictates that the system behaviour can be controlled by intervening with the system structure. That is, a system can be designed to give a desired behaviour and this can be achieved by changing the components or interactions, or both.

The other two premises state about the underlying philosophy in developing the component structure, and interactions respectively.

Fundamental Premise II

"In a discrete system (having discrete components united at a finite number of interfaces) a component/sub-system can be modelled in isolation to characterize its behaviour as an independent entity irrespective of its interactions with other components and the environment."

This means that the components are independent so far as their internal behaviour is concerned. The components can be taken out either physically or conceptually to develop their models representing their internal behaviour and then they can be refitted. The component models can be aggregated to get the model of the overall system. This is essentially a mechanistic viewpoint of the system and is a strong theoretical assumption so far as the managerial and socio-economic systems are concerned. However, in practice this assumption is relaxed to a great extent in PST modelling of these systems, as usually the components in the managerial and social systems are defined uniquely by their placement in the system, and the number of their interfaces depend on their interrelationship with other components. This is unlike the physical real components such as mass, spring, resistor, capacitor, triode, etc. which have well defined terminals irrespective of the system and their interrelationship in a particular system. Whereas, the number of terminals for a transformation process depends on its linkages with other processes.

Fundamental Premise III

"The components of a system interact in a manner such that the principle of conservation holds good within the system for all of its variables."

This takes a view of the system which has strict continuity and compatibility in its relationships. Again this conforms to a mechanistic view point and though true for mechanistic systems it may not always hold good for all managerial and socio-economic systems. The principle of conservation can be better assumed in these systems dealing with more tangible variables, e.g. material flow systems, energy flow systems, and money flow systems. The information flow systems are essentially non-conserved in nature.

In view of the above, it can be said that philosophically PST conforms of the "mechanistic" metaphor of the systems and it can be used to analyze even the managerial and social systems from a mechanistic view point, which is also desired in a wide variety of situations.

Further, as it provides a mechanistic representation of the systems, which can be used as good basis for "optimization" and system design, PST is philosophically inclined towards "hard systems thinking".[21]

2.5 FUNDAMENTAL POSTULATES

As discussed in the previous section, PST is based on the hard and analytic systems philosophy. It follows the basic analysis-synthesis cycle of systems approach, i.e. dividing whole into parts and resynthesizing back into the whole to infer about the whole more rationally, systematically, and objectively. The three fundamental postulates of PST: (i) component postulate, (ii) system postulate, and (iii) interconnection postulate, conform to this analysis-synthesis cycle as shown in Figure 2.2.

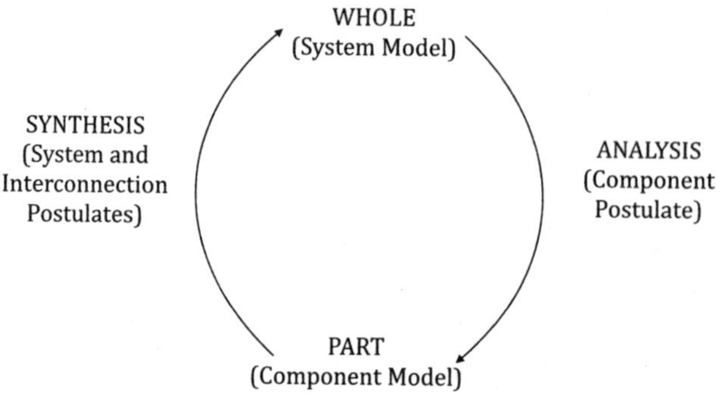

Figure 2.2: Analysis-Synthesis Cycle of Physical System Theory

PST is based on the notion of a linear graph. The first notion of the theory is that a single oriented line segment represents the measurement of two complementary variables (an "across" variable, and a "through" variable). The "across" variable normally exhibits different values at the terminals, whereas, the "through" variable exhibits the same value at both the terminals. The fundamental postulates of PST are as follows:

Component Postulate

"The pertinent performance and behavioural characteristics of every n-terminal component in an identified system structure are completely and uniquely defined by a component terminal graph of (n–1) edges and a set of (n–1) equations relating (n–1) "across" and (n–1) "through" variables."

In order to develop PST model of any system, its components are first defined and the component models are developed for all components separately. The component models are composed of component terminal graphs or free body diagrams, and component terminal equations. The component terminal graphs are dependent upon the number of terminals (interfaces) of the component. The component terminal graph ideally has a tree structure, i.e. it interconnects all the terminals without forming a loop. Thus, n terminals of a component can be connected by a component terminal graph of $(n-1)$ edges. Every edge is associated with an "across" variable and a "through" variable, and thus, there are $2\ (n-1)$ terminal variables for a component having n terminals.

In general, for actually establishing the component model after selecting a terminal graph, it is required to select one set S_1 of $(n-1)$ terminal variables as independent variable functions of time, and the remaining set S_2 of $(n-1)$ terminal variables as dependent variable functions of time. The only requirement on sets S_1 and S_2 is that each contains $(n-1)$ terminal variables. The component model consists of a set of $(n-1)$ relations or more generally, a mapping showing the variable functions of time in S_2 as a function of the variable function of time in set S_1.[22]

System Postulate

"On uniting the component terminal graphs in a one-to-one correspondence with actual interconnections or interrelations of the components in the system, the system graph is obtained."

The system graph is characterized as having 'v' vertices corresponding to interfaces between the identified components in the system, and 'e' edges corresponding to the total number of edges in the component terminal graphs and dummy edges used to close the graph (if any). Thus, the total number of variables associated with the system graph is "$2e$". Usually, some variables are exogenously specified as stimulus to drive the system and are called as drivers. Let n_x be the number of specified across variables or across drivers, and n_y be the number of specified through variables or through drivers, then the total number of unknown variables are $(2e–n_x–n_y)$, which required $(2e–n_x–n_y)$ number of equations to be solved.

Out of these, already $(e–n_x–n_y)$ component equations are developed to specify the performance characteristics of the components; the remaining e linear homogeneous algebraic equations are obtained as structural constraints by the interconnection pattern of the components. The total structure of the system model is shown in Figure 2.3.

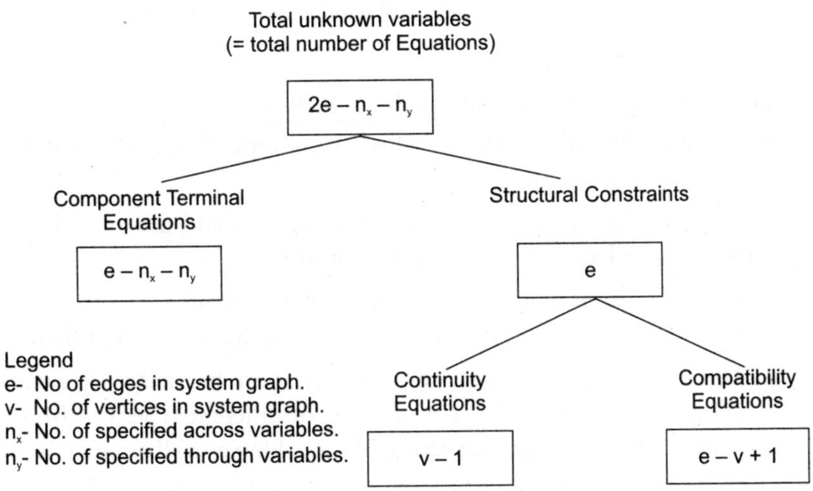

Figure 2.3: Structure of a Physical System Theory Model

Interconnection Postulates

The interconnection postulates synthesize the different components into the system framework on the basis of the ability of the system graph to permit description of the interconnection pattern of the components. As there are two categories of variables, the principle of conservation operating on each one leads to a separate law or sub-postulate.

Cutset Sub-postulate or Continuity Law

"The law states that the flow is conserved in the system, i.e. the algebraic sum of the flow rates into vertices or the algebraic sum of the through variables implied by the oriented edges of any cutset in the system graph is zero."

This law is based and is the generalization of the Kirchhoff's Current Law and also referred to as GKCL. The total number of continuity or fundamental cutset equations are $(v$-$1)$, which are determined by the number of branches in the tree identified on the system graph with "v" vertices. This law defines strict continuity of flow through the system.

Circuit Sub-postulate or Compatibility Law

"The 'across' variable evaluated of whatever path is followed between any two vertices will be the same, i.e. the algebraic sum of the across variables implied by the oriented edges of any circuit in the system graph is zero."

The law is based on the principle of conservation regarding the across variables, is the generalization of the Kirchhoff's Voltage Law, and is also referred to as GKVL. The total number of compatibility or fundamental circuit equations are $(e-v+1)$ which are determined by the number of chords in the co-tree (complementary to tree) in the system graph with "v" vertices and "e" edges. This law defines the compatibility of across variables within the system.

2.6 STRENGTHS AND LIMITATIONS

In order to appreciate the applicability of PST to model real life systems, it is desirable to examine critically from the viewpoint of its strengths and limitations; some major ones are outlined as follows:

Strengths

The significant strengths of PST modelling approach are as follows:

- **i.** It has a graphic input interface, which helps in conceptualizing the system as well as in communicating the system structure.
- **ii.** It uses powerful tools of linear graph theory, matrix algebra, and differential equations so as to make the modelling as well as analysis an efficient process.
- **iii.** There are well-defined steps of modelling and the model structure is transparent. The modelling process is having a high degree of repeatability and is, thus, having a scientific rigour.
- **iv.** Since it is based on the principles of linear graph theory, it can be used as a general system theory; the physical (real as well as conceptual) components of different types can be abstracted in the form of a graph, which facilitates generalization of modelling as well as analysis process.
- **v.** It inbuilds "across" as well as "through" variables in a single model, and thus, provides a more comprehensive framework than other flow modelling approaches, which primarily consider the through or flow variables only.
- **vi.** The dynamic model in the framework of state-space modelling is well established in linear system theory and gives a lot of insight about the dynamic behaviour of the system. The strong points of control system theory are inbuilt in such a model and powerful solution procedures can be used.
- **vii.** It provides an analytical solution to the system models as well as simulation or numeric analysis.
- **viii.** The coefficients are defined for each component separately, and thus, can be easily measured as the attention is to be focused only on one component at a time. It also helps in channelizing the modelling and validation process.
- **ix.** These models can be integrated with other mathematical models and can be used at different levels of planning.

Limitations

Some major limitations of PST modelling approach are as follows:

i. It takes a mechanistic view of the system, which is good for engineering systems, but has a limited utility for a large class of managerial and social systems, which are more organic and cultural in nature.

ii. The formulation is rigid in terms of the definition of variables and the steps of modelling and analysis, which restricts its application to ill defined problems that are common in managerial and socio-economic systems.

iii. The theory is based on the principle of conservation, and thus, cannot be applied directly in the systems where principle of conservation does not hold good. However, with certain modifications, such as introduction of dummy edges, it can handle non conserved situations also, but certainly it increases the number of variables and system equations.

iv. It does not model information system very explicitly and clearly. The information linkages in terms of across variable interactions are modelled implicitly.

v. Mathematical treatment is rigorous, which restrict the application to softer/ill defined managerial and social systems.

vi. It is difficult to apply to degenerate systems.

vii. The model basically assumes linear relationships, and thus, has a limited utility in managerial and social systems, which have high degree of non-linearity. Though it can also handle non-linear relationships with advanced treatment the process becomes complex, which again delimits applicability.

viii. Basically, the model is deterministic and can handle stochasticity only with advanced treatment, which demands higher degree of expertise as well as more extensive mathematical treatment.

ix. The dynamic model is having a rigid structure and demands a lot of mathematical manipulation. Thus, the applicability to policy modelling and analysis is limited.

x. The state-space model may have rate to rate loops, which in managerial and social systems may exhibit false dynamics.

xi. Non-availability of user-friendly specialized software is a major factor deterring its application to managerial and socio-economic systems. The control systems theory based software is available for analysis, which does not provide modelling support.

xii. For the modelling, the variables in the system have to be categorized into "across" and "through" variables, which at times, are difficult to be defined. Moreover, many variables in the system may not fit in the strict definition of "across" and "through" variables and strong assumptions are to be made to handle those variables.

xiii. Simultaneous equations are to be solved for analyzing the model, which for a larger system becomes complex and, at times, indeterminate.

Many of the above mentioned limitations have been overcome by simplifying the modelling in the context of managerial and socio-economic systems, and by enlarging the paradigm to generalize the variables as discussed in the subsequent chapters (chapters four and five). Further, it can be made more applicable by suitably interfacing it with System Dynamics[23] and other modelling methodologies which is presented in the coming sections. This is creating flexibility in its application and enhancing the utility for modelling a vast category of physical systems.

2.7 MODELLING OF MANAGERIAL AND SOCIO-ECONOMIC SYSTEMS

When it is required to have a mechanistic view of the managerial and socio-economic systems, PST offers a sound modelling framework. As has already been discussed, the application of PST requires the following:

i. The system should be discrete.

ii. Variables are measurable and definable in "across" and "through" categories.

iii. Structure of the system is clear in terms of components and their interfaces, which are finite in number.

iv. Principle of conservation holds good for the system.

The managerial and social systems are discrete with finite number of interfaces. In many cases, particularly, when the flow variables are tangible, e.g. material flow, energy flow, money flow, etc. they can be measured quantitatively. The "across" and "through" variables can be defined and the principle of conservation also remains valid. In such situations, the structure is transparent, e.g. manufacturing systems, financial systems, intersectoral interactions, and the components can be defined at different levels of aggregation depending upon the purpose of the study. If the structure is not well defined, then first of all an exercise should be done to clarify the structure, say by using Interpretive Structural Modelling[24] and then the PST model can be developed, e.g. in the case of Management Information Systems[25].

Once the variables are defined and components/sub-systems for modelling are identified the first task is to develop component models. In case of engineering (physical real) systems, the mathematical models of different basic components, e.g. resistor, capacitor, inductor, mass, spring, damper etc. are well established, whereas in the managerial and socio-economic systems there are no such well defined components with established models representing their internal behaviour. In many cases, the component models are based on empirical analysis or assumed behaviour. However, PST modelling of managerial and socio-economic systems is greatly facilitated by developing generalized models of some basic processes that can be identified in these systems.

Experience in modelling managerial and socio-economic systems using PST has shown that the generalized process of modelling leads to some redundancies in PST models of these systems. Thus, owing to the peculiar features of the systems, simplification of modelling as well as analysis can be attempted.

In many practical situations, all the variables cannot be classified according to the strict definitions of "across" and "through" variables. The paradigm, thus, can be enlarged by generalizing the definition of variables, facilitating the application of PST to comparatively softer systems.

Basic Processes

Koenig et al.[26] proposed ecosystem concept to extend the application of PST to engineering-economic systems; according to which, different components of the ecosystem are conceptualized to perform any one or a combination of the three basic processes: material transformation process; material transportation process; and material storage process. These processes can be generalized for any resource, viz. material, energy, money, manpower, information, etc. in the context of various managerial and socio-economic systems. Though theoretically different components of the system under consideration can be conceptualized to perform any one or a combination of these three basic processes, it is worthwhile to define their variants to suit practical situation demanding specialized treatment. Thus, for example, aggregation, disaggregation, and inspection processes are specialized transformation processes and separate models for these could be developed. Hence, it is proposed to consider the following basic processes to help the modelling of managerial and socio-economic systems. The generalized component terminal equations for these basic processes are given in Appendix 2 A, which can be suitably adapted to model a specific system.

i. ***Transformation Process:*** The transformation process can be defined as a transformation of resources to achieve a well-defined change in their physical, chemical, technological, biological or functional characteristics. A transformation process can have multiple inputs and multiple outputs. To facilitate the modelling, a multi-input-multi-output transformation process can be conceptualized as if there is one main output and the remaining outputs are auxiliaries or waste as shown by the free-body diagram of a typical generalized transformation process in Figure 2.4 (a). The process vertex is treated as datum node which acts as reference. The component terminal equations for such a transformation process with n inputs, m wastes and one main output have one cost (across variable) equation for the main output and '$n+m$' flow equations for the n inputs and m wastes. Some examples of transformation processes in managerial and socio-economic systems are: manufacturing stages, information processing or decision-making stages, managerial position, project activities, energy processing, training of manpower, etc. This process is most widely applied in this book.

ii. ***Transportation Process:*** This can be viewed as a special type of transformation process wherein resource is simply moved from one geographic location to another at a cost. This includes collection, translocation and distribution of various kinds of resources with an origin and a destination as shown in Figure 2.4 (b). The component terminal equations for such a process will be exhibiting the internal characteristics of the transportation process, may be the cost-volume relationship. If it is a multi-commodity flow transportation process, then additional equations for aggregating multiple flows into an equivalent flow, and multiple cost equations will be there. Some common examples of transportation process are: materials handling in a plant, transportation/distribution of goods, passenger transportation, energy transmission, information communication, etc. Specific models of transformation process are given in chapters 13 and 23.

iii. ***Storage Process:*** This can be viewed as a special type of transformation process in which the input and output are identical in form, and the resources are carried over time. The free body diagram of a generalized storage process is shown in Figure 2.4 (c). In this case, there are two accumulation or level equations; one for the basic resource flowing through the system and the other for the associated valued flow may be energy or capital accumulation. For the output edge, a cost (across variable) equation is written, and for the input edge a flow equation is written depending upon the ordering policy. Some examples of storage processes are: storage of material, information storage, cadre in a manpower planning system, etc. Specific models of the storage process are presented in chapters 9 and 18.

iv. ***Aggregation Process:*** This is similar to transformation process, but there is no change in characteristics. The process simply aggregates multiple inputs into a single output. The free body diagram of a generalized aggregation process is shown in Figure 2.4 (d). There is one cost (across variable) equation for the aggregated output and n flow equations for n inputs. Some examples of aggregation process are: aggregation of finance from different sources, aggregation of returns from different projects, etc. Specific models of the aggregation process are presented in chapters 12 and 14.

v. ***Disaggregation Process:*** This is also a special class of transformation process with a single input disaggregated into multiple outputs with/without change in characteristics. The free body diagram of a generalized disaggregation process is shown in Figure 2.4 (e). The component terminal equations will involve cost (across variable) and flow equations and can be defined as per the situation. Some examples of disaggregation processes are: classification of a product into different streams; investment of finances to different projects, etc. A specific model of the disaggregation process is presented in chapter 12.

vi. ***Inspection Process:*** This is a special type of transformation process where the input is classified into two categories, viz. conforming and non-confirming by inspecting the input. In the process, there could be two kinds of inspection errors, viz. declaring conforming parts as non-conforming, or declaring non-conforming parts as conforming. The free body diagram of a generalized inspection process is shown in Figure 2.4 (f). The component terminal equations involve the definition of these inspection errors, costing of the conforming output, and flow of input and waste (non conforming output). Some examples of inspection processes are raw material inspection, stage inspection, finished goods inspection, etc. A specific model of the inspection process is presented in chapter 12.

Standard generalized models of many more specialized processes can be developed through research to facilitate model building for managerial and socio-economic systems. One important process that can be of great importance in business systems could be exchange or transaction process where input and output are different resources.

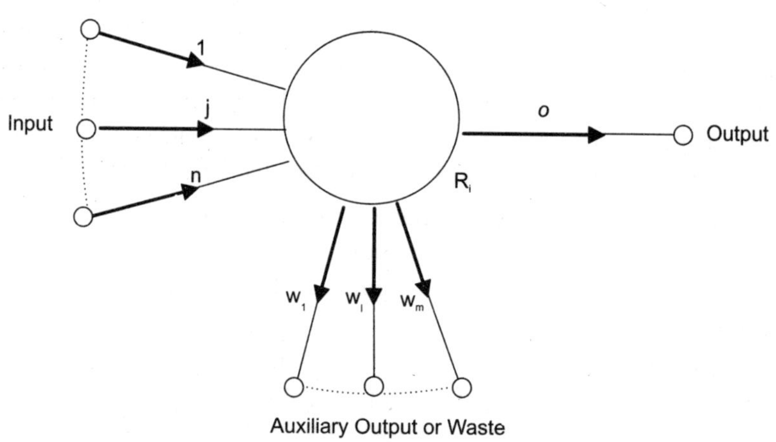

(a) Transformation Process

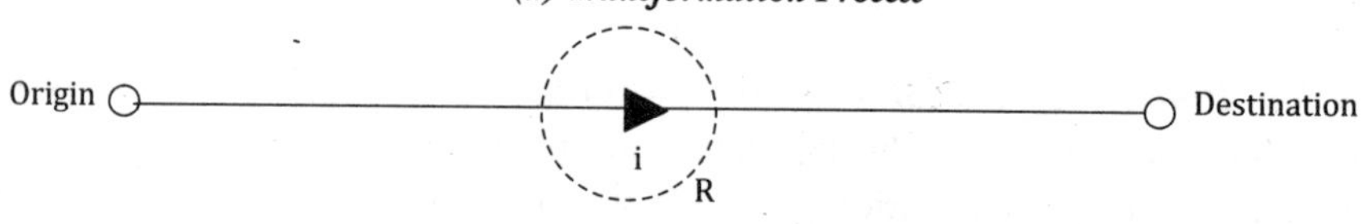

(b) Transportation Process

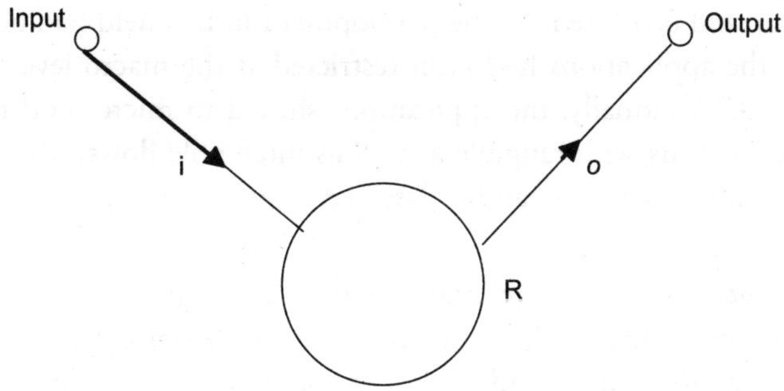

(c) Storage Process

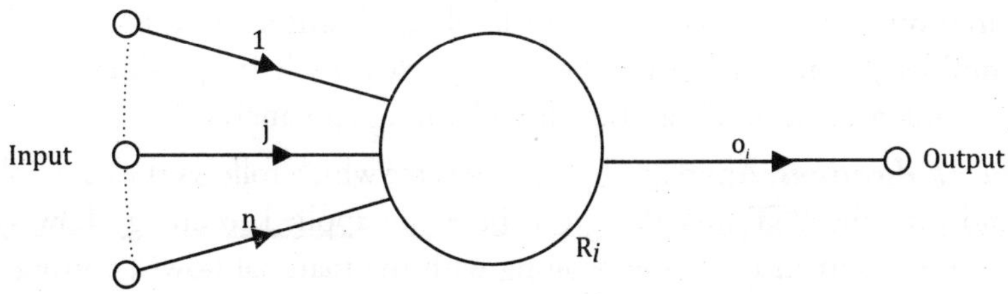

(d) Aggregation Process

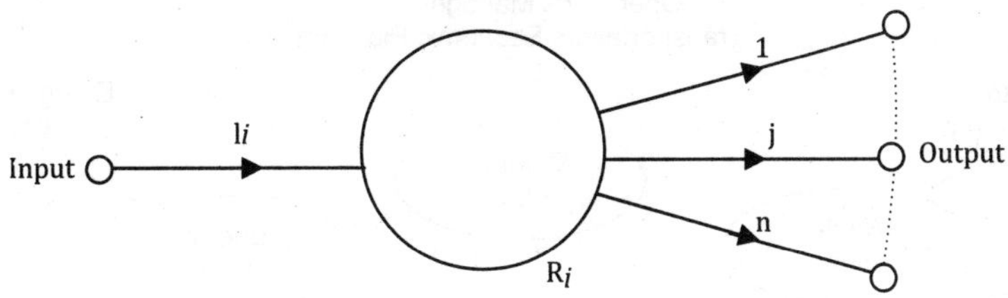

(e) Disaggregation Process

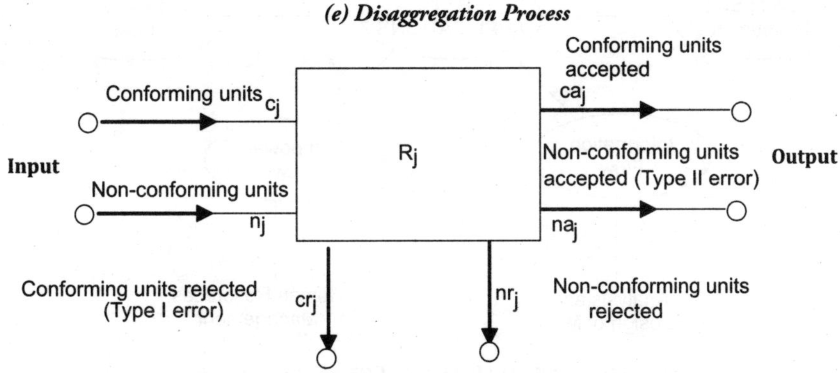

(f) Inspection Process

Figure 2.4: Free Body Diagrams

Applications

PST has been extensively applied in the context of electrical, mechanical, hydraulic and other engineering systems. Though its application to managerial and socio-economic systems was initiated in the context of engineering-economic and ecosystems[27], the development in this field had been comparatively slow and sporadic. Initially, the applications had been restricted at the macro level on the lines similar to proposed by Koenig et al.[28] Gradually, the applications shifted to micro level manufacturing systems and then to a variety of systems with tangible as well as intangible flows. The applications of PST to managerial and socioeconomic systems can be classified as per the resource flow (through variable) as shown in Figure 2.5.

i. ***Material Flow Applications:*** Since material flow is tangible and follows the principal of conservation, the application of PST was first extended from engineering systems to material flow systems both at the macro and micro levels. At the macro level, the applications have successfully been made for modelling of national economic systems[29], water resource systems[30], and transportation systems[31]. At the micro level, applications have been made to productivity analysis[32], multistage manufacturing systems[33], closed chain production systems[34], project systems[35], technology selection[36], and quality planning and analysis[37].

ii. ***Energy Flow Applications:*** Another physical resource which follows the principle of conservation is energy, and thus the PST modelling can be easily applied to energy flow systems. It can be applied to do flow modelling of energy along with the material flow[38], costing of energy, energy planning at the national level[39], as well as the enterprise level and energy audit studies.

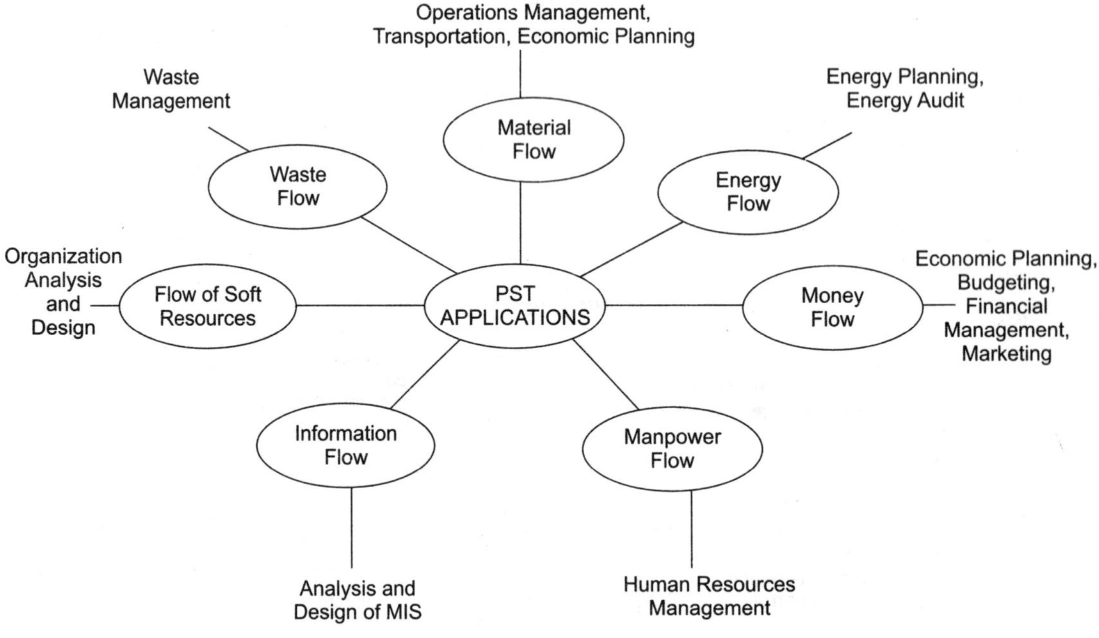

Figure 2.5: Applications of Physical System Theory

iii. ***Money Flow Applications:*** The associated flow with all physical resources is that of money, which is tangible and follows the principle of conservation. At the macro level, the applications in the area of economic planning deal with the flow of monetary resources[40]. It has also been applied in the context of central budget[41] to bring a performance orientation in the budgeting process. Budgeting applications have also been made in the area of marketing management to estimate advertising and sales promotion budget. It has been applied to integrate all the three major financial decisions, viz., financing decision; investment decision; and dividend decision along with mathematical programming[42]. Applications of PST can certainly be made in most of the cash flow problems.

iv. ***Manpower Flow Applications:*** Though manpower related systems are associated with softer and ill defined aspects, there are quite a few areas which are comparatively tangible where the applications of PST have been made, e.g. manpower planning, costing of training systems, project planning, etc. It can be seen that in the conventional rigid framework there is a limited scope of PST in this area. However, by using the enlarged paradigm with generalized variables, and by suitably interfacing it with System Dynamics, and Fuzzy Set Theory many more applications in human resource management can be successfully made.

v. ***Information Flow Applications:*** Flow of information is comparatively less tangible and is not conserved. However, at the architectural level with the application of information technology the information flow has become more tangible. Applications of PST to meta analysis of Management Information System architecture with respect to cost, flow and quality of information to aid various decisions such as budgeting, technology selection, and capacity planning can be made as can be seen in Gupta[43]. Since the flow of information is associated with ambiguities, integration of enlarged PST paradigm with System Dynamics and Fuzzy Sets will help a great deal in the analysis of Information Systems.

vi. ***Flow of Soft Resources:*** Attempts have also been made in the softer areas, such as organizational analysis and design to explore the utility of PST[44]. In the organizational context, the organizational structure is comparatively a more defined area and both the mechanistic and organismic considerations are of interest. The parity of authority and responsibility and a balanced allocation of authority and responsibility are of central concern in organizational design, which can be flexibly modelled by PST in conjunction with Fuzzy Sets, and other qualitative analysis techniques.

vii. ***Waste Flow Applications:*** One significant area which shows the utility of PST in the real life managerial and socio-economic systems, and which could not be effectively handled by other modelling methodologies is of Waste Management. With every resource flow system, there is associated flow of waste and the PST modelling framework can capture this very explicitly. The aspects of waste recycling, rework, and disposal have been modelled by using PST at the macro level[45], municipal waste management[46], and in the manufacturing management[47]. All operations management decisions can incorporate waste considerations in manufacturing by using PST.

2.8 RELATIONSHIPS WITH SYSTEM DYNAMICS

Introduction to System Dynamics

System Dynamics (SD) is yet another methodology which is based on the concept of flow modelling like PST. It originated with the pioneering work of Forrester[48] and developed subsequently during 1960s[49]. Till 1980 it has taken the shape of a discipline and then its further developments in the area of qualitative system dynamics[50], interfaces, use of AI, etc.[51] have been initiated. It is built upon the foundations laid by three background disciplines, viz. traditional management of social systems, feedback theory or cybernetics, and computer simulation. System Dynamics finds its roots essentially in the causal philosophy, assuming that the behaviour of a system is principally governed by its structure, and flow structure is most effective way of viewing an organization. Its main purpose is "gaining deep insight into the working of a system" and thus, governed by the "learning" or soft systems thinking paradigm. It has more applications at the level of policy analysis and is supported by a set of diagramming aids, and specialized user-friendly simulation packages.

A Critical Comparison

PST and SD are critically compared for their similarities and differences for important facets such as, foundations, philosophy, paradigm, metaphor, system structure, variables, modelling, analysis, software, management level, strengths, limitations, and applicability for managerial systems, as shown in Table 2.1. It can be observed through such a critical comparison, that the two modelling methodologies are essentially complementary in nature and more meaningful applications of these methodologies can be made in a "Flexible System Methodology" framework[52] by suitably interfacing the two as per the requirements of the problem situation.

Table 2.1: A Critical Comparison of Physical System Theory and System Dynamics

Basis of Comparison	Physical System Theory	System Dynamics
FOUNDATIONS	Its roots primarily lie in *engineering principles* and disciplines such as graph theory, electrical network theory and linear system theory	It is primarily based on *management principles* and associated disciplines such as traditional management, cybernetics, and computer simulation.
PHILOSOPHY	The philosophy is *implicitly causal*	The philosophy is *explicitly causal*
	Flow structure is determined in terms of components and interrelationships, which primarily deals with the physical flow; the information flow is implicitly considered.	*Flow structure* is determined in terms of with causal feedback chains, which deals with both the physical and information flows explicitly.
	Lays more emphasis on *principle of conservation.*	Lays more emphasis on *feedbacks.*
PARADIGM	Provides a basis for optimization, and thus, oriented towards the *hard systems thinking.*	Essentially focused on learning about systems, and thus, oriented towards *soft systems thinking.*
METAPHOR	Governed by a *mechanistic* view of the system.	Governed by an *organismic* view of the system.

SYSTEM STRUCTURE	The underlying philosophy, paradigm and metaphor requires a well-defined system structure, which makes it more *rigid* in operation.	Problem situations and systems with ill-defined structure can be modelled which makes it more *flexible* in operation.
	Applied basically for the analysis of *discrete* systems.	Applied basically to *continuous* systems.
VARIABLES	Basically two types of variables are considered, i.e. "across", and "through".	The basic variables considered are "levels", and "rates".
	Variables in modelling are treated as *continuous.*	Variables theoretically in modelling are treated as *continuous.*
MODELLING	Based on *flow modelling.*	Based on *flow modelling.*
	Models are *static* as well as *dynamic.*	Models are essentially *dynamic.*
	Modelling is governed by mixed formulation: (i) *Differentiation* in time domain, (ii) Linear time invariant equation in s-domain, and (iii) Integrated differential equations in operation domain.	Governed by the concept of *integration*
	Basically considers *linear* relationships, and *deterministic* in nature.	*Nonlinear* and *deterministic.*
ANALYSIS	The analysis is done *analytically* using powerful tools of matrix algebra and differential equations.	The analysis is done in *simulation* framework.
	Sensitivity analysis is done through differentiation, which is essentially *parametric.*	Sensitivity as well as policy analysis is done through experimentation including both the changes in *parameters,* and *model structure.*
SOFTWARE	Specialized software for managerial applications *not available.* Control system theory based software are used.	Specialized *simulation packages* are developed, which are user friendly with graphic features.
MANAGEMENT LEVEL	In view of the attributes of the methodology, it fits, better to analyze the problems and systems at the *operational management* level (Figure 2.6)	In viewing of basically modelling the policy structures it is best suited at the *strategic management* level (Figure 2.6).

APPLICABILITY	It is applicable very effectively to the *engineering systems,* and the managerial and socio-economic systems with *tangible flows,* e.g. material, energy, money, manpower. Can be adapted to model *intangible flows* as well, e.g. information, responsibility etc.	It is more applicable to *managerial and socio-economic systems.* It takes both the tangible and intangible flows at one place, and can amalgamate multiple flows together.
PRACTICE	*Individual modeller,* demand high degree of expertise, basically applied at quantitative level.	*Group based modelling,* participants need not have high degree of expertise, applied both at qualitative and quantitative levels.
STRENGTHS	• Has scientific rigour. • Graphic interface helps conceptualization • Precision results due to analytical modelling framework • Powerful tools of mathematics can be utilized. • High degree of repeatability in modelling and thus the models can be generated through computer.	• Can accommodate modelling of ill-defined situations. • Diagramming aids are available for conceptualization. • More flexible in application. • The mathematics used is simple and can be easily comprehended by the manager. • It translates mental models and helps in storing as well as communicating managerial experience and knowledge. • Can be used for development of games and decision support systems. • Can capture non-linearity and dynamics.
LIMITATIONS	• The framework is more rigid and can be applied to well-defined situations and systems only. • Demands a high degree of mathematical knowledge and expertise by the user. • Basically linear and deterministic in nature. • Non-linearity and stochasticity can be handled by advanced treatment. • Policy analysis cannot be carried out effectively.	• Lacks scientific rigour. • Models may incorporate bias of managers. • Repeatability is low; each modeller may have a different model of the same problem situation. • Requires more involvement of managers.

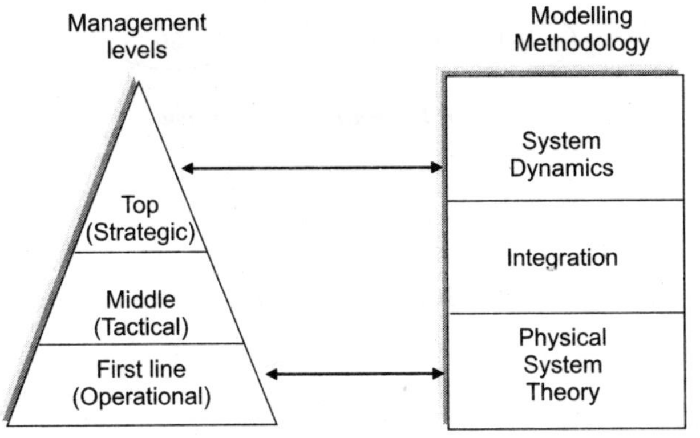

Figure 2.6: Applicability of Physical System Theory and System Dynamics as per Management Levels

Integration or Interfacing

In view of complementarity of PST and SD, as is reflected by the critical comparison in the preceding section, the two modelling methodologies can be suitably integrated or interfaced to cater to the differing requirements of problem situation lying on the whole continuum from well structured to unstructured in a Flexible System Methodology framework. As can be seen in Figure 2.6, PST is covering more the operational level problems, whereas SD is catering more effectively the strategic level problem situations. A suitable integration of the two will help bridging the gap and the problem situations from strategic to tactical management can be analyzed in an integrative and cohesive framework.

Three possible schemes of integration of PST and SD are shown in Figure 2.7, which have been tried in the past and exhibited meaningful application of the integrated models.

i. ***In Succession:*** As shown in Figure 2.7 (a), PST and SD can be effectively integrated in succession, i.e. developing a model in PST framework, and then converting it into a SD simulation model for simulating the behaviour[53]. The PST model in simplified framework can be easily simulated using DYNAMO simulation package.

ii. ***Submerging one into Another:*** As shown in Figure 2.7 (b), the PST model can be made a submodel in a larger SD model so that it can be solved in every time period in a SD simulation framework. In this manner, hierarchical SD models can be developed and the strengths of PST may be used for that part of the problem situation which is well structured. An application of this type of integration is made in the context of energy policy planning in the Indian economy[54]. It has developed an integrated PST-SD model in which a PST subroutine is developed to model the input-output relationships in an intersectoral economy and various policies are modelled in a SD framework.

iii. ***Amalgamation:*** As shown in Figure 2.7 (c), the two techniques can be amalgamated so that within a single model part application of both is made as part requirements leading to a totally new formulation. An attempt in this direction has been made while developing a DSS for unit exchange spares management for Indian Railways[55]. It develops the simplified system graph of various maintenance stages as transformation processes whereas storage processes and their policies are developed in a SD framework owing to the inherent advantages of SD policy analysis. This has resulted into a simulation model of maintenance management in spreadsheet framework in which "what-if" analysis can be effectively done in a DSS environment.

(a) Integration in Succession

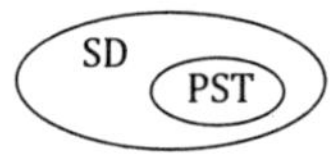

(b) Submerging One into Another

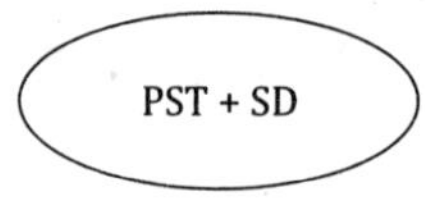

(c) Amalgamation

Figure 2.7: Schemes of Integration of Physical System Theory and System Dynamics

2.9 FLEXIBILITY IN APPLICATION

In view of the requirements of a variety of systems, physical system theory can be applied in a flexible manner on the fronts of systems to be analyzed, paradigm, structure, metaphor, level of application, type of model, variables and relationships, and nature of analysis. The flexibility refers to dynamic interplay on continuum exercising the freedom of choice[56]. In order to exercise the freedom of choice, the first task is to identify the continua. Some important continua on which PST can be flexibly applied are shown in Figure 2.8.

The PST can be applied on one hand to physical real systems, whereas, with suitable adaptation it can be applied, on the other hand, to physical conceptual systems exhibiting its adaptability for variety of system applications.

Though, primarily PST is inclined towards hard systems paradigm and used more as an approach of system design, it can also be used in a soft systems thinking framework, as a learning aid to clarify the system structure in terms of its components and relationships, as well as the system behaviour under varying conditions. When PST is integrated to SD it operates from hard to soft systems thinking.

The basic paradigm of PST applications is analytic, as shown in Figure 2.2. However, it can be applied synthetically in the context of flexible systems management using situation-actor-process (S-A-P) framework[57]. First of all, the context of the system under consideration is clarified and the situation in which it is placed is analyzed. Next, the actors dealing within the system are identified and in the light of the situation and actor the role of the system is defined which is used as a basis for its analysis under the three postulates. Such a synthetic paradigm provides a basis for reengineering the process/system under consideration.

The structure of PST could be traditional or simplified depending upon the type of system under analysis. Further, if the variables are abstract and general, the enlarged framework with quality variables can be applied.

Usually, PST is based on mechanistic metaphor and applied at the operational level of management. However, by interfacing with SD it can be applied with organismic view and the application can be extended to deal with issues of policy at the strategic level.

The PST model can be developed empirically by field data or can be placed in a simulation framework. On the front of variables, it can use specific "across" or "through" variables or can have all the variables as general variables. The variables could be deterministic, stochastic, or combined. Similarly, the variables, as well as relationships can be fuzzified to deal with qualitative or linguistic variables. Traditionally, PST uses linear relationships. However, in case of managerial and socio-economic systems mathematical models can be framed to incorporate non-linearity in a simplified simulation framework.

The PST models can be analyzed analytically or can be simulated in SD or Monte Carlo framework to capture dynamics and stochasticity.

PST, thus, should be applied in a more flexible manner by examining the requirements of the situation, and actors so as to adapt to the problem situation more realistically.

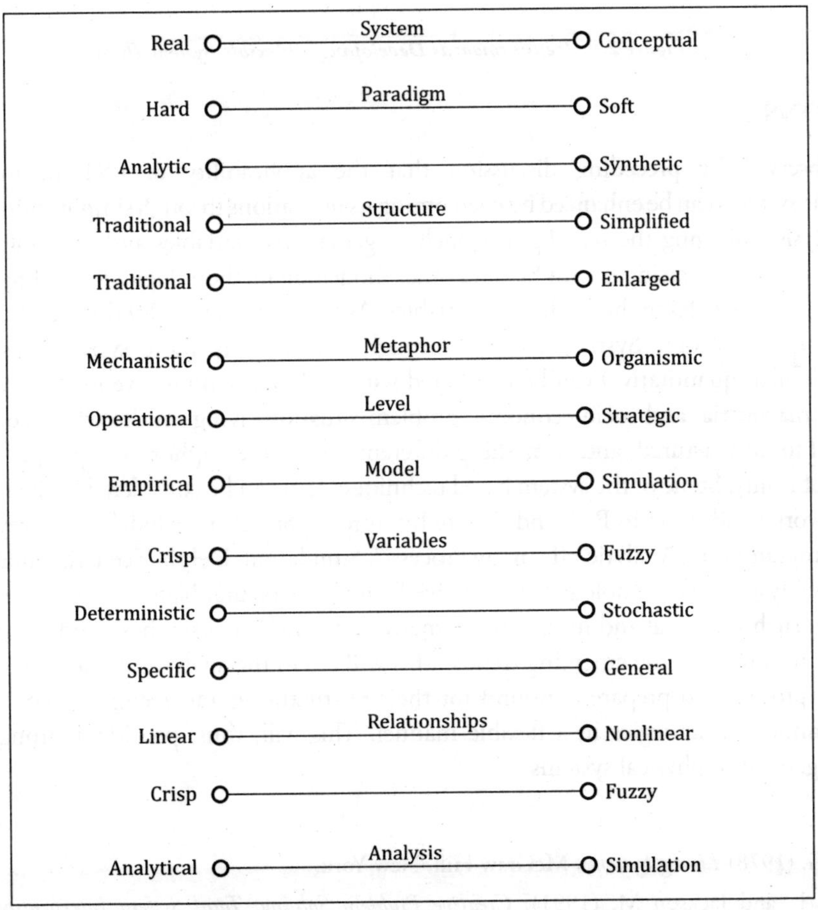

Figure 2.8: Flexibility in Physical System Theory

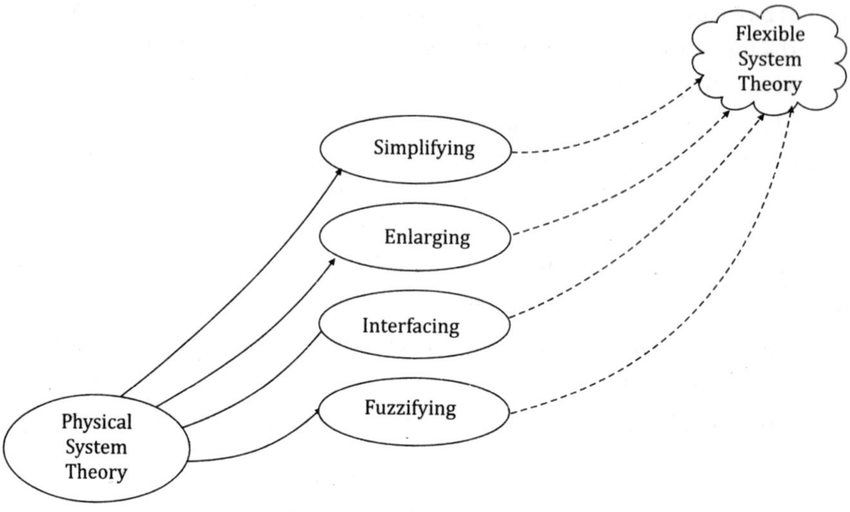

Figure 2.9: Trends towards Developing a Flexible System Theory

REFLECTIONS

It can be observed by preceding discussion that the applicability of PST to managerial and socio-economic systems can be enhanced by overcoming its limitations through simplification of modelling as well as analysis; enlarging the paradigm to include generalized variables and resulting in a less rigid formulation; suitably interfacing it with SD and other modelling methodologies to cater to comparatively softer problems, and fuzzifying the qualitative variables. As shown in Figure 2.9 this trend of development is giving way to a "Flexible System Theory" through which many more systems based techniques (qualitative as well as quantitative) can be integrated with PST in an innovative manner to help develop the models of managerial and socio-economic problem situations lying on the whole continuum from well structured to unstructured, and enmeshing different system methaphors and perspectives as per the requirements of reality. Some of the system based techniques that can be considered in the Flexible System Theory framework in addition to PST and SD are Interpretive Structural Modelling, Fuzzy Set Theory[58], Linear/Goal Programming, Analytic Hierarchy Process[59], Simulation, Delphi/Scenario building, statistical techniques, Soft Systems Methodology[60], Interactive Planning[61], Neural Nets, Expert Systems, and so on. It is hoped that such a spectral and integrative framework will help in a synthesis and convergence of the systems theory in tune with the emerging trends. This will aid in the analysis of manufacturing, business and economic processes to prepare a ground for their restructuring and reengineering, and ultimately regenerating innovative strategies in a flexible manner. This will, thus, provide a strong basis for the effective management of physical systems.

Notes

1. Miller J.G. (1978) *Living Systems,* McGraw Hill, New York.
2. Flood R. L. and Jackson M. (1991), *Creative Problem Solving: Total System Intervention,* John Wiley, New York.

3. Etizioni A. (1968) *The Active Society,* Free Press, New York.
4. Evans F.J. and Dixhoorn J.J. Von (1974), Towards More Physical Structure in Systems Theory, in *Physical Structure in Systems Theory,* Eds. Dixhoorn, J.J. Von and Evans F.J., Academic Press, London.
5. Nickle C.A., (1925) Oscillographic Solution of Electro-Mechanical Systems, *Trans AIEE,* 44, 844-856.
6. Olson H.F. (1943), *Dynamical Analogies,* Von Nostrand, Princeton.
7. Murphy G., Shippy D.J. and Luo H.L. (1963), *Engineering Analysis,* Iowa State University, University Press, USA.
8. Firestone F.A., (1933), A New Analogy Between Mechanical and Electrical Systems, *J. Acoust Soc. A.,* 4, 249-267.
9. Trent H.M. (1955), Isomorphisms Between Oriented Linear Graphs and Lumped Physical Systems, *J. Acoust Soc. A.,* 27, 500-527.
10. Koenig H.E. and Blackwell W.A. (1961) *Electro-Mechanical System Theory,* McGraw Hill Book Co., New York.
11. Koenig H.E., Tokad Y. and Kesavan H.K. (1967) *Analysis of Discrete Physical Systems,* McGraw Hill Book Co., New York.
12. Roe P.H. (1966), *Networks and Systems,* Addison Wesley, New York.
13. Satsangi P.S. and Ellis J.B. (1971a), General Systems from Network Systems: A Philosophy of Modelling, *Int. J. of Systems Science,* 2(1), 1-16.
14. Koenig H.E., Cooper W.E. and Falvey J.M. (1972) Engineering for Ecological, Sociological and Economic Compatibility, *IEEE Trans. Systems, Man and Cybernetics,* 2 (3), 319.
15. Sushil (1992a) Simplification of Physical System Theory in the Modelling of Manufacturing, Organizational and other Socio-economic Systems, *Int. Journal of Systems Science,* 23 (4), 531-543; Sushil (1990a), Modelling of Quality in Physical System Theory: An Extended Framework, *Int. Journal of Systems Science,* 21 (12), 2489-2512.
16. Sushil (1994a), Flexible System Methodology, *Systems Practice,* 7(6), Also Presented as Key Note Address at the Ninth International Congress on Cybernetics and Systems, New Delhi January, 1993.
17. Harary F. Norman R.Z. and Cartwright D. (1965) *Structural Models: An Introduction to the Theory of Directed Graphs,* John Wiley and Sons, New York.
18. Seshu S. and Reed M.B. (1961), *Linear Graphs and Electrical Network,* Addison-Wesley Pub. Co., Reading.
19. Glass C.M. (1976), *Linear Systems–with Application and Discrete Analysis,* West Publishing Co., New York.
20. Zadeh L.A. and Desoer C.A. (1963) *Linear Systems Theory: The State Space Approach,* McGraw Hill Book Co., New York.
21. Checkland P. (1981), *Systems Thinking: Systems Practice,* John Wiley and Sons, New York.
22. Koenig, Tokad and Kesavan (1967), *op cit.*
23. Forrester J.W. (1961), *Industrial Dynamics,* MIT Press, Cambridge Massachusetts.
24. Warfield J.N. (1990), *A Science of Generic Design: Managing Complexity Through Systems Design, Vol. I and II,* Intersystems Publication, USA.
25. Gupta M.P. (1990), *Modelling and Analysis of Management Information System Architecture of National Level,* Ph.D. Thesis, IIT Delhi.

26. Koenig, Cooper and Falvey (1972), *op. cit.*
27. Koenig H.E. and Tammala R.L. (1972), Principles of Ecosystem Design and Management, *IEEE Trans. Systems, Man and Cybernetics,* 2(4), 449-459.
28. Koenig, Cooper and Falvey (1972), *op cit;* Satsangi and Ellis (1971 a), *op. cit.*
29. Bhargava D.K. and Satsangi P.S. (1978), Economic System: A Multidimensional Mode, *Int. J. of System Science,* 9 (6), 603-612; Satsangi P.S. (1969) *A Physical System Theory Modelling Framework Generalized for Large Scale Economic Systems,* Ph.D. Thesis, University of Waterloo Canada; Satsangi P.S. and Ellis J.B. (1971b), A Generalized System Theoretic Framework for Modelling Large Scale National Economic Systems in Dynamic, Structural and Spatial Terms, *Int. J. of Systems Science,* 2(2), 213; Abid H.
30. Agnihotri G. (1984), *Dynamic Modelling Studies of Water Resources Systems for Regional Development Applications: The Case of Punjab Region of India,* Unpublished Ph.D. Thesis, IIT, Delhi.
31. Mathur M.C. (1984), *Physical Systems Theory Research Investigations of Large Scale Transportation System Problems,* Unpublished Ph.D. Thesis, IIT, Delhi; Satsangi P.S. (1977), A Physical System Theory Modelling Framework for Transportation System Studies, *IEEE Trans. Systems, Man and Cybernetics,* 7(11), 763-778.
32. Sushil, Singh N. and Jain B.K., (1988), A Physical System Theory Approach to Material Flow and Productivity Analysis, *Engineering Costs and Production Economics,* 13, 200-215.
33. Singh N. and Sushil (1990 a) A Physical System Theory Framework for Modelling Manufacturing Systems, *Int. Journal of Production Research,* 28 (6), 1067-1082; Kak A. (1993), *Application of Physical System Theory to Operations Management: A Case Study,* M. Tech. dissertation, IIT, Delhi.
34. Sushil (1991) Simplified Physical System Theory Model of a Closed Chain Production System, *Int. Journal of Systems Science,* 22 (6), 987-999.
35. Kanda A. (1986) *Project Planning Studies with Renewable and Non-renewable Resources,* Unpublished Ph.D. Thesis, IIT Delhi.
36. Singh N. and Sushil (1990 b) Technology Selection Models for Multistage Production Systems: Joint Application of Physical System Theory and Mathematical Programming, *European Journal of Operational Research,* 47, 248-261.
37. Sushil and Pankaj (1992), Systems Modelling of Cost and Quality: An Application of Extended Physical System Theory and Simulation, *Systems Research,* 9(1), 43-66.
38. Singh T.P. (1994), *Analysis and Design of Waste Management System for Engineering Industry,* Ph.D. Thesis, TIET Patiala.
39. Vij A.K. (1990), *Energy Policy Modelling and Analysis for the Indian Economy,* Ph.D. Thesis, IIT, Delhi.
40. Ellis J.B. (1965), *The Description and Analysis of Socio-Economic Systems by Physical System Techniques,* Ph.D. Thesis, Michigan State University; Satsangi and Ellis (1971b), *Op. Cit;* Vij A.K. (1990), *Energy Policy Modelling and Analysis for the Indian Economy,* Ph.D. Thesis, IIT, Delhi.
41. Gupta M. (1991) *Systems Modelling for Central Budget,* M. Tech. Dissertation, IIT Delhi.
42. Kulkarni R.G. (1993) *Modelling and Analysis of Financial Decisions Using Physical System Theory and Goal Programming,* M.Tech. Dissertation, IIT Delhi.
43. Gupta (1990) *op. cit.*
44. Sushil and Raghunathan B., (1994), Interactive Decision Support System for Organizational Analysis, *Decision Support Systems,* 11, 283-298.
45. Sushil (1989) *Systems Approach to National Planning: A Study in Waste Management,* Anmol Publications, New Delhi.

46. Sushil, Jalan R.K. and Srivastava V.K. (2000) The Management of Municipal Solid Waste Using Flexible Systems Approach, in V.I. Grover et al. (eds.), *Solid Waste Management,* Oxford and IBH, New Delhi.
47. Singh (1994), *op. cit.*
48. Forrester J.W. (1958), Industrial Dynamics: A Major Breakthrough for Decision Makers, *Harvard Business Review,* 36 (4), 37-66.
49. Forrester J.W. (1961), *op. cit.;* Forrester J.W., (1968), *Principles of Systems,* MIT Press, Cambridge, Massachusetts; Forrester J.W. (1969), *Urban Dynamics,* MIT Press, Cambridge, Massachusetts; Forrester J.W. (1971), *World Dynamics,* MIT Press, Cambridge Massachusetts.
50. Wolstenholme E.F. (1990), *Systems Enquiry: A System Dynamics Approach,* John Wiley and Sons, Sussex, U.K.
51. Sushil (1993a) *System Dynamics: A Practical Approach for Managerial Problems,* Wiley Eastern Ltd., New Delhi; Sushil (ed) (1993b) *System Dynamics for Management Support,* Wiley Eastern, New Delhi.
52. Sushil (1994a), *op. cit.*
53. Kumar S. and Satsangi P.S. (1993) System Dynamics Simulation of Physical Systems with Application to Hopfield Neural Networks, in Sushil (ed) *Systems Dynamics for Management Support,* Wiley Eastern Ltd., New Delhi.
54. Vij A.K., Vrat P. and Sushil (1988 b), Energy Modelling through Integration of System Dynamics and Physical System Theory, *System Dynamics: International Journal of Policy Modelling,* 1(1); Vij A.K., Sushil, and Vrat P. (1993), Energy Policy Analysis for the Indian Economy through Integration of System Dynamics and Physical System Theory, in Sushil (ed.), *System Dynamics for Management Support,* Wiley Eastern Ltd., New Delhi, 371-383.
55. Kumar N. and Vrat P. and Sushil (1994), A Simulation Study of Unit Exchange Spares Management of Diesel Locomotives in the Indian Railways, *International Journal of Production Economics,* 33, 225-236.
56. Sushil (1997), Flexible Systems Management: An Evolving Paradigm, *Systems Research and Behavioral Science,* 14(4), 259-275.
57. Sushil (2001) SAP-LAP Framework, *Global Institute of Flexible Systems Management,* 2 (1), 51-55.
58. Zadeh L.A. (1965), Fuzzy Sets, *Information and Control,* 8, 338-353.
59. Saaty T.L. (1980), *The Analytic Hierarchy Process,* McGraw-Hill, New York.
60. Checkland (1981), *op. cit.*
61. Ackoff R.L. (1981), *Creating the Corporate Future,* John Wiley and Sons, New York.

APPENDIX 2A

GENERALIZED MODELS OF BASIC PROCESSES

The generalized models of the basic processes are developed based on the free body diagrams or component terminal graphs presented in Figure 2.4.

In all the models 'x' represents an 'across' variable and 'y' represents a 'through' variable.

(a) Transformation Process

Across variable or cost equation

$$-x_{oi} = \sum_{j=1}^{n} k_{ji} x_{ji} \pm \sum_{l=1}^{m} kw_{il} xw_{il} + f_i(y_{oi}) \tag{2.1}$$

Cost of input Cost of waste Processing cost

The negative signs for x_{oi} in equation (A.1) represents the flow out of the reference node as reference node is at zero potential. In case of simplified models, this sign is to be removed (chapter four). This negative sign is ignored in other formulations for the sake of simplicity.

The cost of waste can be either way, i.e. either we have to spend cost to dispose waste, which will add to the total cost or we can have salvage value for the waste which will reduce the total cost and thus ± signs are given for cost of waste in equation (A.1).

Through Variable or Flow Equations

$$y_{ji} = k_{ji} y_{oi} \qquad j = 1,\ldots, n \tag{2.2}$$

$$yw_{il} = kw_{il} y_{oi} \qquad l = 1,\ldots, m \tag{2.3}$$

where,

x_{oi} = Unit cost of output of *i*th transformation process

x_{ji} = Unit cost of *j*th input to ith process

xw_{il} = Unit cost/salvage value of *l*th waste from *i*th process

y_{oi} = Flow of output from *i*th process

y_{ji} = Flow of *j*th input to *i*th process

yw_{il} = Flow of *l*th waste from *i*th process

k_{ji} = Technological coefficient for *j*th input to *i*th process (input per unit of output)

$= y_{ji}/y_{oi}$

f_I = Processing cost function for *i*th process

In case of economies of scale

$$f_I(y_{oi}) = c_I - m_i y_{oi}$$

where

c_I = Fixed processing cost for *i*th process

m_i = slope of process cost curve for *i*th process.

(b) Transportation Process

For single product flow, the characteristic equation is

$$x_i = g_i\,(y_i) \tag{2.4}$$

where

$x_i =$ Unit cost of transportation for *i*th transportation link

$y_i =$ Flow on *i*th link

$g_i =$ Transportation cost function for *i*th link

For multi product flow on the *i*th link

$$x_{ji} = g_{ji}\,(y_{ji}) \quad for\ j=1, \ldots, n \tag{2.5}$$

$$x_i = \sum_{j=1}^{n} w_{ji} x_{ji} \tag{2.6}$$

$$y_{ji} = w_{ji} y_i \quad for\ j=1, \ldots, n \tag{2.7}$$

where,

$x_{ji} =$ unit cost of transportation for *j*th type on ith link

$y_{ji} =$ Flow of *j*th type on *i*th link

$w_{ji} =$ Weightage of *j*th type of flow on *i*th link.

(c) Storage Process

$$\frac{dM(t)}{dt} = y_i(t) - y_o(t) \tag{2.8}$$

$$\frac{dE(t)}{dt} = x_i(t) \cdot y_i(t) - x_o(t) \cdot y_o(t) + h(M, y_i, t) \tag{2.9}$$

$$x_o\,(t) = E/M \tag{2.10}$$

Policy equations are different for forward and backward computations.

For forward analysis, the policy for shipment y_o can be of multiple types depending upon inventory level *M(t)*, inflow $y_i(t)$, averages of different kinds, orders etc., i.e.

$$y_o\,(t) = h_1\,(M(t), y_i\,(t), y\text{–}y_i\,(t), \ldots.) \tag{2.11}$$

For backward analysis, the policy of acquiring inventory in the storage can again be of multiple types, depending upon, inventory level *M(t)*, desired inventory DM; output flow $y_o\,(t)$ or average output, $\bar{y}_o\,(t)$, and so on, i.e.

$$y_i\,(t) = h_2\,(M(t), DM(t), y_o\,(t), \bar{y}_o\,(t), \ldots) \tag{2.12}$$

where,

$x_i\,(t) =$ Unit cost of input in time period *t*

$x_o\,(t) =$ Unit cost of output in time period *t*

$y_i\,(t) =$ Flow of input in time period *t*

$y_o(t)$ = Flow of output in time period t

$\bar{y}_i(t)$ = Average input upto time period t

$\bar{y}_o(t)$ = Average output upto time period t

$M(t)$ = Material accumulation or inventory level in time period t

$DM(t)$ = Desired inventory level in time period t

$E(t)$ = Energy or capital accumulation in time period t

h = Maintenance cost function

h_1 = Policy function for shipment policy

h_2 = Policy function for inventory acquisition policy

(d) Aggregation Process

Across variable or cost equation

$$-x_{oi} = \sum_{j=1}^{n} k_{ji} x_{ji} + f_i(y_{oi}) \tag{2.13}$$

Cost of input Aggregation cost

Flow equations

$$y_{ji} = k_{ji} y_{oi} \text{ for } j = 1, \ldots, n \tag{2.14}$$

where

x_{oi} = Unit cost of aggregated output for *i*th process

x_{ji} = Unit cost of *j*th input to *i*th process

y_{oi} = Flow of aggregated output for *i*th process

y_{ji} = Flow of input of *j*th type to *i*th process

k_{ji} = Ratio of *j*th input in aggregated output of process i

= y_{ji}/y_{oi}

f_i = Aggregation cost function for process i

(e) Disaggregation Process

Across variable or cost equation

$$-x_{li} = \sum_{j=1}^{n} k_{ij} x_{ij} \tag{2.15}$$

Flow equations

$$y_{ij} = k_{ij} y_{li} \text{ for } j = 1, \ldots, n \tag{2.16}$$

where,

x_{li} = Unit cost of total input to *i*th disaggregation process

x_{ij} = Unit cost of *j*th disaggregated output from *i*th process

y_{li} = Flow of total input to *j*th process

y_{ij} = Flow of *j*th disaggregated output from *i*th process

k_{ij} = Ratio of *j*th disaggregated output to total input of *i*th process

= y_{ij}/y_{li}

(f) Inspection Process

Cost equation for conforming output

$$x_{caj} = k_{cj} x_{Cj} \quad - k_{crj} x_{Crj} \quad + f_{caj}(y_{Caj}) \tag{2.17}$$

Cost of conforming input — Cost of conforming rejects — Cost of inspection

Cost equation for non-conforming output

$$x_{naj} = k_{nj} x_{nj} \quad - k_{nrj} x_{nrj} \quad + f_{naj}(y_{naj}) \tag{2.18}$$

Cost of non-conforming input — Cost of non-conforming rejects — Cost of inspection

Flow equations

$$y_{cj} = k_{cj} \cdot y_{caj} \tag{2.19}$$

$$y_{crj} = k_{crj}\, y_{caj} \tag{2.20}$$

$$y_{nj} = k_{nj} \cdot y_{naj} \tag{2.21}$$

$$y_{nrj} = k_{nrj} \cdot y_{naj} \tag{2.22}$$

where,

x_{caj} = Unit cost of conforming units accepted from *j*th inspection process

x_{cj} = Unit cost of conforming units input to *j*th process

x_{crj} = Unit cost of conforming units rejected from *j*th process (includes cost of salvaging rejects, and cost of separating rejects into repair/rework/scrap)

x_{naj} = Unit cost of non-conforming units accepted from *j*th inspection process

x_{nj} = Unit cost of non-conforming units input to *j*th process

x_{nrj} = Unit cost of non-conforming units rejected from *j*th process

y_{caj} = Confirming units accepted at *j*th process

y_{cj} = Confirming units input to *j*th process

Y_{crj} = Conforming units rejected from *j*th process (type I error)

y_{naj} = Non-conforming units accepted at *j*th process (type II error)

y_{nj} = Non-conforming units input to *j*th process

y_{nrj} = Non-conforming units rejected from *j*th process

k_{cj} = Ratio of conforming units input to conforming units accepted

= $y_{cj}/y_{caj} = 1/(1-\alpha_j)$

k_{crj} = Ratio of confirming units rejected to conforming units accepted

= $y_{crj}/y_{caj} = \alpha_j/(1-\alpha_j)$

α_j = Probability of conforming units rejected

= y_{crj}/y_{cj}

k_{nj} = Ratio of non conforming units input to non conforming units accepted

$= y_{nj}/y_{naj} = 1/\beta_j$

k_{nrj} = Ratio of non conforming units rejected to non conforming units accepted

$= y_{nrj}/y_{naj} = (1-\beta_j)/\beta_j$

β_j = Probability of non conforming units accepted

$= y_{naj}/y_{nj}$

3 STEPS OF MODELLING AND ANALYSIS

3.1 INTRODUCTION

The basics of the methodology of PST have been discussed in the previous chapter. However, for the purposes of implementation, specific steps of modelling and analysis are to be illustrated, which is the main purpose of this chapter. These steps, by and large, are followed in most of the applications reported in subsequent parts of the book. While attempting modelling and analysis, one may either adopt special matrix based formulations, such as Brand-Chord formulation and State-space formulation, or adopt a simple step-by-step solution procedure amenable to simulation. Keeping in view the simplicity of the latter approach, it has been widely used in various applications. Thus, the special formulations have only been introduced in the chapter. Operational details of these formulations would have been of theoretical interest only and thus not presented here.

The chapter provides specific steps of modelling and analysis, with an illustrative example, introduces special formulations, alternate software packages that may be used, and possible integration with other modelling methodologies. This provides a flexible modelling methodology for managerial and socio-economic systems.

3.2 STEPS OF MODELLING AND ANALYSIS

Keeping in view the fundamental postulates of PST, as described in chapter two, operational steps of modelling and analysis are delineated. The steps along with flexibility domains are shown in Figure 3.1.

Identification of Variables

The first step in developing any physical system model is to identify the variables of interest and classify them in two categories:

Across Variables having different values at the terminals for a component and governed by law of compatibility for the system

Through Variables having same values at the terminals for a component and governed by the law of continuity for the system

Select across and through variables for different types of systems are shown in Table 3.1 covering real physical systems as well as conceptual physical systems (socio-economic systems).

Table 3.1: Examples of 'Across' and 'Through' Variables

Nature of Variable	Real Physical Systems			Conceptual Physical Systems (Socio-economic Systems)		
	Electrical	**Mechanical**	**Hydraulic**	**Transport**	**Economy**	**Organization**
Across	Voltage	Velocity, displacement	Pressure	Traffic intensity	Unit cost	Unit authority
Through	Current	Force	Flow	Flow of traffic	Flow of resources	Responsibility

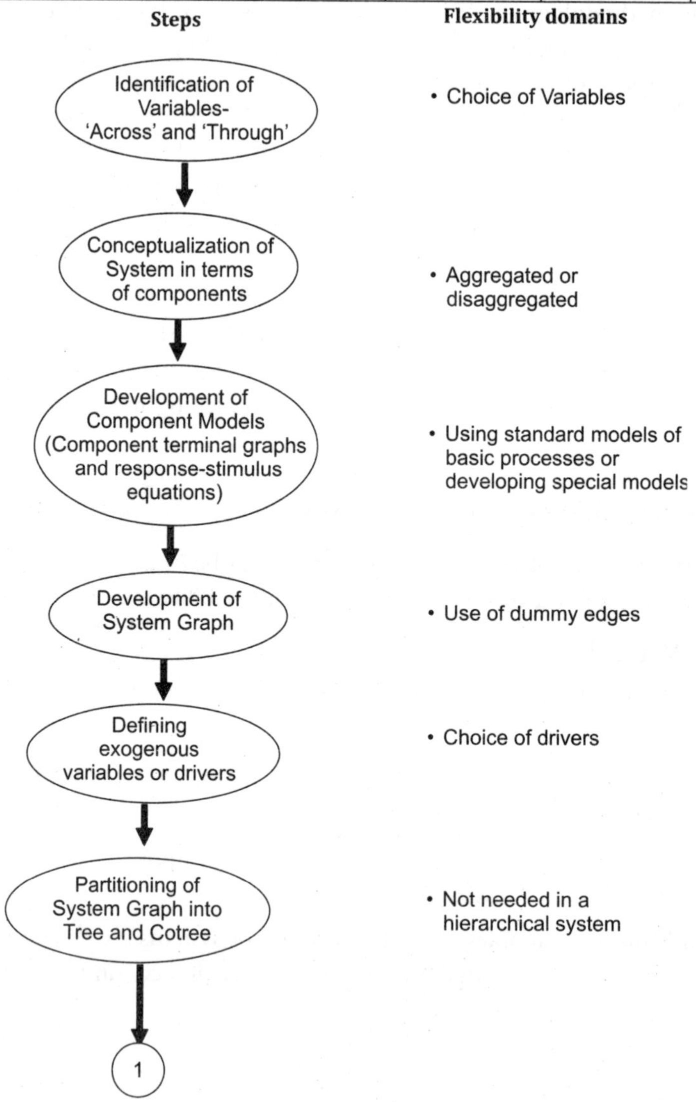

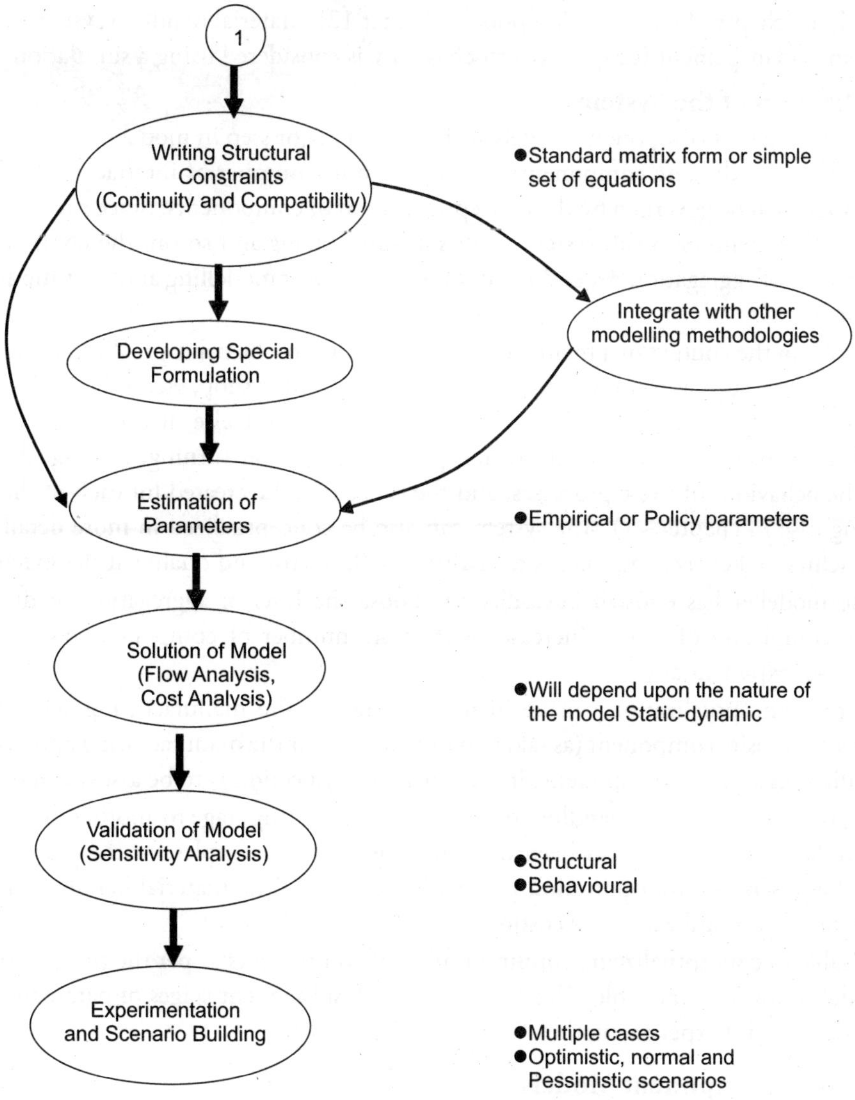

Figure 3.1: Steps of Modelling and Analysis

The definitions of 'across' and 'through' variables may be same throughout the system or vary with different components. For example, in the context of manufacturing system applications, the 'through' variable is taken as flow of material or capacity of manufacturing in terms of weight or number of units, and the 'across' variable is taken as unit cost of the material or manufacturing. Whereas, in the context of financial decision-making system (chapter 14), though the through variable is same throughout the system, i.e. flow of finances, the across variable varies with different financial decisions, such as cost of capital, rate of return, interest rate, and dividend rate.

Further, the variables could be taken as deterministic or stochastic. In most of the applications presented in this book the variables / parameters taken are deterministic. However, only in few cases, such as quality analysis (chapter 11), inspection policy (chapter 12), material handling systems (chapter 13), and maintenance management (chapter 18) stochasticity is considered using a simulation approach.

Conceptualization of the System

Since PST modelling uses component models as the base, a major step in modelling is to conceptualize the system in terms of discrete components united at a finite number of interfaces. The shape of the subsequent model will be governed by the conceptualization of components. Since the system is having hierarchy, i.e. sub-systems, sub-sub-systems, sub-sub-sub-systems and so on, the components can be defined at any level of aggregation depending upon the purpose of modelling and the time and resources available for it.

For example, in the context of a manufacturing system, it may be conceptualized in terms of major shops as components, such as foundry shop, forging shop, machine shop, etc. if the analysis is required to understand the capacities and costing of these major shops. Or else, it can be conceptualized at a lower level in terms of various manufacturing processes, such on turning, milling, drilling etc. to understand the behaviour of these processes, and the capacity to be created for each of them (e.g. gear manufacturing case in chapter six). The system can also be conceptualized in more detail in terms of individual machines taken as components to analyze the flow, cost and quality at the machine level.

Thus, the modeller has enough flexibility to choose the level of aggregation or disaggregation. However, the complexity of model increases with more number of components as the number of variables increases exponentially.

Further, the conceptualization can be in terms of taking the manufacturing stage and in-stage inspection as a composite component (as taken in this chapter) or the manufacturing and inspection can be conceptualized as separate components if the impact of inspection is to be assessed independently.

Similarly, the cost of material handling from one manufacturing stage to another may be combined with the manufacturing stage as has been done in most of the examples, or the materials handling can be treated as a separate transportation process (chapter 13). The material handling system can be analyzed for congestion, utilization, or costing.

It is advisable to conceptualize minimum number of components as per the purpose of analysis to keep the model simple and tractable. This will facilitate all subsequent stages of parameter estimation, solution, validation, and experimentation.

Development of Component Models

Once the system is conceptualized in terms of components, we can apply the component postulate (chapter two) to develop component model of each component separately. The component model consists of the component terminal graph and a set of component terminal or response-stimulus equations relating all the variables on the component terminal graph.

The component, models can be developed using a standard model of a typical component, e.g. in case of engineering systems, we may take standard equations for mass, spring, damper, resistor, capacitor, inductor, etc. In case of managerial and socio-economic systems, we may adapt the standard models of basic processes, such as transformation, transportation, storage etc. as given in chapter two. In case of a unique component, not fitting to the basic processes, an empirical model may be developed.

The response-stimulus equations describe the internal behaviour of the component as measured at its interfaces. In case of engineering systems, the terminals of a component are defined irrespective of its placement in the system, e.g. a resistor will always have two terminals (and thus one equation) whether connected in series or parallel with other components, and a triode will have three terminals (and thus two equations).

However, this is not true with respect to components in a managerial system. For example, the number of terminals on a transformation process as a manufacturing stage will depend upon the number of inputs and outputs of the manufacturing stage.

Development of System Graph

Once all the component models are developed, the interaction of the components is to be modelled. This is achieved by following the systems postulate (chapter two). All the component terminal graphs are united in one-to-one correspondence to the manner in which components are united in the real system. This results into the generation of system graph.

The system graph should be a closed graph, i.e. minimum two edges should join each vertex. If any vertex is left open, circuit will not be completed. In case of any open vertices, dummy edges are to be introduced to connect the vertex with the reference node. The reference node is the datum with reference to which all the across variables (e.g. voltage) are measured. The potential of the reference node is zero.

In case of various basic processes, the process node is the reference node marked as R as shown in Figure 2.4. In case of manufacturing systems modelled based on transformation processes only (as given in illustrative example in this chapter and in Part II) the system graph remains open after joining the component terminal graphs. The end vertices are connected to reference node by dummy edges to close the graph The reference node act as the market from which direct input is taken into the manufacturing system and direct output and waste is given back. Through physically this graph may appear to be open (Figure 3.4) the graph is practically closed as all reference nodes are the same node having zero potential.

Defining Exogenous Variables

The exogenous variables are the drivers which drive the system. For example, in an electrical system we may have a voltage driver or a current driver. A driver is an exogenously specified variable and, thus, does not require a separate equation.

In case of a system graph having e edges, there will be in all $2e$ variables associated with it, out of which n_x could be the number of specified across variables, and n_y could be the number of specified through variables. Thus, the total unknown variables in a system graph with e edges are $(2e-n_x-n_y)$. The total number of equations for these unknown variables are shown in Figure 2.3 (chapter two).

In order to make a practical sense of the model, at least one variable is to be defined exogenously, whose impact we want to assess. For example, in case of a manufacturing system, the across drivers could be cost of direct input material and the salvage cost of waste disposed, and the through driver could be total direct output from the manufacturing system (refer models in chapters four, six and seven).

Partitioning of System Graph

In order to develop constraint equations, the system graph may be partitioned into two parts, viz. 'tree' and 'cotree'.

Tree has the following characteristics:

i. It is an interconnected sub-graph interconnecting all the vertices.

ii. It has no closed path.

iii. All the edges associated with across drivers should be in the tree.

iv. All the edges associated with through drivers should not be in the tree.

The part of the system graph complimentary to 'tree' is 'cotree'.

'Tree' is used to write the continuity constraints or fundament-cutset equations (*f*-cutset equations), and 'cotree' is used to write the compatibility constraints or fundamental–circuit equations (*f*-circuit eq.). Since there are v vertices in a system graph the number of 'branches' in a tree are $(v\text{-}1)$ and thus the remaining edges, i.e. number of 'chords' in cotree are $(e\text{–}v\text{+}1)$. Since the simplification of the modelling methodology is proposed in chapter four, most of the applications presented in this book do not follow this step of partitioning the system graph into tree and cotree.

Writing Structural Constraints

As there are two types of variables in the physical systems model, i.e. across and through, there are two types of structural constraints, i.e. compatibility and continuity as presented in the interconnection postulate in chapter two[1].

f-cutset equations

To write the continuity constraints either we can write the continuity equations on each vertex in the system graph except the reference node, or we may identify a *f*-cutset associated with each branch of the tree.

A 'cutset' is a set of edges, which if partitioned will divide the system graph into two parts. A fundamental-cutset (*f*-cutset) is that cutset which has only one tree branch in it. Since there are $(v\text{–}1)$ branches, there are $(v\text{–}1)$ *f*-cutset equations.

The *f*-cutset or continuity equations can be written in simple equation form as a set of equations, or in standard matrix form as given below:

$$[U \mid C_c]\left[\frac{Y_b}{Y_c}\right] = 0$$

or,

$$\begin{array}{r} \\ n_x \\ v-1-n_x \end{array}\begin{array}{c} \text{Branches} \quad \text{Chords} \\ \left[\begin{array}{cc|cc} U & 0 & C_{11} & C_{12} \\ 0 & U & C_{21} & C_{22} \end{array}\right] \end{array}\begin{bmatrix} Y_{bd} \\ Y_{b2} \\ \hline Y_{c1} \\ Y_{cd} \end{bmatrix} = 0$$

where,

Y_b – vector of through variables with branches $(v\text{–}1)$

Y_{bd} – vector of through variables with specified branches (n_x)

Y_{b2} – vector of through variables with unspecified branches $(v\text{–}1\text{–}n_x)$

Y_c – vector of through variables with chords $(e\text{–}v\text{+}1)$

Y_{c1} – vector of through variables with unspecified chords $(e\text{–}v\text{+}1\text{–}n_y)$

Y_{cd} – vector of through variables with specified chords (n_y)

f-circuit equations

The compatibility constraints may be either written for all the circuits in the system graph in the equation form or we may identify only fundamental-circuits (*f*-circuits) and write the equations in standard matrix form. If the compatibility is defined for *f*-circuits it is defined for all the circuits in the system graph.

A *f*-circuit is that circuit in which there is only one cotree chord and rest all are branches. Since there are $(e-v+1)$ chords in cotree, there are $(e-v+1)$ *f*-circuits and there are $(e-v+1)$ *f*-circuit equations as given below:

$$\left[B_b \,\middle|\, U\right]\left[\frac{X_b}{X_c}\right]=0$$

$$\text{or, } \begin{matrix} e-v+1-n_y \\ n_y \end{matrix}\overset{\text{Branches}\quad\text{Chords}}{\begin{bmatrix} B_{11} & B_{12} & \big| U & 0 \\ B_{21} & B_{22} & \big| 0 & U \end{bmatrix}}\begin{bmatrix} X_{bd} \\ X_{b2} \\ \hline X_{c1} \\ X_{cd} \end{bmatrix}=0$$

where,

X_b – vector of across variables with branches $(v-1)$

X_{bd} – vector of across variables with specified branches (n_x)

X_{b2} – vector of across variables with unspecified branches $(v-1-n_x)$

X_c – vector of across variables with chords $(e-v+1)$

X_{c1} – vector of across variables with unspecified chords $(e-v+1-n_y)$

X_{cd} – vector of across variables with specified chords (n_y)

Principle of Conservation

According to the property of abstract graphs, the cutset submatrix and circuit sub-matrix are negative transpose of each other, i.e.

$$C_c = -B_b^T$$

$$\text{or } \begin{bmatrix} C_{11} & C_{12} \\ C_{21} & C_{22} \end{bmatrix} = -\begin{bmatrix} B_{11}^T & B_{12}^T \\ B_{21}^T & B_{22}^T \end{bmatrix}$$

Thus, we can derive one submatrix from the graph and the other is derived by the transpose operation.

Formulations for Solution

The set of system equations, i.e. component terminal equations and structural constraints can be adapted to develop formulations for solution in the following manner:

i. Solve the set of system equations either simultaneously or in a step-by-step manner using any programming language, data base package, or electronic spreadsheet.

ii. Develop special formulations to reduce the number of simultaneous equations to be solved.

iii. Develop integrative models by integrating PST model with other modelling methodologies.

i. ***Solving the Set of System Equations:*** The total number of system equation are $(2e-n_x-n_y)$ consisting of $(e-n_x-n_y)$ component terminal equations and *e* structural constraints. If we solve the system equations as it is, then set of $(2e-n_x-n_y)$ equations are to be solved simultaneously. This number can be reduced by either resorting to special formulations or in case of hierarchical systems solving them in a step-by-step manner (as given in chapter four). In most of the applications reported in this book, a step-by-step solution procedure is used.

ii. ***Special Formulations:*** Depending upon the nature of component terminal equations we can use three types of formulations[2]

- Branch Formulation
- Chord Formulation
- Branch-Chord Formulation

The nature of component terminal equations, starting point of solution, and number of simultaneous equations to be solved in each case is given in Table 3.2.

The standard form of component terminal equations in all the three formulations are given here. The detailed derivations are not discussed as these formulations are not applied in the book except in chapter 21, in which branch-chord formulation is used.

Table 3.2: Specification of Special Formulations

Formulation	Type of Component terminal Equations	Starting Point of solution	Number of simultaneous equations to be solved
Branch	Explicit in through variables (short-circuit parameter form)	*f*-cutset equations	$(v-1-n_x)$
Chord	Explicit in across variables (open circuit parameter form)	*f*-circuit equations	$(e-v+1-n_y)$
Branch-Chord	Hybrid form	Component-terminal equations	$(e-n_x-n_y)$

Branch Formulation

$$\begin{bmatrix} Y_{b2} \\ Y_{c1} \end{bmatrix} = \begin{bmatrix} W_{11} & W_{12} \\ W_{21} & W_{22} \end{bmatrix} \begin{bmatrix} X_{b2} \\ X_{c1} \end{bmatrix} + \begin{bmatrix} F_1 \\ F_2 \end{bmatrix}$$

Chord Formulation

$$\begin{bmatrix} X_{b2} \\ X_{c1} \end{bmatrix} = \begin{bmatrix} Z_{11} & Z_{12} \\ Z_{21} & Z_{22} \end{bmatrix} \begin{bmatrix} Y_{b2} \\ Y_{c1} \end{bmatrix} + \begin{bmatrix} G_1 \\ G_2 \end{bmatrix}$$

Branch-Chord Formulation

$$\begin{bmatrix} W_{11} & W_{12} \\ W_{21} & W_{22} \end{bmatrix} \begin{bmatrix} X_{b2} \\ X_{c1} \end{bmatrix} = \begin{bmatrix} Z_{11} & Z_{12} \\ Z_{21} & Z_{22} \end{bmatrix} \begin{bmatrix} Y_{b2} \\ Y_{c1} \end{bmatrix} + \begin{bmatrix} F_1 \\ F_2 \end{bmatrix}$$

In case of dynamic systems, the state-space formulation is used.

State-Space Model

State-space model is a standard model in linear systems theory in which all the system equations are reduced to two sets of equations, viz. state equations, and output equations. The state equations are first order differential equations and output equations are simple algebraic equations as given below:

$$\dot{S}\ (t) = S\ (t) + I\ (t)$$

$$O\ (t) = S\ (t) + I\ (t)$$

where,

$\dot{S}$ — first order differential of state vector

S — state vector

O — output vector

I — input vector

The detailed steps of state-space modelling are not provided as this model is not applied in this book. For dynamic modelling, PST model is integrated with system dynamics models[3].

iii. ***Integrative Models:*** In this case, flexible systems methodology[4] is applied to suitably integrate PST models with other modelling methodologies to match the modelling attributes with the problem attributes as closely as possible. The strengths and weaknesses of various modelling methodologies are critically examined in the light of the problem requirements and a suitable integration is selected following the steps of flexible systems methodology as discussed in chapter one.

A summary of the type of integrative flexible models used in various chapters is provided in Table 3.3. Other chapters used only PST modelling in different forms.

The details of integration are provided in respective chapters.

Estimation of Parameters

Once the model is prepared in a solvable form, there is need of parameter values to generate the behaviour of the model. Most of the parameters in PST model are empirical in nature, which are to be measured based on past data.

For example, in case of electrical systems the parameters of various components could be resistance, capacitance, and inductance.

In case of manufacturing systems, the parameters are in terms of technological coefficients (dependent upon the amount of waste generated and input required) and processing cost co-efficients. These can be measured based on past performance or specifications of the system.

The sensitivity analysis will help in identifying the sensitive parameters, which need to be measured more accurately.

In some cases, the parameters can be either normative or reference parameters which can be either based on benchmarking data or based on other models.

A third type of parameters is policy parameters, i.e. the parameters the management want to specify such as rejection rate, cost reduction parameters, quality parameters, and so on.

If PST model is integrated with other models, such as goal programming, then the parameters are optimized. In case of stochastic parameters, simulation can be used for parameter estimation as has been done in chapter 11 to estimate quality parameters. In case of ambiguous parameters, such as information quality, fuzzy set theory may be applied (chapter 17) for parameters estimation.

Table 3.3: Integrative Flexible Models in Various Applications

Chapter	Model Integrated with PST Model	Type of Integration
10. Technology Selection	Non-linear 0-1 programming	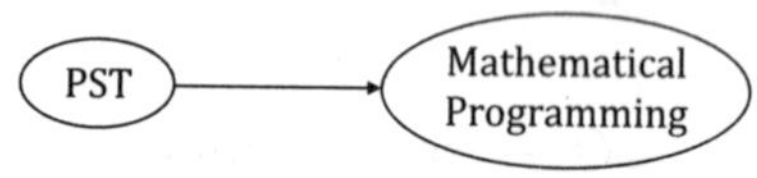
11. Cost and Quality Analysis	Simulation	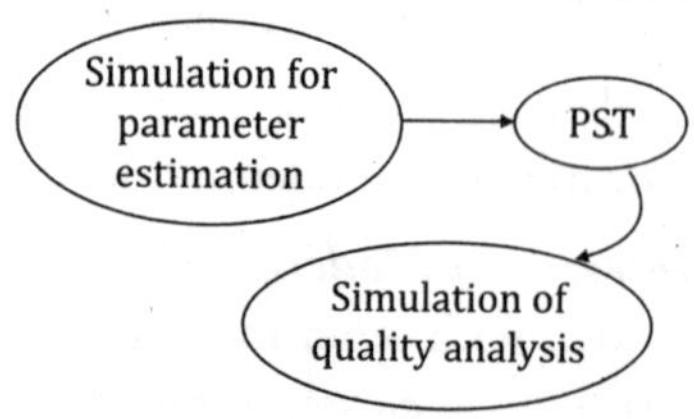
12. Inspection Policy	Simulation, Heuristics	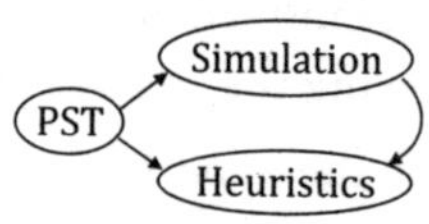
13. Material Handling	Simulation	
14. Financial Decision-Making	Goal Programming (GP) Net Present Value (NPV)	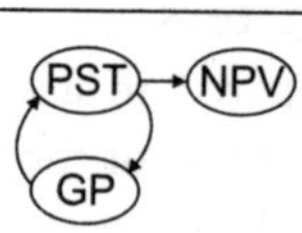
17. Management Information System	Fuzzy Set Theory (FST), Analytic Hierarchy Process (AHP)	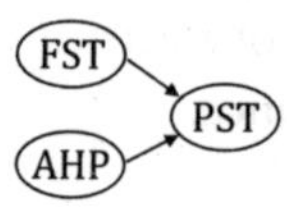

18. DSS for Maintenance Management	System Dynamics (SD)	
20. National Budget	Analytic Hierarchy Process (AHP)	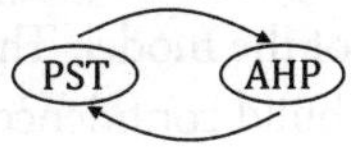
21. Energy Planning	System Dynamics (SD)	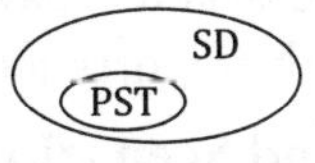
22. Steel Planning	System Dynamics (SD)	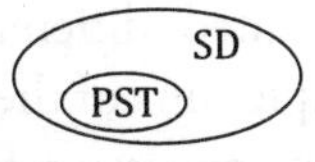
24. Waste Management	System Dynamics (SD)	

Solution of Model

The values of exogenous variables and parameters are fed as input data and the model can be solved using a software package. Various modes of software support are discussed in a separate section at the end of the chapter. Two broad approaches for solving the model are: analytical mode, and simulation mode. In case a step-by-step solution procedure is used, the model can be simulated over time using electronic spreadsheets, system dynamics software packages, or any equation solver package.

Validation of the Model

Validation of the model is predominantly a process of building confidence in the model. It is a relative concept, i.e. the model inspires greater confidence them its alternatives[5]. The two major types of tests for validating physical systems models are for

- structural validation, and
- behavioural validation.

Structural Validation

Some important tests for structural validation are:

i. ***Structure verification test:*** The structure of the model is reviewed to examine the components, and their interrelationship as it exists in the real system.

ii. ***Dimensional consistency test:*** The system equations are checked for dimensional consistency.

iii. ***Boundary adequacy test:*** The boundary of the model structure is examined, whether the components included are adequate to address the problem.

iv. ***Face validity test:*** The validity of the model on the face of it is examined, that if the model structure looks like the real system.

v. ***Parameter verification test:*** The parameters are reexamined both for their definition and value.

Behavioural Validation

The structural validation leads to behavioural validation, where the results or model behaviour is examined for its suitability and consistency.

i. ***Behaviour–reproduction test:*** The model is run for a known set of output variables. If the model results are close to the actual values the model is verified.

ii. ***Sensitivity analysis:*** Sensitivity analysis of model parameters is carried out to assess the sensitivity and robustness of the model. This test is shown in most of the applications reported in this book. It is always easier to build confidence in models exhibiting low sensitivity to a range of parameter values.

Most of the structural and behavioural validation tests were performed in the applications reported in different parts, but the reporting of the same is not done in each case to avoid repetition.

Experimentation and Scenario Building

Once the model is validated, different test data (in terms of exogenous variables or policy parameters) are fed to obtain multiple runs of the model. For example, different cases are reported in chapter 13 for simulating scenarios on congestion factor and utilization levels.

The model can also be experimented by marginally changing the structure. For example, chapter six on multistage manufacturing systems reports these case experiments (Table 6.8): (i) rework at stages two and five, (ii) rework at stage two only; and (iii) rework at stage five only.

The experimentation and resulting scenarios act as a useful guide for decision-making about capacities, re-work, inspection levels, cost control, technology selection, etc.

3.3 ILLUSTRATIVE EXAMPLE

An example of a serial manufacturing system is taken to illustrate the steps of modelling and analysis. The same example is taken in chapter four for simplification and thus the parameters and results of the example are given in chapter four and not repeated here. The system is designed to produce shafts in three operations: turning, milling, and drilling. An incoming inspection is made on bar stock and unacceptable material is scrapped. After each operation, a cursory inspection is performed and the parts not meeting the specifications are scrapped.

Various steps of modelling and analysis are illustrated below.

i. ***Identification of Variables:*** The following variables have been identified for the system:
   a. *Through variables:* Flow of parts through the manufacturing system in terms of number of parts.
   b. *Across variables*: Unit cost of parts in terms of $ per part.

ii. ***Conceptualization of the System:*** The system is conceptualized to be consisting of four components:
   a. Initial Inspection
   b. Turning and Inspection
   c. Milling and Inspection
   d. Drilling and Inspection

Inspection at each manufacturing stage is clubbed with the manufacturing as the main thrust was to analyze the manufacturing process and not the efficacy of inspection.

iii. ***Development of Component Models:*** Each manufacturing stage is considered as a transformation process taking input, processing it and giving the output and rejects. The component terminal graphs of the four components are shown in Figure 3.2.

Using the standard model of the transformation process (chapter two) the following component terminal equations are developed. Each component has one cost equation for the output, and two flow equations–one for input and the other one for waste.

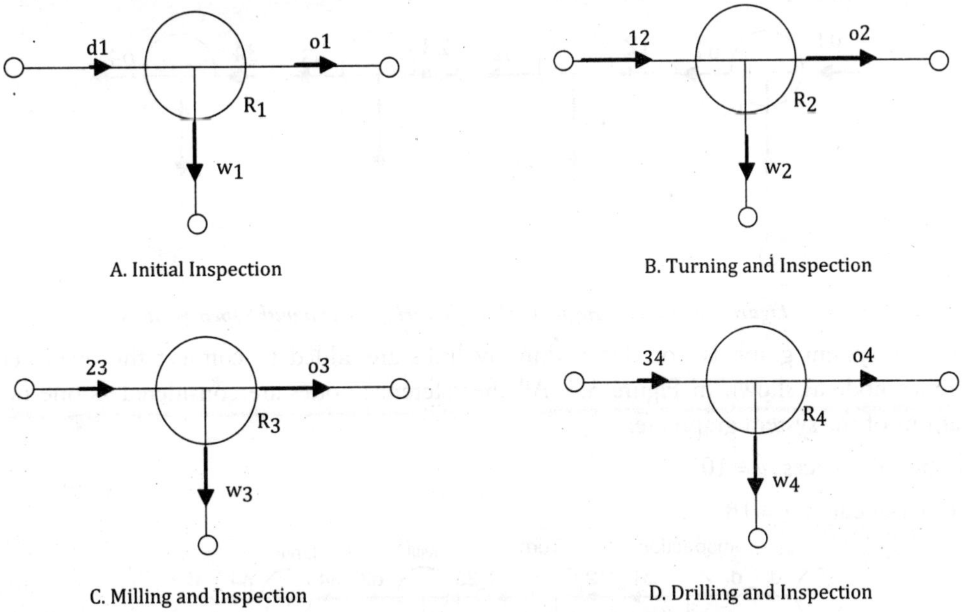

Figure 3.2: Component Terminal Graphs

a. *Initial inspection:*

$$-x_{o1} = k_{d1}\,x_{di} - kw_1\,xw_1 + f_1$$
$$y_{d1} = k_{d1}\,y_{o1}$$
$$yw_1 = kw_1\,y_{o1}$$

b. *Turning and inspection:*

$$-x_{o2} = k_{12}\,x_{12} - kw_2\,xw_2 + f_2$$
$$y_{12} = k_{12}\,y_{o2}$$
$$yw_2 = kw_2\,y_{o2}$$

c. *Milling and inspection:*

$$-x_{o3} = k_{23}\,x_{23} - kw_3\,xw_3 + f_3$$
$$y_{23} = k_{23}\,y_{o3}$$
$$yw_3 = kw_3\,y_{o3}$$

d. *Drilling and inspection:*

$$-x_{o4} = k_{34}\,x_{34} - kw_4\,xw_4 + f_4$$
$$y_{34} = k_{34}\,y_{o4}$$
$$yw_4 = kw_4\,y_{o4}$$

iv. ***Development of System Graph:*** On uniting the component terminal graphs the system graph is developed as shown in Figure 3.3.

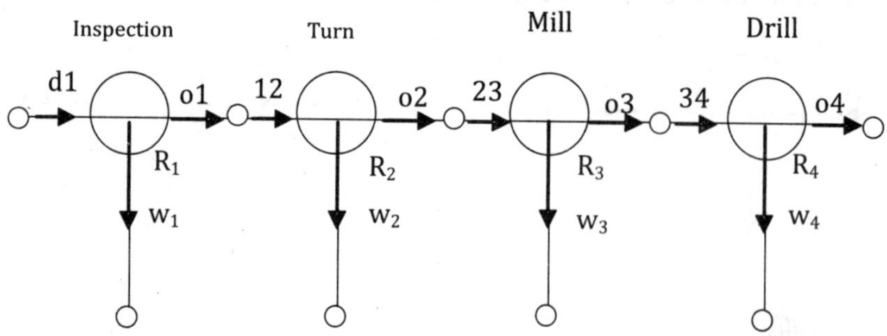

Figure 3.3: System Graph of Manufacturing System with Open Vertices

Since the system graph is not closed dummy links are added to connect the open vertices to the reference node as shown in Figure 3.4. All the reference nodes are considered as one node. The specifications of the system graph are:

Number of vertices $v = 10$

Number of edges $e = 18$

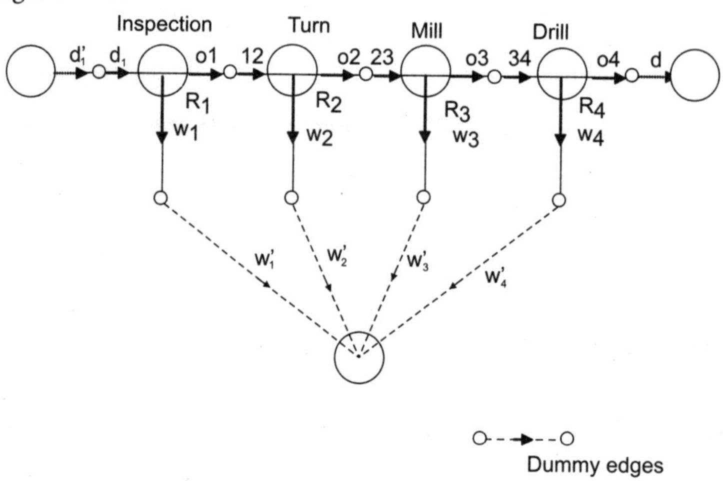

Figure 3.4: System Graph of Manufacturing System— Closed with Dummy Edges

v. ***Defining Exogenous Variables:*** The exogenous variables for the system are defined as follows:

Specified across variables $n_x = 5$

- Unit cost of input material (x_{di}')
- Unit salvage value of rejects at all the four stages.
  (xw'_1, xw'_2, xw'_3, xw'_4)

Specified through variables $n_y = 1$

- Output of the manufacturing system (y_d)

The values of these variables are given in Table 4.1 in the next chapter.

vi. Partitioning of System Graph: The system graph is divided into 'tree' and 'cotree' as shown in Figure 3.5.

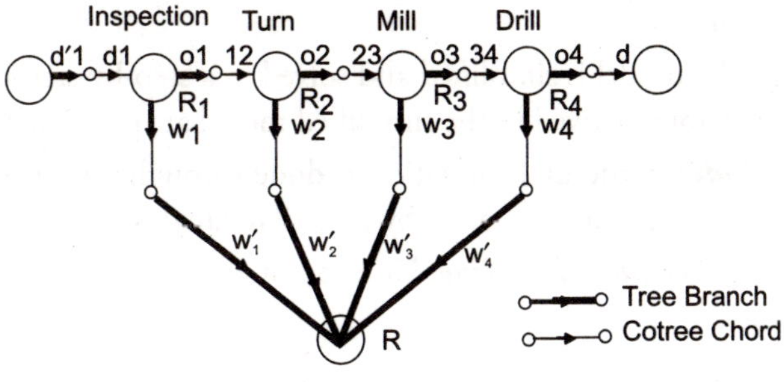

Figure 3.5: System Graph of Manufacturing System— Partitioned into Tree and Cotree

Number of tree branches = v–1 = 10 – 1 = 9

(all specified across variables are included in the tree and specified through variable kept outside the tree)

Number of cotree chords = $e - v + 1 = 18 - 10 + 1 = 9$

vii. Writing Structural Constraints: The structural constraints are written in simple equation form as given below. This is a special case where every cutset is a *f*-cutset, and every circuit is a *f*-circuit. Thus, the continuity equations are written on all interconnecting nodes, and compatibility equations are written on all circuits connecting one interconnecting node and reference node.

a. *Continuity equations:*

$$y'_{d1} - y_{d1} = 0, \quad y_{o3} - y_{34} = 0, \quad yw'_2 - yw_2 = 0$$
$$y_{o1} - y_{12} = 0, \quad y_{o4} - y_d = 0, \quad yw'_3 - yw_3 = 0$$
$$y_{o2} - y_{23} = 0, \quad yw'_1 - yw_1 = 0, \quad yw'_4 - yw_4 = 0$$

b. *Compatibility equations:*

$$x_{d1} + x'_{d1} = 0, \quad x_{34} + x_{o3} = 0, \quad xw_2 + xw'_2 = 0$$
$$x_{12} + x_{o1} = 0, \quad x_d + x_{04} = 0, \quad xw_3 + xw'_3 = 0$$
$$x_{23} + x_{o2} = 0, \quad xw_1 + xw'_1 = 0, \quad xw_4 + xw'_4 = 0$$

c. *Mass balance equations:* The mass balance equations help in relating the parameters of the system.

$$y_{d1} = y_{o1} + yw_1, \quad y_{23} = y_{o3} + yw_3$$
$$y_{12} = y_{o2} + yw_2, \quad y_{34} = y_{o4} + yw_4$$

or expressed in terms of technological coefficients

$$k_{d1} = 1 + kw_1, \quad k_{23} = 1 + kw_3$$
$$k_{12} = 1 + kw_2, \quad k_{34} = 1 + kw_4$$

viii. Formulation for Solution: No special formulation is needed, as this a pure hierarchical system and a step-by-step solution procedure can be used as given in chapter four.

ix. ***Estimation of Parameters:*** The parameters of the system are technological coefficients *k* and the processing cost coefficients *f*. The values of these parameters for the simplified model are given in Table 4.1 (chapter four).

x. ***Solution of Model:*** The model is simplified and solved in a step-by-step manner and the values of through and across variables derived in the simplified model are given in Table 4.3 (chapter four).

xi. ***Validation of the Model:*** Structural validation is done through reviewing the structure and face validity, and behavioural validation is done by sensitivity analysis given in chapter four.

xii. ***Experimentation and Scenario Building:*** No experiments were performed on this model as this was not a real life case.

3.4 SOFTWARE SUPPORT

The PST models developed by using the steps given in the previous sections can be solved either manually (for small models) or with the help of a software package. Multiple types of software support is possible for solving PST models:

i. Using programming language
ii. Using electronic spreadsheet and DBMS
iii. Using system dynamics software
iv. Developing specialize packages

Using Programming Language

The PST models can be programmed in standard programming languages, such as C^{++} etc. If we use special formulations, this would be an appropriate mode. However, this is not advocated for simpler step-by-step procedure. If the PST model is combined with mathematical programming or simulation model then a specific programme will be needed to solve the integrated model. This is the expert mode and is not recommended for a normal user in management.

Using Electronic Spreadsheet and DBMS

If we are modelling a hierarchical system, in which only one equation is to solved at a time in a step-by-step manner, the equations can be inputed as formulae in electronic spreadsheets such as 'Excel'. This is a very convenient mode of solving the model as the sequencing of equations is done by the package and many 'what-if' experiments can be conducted. The model can also be solved dynamically overtime. This also has the advantage of displaying the results graphically. In essences, the PST model gets converted into a Decision Support System (DSS). This mode is used in Chapter 18.

The PST models can also be conveniently made in Database Management Systems (DBMS). An example of this kind is made in Chapter 15 to develop a DSS for organizational analysis.

Using System Dynamics Software

System dynamics (SD) is a dynamic modelling methodology which simulates the equations over time. The most fundamental software for SD modelling is Dynamo[5]. The PST models as a set of equations can be written as Dynamo equations, and interactive simulation with graphic display can be carried out on Stella.

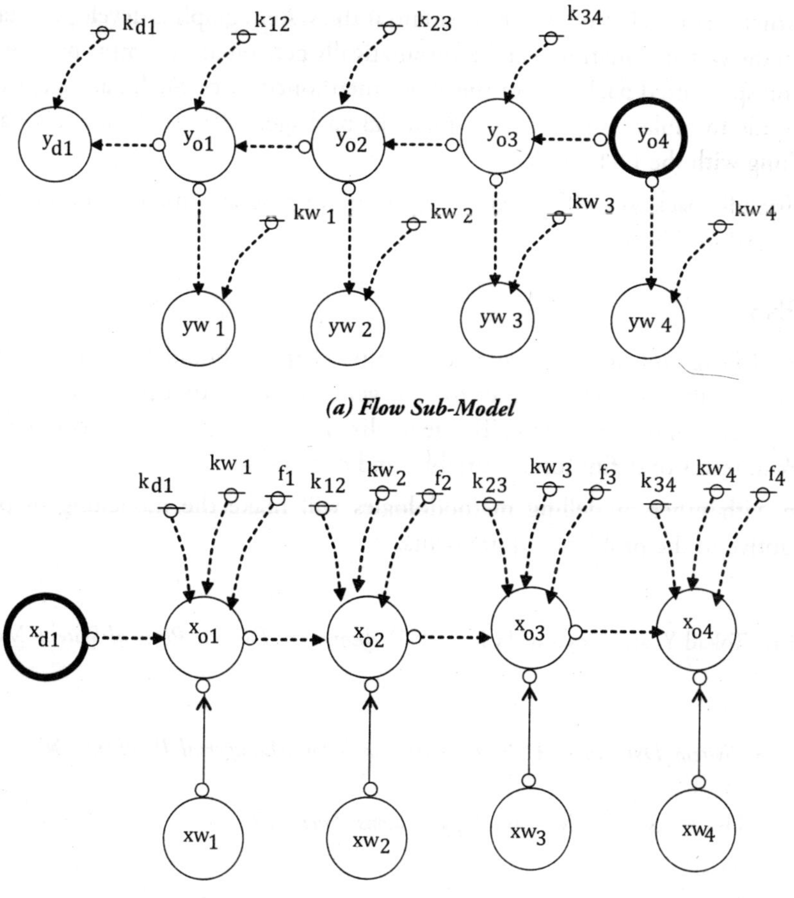

(a) Flow Sub-Model

(b) Cost Sub-Model

Figure 3.6: Systems Dynamics Flow Diagram of Four Stage Serial Manufacturing System
(For simplified system graph and equations refer chapter four)

The simplified system model of the illustrative example given in this chapter and chapter four is converted into a SD model as an illustration. Since this is a static model, there are no levels and rates in the SD model. It is purely a model of auxiliaries that can be solved hierarchically multiple times as per the exogenous flow variable. The flow diagram of the model is given in Figure 3.6 and the Dynamo equations are given in Appendix 3A.

This mode is very useful if the model is dynamic and has an integration with SD model for some purposes. This is applied in case of applications in chapters 22 and 24.

Developing Specialized Packages

It would be a worthwhile exercise to develop specialized packages for modelling special classes of physical systems. For example, there can be separate packages for manufacturing systems, organizational systems, financial decision-making, and so on.

Since the structure of PST model is transparent, if the system graph is developed graphically on the computer screen the system equations can be automatically generated. Attempts have been made by the author to develop specialized packages for the above mentioned cases. Such packages can be tailored to inbuild area specific formulae, e.g. in case of financial package, NPV, IRR, financial ratios etc. can be incorporated along with the PST model.

Such user-friendly packages will be extremely useful for practical implementation of the modelling framework proposed in this book.

REFLECTIONS

The steps of modelling and analysis presented in this chapter should be seen in conjunction with chapter four for simplification and chapter five for generalization. In a practical sense, the simplified modelling framework will be very useful. The generalized framework will facilitate to give flexibility to PST framework in terms of definition of variables and constraints.

Integration with other modelling methodologies will make the modelling of physical systems flexible and adaptive to the problem requirements.

Notes

1. Koenig H.E., Tokad Y. and Kesavan H.K. (1967) *Analysis of Discrete Physical System,* McGraw Hill Book Co., New York.
2. *Ibid.*
3. Sushil (1993) *System Dynamics: A Practical Approach for Managerial Problems,* Wiley Eastern Limited, New Delhi.
4. Sushil (1994) Flexible Systems Methodology, *Systems Practice,* 7 (6).
5. Sushil (1993) *op. cit.*

APPENDIX 3A

DYNAMO EQUATIONS OF PST MODEL OF SERIAL MANUFACTURING SYSTEM

```
* PST MODEL – FLOW SUB-MODEL
A  YO4 .K = 100
A  YO3.K = K34 * YO4.K
C  K34 = 1.0204
C  KW4 = 0.0204
A  YO2.K = K23 * YO3.K
A  YW3.K = KW3 * YO3.K
C  K23 = 1.0526
C  KW3 = 0.0526
A  YO1.K = K12 * YO2.K
A  YW2 = KW2 * YO2.K
C  K12 = 1.0309
C  KW2 = 0.0309
A  YD1.K = KD1 * YO1.K
A  YW1.K = KW1 * YO1.K
C  KD1 = 1.002
C  KW1= 0.002
* PST MODEL – COST SUB-MODEL
A  XD1.K = 6.4
A  XO1.K = KD1 * XD1.K – KW1 * XW1.K + F1
A  XW1.K = 3.6
C  F1 = 0.1
A  XO2.K = K12 * XO1.K – KW2 * XW2.K. + F2
A  XW2.K = 3.6
C  F2 = 1.5
A  XO3.K = K23 * XO2.K – KW3 * XW3.K + F3
A  XW3.K = 3.6
C  F3 = 3.0
A  XO4.K = K34 * XO3.K – KW4 * XW4.K +F4
A  XW4.K = 3.6
C  F4 = 0.5
SPEC  DT = 1, LENGTH = 1
```

4 SIMPLIFICATION IN MODELLING AND ANALYSIS

4.1 INTRODUCTION

The growing scale and complexity of systems requires the solutions to be provided more systematically, rationally, and objectively. Systems theory has contributed greatly towards the modelling and analysis of large-scale socio-economic systems. It has been observed that a significant feature common to all types of systems, whether physical or socio-economic, is that these are viewed as a collection of discrete components united at a finite number of interfaces through which they interact with each other. These interactions, in a physical system theory (PST) framework, are represented by a system graph, which is governed by the principle of conservation.

With increase in the size and complexity of a system, the complexity of system graph in the PST model increases multiplicatively. The number of edges in the system graph, and consequently the associated variables and equations, pose a serious limitation on its applicability to the real life large-scale socio-economic systems.

In order to make the framework of PST more practically applicable, it is necessary to resort to simplification of PST models. The purpose of simplification is to create a balance between the variety of the system to be handled by the analyst and the analyst's resources. According to Hare[1], "as such, simplification of system definition is necessarily related to analyst's objectives and resources, not some historical concept of elegance, or neatness". For our discussion, we can divide simplification methods into the two main categories: (i) first-order methods, which directly reduce complexity by involving actions of 'elimination' and 'grouping'; and (ii) higher order methods, which simplify a system indirectly, through a series of steps. In this chapter, both the approaches of simplification are employed in the context of the modelling of manufacturing, organizational and other socio-economic systems.

The experience gained by the author in the modelling of these socio-economic systems using PST has indicated the need for such simplifications. Typical features that are commonly exhibited by the models of these systems, irrespective of their size and complexity, have shown the way towards the simplification of these models. The first-order methods of elimination and grouping are used in simplifying the PST models of these systems. The higher order methods of step-wise simplification are employed in the simplification of the analysis procedure in a class of socio-economic systems, which are forward flow or hierarchical in nature. The approach of simplification suggested in this chapter is illustrated with the example of the modelling of a multi-stage manufacturing system.

The flexible system methodology used for simplification of PST models is depicted in Figure 4.1. The proposed simplification will open new options both on the fronts of modelling and analysis. In view of the simplified framework of modelling, specialized software can be developed to autogenerate

PST models. The simplification in analysis generates new models of solution such as use of electronic spreadsheets, system dynamics, and other simulation techniques. The flexibilities are created to handle non-linearity and stochasticity in systems.

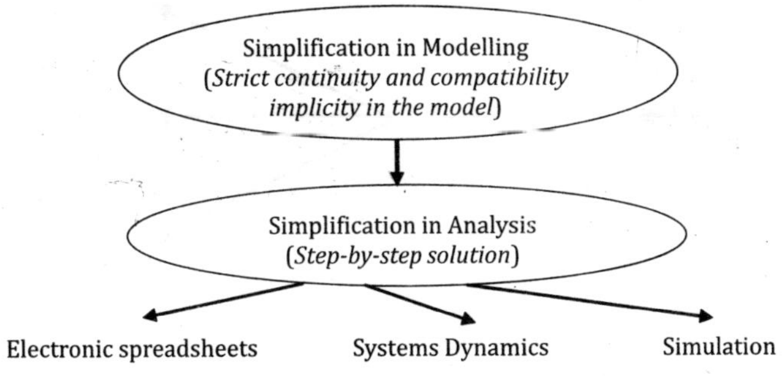

Figure 4.1: Flexible System Methodology for Simplification

4.2 SIMPLIFICATION

There are certain peculiar features that can be observed in PST models of managerial and socio-economic systems, particularly based on the model of the transformation process. Two significant features of these systems are as follows:

i. In most of the interconnections of components there is one-to-one correspondence, i.e. what is the output of one component, is the input to some other component as shown in Figure 4.1 (a). The branching at the connecting node takes place in very few cases as shown in Figure 4.1 (b).

ii. A large class of these systems is forward flow or hierarchical in nature, i.e. the input is processed in stages, one after the other.

The first feature results in simplification in modelling, whereas the second feature results in simplification in solution of the model[2].

Simplifications in Modelling

As PST provides a generalized framework to model a variety of systems, the direct application of the PST approach to the modelling of managerial and socio-economic systems leads to redundancy in the original model owing to the above mentioned peculiarities of these systems. In view of these peculiar features, the PST models can be considerably simplified resulting in reduction of the number of variables and number of system equations. It is redundant to write the continuity and compatibility equations for the type of interconnections shown in Figure 4.2 (a), which are very common in such systems. This calls for a simplification of the modelling approach in the context of these systems.

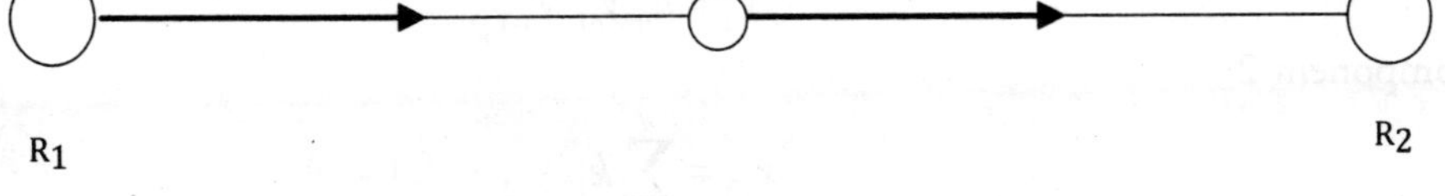

(a) A Typical Interconnection

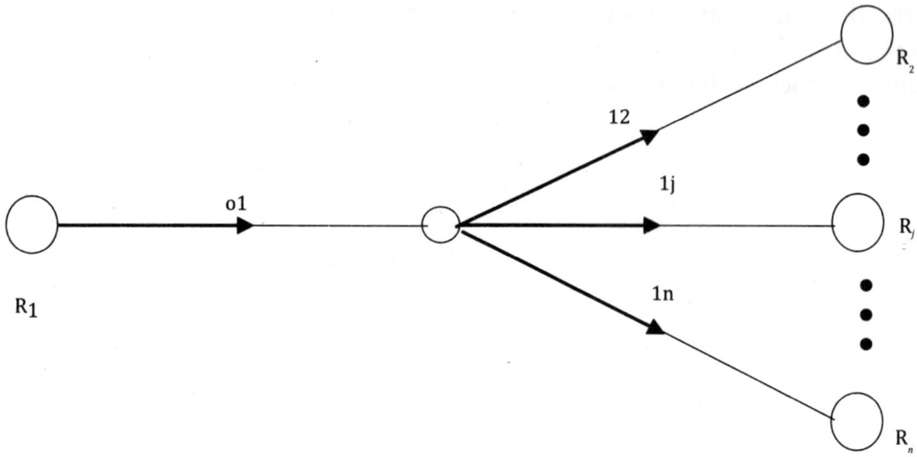

(b) An Interconnection with Branching

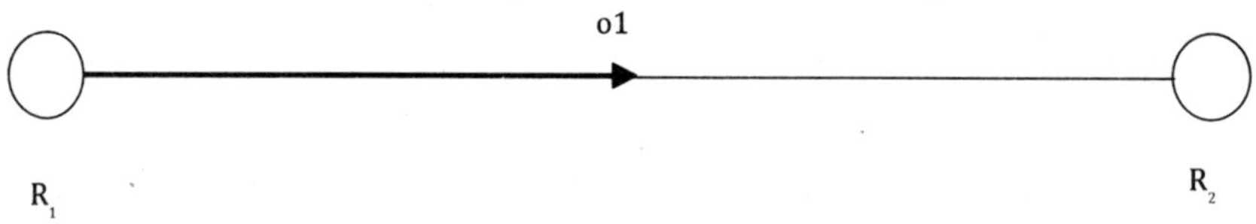

(c) A Simplified Interconnection

Figure 4.2: Types of Interconnections and Simplification

As in Figure 4.2 (a) the continuity equation is as follows:

$$y_{o1} - y_{12} = 0 \tag{4.1}$$

or

$$y_{o1} = y_{12}$$

The compatibility equation for this interconnection is as follows:

$$x_{o1} + x_{12} = 0$$

or

$$-x_{o1} = x_{12} \tag{4.2}$$

The component models of these two components are

For Component 1:

$$-x_{o1} = \sum_{j=1}^{n} k_{j1}\, x_{j1} + f_1(y_{o1}) \tag{4.3}$$

$$y_{j1} = k_{j1}\, y_{o1,}\ \forall_j \tag{4.4}$$

For Component 2:

$$-x_{o2} = \sum_{j=1}^{n} k_{j2} x_{j2} + f_2(y_{o2}) \tag{4.5}$$

$$y_{j2} = k_{j2}\, y_{o2,}\ \forall_j \tag{4.6}$$

As there is one-to-one correspondence, the links o1 and 12 can be replaced by a single link, say o1, as shown in Figure 4.2 (c). The continuity and compatibility equations can be absorbed in the component terminal equations. The revised equations are:

For Component 1: the negative sign of cost x_{o1} is removed

$$x_{o1} = \sum_{j=1}^{n} k_{j1}\, x_{j1} + f_1(y_{o1}) \tag{4.7}$$

$$y_{j1} = k_{j1}\, y_{o1,}\ \forall_j \tag{4.8}$$

For Component 2: x_{12} is replaced by x_{o1} in component model 2 and the negative sign of x_{o1} is absorbed in the component model 1.

$$x_{o2} = k_{12}\, x_{o1} + \sum_{j=2}^{n} + k_{j2}\, x_{j2} + f_2(y_{o2}) \tag{4.9}$$

$$y_{o1} = k_{12}\, y_{o2} \tag{4.10}$$

$$y_{j2} = k_{j2}\, y_{o2} \qquad \text{for } j = 2, \ldots\ldots, n \tag{4.11}$$

The rules for simplification for such a system are as follows:

i. Eliminate the interconnection node (Figure 4.2 (a)) and merge the links which are in one-to-one correspondence by maintaining the notation of the input links, as shown in Figure 4.2 (c).

ii. Merge the continuity and compatibility equations in the component models by changing the notation of relevant variables and by removing the negative sign of cost (across variable) of output from the component equation of the previous component.

iii. Eliminate all the dummy edges interconnecting the exogenous input/output/waste with the reference node.

The above simplification will result in halving of the number of edges in the system graph (if it has no branching nodes); the number of variables and the number of system equations also reduce to half of the original. It may be worth noting that in the process of simplification, nothing is lost regarding the details of the system. The simplified model will only have the component equations with inbuilt structural constraints; there may be a few structural constraints in the simplified models in case if branching at the interconnection nodes takes place.

The simplified system graph is not governed by the rules of tree selection, etc. that are applicable for the original system graph. This simplified system graph should not be compared on those theoretical considerations, which are applicable to the system graph in the original PST model.

Simplification in Analysis

In case of forward flow or hierarchical systems, the solution procedure of a PST model can be considerably simplified by using a step-by-step solution procedure which facilitates the application of PST to large-scale managerial and socio-economic systems.

On joining all component graphs into 1:1 correspondence, a system graph is obtained which has e edges and v vertices. If n_x is the number of specified across variables and n_y is the number of specified through variables then a total of $(2e-n_x-n_y)$ equations have to be simultaneously solved for analyzing the system model to obtain the values of all the unknown across and through variables.

The complexity can be reduced by making use of the Branch-Chord Technique[3] as the component terminal equations are in hybrid form. In this formulation, the total number of simultaneous equations to be solved will be $(e-n_x-n_y)$.

In the case of the system that processes the product in successive stages with only forward flows and having no backward interaction, the solution procedure can be considerably simplified. It may be possible to obtain the value of all the unknown variables by solving only one equation at a time and making appropriate substitutions in a step-wise procedure as discussed below. As many of the multi-stage production systems are forward flow systems, this simplified solution procedure is helpful in applying physical system theory to analyze large scale manufacturing systems.

The steps for obtaining the values of "through" variables are as follows:

- **i.** Solve the flow equations for the last process to determine the values of flow of input/waste.
- **ii.** Solve the continuity equations to obtain the corresponding values of flows from other processes; only in case of interconnection nodes with branching.
- **iii.** Repeat steps (i) and (ii) successively until the value of flow of terminal input/waste is obtained.

The steps for obtaining the values of "across" variables are as follows:

- **i.** Solve the cost (across variable) equations for the first process having only exogenous inputs/wastes.
- **ii.** Solve the compatibility equation to obtain the value of the corresponding across variable going to next process, only in case of interconnection nodes with branching.
- **iii.** Solve the cost (across variable) equation of the next process.
- **iv.** Repeat steps (ii) and (iii) successively until the final cost (across variable) is calculated.

4.3 ILLUSTRATIVE EXAMPLE

In order to illustrate the concepts of simplification we now take up an example of a multistage manufacturing system.

Modelling and Analysis of a Multi-Stage Manufacturing System

The material flow is the basic characteristic of manufacturing systems. Each stage of a multi-stage manufacturing system is conceptualized as a material transformation process. In general, every process transforms input into output and generates some waste. The exogenous output is given in the form of finished product, and the exogenous waste in the form of scrap.

The across variable is taken to be the unit cost of material, while the flow of material is a through variable. The technological coefficient represents the input requirement per unit of output.

For the purpose of illustration, a serial manufacturing system[4] (already modelled in chapter three) is described. This system is designed to produce a shaft in three operations, viz. turn, mill, and drill. An incoming inspection is made on bar stock and unacceptable material is scrapped. After each operation, a cursory inspection is performed and the parts not meeting the specifications are scrapped. The values of different parameters and exogenous variables are given in Table 4.1.

Table 4.1: Values of Parameters and Exogenous Variables

Stage	Technological co-efficients of inputs	Values	Technological co-efficients of wastes	Values	Process cost co-efficients ($)	Values
Initial Inspection	k_{d1}	1.002	kw_1	0.002	f_1	0.1
Turning	k_{12}	1.0309	kw_2	0.0309	f_2	1.5
Milling	k_{23}	1.0526	kw_3	0.0526	f_3	3.0
Drilling	k_{34}	1.0204	kw_4	0.0204	f_4	0.5

Unit cost of incoming material x_{d_1} = \$6.4 per piece.

Unit cost of scrap = $xw_1 = xw_2 = xw_3 = xw_4$ = \$ 3.6 per piece.

Final output $y_{o4} = y_d$ = 100 parts.

The manufacturing system is considered to be composed of four stages, namely, initial inspection; turning and inspection; milling and inspection, and drilling and inspection. The component models, system graph, continuity equations, compatibility equations, and mass balance equations for this system are given in chapter three.

The system model is simplified using the rules previously outlined. The simplified system graph is given in Figure 4.3. The specifications of the PST model and the simplified model are compared in Table 4.2. While counting the number of vertices, all the reference nodes *(R)* are counted as a single node.

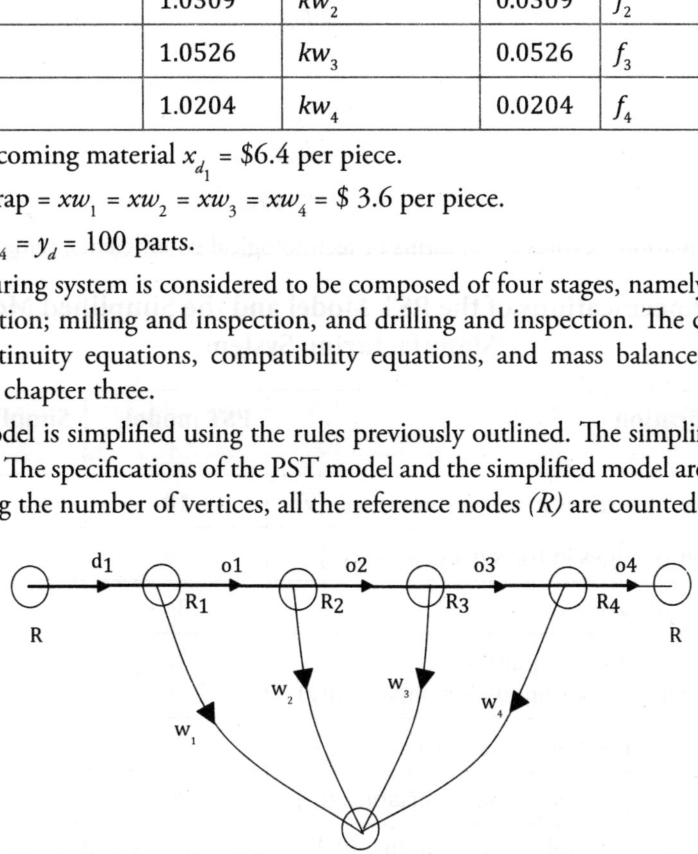

Figure 4.3: Simplified System Graph of Serial Manufacturing System

The equations of the simplified model, in which the structural constraints are merged into the component terminal equations, are as follows:

Initial inspection:

$$x_{o1} = k_{d1}x_{d1} - kw_1 xw_1 + f_1 \tag{4.12}$$

$$y_{d1} = k_{d1}y_{o1} \tag{4.13}$$

$$yw_1 = kw_1 y_{o1} \tag{4.14}$$

Turning and inspection:

$$x_{o2} = k_{12}x_{o1} - kw_2xw_2 + f_2 \tag{4.15}$$

$$y_{o1} = k_{12}y_{o2} \tag{4.16}$$

$$yw_2 = kw_2\,y_{o2} \tag{4.17}$$

Milling and inspection:

$$x_{o3} = k_{23}x_{o2} - kw_3xw_3 + f_3 \tag{4.18}$$

$$y_{o2} = k_{23}y_{o3} \tag{4.19}$$

$$yw_3 = kw_3\,y_{o3} \tag{4.20}$$

Drilling and inspection:

$$x_{o4} = k_{34}x_{o3} - kw_4xw_4 + f_3 \tag{4.21}$$

$$y_{o3} = k_{34}y_{o4} \tag{4.22}$$

$$yw_4 = kw_4\,y_{o4} \tag{4.23}$$

The mass balance equations, expressed in terms of technological co-efficients, remain as they are.

Table 4.2: Specifications of the PST Model and the Simplified Model of the Manufacturing System

S. No.	Specification	PST model	Simplified model
1.	Number of vertices in the system graph (v)	10	1
2.	Number of edges in the system graph (e)	18	9
3.	Number of variables	36	18
4.	Number of system equations (components + continuity + compatibility)	30	12
5.	Number of specified across variables (n_x)	5	5
6.	Number of specified through variables (n_y)	1	1

There are a total of 36 variables in the original PST model, out of which six variables (x_{d1}, xw_1, xw_2, xw_3, xw_4, and y_{o4}) are specified, i.e. number of specified across variables (n_x) = 5 and number of specified through variables (n_y) = 1. If the branch-chord formulation is used, then (e–n_x–n_y), i.e. 12 simultaneous equations will have to be solved. The solution procedure can be considerably simplified as this is a pure forward flow system. By using the proposed simplified solution procedure, the values of all the unknown variables are determined by solving only one equation at a time.

The values of all the flow variables are determined by following the steps given earlier for a desired output y_{o4} = 100 parts, as shown in Table 4.3. The values of all the cost variables are determined by following the steps given earlier, as shown in Table 4.3.

Table 4.3: Values of Through and Across Variables

Sr. No.	Through Variable	Value (no. of parts)	S. No.	Across Variable	Value (in $)
1.	y_{d1}	110.95	9.	x_{o1}	5.52
2.	y_{o1}	110.73	10.	x_{o2}	8.3327
3.	y_{o2}	107.41	11.	x_{o3}	11.9604
4.	y_{o3}	102.04	12.	x_{o4}	12.778
5.	yw_1	0.22			
6.	yw_2	3.32			
7.	yw_3	5.37			
8.	yw_4	2.04			

The system equations are solved to derive the output cost per unit, i.e. x_{o4} in terms of various co-efficients and the exogenous variables, as follows:

$$\begin{aligned} x_{o4} = {} & k_{34}\, k_{23}\, k_{12}\, k_{d1} x_{d1} - k_{34} k_{23} k_{12} kw_1 xw_1 \\ & + k_{34} k_{23} k_{12} f_1 - k_{34} k_{23} kw_2 xw_2 + k_{34} k_{23} f_2 \\ & - k_{34} kw_3 xw_3 + k_{34} f_3 - kw_4 xw_4 + f_4 \end{aligned} \tag{4.24}$$

On partially differentiating equation (4.24) with respect to the technological co-efficients of various stages, and then substituting the values of various coefficients and exogenous variables, the change in output cost for a unit change in technological co-efficient can be computed. The various manufacturing stages are ranked according to the sensitivity values as follows:

$$\frac{\partial x_{o4}}{\partial kw_4} = 7.73184 \tag{4.25}$$

$$\frac{\partial x_{o4}}{\partial kw_3} = 4.58709 \tag{4.26}$$

$$\frac{\partial x_{o4}}{\partial kw_2} = 3.12083 \tag{4.27}$$

$$\frac{\partial x_{o4}}{\partial kw_1} = 3.10033 \tag{4.28}$$

It is observed that for the given data, the unit cost of the output is most sensitive to the waste generated at stage 4, whereas the unit cost is least sensitive to the waste generated at stage 1. Since the input material to stage 4 is processed at a number of previous stages, it has a higher value than that for stage 1 and the outcome of the model is logical.

4.4 APPLICABILITY OF THE SIMPLIFIED FRAMEWORK

The simplified framework of the PST presented in previous sections provides a methodology to model manufacturing, organizational and other socio-economic systems with a reduced number of variables and system equations. Usually, the number of variables reduces to almost half of the existing theoretical counterpart as presented in chapter three. Further, the analysis of forward flow systems is considerably simplified to solve only a single (or very few) equations at a time.

Such a simplified framework has wide applicability in a variety of socio-economic systems and this will promote the modelling and analysis of these systems using PST, which was otherwise either not possible or very difficult in the existing framework. The proposed simplified framework can be further extended to cater for the needs of a larger class of systems. This framework can be adopted both for the study of system behaviour and system design. Particularly in cases of system design, e.g. technology selection, etc., where the PST model is to be integrated with some mathematical programming model and the degree of theoretical sophistication required on the PST front is less because the ultimate solution is to be arrived at by the mathematical programming technique, such a simplified framework will be of high utility.

Moreover, the aspects of non-linearity and stochasticity, which could not be extensively explored in PST can now be examined more thoroughly in view of the reduced complexity of the existing models.

REFLECTIONS

A simplified framework in terms of modelling and analysis of manufacturing, organizational and other socio-economic systems using PST has been proposed. The simplified framework has been illustrated using the example of a manufacturing system. Simplified models in other areas such as marketing, finance, information systems, etc., can also be developed. If the spirit of simplification is taken up in a positive manner, it is hoped that more avenues of simplified modelling of socio-economic systems will be opened up and practical implementation of the PST will be facilitated. This simplified framework has been largely used for modelling in subsequent parts of the book.

Notes

1. Hare Jr V.C. (1967) *Systems Analysis: A Diagnostic Approach,* Harcourt, Brace and World, New York.
2. Sushil (1992) Simplification of Physical System Theory in the Modelling of Manufacturing, Organizational and other Socio-Economic Systems, *Int. Journal of Systems Science,* 23 (4), 531-543.
3. Koenig H.E., Tokad Y., and Kesavan H.K. (1967) *Analysis of Discrete Physical Systems*, McGraw Hill, New York.
4. Davis R.P. and Kennedy Jr W.J., (1987) *Int. J. Prod. Res.* 25, 337.

5 EXTENDED AND GENERALIZED FRAMEWORK

5.1 INTRODUCTION

The PST does not use the notion of analogies, but is based on the concept of a linear graph. A two-terminal component is represented as a single oriented line segment associated with a pair of complementary variables: an 'across' variable *x*, and a 'through' variable *y*. In the case of a component, the 'across' variable takes different values at the terminals, whereas the 'through' variable takes the same values at the terminals. In the case of a system, the 'across' variable is governed by the law of compatibility, whereas the 'through' variable is governed by the law of continuity.

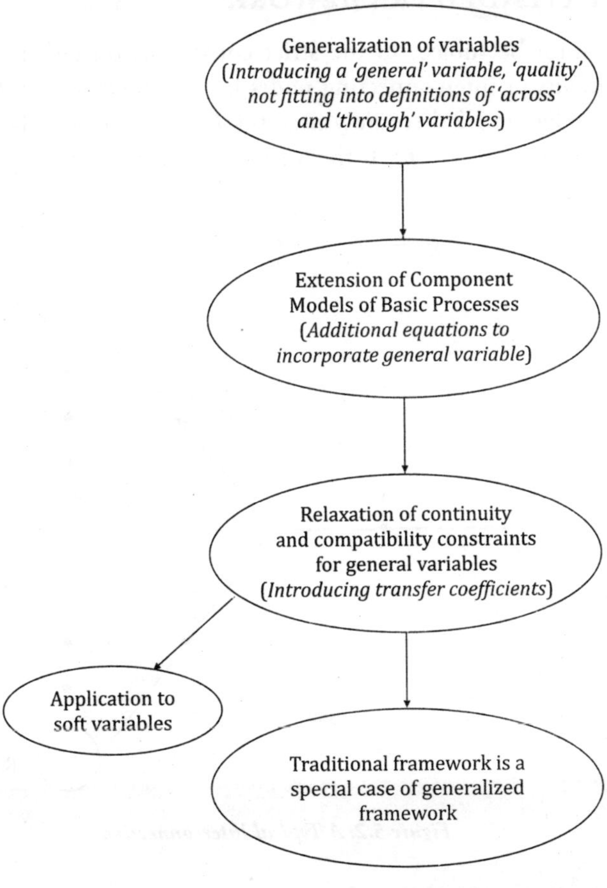

Figure 5.1: Flexible System Methodology for Generalization of Framework

All the variables in managerial and socio-economic systems usually need not conform to the previously mentioned strict definitions of "across" and "through" variables. For example, quality variables of a component may take different values at both the terminals for some components, whereas they may take same values at both the terminals for certain other components of the same system, depending upon characteristics of different components. Further, the quality variables cannot be subject to strict constraints on either the compatibility or the continuity for all the circuits or interconnecting nodes in the system graph. Thus, it is not possible to model quality or other soft variables by treating them either as "across" variables or as "through" variables, as the nature of the constraints that are imposed on such variables are usually not met by quality and other soft variables in real-life systems.

In order to take care of such soft variables, the "across" and "through" variables have been generalized and an enlarged paradigm of PST has been developed by Sushil[1]. The flexible system methodology for generalization of PST framework is shown in Figure 5.1. At the first stage, a general variable is introduced. Based on this variable the component models are extended and the structural constraints are relaxed. The generalized framework can be applied to soft variables, such as quality, managerial effectiveness, risk, and so on. The traditional framework is found to be a special case of the generalized framework.

5.2 NEED FOR AN EXTENDED FRAMEWORK

The quality variables cannot be subject to the strict constraints on either the compatibility or the continuity for all the circuits or interconnecting nodes in the system graph. For example, for a set of interconnections shown in Figure 5.2 there are n circuits $(i1,\ldots ij\ldots,in)$ and there is one interconnection node linking $(n + 1)$ edges $(i, j = 1, \ldots, n)$. If the quality variables are treated as 'across' variables, then the compatibility equations are:

$$x_j = -x_i, \quad \forall j \tag{5.1}$$

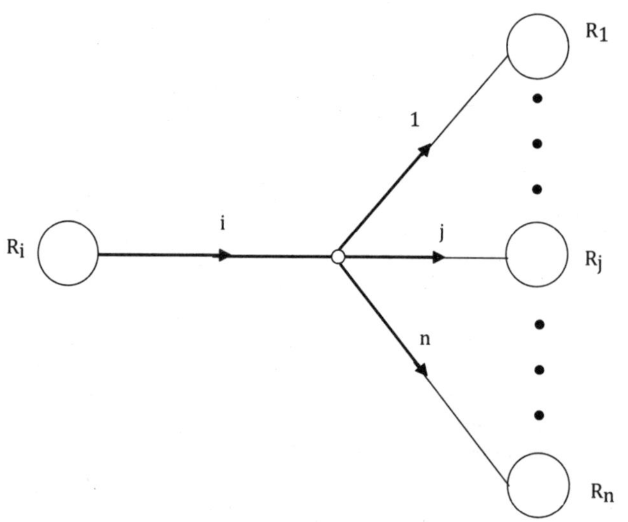

Figure 5.2: A Typical Interconnection

These equations show that quality is transferred similarly to all the n links (j = 1, …, n) from the link number i. In practice, this may not be true for all the interconnections in the system. Usually, there are certain perception gaps, and the quality of the product as perceived by the producer differs from the quality perceived by the receiver of the product, as there may be different needs or different perceptions of the same thing. If there is no perception gap of component i with all the n receivers, i.e. j = 1,…,n then (5.1) is true, but usually this is not the case.

If we treat the quality as a 'through' variable, then the continuity equation to be satisfied is

$$y_i = \sum_j y_j \tag{5.2}$$

This indicates that the quality of the product is transferred in parts. The sum of the quality received by all the n receivers is equal to the total quality supplied. This never happens in reality. If a product goes to n customers it does not mean that quality is apportioned, it may be equal to the original quality or may be different if there are perception gaps.

The quality variables can be treated as across variables only with the following assumptions:

i. the value of quality is different at both the terminals of all the component edges;

ii. there are not perception gaps in transferring the quality, and the compatibility laws are followed.

The quality variables can be treated as 'through' variables only with the following assumptions:

i. the values of the quality are same at the both the terminals of all the component edges,

ii. the quality is divided in the transfer, and the continuity laws are followed.

In both the cases, the assumptions made are unrealistic and their validity is questionable in real life. Thus, on the basis of the above discussion it may be concluded that it is not possible to model quality by treating it either as an 'across' variable or as a 'through' variable, as the nature of the constraints that are imposed on such variables usually not met by the quality variables in real-life systems.

Thus, it can be said that the existing framework of PST is not offering sufficient scope to model the aspect of quality along with other aspects of flow, cost, etc. Further, as has been seen from the overview of PST, the PST modelling framework fails to capture the impact of quality on cost. There are various categories of costs that are associated with identifying avoiding, producing, maintaining and repairing products that do not meet specifications. These could be categorized into prevention, appraisal, internal failure and external failure costs[2].

Leaving aside the cost aspect, it would be much better if the quality and output can be predicted on the basis of inputs and the knowledge of the quality processing at various stages and the perception gaps between various stages, or the required quality of the inputs can be determined for a desired quality of output.

Moreover, since with the application of PST it is possible to pinpoint the source of trouble and inefficiency in terms of cost and productivity a need was felt to make it more meaningful so that the quality can also be incorporated along with the other variables by extending the framework. This would facilitate quality planning and the effect of various processes on quality so that the source of trouble in terms of the adverse effects on quality can be pinpointed. This would facilitate the corrective actions well in advance to prevent the recurrence of such causes in the future.

It is emphasized that the modelling for quality is no less important, and since PST offers a good structural framework, it would be a worthwhile exercise if its framework could be extended to incorporate quality variables. Such an extension could be possible by generalizing the existing set of variables ('across' and 'through') by relaxing the constraints imposed and introducing a third category of variables typified as "quality" variables in the extended PST framework. Such a framework will overcome the limitations cited above and will be more pragmatic.

5.3 EXTENDED FRAMEWORK

In order to take care of quality and other generalized variables in the system, an extended framework of PST has been developed in this section. As already outlined in the original framework of the PST, there are only two variables associated with every edge of the component model/system graph; an 'across' variable x and a 'through' variable y.

In the extended framework, every edge in the system graph is considered to be associated with an additional generalized variable typified as "quality", i.e. there are three variables associated with every edge: an "across" variable x, a "through" variable y, and a "quality" variable z. The quality variable is governed by the following considerations.

i. For components, the quality variable can take either the same value at both the terminals of an edge or different values at both the terminals depending upon the characteristics of the component.

ii. In the system, the quality variable is governed by the quality transfer constraint which takes care of the perception gap at every interconnection by introducing a quality transfer coefficient or function.

The quality transfer equations are

$$z_j = t_{ij} z_i, \quad \forall_j \tag{5.3}$$

where t_{ij} is the quality transfer co-efficient from edge i to edge j. If $t_{ij} > 1$, then the increased quality is perceived in the transfer by the edge j; if $t_{ij} < 1$, the decreased quality is perceived in the transfer by edge j; but if $t_{ij} = 1$ there is no change in the perceived quality in the transfer.

The 'quality' variable is the generalized variable, whereas the conventional 'across' and 'through' variables are specific cases of the 'quality' variables.

The 'across' variable is a special case of quality variable when it takes different values at both the terminals for all the components, and $t_{ij} = 1$ for all the interconnections.

The 'through' variable is a special case of quality variable when it takes the same values at both the terminals for all the components, and $\sum_j t_{ij} = 1$ for all the interconnections.

Examples of 'across', 'through' and 'quality' variables, for different physical and socio-economic systems, are given in Table 5.1.

The revised component equations for a two terminal component are:

$$x_i = f_{1i}\,(y_i, z_i) \tag{5.4}$$

or

$$y_i = f_{2i}\,(x_i, z_i) \tag{5.5}$$

or

$$z_i = f_{3i}\,(x_i, y_i) \tag{5.6}$$

Table 5.1: Examples of 'Across', 'Through' and 'Quality' Variables

Nature of variable	Physical System			Socio-economic System		
	Electrical	Mechanical	Hydraulic	Transport	Economy	Organization
Across	Voltage	Velocity, displacement	Pressure	Traffic intensity	Unit cost	Unit authority
Through	Current	Force	Flow	Flow of traffic	Flow of resources	Responsibility
Quality	Power factor	Quality of force (impact or gradual)	Turbulence	Level of service	Quality of resources	Managerial effectiveness

The system graph is obtained in a similar manner as in the original framework and it will have e edges and v vertices. Then the total number of variables in the system is $3e$, as in the extended framework there are three variables with each edge.

The structural constraints in the extended framework are as follows:

i. compatibility equations for x variables, which are $e–v+1$ in number;

ii. continuity equations for y variables which are $v–1$ in number;

iii. quality transfer equations for z variables.

The number of equations for the existing framework and the extended framework of the PST are compared in Table 5.2.

Table 5.2: Number of Equations

No.	Type of Equation	Number of Equations in Framework of PST	
		Existing	Extended
1.	Component terminal equation	e	z*
2.	Continuity equation	v – 1	v – 1
3.	Compatibility equation	e – v + 1	e – v + 1
4.	Quality transfer equation	-	e – z**
	Total number of equations	2e	3e

* z is the number of component terminal equations, which is also different for the models used for forward and backward calculation of quality in the case of forward flow systems, as illustrated in the next section.

** The number of quality transfer equations may be different in the case of forward or backward calculation of quality in the case of forward flow systems. In the case of backward calculation, the number is increased by the number of vertices where branching is taking place, as given in the illustrative example.

5.4 MODELLING OF BASIC PROCESSES IN THE EXTENDED FRAMEWORK

The models of the basis process, i.e. transformation, transportation, and storage, in the extended framework are as follows:

Transformation Process

The free body diagram or component terminal graph of *i*th transformation process with *n* different inputs and *m* types of wastes or auxiliary outputs is shown in Figure 5.3.

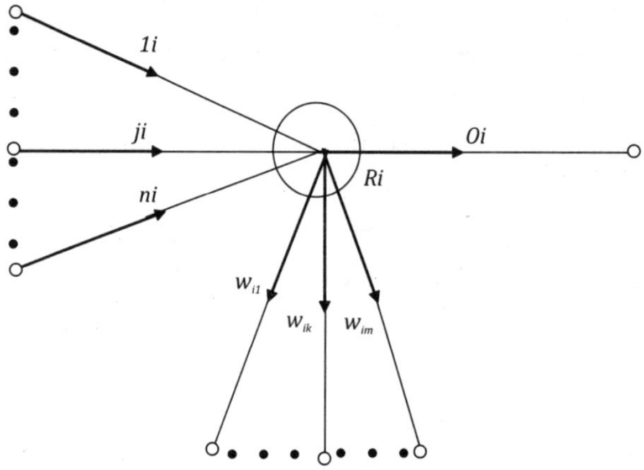

Figure 5.3: Free Body Diagram of ith Transformation Process

The component terminal equations comprise a cost equation, flow equations, and quality equations. The quality equations are different for the backward analysis (i.e. calculation of quality of inputs and wastes by knowing the quality of output) as well as forward analysis (i.e. calculation of the quality of output and wastes by knowing the quality of inputs).

Let

x_{oi} = unit cost of output

x_{ji} = unit cost of input *j*

xw_{ik} = unit cost of waste *k*

y_{oi} = flow of output

y_{ji} = flow of input *j*

yw_{ik} = flow of waste *k*

z_{oi} = quality of output

z_{ji} = quality of input *j*

zw_{ik} = quality of waste *k*

Then the component equations are as follows:

Modified Cost Equation

The cost equation would have to be modified in the light of the impact of the quality on the cost. The modified cost equation is

$$-x_{oi} = \sum_{j=1}^{n} k_{ji}x_{ji} - \sum_{k=1}^{m} kw_{ik}xw_{ik} + f_i(y_{oi}) + p_i(z_{oi}) \tag{5.7}$$

where

k_{ji} = technological co-efficient for input j (defining input j required per unit of output)

kw_{ik} = technological co-efficient for waste k

f_i = processing cost function that represents the economies of scale $f_i(y_{oi}) = c_i - m_i y_{oi}$

p_i = quality-cost function representing the variation in processing cost with the quality

In a linear model all the coefficients are assumed to be linear. The cost equation for any kind of coefficients/functions is generalized as follows:

$$-x_{oi} = F_i(k_{ji}, kw_{ik}, f_I(y_{oi}), p_i(z_{oi}), x_{ji}, xw_{ik}) \tag{5.8}$$

Flow Equations

There is no change in the flow equations in the extended framework. The flow equations for the ith component are as follows:

$$y_{ij} = k_{ji}\, y_{oi}, \quad \forall j \tag{5.9}$$

$$yw_{ik} = kw_{ik}\, y_{oi}, \quad \forall k \tag{5.10}$$

Quality Equations (for backward analysis)

The quality equations are added to the existing PST model which will help in the analysis of the quality, as follows:

$$z_{ji} = q_{ji}\, z_{oi}, \quad \forall j \tag{5.11}$$

$$y_{wik} = qw_{ik}\, z_{oi}, \quad \forall k \tag{5.12}$$

where q_{ji} is the equality-processing coefficient for input j, and qw_{ik} is the quality-processing coefficient for waste k.

Quality Equations (for forward analysis)

The quality of output can be calculated by knowing the quality of inputs, and then the quality of wastes can be calculated as

$$z_{oi} = \sum_{j=1}^{n} a_{ji} z_{ji} \tag{5.13}$$

$$zw_{ik} = qw_{ik}\, z_{oi}, \quad \forall k \tag{5.14}$$

where a_{ji} is the modified quality processing co-efficient:

$$a_{ji} = \frac{k_{ji}}{\sum_j k_{ji}} q_{ji} \quad (5.15)$$

a_{ji} takes care of both quality and flow in the aggregation process.

Transportation Process

The free body diagram of the *i*th transportation process is shown in Figure 5.4. Let x_i be the unit cost of transportation, y_i be the flow in *i*th transportation, and z_i be the quality of *i*th transportation. Then the cost equation is modified to incorporate the aspects of quality, whereas the quality of transportation is dependent upon the flow volume and the unit cost of transportation, as expressed in the following equations:

$$-x_i = g_i(y_i) + p'_i(z_i) \quad (5.16)$$

$$z_i = r_i(x_i, y_i) \quad (5.17)$$

where g_i is the function relating flow of transportation to unit cost, p'_i is the function relating quality of transportation to unit cost, and r_i is the function of flow and unit cost of transportation determining the quality of transportation.

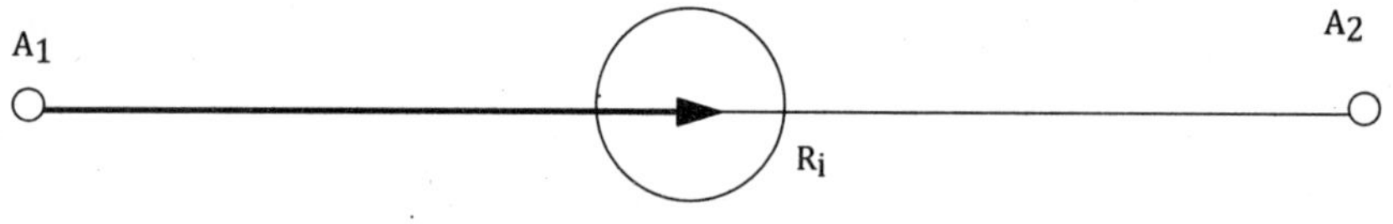

Figure 5.4: Free Body Diagram of ith Transportation Process

Storage Process

The free body diagram of a storage process is shown in Figure 5.5.

Let x_i be the unit cost of input to storage, x_o be the unit cost of output from storage, y_i be the flow of input, y_o be the flow of output, z_i be the quality of input, z_o be the quality of output, M be the material or resource accumulation, S be the capital accumulation, Z be the quality accumulation, and t be the time period.

Then the component-terminal equations are as follow:

a. The rate of material resource accumulation is not affected by quality.

$$\frac{dM}{dt} = y_i - y_o \quad (5.18)$$

b. The rate of capital accumulation is dependent upon the quality of input and output. The modified equation is

$$\frac{dS}{dt} = (x_i y_i - x_o y_o) + h(M_i, y_i, y_o, z_i, z_o, t) \quad (5.19)$$

where h is a function representing maintenance capital required.

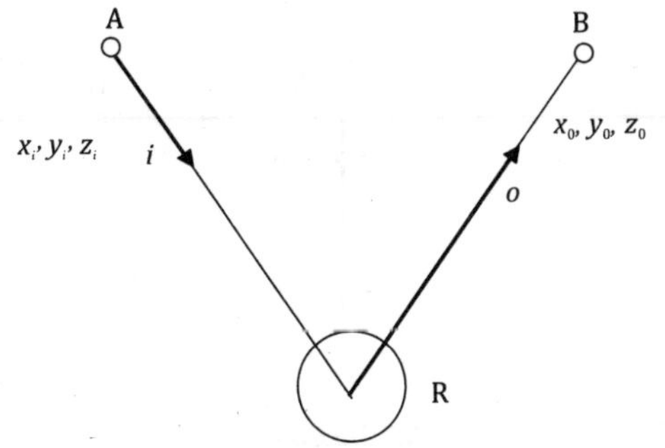

Figure 5.5: Free Body Diagram of Storage Process

c. The rate of quality accumulation is expressed as

$$\frac{dZ}{dt} = (y_i z_i - y_o z_o) + s(M, y_i, y_o, z_i, z_o, t) \tag{5.20}$$

where s is a function representing instantaneous quality loss is storage because of degradation.

d. The unit cost of output is dependent upon the quality of output

$$x_0 = \frac{S}{M} + P^{11}(z_0) \tag{5.21}$$

where p^{11} is a function representing the effect of quality on unit cost of output.

e. Output quality can be determined by knowing the order in which the output is taken from the storage, e.g. LIFO (last in first out), FIFO (first in first out), etc. As z_i is known for all times, the corresponding z_o can be assessed. In the case of a random output system, the average quality can be calculated as

$$z_o = \frac{Z}{M} \tag{5.22}$$

5.5 ILLUSTRATIVE EXAMPLE

In order to illustrate the principles and concepts of modelling quality in an extended framework of PST, a simple example of a manufacturing system with three manufacturing stages, namely turning, milling and assembly, is chosen. The raw material input comes to the turning stage. The output of the turning stage is divided into two parts; one part goes directly to the assembly, whereas the other part is further processed at the milling stage. The output of milling also goes to the assembly which gives the finished product. The waste generated at all three stages is declared as scrap and directly disposed by salvaging. For the sake of the illustration, only the transformation processes are considered. The system model in terms of component terminal equations and structural constraints in the extended framework is given in Appendix 5 A. The component terminal graphs are shown in Figure 5.6 and the system graph is shown in Figure 5.7.

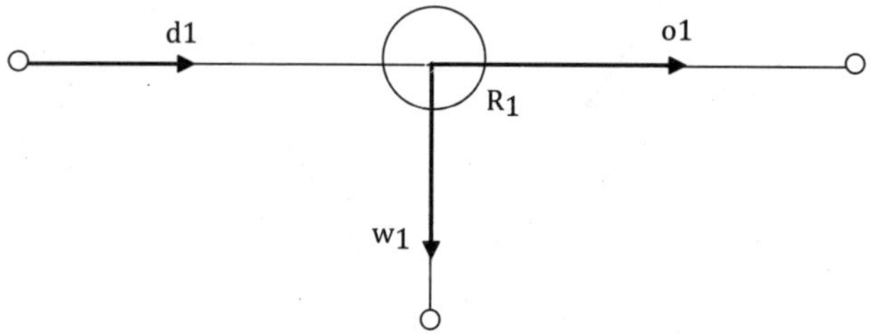

(a) Turning Stage

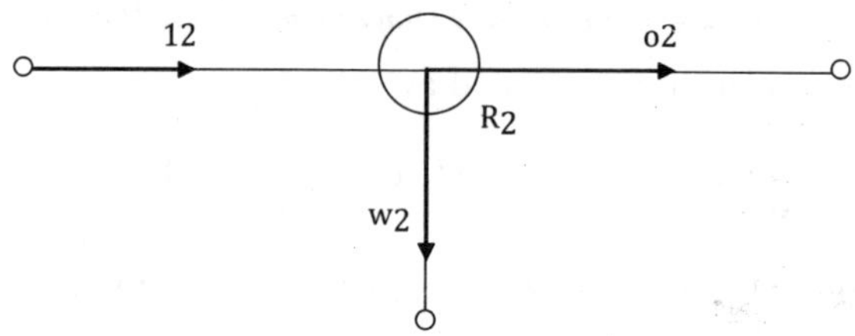

(b) Milling Stage

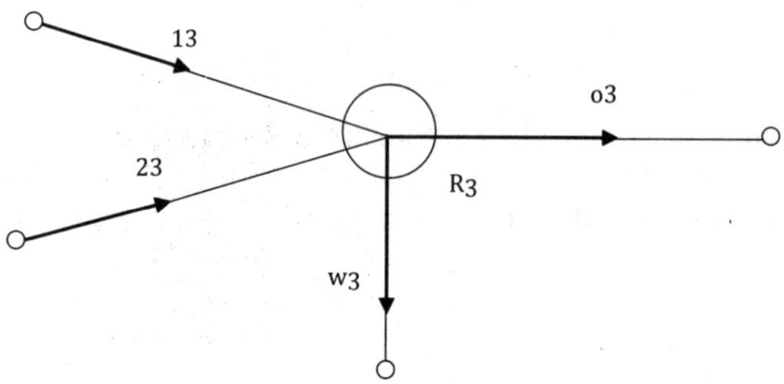

(c) Assembly Stage

Figure 5.6: Component Terminal Graphs

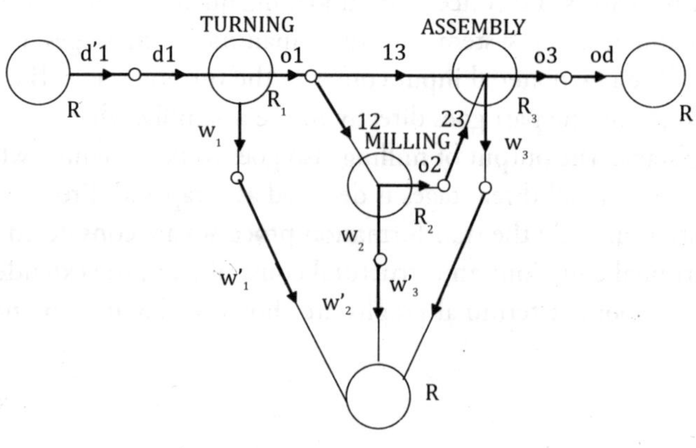

Figure 5.7: System Graph of the Manufacturing System

The parameters of the illustrative system are estimated as shown in Table 5.3.

Table 5.3: Values of Different Parameters

Sr. No.	Stage	Techno-logical coefficients	(Modified) quality processing coefficients	Quality transfer coefficients	Process cost function	Quality cost function
1	Turning	k_{d1}=1.05 kw_1=0.05	q_{d1}=1.077 qw_1=0.077 a_{d1}=0.929	$t_{d'1,d1}$=1 $t_{w1,w1'}$=1	c_1=3 $m_{1,}$=0.01	q_1=0.1
2	Milling	k_{12}=1.03 kw_2=0.03	q_{12}=0.92 qw_2=0.17 a_{12}=1.087	$t_{o1,12}$=0.861 $t_{w2,w2'}$=1	c_2=4 m_2=0.015	q_2=0.2
3	Assembly	k_{13}=0.22 k_{23}=0.80 kw_3=0.02	q_{13}=1.2 q_{23}=1.1 qw_3=0.3 a_{13}=0.178 a_{23}=0.713	$t_{o1,13}$=0.923 $t_{o2,23}$=0.917 $t_{w3,w3'}$=1 $t_{o3,0d}$=1	c_3=2 m_3=0.002	q_3=0.15

Analysis and Results

All the system equations are solved in the order to calculate first the values of all the quality variables (by solving the quality equations and quality transfer equations, backwards as well as forwards, then to calculate the values of all the flow variables (by solving the flow equations and continuity equations backwards by specifying y_{od}), and then to calculate the values of all the cost variables (by solving the cost equations and compatibility equations knowing the exogenous costs of input and wastes). Below are different sets of calculations that have been performed for this illustrative example.

i. Calculation of quality backwards by specifying the desired quality of output as low (z_{od} = 3), medium (z_{od} = 5), high (z_{od} = 8), and perfect (z_{od} = 10). These calculations are performed both with and without perception gaps.

ii. Similarly, the quality is analyzed forward by specifying the available quality of input z_{d1} as low (3) medium (5), high (8) and perfect (10), with and without perception gaps.

iii. The flow on different edges is calculated.

iv. The values of cost variables are determined both by considering the effect of quality on cost (by taking quality cost function q_i) and by ignoring the effect of quality. The cost is calculated for all the four levels of the desired quality of the output.

The results of the analysis are summarized in Tables 5.4, 5.5, 5.6, and 5.7. Table 5.4 gives the values of various quality variables for four different levels of the desired quality of output, namely, low (3 points), medium (5 points), high (8 points), and perfect (10 points). The calculations have been presented both when the system has perception gaps and when no perception gaps are there, i.e. the values of all quality transfer co-efficients are taken to be unity. The results in Table 5.4 indicate that the required quality of input increases proportionately with the quality of output, and this is higher with the perception gaps in the system. It is worth noting that the desired quality of input ($z_{d'1}$) is more than 10 for the high quality output (i.e. 11.2) and for the perfect quality of output (i.e. 14.0). This means that it is not possible to produce high-quality and perfect-quality output with the existing system. The best quality output that can be obtained with perception gaps is 7.22 points as shown in Table 5.5.

Table 5.4: Values of Quality Variables (Backward Calculation)

Sr. No.	Quality variable	Values of quality variables for desired quality of output (z_{od})							
		Low (3)		Medium (5)		High (8)		Perfect (10)	
		With perception gap	Without perception gap	With perception gap	Without perception gap	With perception gap	Without perception gap	With perception gap	Without perception gap
1.	z_{o3}	3.0	3.0	5.0	5.0	8.0	8.0	10.0	10.0
2.	zw_3	0.9	0.9	1.5	1.5	2.4	2.4	3.0	3.0
3.	zw'_3	0.9	0.9	1.5	1.5	2.4	2.4	3.0	3.0
4.	z_{23}	3.3	3.3	5.5	5.5	8.8	8.8	11.0	11.0
5.	z_{13}	3.6	3.6	6.0	6.0	9.6	9.6	12.0	12.0
6.	z_{o2}	3.6	3.3	6.0	5.5	9.6	8.8	12.0	11.0
7.	zw_2	0.61	0.56	1.02	0.94	1.63	1.5	2.04	1.87
8.	zw'_2	0.61	0.56	1.02	0.94	1.63	1.5	2.04	1.87
9.	z_{12}	3.31	3.04	5.52	5.06	8.83	8.1	11.04	10.12
10.	z_{o1}	3.9	3.6	6.5	6.0	10.4	9.6	13.0	12.0
11.	zw_1	0.3	0.28	0.5	0.46	0.8	0.74	1.0	0.92
12.	zw'_1	0.3	0.28	0.5	0.46	0.8	0.74	1.0	0.92
13.	$z_{d'1}$	4.2	3.88	7.0	6.46	11.2	10.34	14.0	12.92
14.	$z_{d'1}$	4.2	3.88	7.0	6.46	11.2	10.34	14.0	12.92

Table 5.5: Values of Quality Variables (Forward Calculation)

Sr. No.	Quality variable	Values of quality variables for desired quality of input ($z_{d'1}$)							
		Low (3)		Medium (5)		High (8)		Perfect (10)	
		With perception gap	Without perception gap	With perception gap	Without perception gap	With perception gap	Without perception gap	With perception gap	Without perception gap
1.	z_{d1}	3.0	3.0	5.0	5.0	8.0	8.0	10.0	10.0
2.	z_{o1}	2.79	2.79	4.65	4.65	7.43	7.43	9.29	9.29
3.	zw_1	0.21	0.21	0.36	0.36	0.57	0.57	0.72	0.72
4.	zw'_1	0.21	0.21	0.36	0.36	0.57	0.57	0.72	0.72
5.	z_{12}	2.40	2.79	4.0	4.65	6.4	7.43	8.0	9.29
6.	z_{13}	2.58	2.79	4.29	4.65	6.86	7.43	8.57	9.29
7.	z_{o2}	2.61	3.03	4.35	5.05	6.96	8.08	8.7	10.1
8.	zw_2	0.44	0.52	0.74	0.86	1.18	1.37	1.48	1.72
9.	zw'_2	0.44	0.52	0.74	0.86	1.18	1.37	1.48	1.72
10.	z_{23}	2.39	3.03	4.0	5.05	6.38	8.08	7.98	10.1
11.	z_{o3}	2.16	2.66	3.62	4.43	5.77	7.08	7.22	8.85
12.	zw_3	0.65	0.8	1.09	1.33	1.73	2.12	2.17	2.66
13.	zw'_3	0.65	0.8	1.09	1.33	1.73	2.12	2.17	2.66
14.	z_{od}	2.16	2.66	3.62	4.43	5.77	7.08	7.22	8.85

Table 5.6: Values of Flow Variables for Desired Output y_{od} = 100 units

No.	Flow Variable	Value
1.	y_{o3}	100
2.	yw_3	2
3.	yw'_3	2
4.	y_{23}	80

5.	y_{13}	22
6.	y_{o2}	80
7.	yw_2	2.4
8.	yw'_2	2.4
9.	y_{12}	82.4
10.	y_{o1}	104.4
11.	yw_1	5.22
12.	yw'_1	5.22
13.	y_{d1}	109.62
14.	$y'_{d'1}$	109.62

As shown in Table 5.6 the input required is 9.62% more than the flow of output, which is due to the waste generated at all the stages.

The impact of the quality on the cost is shown in Table 5.7. The unit cost of output without quality consideration (calculated by the conventional PST model) is $16.788, which is continuously increasing with the increase in the desired quality of the output. For perfect quality of output, the unit of cost is as high as $21.565. It may be further noted that the costs with perception gaps are higher than those without perception gaps in this particular system. This further substantiates the point that the existing PST framework is inadequate to account for the quality-cost relationships, which are no less important.

Table 5.7: Values of Cost Variables for $x_{d'1}$ = \$10.00, $xw'_1 = xw'_2$ = \$3.00, xw'_3=\$4.00, y_{od} = 100 units, z_{od} = [3,5,8,10]

		Values of cost variables for desired quality of output (z_{od}) (in \$)								
			Low (3)		Medium (5)		High (8)		Perfect (10)	
Sr. No.	Cost variable	Value without Quality consideration (in\$)	With perception gap	Without perception gap	With perception gap	Without perception gap	With perception gap	Without perception gap	With perception gap	Without perception gap
1.	x_{d1}	10.0	10.0	10.0	10.0	10.0	10.0	10.0	10.0	10.0
2.	xw_1	2.0	2.0	2.0	2.0	2.0	2.0	2.0	2.0	2.0
3.	x_{o1}	12.356	12.746	12.716	13.006	12.906	13.396	13.236	13.656	13.456

4.	x_{13}	12.356	12.746	12.716	13.006	12.906	13.396	13.236	13.656	13.456
5.	x_{12}	12.356	12.746	12.716	13.006	12.906	13.396	13.236	13.656	13.456
6.	xw_2	3.0	3.0	3.0	3.0	3.0	3.0	3.0	3.0	3.0
7.	x_{o2}	15.437	16.558	16.467	17.306	17.103	18.428	18.103	19.176	18.77
8.	x_{23}	15.437	16.558	16.467	17.306	17.103	18.428	18.103	19.176	18.77
9.	xw_3	4.0	4.0	4.0	4.0	4.0	4.0	4.0	4.0	4.0
10.	x_{o3}	16.788	18.221	18.141	19.176	18.992	20.61	20.314	21.565	21.196
11.	x_{od}	16.788	18.221	18.141	19.176	18.992	20.61	20.314	21.565	21.196

Sensitivity Analysis

The sensitivity analysis has also been performed, to assess the impact of changes in the parameters on the quality of output and on the cost of output.

In order to perform the sensitivity analysis for quality, the quality equations have been solved backwards, starting from the output quality z_{od}, in terms of exogenous variables and the following final relationship is obtained.

$$z_{od} = (t_{od,o3}\, a_{13}\, t_{13,o1}\, a_{d1}\, t_{d1,d'1} + t_{od,o3}\, a_{23}\, t_{23,o2}\, a_{12}\, t_{12,o1}\, a_{d1}\, t_{d1,d'1}\, z_{d'1})$$

This equation is partially differentiated with respect to various modified quality processing coefficients *a* and the quality transfer coefficients *t*. The values of these partial differentials are shown in Table 5.8, indicating that the output quality is most sensitive to processing at stage 3 with respect to the input from stage 2, i.e. a_{23}, and least sensitive to the processing at stage2, i.e. a_{12}. In terms of perception gaps, the output quality is affected maximum by the perception gap between stage 3 and stage 1, i.e. $t_{13,o1}$. The impact of quality processing is higher than that of the perception gaps in this particular system.

Table 5.8: Sensitivity Analysis for Quality

Sr.No.	Partial differential for quality processing	Value	Sr. No.	Partial differential for quality transfer	Value
1.	$\frac{\partial z_{od}}{\partial a_{23}}$	$= 1.279\, z_{d'1}$	5.	$\frac{\partial z_{od}}{\partial t_{d,1d'1}}$	$= 1.0911\, z_{d'1}$
2.	$\frac{\partial z_{od}}{\partial ad_1}$	$= 1.1745\, z_{d'1}$	6.	$\frac{\partial z_{od}}{\partial t_{23,02}}$	$= 0.8362\, z_{d'1}$
3.	$\frac{\partial z_{od}}{\partial a_{13}}$	$= 1.0065\, z_{d'1}$	7.	$\frac{\partial z_{od}}{\partial t_{12,01}}$	$= 0.7852\, z_{d'1}$

4.	$\frac{\partial z_{od}}{\partial a_{12}}$	$= 0.8389\, z_{d'1}$	8.	$\frac{\partial z_{od}}{\partial t_{13,01}}$	$= 0.1654\, z_{d'1}$

The sensitivity analysis for the cost is also performed by solving the cost equations in a backward manner in order to determine the output cost x_{od} in terms of exogenous costs, flow, and quality, as given by the following expression:

$$
\begin{aligned}
x_{od} = (1 + kw_3 + k_{23}\, kw_2)\, [&-(1 + kw_1)\, x_{d'1} + kw_1\, xw'_1 \\
&- (c_{1\text{-}}\, m_1(1 + kw_3 + k_{23}\, kw_2)\, y_{od}) \\
&- p_1(t_{o1,\,13}\, q_{13}\, t_{o3,od} z_{od})] \\
&+ k_{23}\, kw_2\, kw'_2 + k_{23}\, (c_{2\text{-}}\, m_2\, k_{23}\, y_{od}) \\
&+ k_{23}\, p_2(t_{o2,\,23}\, q_{23}\, t_{o3,od} z_{od}) \\
&+ kw_3\, xw'_3 + (c_{3\text{-}}\, m_3 y_{od}) + p_3(t_{o3,od} z_{od})
\end{aligned}
$$

The above cost equation is partially differentiated with respect to technological coefficients k, processing cost co-efficients c and m, quality-cost co-efficients p, quality processing coefficients q and quality transfer co-efficients t. The values of the partial differentials are shown in Table 5.9. The results in this table indicate that the impact of technological coefficient of waste on the cost is almost the same for all three stages, and it is highest for the first stage, kw_1. The impact of processing co-efficients on the final cost is comparatively lower, but is again highest for stage 1, c_1. The impact of the quality-cost coefficients on the final cost is found to be the highest and this is a maximum for stage 1, p_1. The impact of the perception gaps on the output cost is comparatively less.

Table 5.9: Sensitivity Analysis for Cost

No.	Partial differential	Value
A. Effect of technological coefficients (waste)		
1.	$\frac{\partial x_{od}}{\partial kw_1}$	8.352
2.	$\frac{\partial x_{od}}{\partial kw_3}$	8.336
3.	$\frac{\partial x_{od}}{\partial kw_2}$	8:332
4.	$\frac{\partial x_{od}}{\partial kw_{23}}$	3.931

B. Effect of processing cost coefficients		
5.	$\frac{\partial x_{od}}{\partial c_1}$	1.044
6.	$\frac{\partial x_{od}}{\partial c_3}$	1.0
7.	$\frac{\partial x_{od}}{\partial c_2}$	0.8
8.	$\frac{\partial x_{od}}{\partial m_2}$	-64.0
9.	$\frac{\partial x_{od}}{\partial m_3}$	–100.0
10.	$\frac{\partial x_{od}}{\partial m_1}$	–108.9936
C. Effect of quality coefficients		
11.	$\frac{\partial x_{od}}{\partial p_1}$	11.5633
12.	$\frac{\partial x_{od}}{\partial p_3}$	10.0
13.	$\frac{\partial x_{od}}{\partial p_2}$	8.0696
14.	$\frac{\partial x_{od}}{\partial t_{o2,23}}$	1.76
15.	$\frac{\partial x_{od}}{\partial q_{23}}$	1.4672
16.	$\frac{\partial x_{od}}{\partial t_{01,13}}$	1.2528
17.	$\frac{\partial x_{od}}{\partial q_{13}}$	0.9636

REFLECTIONS

An extended framework of PST has been developed by introducing 'quality' as a third category of variables in addition to 'across' and 'through' variables; in fact, the quality variable is a generalization of the 'across' and 'through' variables. By introducing such a variable in the PST framework, it would be possible to model quality and other soft variables, in addition to the conventional aspects being modelled in the PST framework. As the concept of quality is such that it cannot be bound by strict definitions of the 'across' and 'through' variables, the extended framework will be of great help in quality planning. The desired quality of inputs and processing can be known for the desired quality of output and subsequently the impact of the quality on the cost can be analyzed, or the quality of output and the related cost impacts can be calculated for the available quality of inputs. At a second level of analysis, the model can be linked with an optimization model to find the optimum quality and cost of processing. Further, the impact of the perception gaps can also be assessed which will be helpful in identifying the areas where more synchronization in terms of perceptions is needed if better quality is desired.

In the illustrative example, all the parameters and functions were assumed to be deterministic and linear, whereas in real life it need not necessarily be so; in particular, the quality processing coefficients are stochastic in character, and the processing and quality cost functions are usually non-linear in character. If the framework developed in this chapter is explored further it would be possible to account for the stochasticity and non-linearity in the system.

The introduction of quality as a generalized variable will not only facilitate the modelling of quality, but it will be possible to model any other soft variable in the system which does not follow the definitions of across or through variables and the constraints of continuity or compatibility. It is hoped that if the framework proposed in this chapter is properly explored, it will be helpful in quality planning and the modelling of other soft variables.

Notes

1. Sushil (1990) Modelling of Quality in Physical System Theory: An Extended Framework, *Int. Journal of Systems Science,* 21 (12), 2489-2512.
2. Edosomwan J.A. (1987) *Integrating Productivity and Quality Management*, Marcel Dekker, New York.

APPENDIX 5A

SYSTEM MODEL IN EXTENDED FRAMEWORK FOR THE ILLUSTRATIVE EXAMPLE

Component Models

All three stages are considered as transformation processes. By using the generalized model of the *i*th transformation process, the component models of all three manufacturing stages are developed by developing the component terminal graphs (Figure 5.6) and the component terminal equations are as follows:

Turning Stage

Cost equation

$$-x_{o1} = k_{d1}\, x_{d1} - kw_1 xw_1 + f_1\,(y_{o1}) + p_1\,(z_{o1}) \tag{5.23}$$

Flow equations

$$y_{d1} = k_{d1} y_{o1} \tag{5.24}$$

$$yw_1 = kw_1 y_{o1} \tag{5.25}$$

Quality equations (backward analysis)

$$z_{d1} = q_{d1}\, z_{o1} \tag{5.26}$$

$$zw_1 = qw_1\, z_{o1} \tag{5.27}$$

Quality equations (forward analysis)

$$z_{o1} = a_{d1}\, z_{d1} \tag{5.28}$$

$$zw_1 = qw_1\, z_{o1} \tag{5.29}$$

In this case $a_{d1} = 1/q_{d1}$ as there is only one input. Hence, the quality equations for both the backward and forward analysis will be the same.

Milling Stage

Cost equation

$$-x_{o2} = k_{12}\, x_{12} - kw_2 xw_2 + f_2\,(y_{o2}) + p_2\,(z_{o2}) \tag{5.30}$$

Flow equations

$$y_{12} = k_{12} y_{o2} \tag{5.31}$$

$$yw_2 = k_{w_2} y_{o2} \tag{5.32}$$

Quality equations (backward analysis)

$$z_{12} = q_{12} z_{o2} \tag{5.33}$$

$$zw_2 = qw_2 z_{o2} \tag{5.34}$$

Quality equations (forward analysis)

$$z_{o2} = a_{12} z_{12} \tag{5.35}$$

$$zw_2 = qw_2 z_{o2} \tag{5.36}$$

In this case $a_{12} = 1/q_{12}$ as there is only one input. Hence, the quality equations for both the backward and forward analysis will be the same.

Assembly Stage

Cost equation

$$-x_{o3} = k_{13}\, x_{13} + k_{23} x_{23} - kw_3\, xw_3 + f_3\,(y_{o3}) + p_3\,(z_{o3}) \tag{5.37}$$

Flow equations

$$y_{13} = k_{13} y_{o3} \tag{5.38}$$

$$y_{23} = k_{23} y_{o3} \tag{5.39}$$

$$yw_3 = kw_3 y_{o3} \tag{5.40}$$

Quality equations (backward analysis)

$$z_{13} = q_{13}\, z_{o3} \tag{5.41}$$

$$z_{23} = q_{23}\, z_{o3} \tag{5.42}$$

$$zw_3 = qw_3\, z_{o3} \tag{5.43}$$

Quality equations (forward analysis)

$$z_{o3} = a_{13}\, z_{13} + a_{23}\, z_{23} \tag{5.44}$$

$$zw_3 = qw_3\, z_{o3} \tag{5.45}$$

System Graph

On joining the component terminal graphs in 1:1 correspondence, the system graph is obtained as shown in Figure 5.7. In the system graph, dummy linkages, d'_1, w'_1, w'_2, w'_3, and *od* are introduced to make the graph closed. Though the reference nodes R, R_1, R_2 and R_3 are drawn separately for the sake of clarity, these are considered as a single vertex. The total number of vertices v in the system graph is 8, and the total number of edges e is 15.

Structural Constraints

The structural constraints of continuity, compatibility and quality transfer for the given system, are written as follows:

Continuity equations

$$y_{d'1} - y_{d1} = 0 \tag{5.46}$$

$$y_{o1} - y_{12} - y_{13} = 0 \tag{5.47}$$

$$y_{o2} - y_{23} = 0 \tag{5.48}$$

$$y_{o3} - y_{od} = 0 \tag{5.49}$$

$$yw'_1 - yw_1 = 0 \tag{5.50}$$

$$yw'_2 - yw_2 = 0 \tag{5.51}$$

$$yw'_3 - yw_3 = 0 \tag{5.52}$$

Compatibility equations

$$x'_{d1} + x_{d1} = 0 \tag{5.53}$$

$$x_{o1} + x_{12} = 0 \tag{5.54}$$
$$x_{o1} + x_{13} = 0 \tag{5.55}$$
$$x_{o2} + x_{23} = 0 \tag{5.56}$$
$$x_{o3} + x_{od} = 0 \tag{5.57}$$
$$xw_1 + xw'_1 = 0 \tag{5.58}$$
$$xw_2 + xw'_2 = 0 \tag{5.59}$$
$$xw_3 + xw'_3 = 0 \tag{5.60}$$

Quality transfer equations (forward calculation)

$$z_{d1} = t'_{d'1,d1}\, z'_{d'1} \tag{5.61}$$
$$z_{13} = t_{o1,13}\, z_{o1} \tag{5.62}$$
$$z_{12} = t_{o1,12}\, z_{o1} \tag{5.63}$$
$$z_{23} = t_{o2,23}\, z_{o2} \tag{5.64}$$
$$z_{od} = t_{o3,od}\, z_{o3} \tag{5.65}$$
$$zw_1 = t'_{w'1,\, w1}\, zw'_1 \tag{5.66}$$
$$zw_2 = t'_{w2,\, w2}\, zw'_2 \tag{5.67}$$
$$zw_3 = t'_{w3,\, w3}\, zw'_3 \tag{5.68}$$

Quality transfer equations (backward calculation)

$$z_{o3} = t_{od,o3}\, z_{od} \tag{5.69}$$
$$zw'_3 = t_{w3,\, w'3}\, zw_3 \tag{5.70}$$
$$z_{o2} = t_{23,\, o2}\, z_{23} \tag{5.71}$$
$$zw'_2 = t'_{w2,\, w'2}\, zw_2 \tag{5.72}$$
$$z'_{o1} = t_{13,\, o1}\, z_{13} \tag{5.73}$$
$$z''_{o1} = t_{12,\, o1}\, z_{12} \tag{5.74}$$
$$z_{o1} = max\,(z'_{o1}, z''_{o1}) \tag{5.75}$$
$$zw'_1 = t_{w1,\, w1'}\, zw_1 \tag{5.76}$$
$$z_{d'1} = t_{d1,\, d1'}\, z_{d1} \tag{5.77}$$

where

$$t_{od,\, o3} = 1/t_{o3,\, od} \tag{5.78}$$
$$t_{w3,\, w3'} = 1/t_{w3,\, w3} \tag{5.79}$$
$$t_{23,\, o2} = 1/t_{o2,\, 23} \tag{5.80}$$
$$t_{w2,\, w2'} = 1/t'_{w2',\, w2} \tag{5.81}$$
$$t_{13,\, o1} = 1/t_{o1,\, 13} \tag{5.82}$$
$$t_{12,\, o1'} = 1/t_{o1,\, 12} \tag{5.83}$$
$$t_{w1,\, w1'} = 1/t_{w1',\, w1} \tag{5.84}$$
$$t_{d1,\, d1'} = 1/t_{d1',\, d1} \tag{5.85}$$

PART III

APPLICATIONS IN MANUFACTURING MANAGEMENT

6 INTEGRATED OPERATIONS MANAGEMENT OF MULTI-STAGE MANUFACTURING SYSTEMS

6.1 INTRODUCTION

The modelling of manufacturing systems helps in studying the implications of various planning and operating decisions. A number of modelling approaches have appeared in literature. However, there is always a quest for simple modelling methodologies. In this chapter, a simple approach to model multistage manufacturing systems based on physical system theory (PST) is presented.

The objective of this chapter is to develop generalized analytical models for multistage manufacturing systems with rework options using PST concepts and then to show how these models can help in operations management. Specifically, for a specified desired output from the last stage the following analytical relationships at the *i*-th stage are developed:

- the aggregate input requirement;
- the number of units rejected from the basic process;
- the number of units reworked at the rework process;
- the number of units rejected at the rework process;
- the unit output cost from the last stage as a function of various costs and scrap rates at the preceding basic and rework processes up to the *i*-th stage.

These basic analytical relationships can then be used as a management planning tool to address many production planning issues such as material requirement, wastes generated at each stage, machine requirements, justification to acquire precision machines, etc.

There are three levels of planning that concern the operations manager: long-range planning (strategic decisions, such as introducing new products or processes and location of facilities); short-term planning (scheduling); and between these two extremes, intermediate planning or aggregate planning. Aggregate planning involves structuring of a general plan for responding to forecasted demand through some combination of workforce, output, and inventory loadings. One can perceive aggregate planning as a macro-approach to the development of a production plan. This plan allows upperlevel management to make decisions regarding the use of resources without getting into the details of daily schedule and job assignments.

Aggregate planning is the term coined to describe the formulation of general strategies by which capacity can be made to satisfy demand in a most economical way during a specific time period. The common time period is upto a year into the future[1].

For the purpose of generalization, we derive relationships assuming that each stage has a basic process as well as a rework process (precision process). However, when the rework processes are not present at any or some of the stages, results are obtained for such a system by substituting a value of one in the generalized relationships for the rework factors for those stages.

The models developed are illustrated by considering a spur gear manufacturing example. Results with and without rework are analyzed. The sensitivity analysis is carried out to identify the critical machines/operations in a multi-stage manufacturing system of spur gear manufacture.

6.2 AGGREGATE PLANNING METHODS

Graphic and Charting Methods

Some useful planning methods of aggregate planning are graphic and charting methods. These methods consist of the development of simple tables and graphs that enable the operations manager to compare and contrast forecasted demand with existing capacity. This comparison gives the manager the basis for developing plans to achieve intermediate range objectives. The alternative plans are evaluated in terms of their overall cost. The major disadvantage of graphic and charting methods is that they do not necessarily result in the optimum aggregate plan.

These methods involve the following steps[2]:

- **i.** Determine demand for each period.
- **ii.** Determine capacity for each period of regular time,
- **iii.** Determine unit cost for regular time, overtime, subcontracting, and inventory holding cost.
- **iv.** Determine company wide or departmental policies that are followed (e.g. level of safety stock, maximum level of inventory, and preferred work force size).
- **v.** Develop alternative plans and compute the cost of each.
- **vi.** If satisfactory plans emerge, select the best one, otherwise return to step v.

Mathematical Models[3]

Linear Programming

It is possible to formulate aggregate planning in a linear programming framework. The linear programming procedure then identifies the optimal plan for minimizing costs. This is not a trial and error procedure like the graphical method. This plan specifies the number of units of output to produce in each time period, how many shifts the manufacturing facility should operate each time period, and how many units of inventory should be carried each period. It does all of this by taking into account the operating capacity of the facility. One limitation of linear programming is its assumption of linear costs. The linear cost relationships are not always accurate representations of actual costs.

Linear Decision Rules

This procedure results in a set of equations that can be used to calculate the best work force size, production rate, and inventory level for each time period in the planning horizon. This set of equations has been known as the Linear Decision Rules. The advantage of this procedure is that, like linear programming, it guarantees an optimum solution and saves trial and error computations. In addition, it recognizes that some cost relationships may be nonlinear rather than linear. The Linear Decision Rule model has an advantage, as it can use both linear and quadratic cost relationships.

A disadvantage of Linear Decision Rule model is that it must be tailor made for each organization. The procedure requires a careful study of cost structure of a company, which must then be expressed in mathematical form. An extensive mathematical analysis must be made to come up with the proper Linear Decision Rules for that particular company. Whenever, the company's cost relationships change, for example when salaries increase, the mathematical derivation of the Linear Decision Rules must be redone.

Heuristic Approach

A final class of aggregate planning models has evolved over time. These methods apply in situations where management has done aggregate output planning on an intuitive basis.

a. *Management Co-efficient Model:* This procedure requires obtaining records of past work force, production, and inventory decisions. These data are analyzed by multiple regression techniques to find those regression equations that best fit the historical data. These regression equations are then used to make future planning decisions in much the same way that the Linear Decision Rules are used.

b. *Parametric Production Planning:* In this heuristic approach, a rough estimation is made of four variables. These four variables are searched in combination to provide a least cost combination. Since the four variables are inserted into work force and production rate decision rules for actual planning, this approach approximates the firm's aggregate planning environment.

The advantages of these empirical models are that they are easy to obtain if sufficient historical data are available and that by reducing the variability in decision making, they can reduce costs. They must be applied, however, with great caution. The fact that past decision making tendencies have been successful does not necessarily mean that they are applied mechanically to future circumstances. Furthermore, this procedure may provide plans that are no where near optimal.

6.3 INTEGRATED APPROACH TO OPERATIONS MANAGEMENT

The principles of PST can be applied to operations management. The various production stages can be conceptualized as material transformation processes. In general, every process transforms input into output and generates some waste. Physical System Theory can be used as a base to solve the problems of operations management, like, aggregate production planning (APP), materials requirement planning (MRP), capacity requirement planning (CRP), scheduling, costing etc., which will have an additional feature of accounting of waste at each stage.

A generalized model is developed which helps in studying the implications of various planning and operating decisions.

The PST block diagram is prepared with the help of basic data supplied by the concerned operations manager. The main phases of flexible system methodology (as shown in Figure 6.1) are:

- Basic data
- Input to PST model
- PST model
- Output from PST model
- Final results.

Basic Data Required

The basic data required for the development of PST model is given as under:

i. Forecast of demand of a specific period of time.

ii. Scrap rate value and weight of raw material required at different production stages while manufacturing the product.

iii. Production cost at lowest and highest production levels.

iv. Exogenous costs of various raw materials/components.

Input to PST Model

The model maker converts the basic data into the requirements of the model.

i. The forecast of demand gives the flow of material input for that period of time.

ii. The scrap rate value and weight of raw material helps in calculating the various technological coefficients.

iii. The production cost at highest and lowest production levels gives the processing costs.

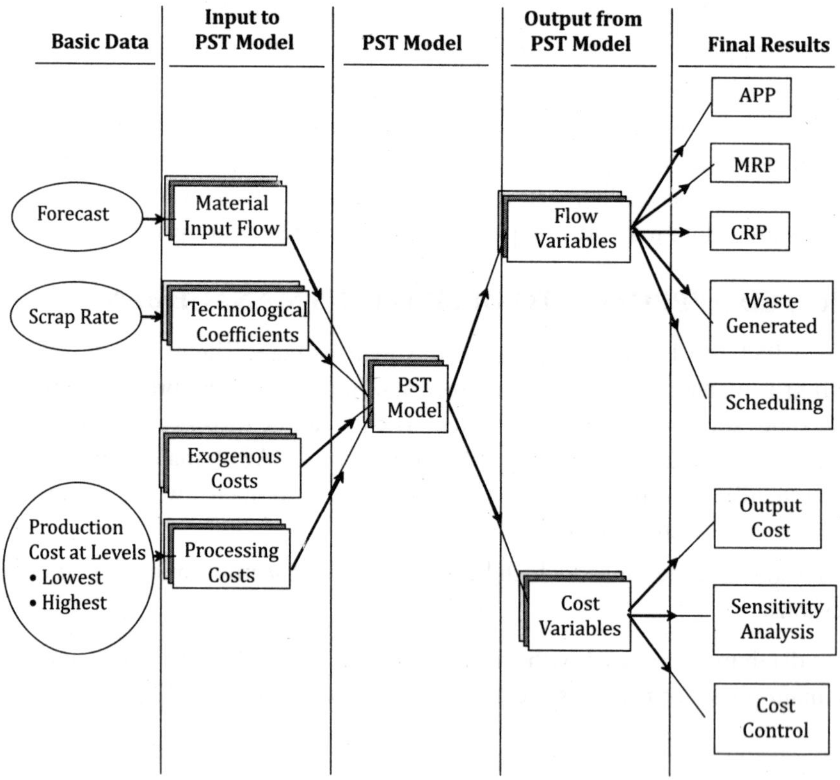

Figure 6.1: Flexible System Methodology for Integrated Operations Management

The various inputs to PST model are:

- Flow of material input
- Technological coefficients
- Processing costs
- Exogenous costs.

Physical System Theory Model

The PST model is developed with the help of input data. The input data has to be accurate and sufficient. The model is based on the principles of material and energy balance. The material transformation processes taking place in a manufacturing unit are modelled. The output from one or more processes acts as an input to other processes. Every process transforms inputs into outputs and generates some waste. The generalized model for multi-stage manufacturing is presented in subsequent section.

Output from PST Model

There are two main outputs from the PST model

i. "Across" variables

ii. "Through" variables

The "across" variable is taken to be the unit cost of material and is denoted by x, while the "through" variable is the flow of material and is denoted by y. The analysis of "across" variables serves as a base for costing while the analysis of "through" variables assists in the production planning. It can also include quality analysis if the extended framework of PST is used as discussed in chapter five.

Final Results from PST Model

The results from the model are listed as under:

i. ***Aggregate Production Plan (APP):*** Aggregate planning is the process of planning the quantity and timing of output over the intermediate time horizon. Within that time frame, the maximum capacity of a production facility is relatively fixed. Given a forecast, planners are concerned with making the best possible use of the organization's labour, materials and capital resources to respond to expected demand which might be either higher or lower than expected.

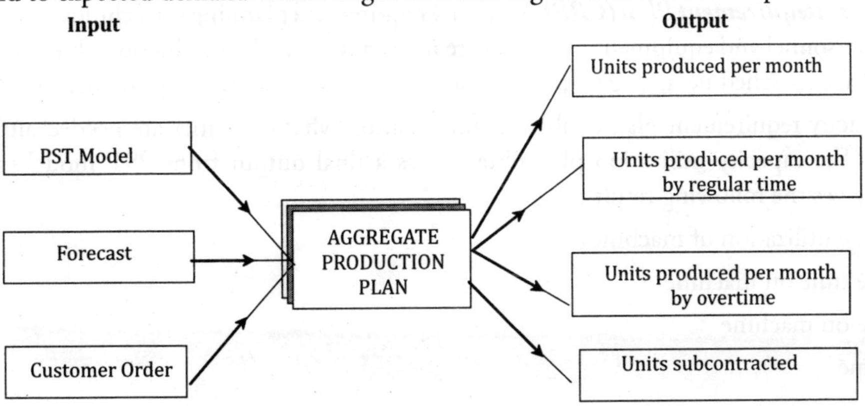

Figure 6.2: Block Diagram for Aggregate Production Plan

The Aggregate Production Plan obtained from PST model, (as shown in Figure 6.2) depicts the following:

- Total number of units produced per month
- Number of units produced by regular time
- Number of units produced by overtime
- Number of units produced per month by subcontracting.

ii. ***Materials Requirement Plan (MRP):*** Materials requirement planning is a technique for determining the quantity and timing for the acquisition of dependent demand items needed to satisfy master schedule/aggregate plan requirements.

The materials requirement plan establishes, specially, what materials are needed. It focuses on priority of materials. The materials requirement plan obtained from PST model (as shown in Figure 6.3) furnishes the following results:

- Bill of materials
- Planned order release
- Measurement of performance of vendors, quality, cost, accuracy, etc.

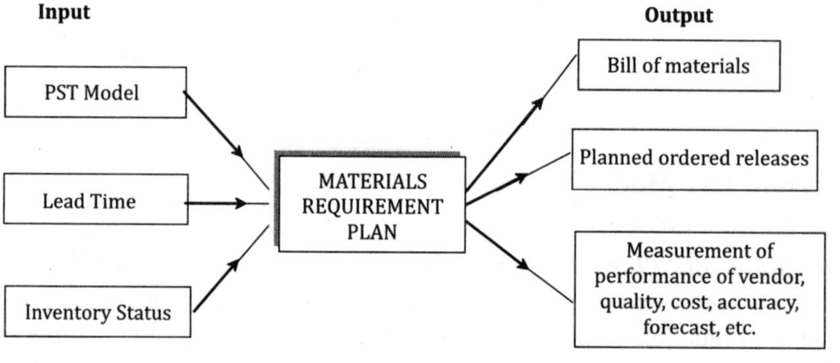

Figure 6.3: Block Diagram for Materials Requirement Plan

iii. ***Capacity Requirement Plan (CRP):*** Capacity requirement planning is a technique for determining what personnel and equipment capacities are needed to meet the production objectives-embodied in the master schedule/aggregate production plan and the materials requirement plan.

The capacity requirement plan establishes information what capacities are needed and when they are needed. The capacity utilization plan obtained as a final output from PST model (as shown in Figure 6.4) gives the following results:

- Percentage utilization of machines
- Available time on machine
- Idle time on machine
- Cycle time.

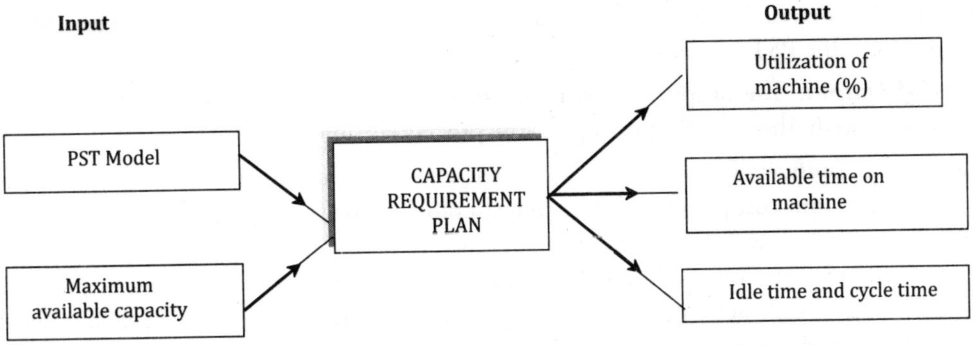

Figure 6.4: Block Diagram for Capacity Requirement Plan

iv. ***Production Schedules:*** The production quantities embodied in the aggregate plan are translated into specific job assignments and short-term or detailed scheduling. At this point, individual jobs are assigned to specific work centres. A realistic schedule is essential for getting the job done and maintaining credibility in the system. If planners call for twice as much work as is reasonable, the schedules will look stupid and will be disregarded. Good scheduling system will release work to a reasonable rate that will keep unnecessary backlogs from the production floor. Gantt charts have been extensively used for scheduling in the past. These charts are easy to understand, and can quickly reveal the current or planned situation to all concerned.

The production schedule can also be obtained from the PST model (as shown in Figure 6.5). The plan thus obtained gives the following details:

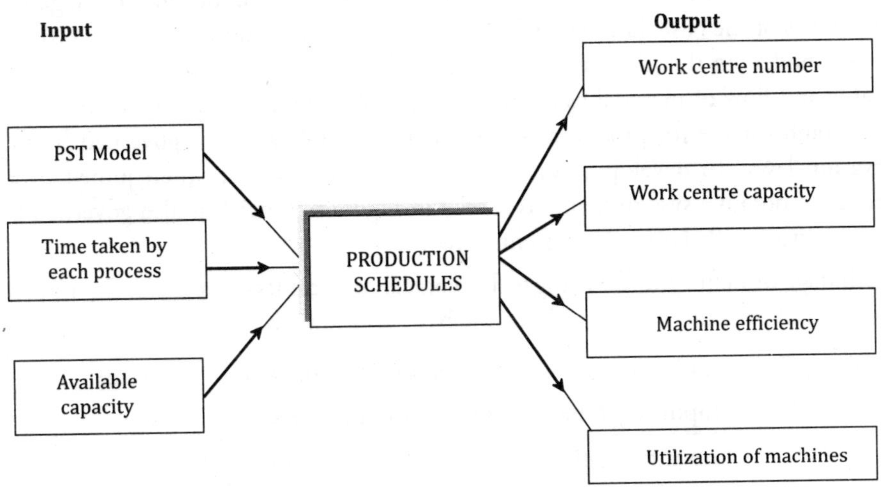

Figure 6.5: Block Diagram for Production Schedules

- Work centre number
- Work centre capacity
- Machine efficiency
- Waste generation at different stages.

v. ***Waste Generation:*** The waste generated at different manufacturing stages can also be got as an output from the PST model.

vi. ***Cost of Output:*** The output cost per unit of the product can be calculated with the help of Physical System Theory. The model gives the process costing.

vii. ***Sensitivity Analysis:*** The sensitivity analysis is also a final output from PST model. It refers to the change in unit cost per unit change in technological coefficient of waste (kw) at a stage, or per unit change in processing cost parameter (c), or the economies of scale. The analysis enables to identify the area or stage that is most or least sensitive to the waste generated, or processing cost or economies of scale at that particular stage. Maximum attention needs to be paid to those stages of production where the sensitivity value is high. This will aid in increasing productivity.

6.4 MODELLING OF MULTI-STAGE MANUFACTURING SYSTEMS

The material flow is the basic characteristic of manufacturing systems. The principles of physical system theory can be applied to manufacturing systems as they obey the principle of conservation. The various production stages can be conceptualized as material transformation processes. In general, every process transforms input into output and generates some waste (scrap). In some cases, the scrap can be classified into reworkable and disposable categories. The rework process may require precision machines. Therefore, each stage is conceptualized to have a basic transformation process (basic machines) and a rework transformation process (precision machines).

Consider a multi-stage manufacturing system consisting of m stages ($j = 1,2,...,m$). Further, assume that each stage consists of a basic and a rework process. We first model the jth stage by developing component models of the basic and rework processes separately as shown in Figure 6.6 and Figure 6.7 respectively. In modelling the jth transformation process, the unit cost of the material is taken as an 'across' variable and flow of material as a 'through' variable. The components are then joined to form a sub-system graph for the jth process as shown in Figure 6.8, which is governed by the continuity and conservation laws. To develop a system model, all the stages are then joined in a one-to-one correspondence to obtain a system graph as shown in Figure 6.9 which is also governed by continuity and energy conservation (balance of payments) laws.

For the jth stage we define the following technological coefficients to represent the input requirements (or waste generated) per unit output for both the basic and rework processes:

Technological co-efficient of input for basic transformation process at the jth stage:

$$kb_{ij} = \frac{\text{Input to } j\text{th basic transformation process}}{\text{Output from } j\text{th basic transformation process}} \tag{6.1}$$

Technological coefficient of waste for basic transformation process at the jth stage:

$$kb_{wj} = \frac{\text{Waste generated at the } j\text{th basic transformation process}}{\text{Output from the } j\text{th basic transformation process}} \tag{6.2}$$

Technological co-efficient of rework at the *j*th process:

$$kb_{rj} = \frac{\text{Rework at the } j\text{th basic transformation process}}{\text{Output from the } j\text{th basic transformation process}} \tag{6.3}$$

Technological co-efficient of input for the rework transformation process at the *j*th stage:

$$kr_{ij} = \frac{\text{Input to the } j\text{th rework process}}{\text{Output from } j\text{th rework process}} \tag{6.4}$$

Technological coefficient of waste for the *j*th rework transformation process:

$$kr_{wj} = \frac{\text{Waste generated at the } j\text{th rework process}}{\text{Output from } j\text{th rework process}} \tag{6.5}$$

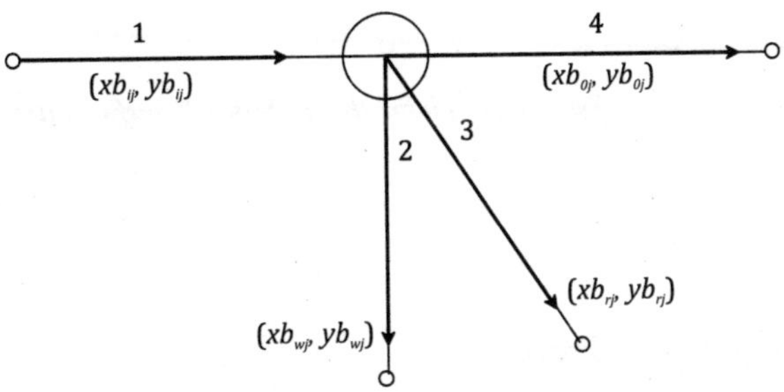

Figure 6.6: Basic Transformation Process with Rework at the jth Stage

Legend

1: Input to the Basic Process

2: Waste from the Basic Process

3: Reworkable Output from the Basic Process

4: Output from the Basic Process.

(): represent the 'across' and 'through' variables.

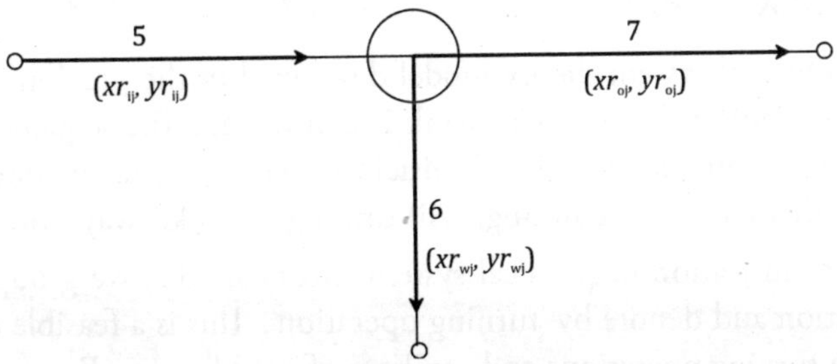

Figure 6.7: Rework Transformation Process at the jth Stage

Legend

5: Input to the Rework Process

6: Waste from the Rework Process

7: Output from the Rework Process

(): represents the 'across' and 'through' variables.

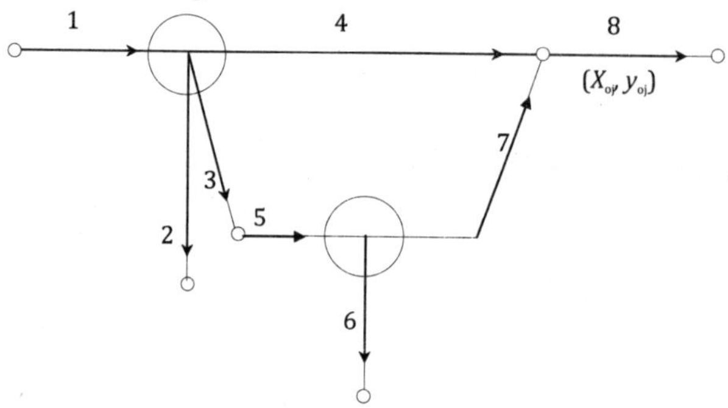

Figure 6.8: Sub-System Graph for the jth Stage Transformation Process

Legend

1-4: Defined in Figure 6.6

5-7: Defined in Figure 6.7

8: Aggregate Output from the *j*th Process.

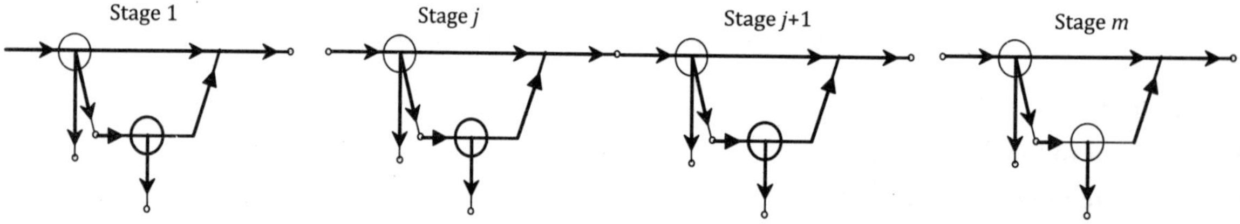

Figure 6.9: System Graph for the m Stage Manufacturing System

The generalized model of the multistage manufacturing system is presented in Appendix 6A.[1]

6.5 ILLUSTRATIVE EXAMPLE

For illustration of the physical system theory model developed in the previous section for multi-stage manufacturing, the machining of a spur gear blank is considered. The sequence of operations for the manufacture of a spur gear from bar stock is: (i) initial inspection; (ii) facing one end; (iii) turning; (iv) parting off; (v) facing other end; (vi) centering; (vii) drilling; (viii) keyway slotting; and (ix) milling.

For the purpose of illustration of physical system theory model, we group the operations (ii) to (vi) into a single operation and denote by 'turning operation'. This is a feasible assumption as all these operations are basically turning operations and can be performed say on Engine Lathe in the sequence outlined. Therefore, the revised sequence of operations after grouping to manufacture a spur gear from a bar stock is: (i) initial inspection; (ii) turning; (iii) drilling; (iv) keyway slotting; and (v) milling.

During the manufacturing process, inspections are made at each stage and the parts not meeting the specifications are scrapped. The information about percentage scrap generated at different stages is used to calculate the technological co-efficients. The following scenarios are examined.

Scenario 1: Gear Manufacturing without Rework

It is assumed that we need to manufacture 1000 finished gears. Rework is not permitted at any of the stages. The input data regarding the scrap rates, processing rates, processing costs at different stages and unit cost of incoming material and salvage value of the scrap are given in Table 6.1. The technological coefficients calculated using the scrap rates information are given in Table 6.2. The data in Tables 6.1 and 6.2 are used to make capacity, cost and sensitivity analyses as follows:

Table 6.1: Input Data for Scenario 1

Sr. No.	Stage	Operating rate (parts per hour)	Scrap rate (%)	Unit processing cost ($)
1.	Initial inspection	25	0.15	0.50
2.	Turning operation	50	4.00	4.50
3.	Drilling operation	50	2.00	2.50
4.	Keyway slotting operation	50	3.00	3.50
5.	Milling operation	30	4.50	4.50

Unit Cost of Incoming Material, xb^i_1 = \$6.4; The Salvage Value of the Scrap per Unit $xb^w_1 = xb^w_2$ = \$3.2.

Table 6.2: Values of the Technological Coefficients (Scenario 1)

Sr No.	Stage	Technological Coefficients of Inputs	Values	Technological Coefficients of Wastes	Values
1.	Initial inspection	kb_{i1}	1.0015022	kb_{w1}	0.0015022
2.	Turning	kb_{i2}	1.0416666	kb_{w2}	0.0416666
3.	Drilling	kb_{i3}	1.0204081	kb_{w3}	0.0204081
4.	Keyway slotting	kb_{i4}	1.0309278	kb_{w4}	0.0309278
5.	Milling	kb_{i5}	1.0471204	kb_{w5}	0.0471204

Capacity Analysis

The capacity analysis can be used for aggregate production planning, materials requirement planning, and capacity requirement planning.

There are five stages and rework is not allowed. Therefore, m = 5; $rf_1 = rf_2 = \dots rf_5 = 1$; final desired output y_d =1000. Using the relationships 31,32 and 33, we obtain the following:

Input to stage 5 (yb^i_5)= $kb^i_5 y_d$ = 1.0471204*1000 = 1047.1204

Waste at stage 5 (yb^w_5)= $kb^w_5 y_d$ = 0.0471204*1000 = 47.1204

Input to stage 4 (yb^i_4)= $kb^i_4 yb_{i5}$ = 1.0309278*1047.1204 = 1079.50553

Waste at stage 4 (yb^w_4)= $kb^w_4 yb_{i5}$ = 0.0309278*1079.50553 = 32.38513031

Number of machines required at stage 5

= yb_{i5}/Procesing rate per machine per day

= 1047.1204/240 = 5.3975 ≅ 6 are needed

Calculations for other stages can be similarly done; the final results are given in Table 6.3.

Table 6.3: Capacity Analysis

Sr. No.	Stage	Number of parts processed	Scrap (units)	Processing rate for 8 hours	Number of machines required	
1.	Initial inspection	1149.1571	1.7236	200	-	
2.	Turning	1147.4334	45.8972	400	2.8685	3
3.	Drilling	1101.5361	22.0306	400	2.7538	3
4.	Keyway slotting	1079.5055	32.3851	400	2.6987	3
5.	Milling	1047.1204	47.1204	240	5.3975	6

Cost Analysis

The cost of incoming material and scrap are exogenously specified. Unit costs calculated using the equation 38 are given in Table 6.4.

Table 6.4: Unit Output Cost at Each Stage

Sr No.	Stage	Unit cost value in $
1.	Initial inspection	6.904807
2.	Turning	11.559173
3.	Drilling	14.229767
4.	Keyway slotting	18.044172
5.	Milling	23.243636

Sensitivity Analysis

The total unit cost equation 38 is partially differentiated with respect to technological coefficients of wastes at different stages. Various production stages are ranked according to their sensitivity values (it refers to a change in unit cost per unit change in the technological coefficient of waste at a stage) as follows:

i. $\partial x_d / \partial kb_{w1} = 3.671787$

ii. $\partial x_d / \partial kb_{w2} = 4.08097896$

iii. $\partial x_d / \partial kb_{w3} = 9.02377402$

iv. $\partial x_d / \partial kb_{w4} = 11.5494561$

v. $\partial x_d / \partial kb_{w5} = 14.991145$

We, therefore, infer that the unit cost of output is most sensitive to the waste generated at the final stage as compared with the previous stages. This is a logical conclusion because of the value addition after being processed at the previous stages.

Scenario 2: Gear Manufacture with Rework

In general, as the variations in the specifications are known to occur, it may often be advantageous to have rework as an alternative to scrapping. The rework factors permit the use of the generalized models for cost and input material requirements to analyze multi-stage manufacturing systems with rework possibilities at any stage. The rework factor assumes a value of one if no rework is permitted at a stage, otherwise its value is determined using the information about the percentage rework at the basic process and the scrap at the rework process itself.

We study three cases of rework for the gear-manufacturing example. In case 1, we permit rework at stags 2 through 5; in case 2 the rework is allowed only at stage 2; whereas in case 3 the rework is permitted at stage 5 only. The data is given in Table 6.5. The technological coefficients calculated are given in Table 6.6. The rework factors for stages 2 to 5 are 0.979791825, 0.98999899, 0.979996012 and 0.974738463 respectively.

Table 6.5: Input Data for Scenario 2

Sr. No.	Stage	Operating rate (parts/hour)	Scrap rate (%)	Rework rate (%)	Processing cost/unit ($/unit)
1.	Initial inspection	-	0.15	-	0.50
2.	Turning	50	2.00	2.00	4.50
3.	Precision turning	10	1.00	-	6.00
4.	Drilling	50	1.00	1.00	2.50
5.	Precision drilling	10	1.00	-	4.00
6.	Keyway slotting	50	1.00	2.00	3.50
7.	Precision keyway	10	1.00	-	6.00

8.	Milling	30	2.00	2.50	4.50
9.	Precision milling	10	1.00	-	7.00

We assume that 1000 finished gears are required. Using the values of rework factors and data from Table 6.5 and Table 6.6, the capacity analysis for case 1 is given in Table 6.7. For this case, we observe that the incoming material requirement is 1064.7886. However, for case 2 and case 3 the incoming material requirements are 1125.9347 and 1118.1781 respectively. For all the three cases the unit output cost analysis is given in Table 6.8. It is interesting to note that for the data assumed, it is economical to permit rework and it is most economical to allow rework at stages 2 to 5 as the unit output costs without rework and for cases 1, 2 and 3 are \$23.243636, \$22.293326, \$23.1471496, and \$22.832480.

Table 6.6: Values of Technological Coefficients (Scenario 2)

Sr. No.	Stage	Technological coefficients of inputs	Values	Technological coefficients of wastes	Values
1.	Initial inspection (Basic)	kb_{i1}	1.0015022	kb_{w1}	0.0015022
2.	Turning (Basic)	kb_{i2}	1.0416666	kb_{w2} kb_{r2}	0.0208333 0.0208333
3.	Precision turning (Rework)	kr_{i2}	1.0101010	kr_{w2}	0.0101010
4.	Drilling (Basic)	kb_{i3}	1.0204081	kb_{w3} kb_{r3}	0.01020408 0.01020408
5.	Precision drilling (Rework)	kr_{i3}	1.0101010	kr_{w3}	0.0101010
6.	Keyway slotting (Basic)	kb_{i4}	1.0309278	kb_{w4} kb_{r4}	0.0103092 0.0206185
7.	Precision keyway (Rework)	kr_{i4}	1.0101010	kr_{w4}	0.0101010
8.	Milling (Basic)	kb_{i5}	1.0471204	kb_{w5} kb_{r5}	0.0206178 0.0265126
9.	Precision milling (Rework)	kr_{i5}	1.0101010	kr_{w5}	0.0101010

The sensitivity analysis of the unit output cost with respect to waste generated at the basic and rework processes is carried out in a similar way as in scenario 1. The results indicate that the successor stages are more critical than the predecessor stages. Therefore, any attempt to improve the technology must start at the last stage. The unit output cost analysis for cases 2 and 3 proves that the rework (improved technology) at the last stage is more economical than allowing rework at the second stage. These analyses would help in making machine (technology) acquisition decisions.

Table 6.7: Capacity Analysis (Rework Allowed on Stages 2 to 5)

Sr. No.	Stage	Number of parts processed	Scrap (units)	Processing rate	Number of machines required	
1.	Incoming inspection	1064.780636	1.597114	-	-	
2.	Turning	1063.183522	21.263650	400	2.6579	3
3.	Precision turning	21.263650	0.212636	80	0.2657	1
4.	Drilling	1041.707849	10.417078	400	2.6042	3
5.	Precision drilling	10.417078	0.104174	80	0.1302	1
6.	Keyway slotting	1031.186664	10.311788	400	2.5779	3
7.	Precision keyway slotting	20.623677	0.206236	80	0.2578	1
8.	Milling	1020.668539	20.413362	240	4.2578	5
9.	Precision milling	25.516703	0.255166	80	0.3189	1

Table 6.8: Unit Output Cost Analysis for Cases with Rework

		Unit output cost at various stages for different cases ($)		
		CASE 1	CASE 2	CASE 3
Sr. No.	Stage	Rework at stages 2 to 5	Rework at stage 2 only	Rework at stage 5 only
1.	Initial inspection	6.904807	6.904807	6.904807
2.	Turning	11.325584	11.325584	11.559173
3.	Precision turning	11.446179	11.446179	-
4.	Drilling	13.973309	14.1144678	14.229767
5.	Precision drilling	14.012990	-	-
6.	Keyway slotting	17.396559	17.95028	18.044172

7.	Precision keyway slotting	17.515936	-	-
8.	Milling	22.117312	23.1471496	22.656466
9.	Precision milling	22.293326	-	22.832480

Other scenarios for the capacity and unit output cost with additional information on tool replacement times, more overtime and fewer machines, etc. can be developed using an approach similar to Davis and Kennedy[3].

6.6 APPLICABILITY OF PST MODELS TO INTEGRATED OPERATIONS MANAGEMENT

The physical system theory models have certain characteristics which makes the modelling and analysis task flexible and efficient as explained below.

i. These models simultaneously consider both the flow and cost variables in an integrated manner in the same model with one set of parameters. The cost models along with flow models thus developed provide an effective quantitative management-planning tool for production planning and control purposes.

ii. The technological coefficient is defined at a single stage irrespective of how it is interconnected to other stages. Thus, the components (production stages) can be modelled independently. However, the transition probability by its very definition involves two stages. Therefore, the parameters of the physical system theory are more pragmatic and comprehensible.

iii. These models are computationally more convenient as in pure forward flow systems, only one equation is to be solved at a time. In case of some recycling in between, some more equations may have to be solved simultaneously.

iv. The sensitivity analysis with respect to changes in technology can be carried out.

v. These models can be integrated with mathematical models and can be used for strategic planning such as system design (technology selection) and tactical/operational planning such as production planning and control.

Physical System Theory can be widely applied to the modelling of manufacturing systems for the purposes of engineering costing, production economics, productivity analysis, waste management, technology selection, etc. In this chapter, only transformation processes were considered. However, the models can be further extended to include transportation and storage processes also. This will facilitate the analysis of material handling and inventory control problems in manufacturing systems.

The concept of physical system theory presented here deals with linear systems. New theoretical developments can be attempted to include non-linearities inherent in production systems. Further, the concepts of physical system theory can be integrated with other modelling approaches such as simulation, optimization, system dynamics, etc., do develop new areas of theoretical research that may be applied to investigate manufacturing systems from altogether new angles.

REFLECTIONS

The physical system theory framework for modelling multi-stage manufacturing systems was presented in this chapter. The modelling approach makes use of parameters which are easily comprehensible and measurable. The generalized models for determining the costs and flows in multi-stage manufacturing systems with rework possibilities at any of the stages were developed. The capacity, cost and sensitivity analysis were carried out for both with and without rework options. A number of scenarios can be developed using the basic results derived in this chapter to help make production-planning decisions. Since the graph theoretic system approach provides a custom tailored fit to each individual system. It can be potentially used for modelling real life manufacturing systems by incorporating realistic system constraints for the purposes of evaluation of alternatives, sensitivity and optimal design studies, and production scheduling.

Notes

1. Singh N. and Sushil (1990) A Physical System Theory Framework for Modelling Manufacturing Systems, *International Journal of Production Research,* 28(6), 1067-1082.
2. Ibid.
3. Davis R.P. and Kennedy Jr. W.J. (1987), Markovian Modelling of Manufacturing Systems, *International Journal of Production Research,* 25, 337.

APPENDIX 6A

GENERATED MODEL OF A MULTI-STAGE MANUFACTURING SYSTEM

Component Model for Basic Transformation Process

The component model for the basic transformation process at the *j*th stage is a set of the following component terminal equations:

Material flow equations

These equations follow directly from equations 6.1 to 6.3.

$$yb_{ij} = kb_{ij}\, yb_{oj} \tag{6.6}$$

$$yb_{wj} = kb_{wj}\, yb_{oj} \tag{6.7}$$

$$yb_{rj} = kb_{rj}\, yb_{oj} \tag{6.8}$$

where:

kb_{ij}, kb_{wj}, *and* kb_{rj} are the technological co-efficients of input, waste and rework at the *j*th basic process respectively; yb_{ij}, yb_{wj}, yb_{rj} and yb_{oj} are the through variables representing input material required, waste generated, material to be reworked and output from the *j*th basic process respectively.

Mass balance equation

$$yb_{ij} = yb_{wj} + yb_{rj} + yb_{oj} \tag{6.9}$$

Cost equation

The following equation is obtained by analyzing the energy conservation law:

$$xb_{ij}\, yb_{ij} + fb_j\, yb_{oj} = xb_{oj}\, yb_{oj} + xb_{rj}\, yb_{rj} + xb_{rj}\, yb_{rj} + xb_{wj}\, yb_{wj} \tag{6.10}$$

where:

xb_{ij} xb_{wj}+xb_{rj}, xb_{oj} and fb_j are the across variables representing unit costs of input, waste, rework, output, and processing at the *j*th basic transformation process. The processing cost fb_j are assumed to be constant.

Component Model for Rework Transformation Process

Similarly, the component model for the rework process at the *j*th stage consists of the following component terminal equations:

Material flow rate equations

These equations follow directly from equations 6.4 and 6.5

$$yr_{ij} = kr_{ij}\, yr_{oj} \tag{6.11}$$

$$yr_{wj} = kr_{wj}\, yr_{oj} \tag{6.12}$$

Mass balance equation

$$yr_{ij} = yr_{wj} + yr_{oj} \tag{6.13}$$

where:

yr_{ij}, yr_{wj} and yr_{oj} are the through variables representing input, waste and output at the rework transformation process at the *j*th stage respectively.

Cost equation

$$xr_{ij}\,yr_{ij} + fr_j\,yr_{oj} = xr_{oj}\,yr_{oj} + xr_{wj}\,yr_{wj} \tag{6.14}$$

where:

xr_{ij}, xr_{wj}, xr_{oj} and fr_j are the across variables representing unit input, waste, output and processing costs respectively at the rework transformation process at *j*th stage.

Model for jth Transformation Stage

Now the basic and rework processes are joined in a one-to-one correspondence to obtain the sub-system graph for the *j*th manufacturing stage as shown in Figure 6.8. The resulting continuity and balance of payments (dollar inflow rate is equal to dollar outflow rate at the joining nodes) equations are:

Continuity equations

$$yb_{rj} = yr_{ij} \tag{6.15}$$

$$y_{oj} = yb_{oj} + yr_{oj} \tag{6.16}$$

where:

y_{oj} is the through variable representing final aggregate output from the *j*th stage.

Balance of payment equations

$$xb_{rj}\,yb_{rj} = xr_{ij}\,yr_{ij} \tag{6.17}$$

$$xb_{oj}\,yb_{oj} + xr_{oj}\,yr_{oj} = x_{oj}\,y_{oj} \tag{6.18}$$

where:

x_{oj} is the across variable representing unit aggregate output cost at the *j*th stage.

On solving the system of equations 6.6–6.9, 6.11–6.13 and 6.15–6.16 we obtain:

$$yb_{ij} = kb_{ij}\,(rf_j)\,y_{oj} \tag{6.19}$$

$$yb_{wj} = kb_{wj}\,(rf_j)\,y_{oj} \tag{6.20}$$

$$yb_{rj} = kb_{rj}\,(rf_j)\,y_{oj} \tag{6.21}$$

$$yr_{wj} = kr_{wj}\,(1-rf_j)\,y_{oj} \tag{6.22}$$

$$yb_{oj} = (1-rf_j)\,y_{oj} \tag{6.23}$$

$$yr_{oj} = (rf_j)\,y_{oj} \tag{6.24}$$

where:

$$rf_j = (kr_{ij} / (kr_{ij} + kb_{ij}))$$

Notice that rf_j represents the rework factor. If kb_{rj} is equal to zero then rf_j is equal to one. That means no rework process exists at the *j*th stage.

On simplification, the equation 6.10 and 6.14 yield the following:

$$xb_{oj} = kb_{ij}\, xb_{ij} - kb_{wj} xb_{wj} - kb_{rj}\, xb_{rj} + fb_j \tag{6.25}$$

$$xr_{oj} = kr_{ij}\, xr_{ij} - kr_{wj} xr_{wj} + fr_j \tag{6.26}$$

Using the equations 25, 26 and 18 we obtain:

$$x_{oj} = (kb_{ij}\, xb_{ij} - kb_{wj} xb_{wj} + fb_j)\,(rf_j) + (fr_j - kr_{wj}\, xr_{wj})\,(1-rf_j) \tag{6.27}$$

The factor $(1-rf_j)$ reduces the second term of equation 27 to zero if rework is not permitted at the *j*th stage. Therefore, the cost and material flow equations for the *j*th stage without rework process are obtained as a special case of a more generalized transformation process involving both basic and rework processes.

The Generalized System Model

To model the multi-stage manufacturing system, the stages are joined in a one-to-one correspondence to form a system graph as shown in Figure 6.9. The system graph is governed by the laws of conservation (continuity as well as balance of payments). Accordingly, the interaction constraints are:

$$y_{o1} = yb_{i2}; \qquad x_{o1}\, y_{o1} = xb_{i2}\, yb_{i2} \qquad \text{(between stage 1 and 2)} \tag{6.28}$$

... ...

... ...

$$y_{oj} = yb_{ij+1}; \qquad x_{oj}\, y_{oj} = xb_{ij+1}\, yb_{ij+1} \qquad \text{(between stages } j \text{ and } j+1) \tag{6.29}$$

... ...

... ...

$$y_{om-1} = y_{im}; \qquad x_{om-1}\, y_{om-1} = xb_{im}\, yb_{im} \qquad \text{(between stages } m-1 \text{ and } m) \tag{6.30}$$

x_d and y_d are the unit output cost and desired output at the *m*th stage.

We obtain the following generalized equations for a multi-stage manufacturing system at the *l*th stage $(j=1,2,\ldots\ldots,l,\ldots\ldots,m)$ when l to m stages are considered together using linking equations 28–30 and material and unit output cost equations 6.19–6.24 and 6.27: Input material required at the *l*th stage:

$$yb_{il} = \prod_{j=1}^{m} kb_{ij} rf_j\, y_d \tag{6.31}$$

Waste generated by the basic process at the *l*th stage:

$$yb_{wl} = kb_{wi} rf_l \prod_{j=l+1}^{m} kb_{ij} rf_j\, y_d \qquad l = 1,2\ldots\ldots\ldots\ldots, m-1; \tag{6.32}$$

$$yb_{wl} = kb_{wl} rf_l y_d \qquad l = m \tag{6.33}$$

Material for rework by the basic process at the *l*th stage:

$$yb_{wj} = kb_{rl} rf_l \prod_{j=l+1}^{m} kb_{ij} rf_j\, y_d \qquad l = 1,2\ldots\ldots\ldots\ldots, m-1; \tag{6.34}$$

$$yb_{rl} = kb_{rl} rf_l y_d \qquad l = m \tag{6.35}$$

Material generated by the rework basic process at the *l*th stage:

$$yr_{wl} = kr_{wl}(1 - rf_l)\prod_{j=l+1}^{m} kb_{ij}rf_j y_d \qquad l = 1, 2............, m-1; \tag{6.36}$$

$$yr_{wl} = kr_{wj}(1 - rf_l)\, y_d \qquad l = m \tag{6.37}$$

The unit output cost at the *m*th stage (x_d) for a coupled multi-stage system (j=1,2,......l,.......,m) is:

$$x_d = \prod_{j=1}^{m}(kb_{ij}rf_j)xb_{il}$$

$$+\prod_{n=l+1}^{m}\prod_{j=n}^{m}[(kb_{ij}rf_j)[(rf_{n-1})(fb_{n-1} - kb_{wn-1}xb_{wn-1})$$

$$+ (1 - rf_{n-1})(fr_{n-1} - kr_{wn-1}\, xr_{wn-1})]\,]$$

$$+ (rf_m)(fb_m\ kb_{wm}\ xb_{wm})$$

$$+ (1 - rf_m)(fr_m - kr_{wm}xr_{wm}) \tag{6.38}$$

Relationship between Scrap Rates, Rework Rates and Technological Coefficients

The main parameters of the physical system theory model are the technological coefficients. In this section we derive expressions for determining the technological coefficients from the information about scrap rates at both the basic and rework processes and rework rate at the basic process. Consider the *j*th transformation process involving both scrap and rework as shown in Figures 6.6 and 6.7. Assume that cb_{wj} and cb_{rj} represent the per cent scrap rate and rework rate at the *j*th basic process respectively and cr_{wj} represents the scrap rate at the *j*th rework process.

$$cb_{wj} = yb_{wj} / yb_{ij} \tag{6.39}$$

$$cb_{rj} = yb_{rj} / yb_{ij} \tag{6.40}$$

$$cr_{wj} = yr_{wj} / yr_{ij} \tag{6.41}$$

These equations along with mass balance equations 6.9 and 6.13 yield the following:

$$kb_{wj} = cb_{wj} / (1 - cb_{wj} - cb_{rj}) \tag{6.42}$$

$$kb_{rj} = cb_{rj} / (1 - cb_{wj} - cb_{rj}) \tag{6.43}$$

$$kb_{ij} = 1 + kb_{wj} + kb_{rj} \tag{6.44}$$

$$kr_{wj} = cr_{wj} / (1 - cr_{wj}) \tag{6.45}$$

$$kr_{ij} = 1 + kr_{wj} \tag{6.46}$$

7 COSTING AND PRODUCTION PLANNING FOR MANUFACTURING OF COMPRESSION SPRINGS

7.1 INTRODUCTION

The objective of this chapter is to show how the multistage production system can be modelled using physical system theory, and how the analysis process can be simplified in forward flow systems to assist in resolving the issues of costing and production planning. The approach has been exemplified by a case study of the manufacturing of compression springs.[1] The system model, consisting of component models of various manufacturing stages, system graph and structural constraints of continuity and compatibility, has been developed and the analysis has been performed for both the cost and flow variables.

7.2 APPLICABILITY FOR COSTING AND PRODUCTION PLANNING

The physical system theory model proposed here deals with two classes of variables, i.e. the unit cost, and the flow of material. Hence, it can be suitably applied to cater for the issues of costing and production planning as shown in Figure 7.1. The unit cost of the product, whether intermediate or final, can be easily calculated considering both the direct and indirect costs of production. Only the exogenous costs of raw material and waste disposal should be known as direct costs. The indirect costs such as labour, energy, overheads, etc., are dealt with by the processing cost function 'f'. This function also takes into consideration the economies of scale. Thus, the proposed model can be easily applied for the purposes of costing in a multi-stage production system, which takes care of possible interactions of processing centres.

The model can be separately run for standard exogenous costs and standard processing coefficients and the actual costs can then be compared with the standard costs for the purposes of cost monitoring. The sensitivity analysis will help in identifying the critical stages for cost control and technology upgradation.

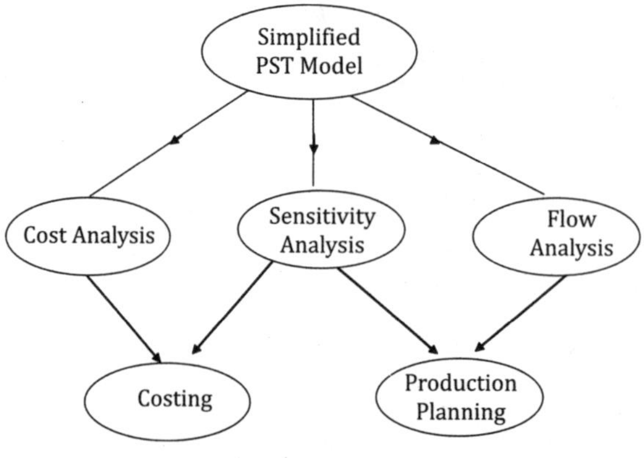

Figure 7.1: Flexible System Methodology for Costing and Production Planning

The value of 'through' variables determined at all stages gives the load on various processing centres. This can be taken as a basis for calculating the number of machines or the capacity of the processing unit to be selected at different processing centres. If the capacity of the machines at a processing centre is known, the model will help in identifying whether or not the load can be catered for by overtime or if a new machine is to be procured. The load calculation at various processing centres is a major input to the production planning process. Further, it also gives the value of waste generated at different stages which can be used in the design of waste disposal systems.

7.3 A CASE STUDY- MANUFACTURING OF COMPRESSION SPRINGS

System Description

The economic model of the manufacture of compression springs used in the clutch plate of a two wheeler, has been considered. The springs are manufactured by a small-scale unit. Only transformation processes have been considered. The production of these compression springs involves the following basic processes:

- **a.** Spring coiling
- **b.** Heat treatment-stress relieving
- **c.** End-grinding
- **d.** Load testing
- **e.** Phosphating

Spring Coiling

The spring coiling is carried out on an automatic universal spring coiling machine. It can coil spring wires from 1.5 to 4.5 mm diameter. It can produce any shape and size of springs within this range. The process of coiling involves feeding the spring wire into the machine through a wire straightened and feeding rollers. The springs are coiled automatically and cut to a pre-set size. The machine has the capacity to coil 60 springs per minute.

Heat Treatment

This is carried out in an electric furnace with air circulation at a temperature of 300°C. One lot of springs is kept for 30 minutes at this temperature. The size of chamber in the furnace is 18"×18"×24" and in one lot, 3000 springs can be heat-treated. The aim of heat-treatment is to relieve stresses induced in the spring during coiling.

End Grinding

This process involves grinding the spring ends from both sides simultaneously and maintaining the ends at right angle to the axis of the spring. In other words, this process ensures squareness of the springs. The process is carried out on an automatic universal springs end grinding machine fitted with two grinding heads. The grinding capacity of this machine is 30 springs per minute.

Load Testing

The process involves the application of a test load, compressing the springs to specified height as defined in the drawing of the spring. The spring is compressed to the specified height by mechanical or hydraulic machines and the load is displayed by a digital system. The machine has a minimum division of 0.5 mg and a maximum spring load capacity of 300 kg. Springs are tested in this machine by random sampling.

Phosphating

This process provides a protective rust-preventive layer of black colour on the surface of the spring. One basket carried 500 compression springs for phosphating and it takes 15 minutes to phosphate this lot of springs.

Simplified System Model

The model is based on the principles of material and energy balance. The simplified system model is composed of component models, and simplified system graph with inbuilt structural constraints.

Component Models

Every production stage is treated as a component which takes some input(s), gives a main output and generates waste. The generalized model of a transformation process, as outlined in chapter three has been utilized to develop component models for all five production stages. The component terminal graphs of the five production processes are shown in Figure 7.2. A lot of 20,000 springs has been considered as the unit of production. The material flows taken in terms of kg are treated as 'through' variables, whereas the costs of production per kg are treated as 'across' variables. The cost and flow equations for different components are given in Appendix 7 A for the simplified model.

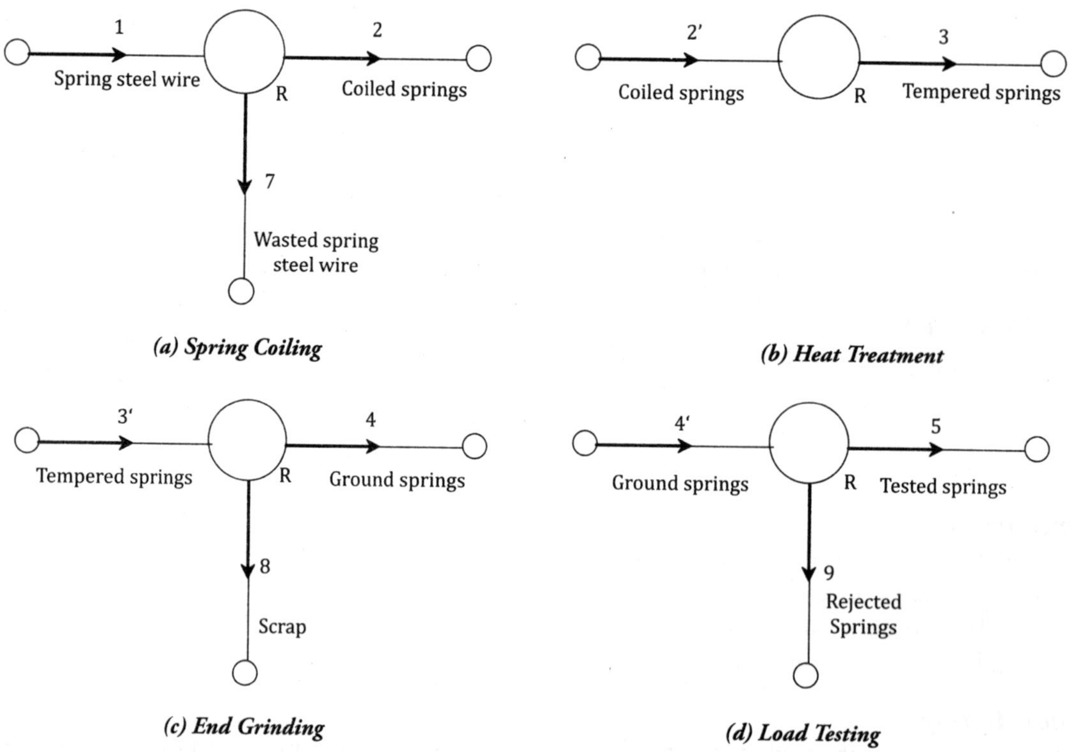

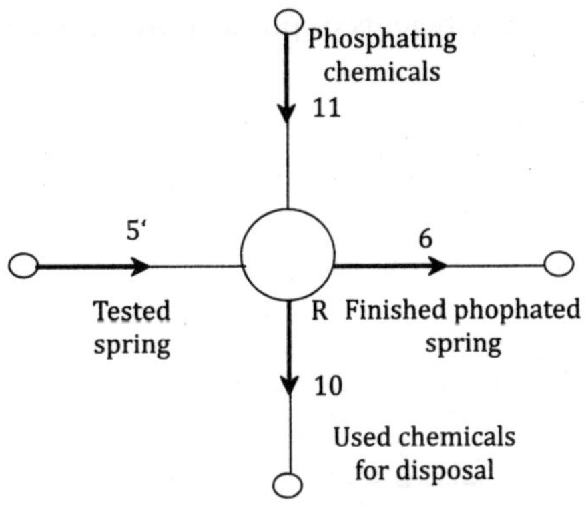

(e) Phosphating

Figure 7.2: Component Terminal Graphs

System Graph

On joining the component terminal graphs in one-to-one correspondence the material flow network is obtained. This network is closed by introducing dummy links. The system graph is shown in Figure 7.3. All the reference nodes (represented by R) in the system graph are treated as a common vertex. The simplified system graph is shown in Figure 7.4.

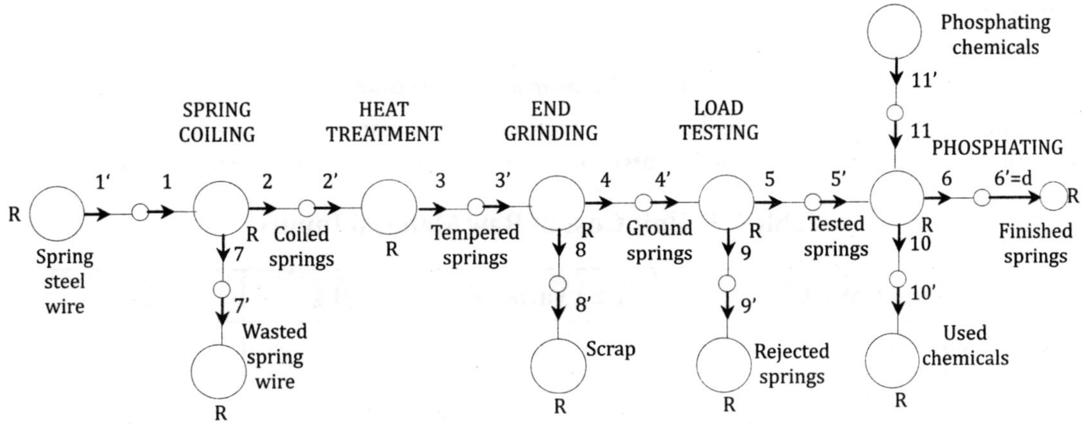

Figure 7.3: System Graph

Structural Constraints

The structural constraints for continuity and compatibility are merged into component terminal equations as per the rules of simplification given in chapter four. The mass balance equations are given in Appendix 7 A.

Input Data

The raw material input at the spring coiling machine is hardened and tempered spring steel wire to ISI:4454. The other raw materials are chemicals for phosphating, which include hydrochloric acid, sodium nitrite, amorphous zinc and water. In one lot 20,000 springs are produced and the weight of one lot (y_6) is 9.35 kg. The costs of raw material inputs are shown in Table 7.1.

Values of various technological coefficients of production are as shown in Table 7.2.

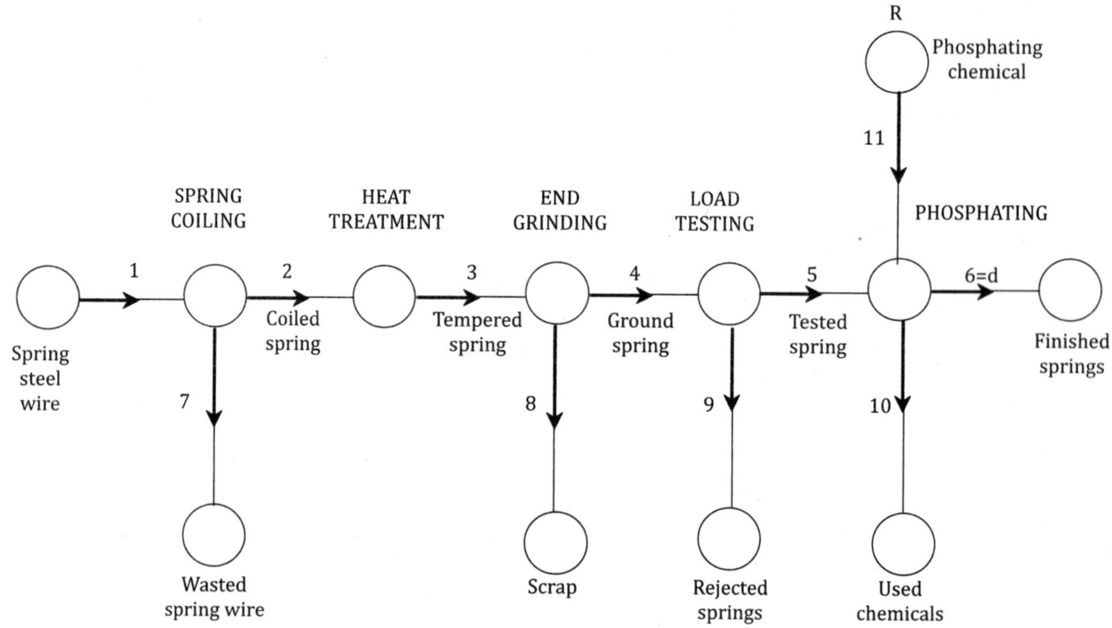

Figure 7.4: Simplified System Graph

Assuming processing costs to be linear functions of the amount of production, the ratios of proportionality (processing cost coefficients) have been worked out and are given in Table 7.3.

Table 7.1: Unit Cost of Raw Material Inputs

Input/Waste	Cost Variable	₹/kg	₹/lot
Spring wire	x_1	18	168.3
Wasted spring wire (can be sold)	x_7	- 0.5	- 4.675
Scrap (can be sold)	x_8	- 0.5	- 4.675
Rejected springs (can be sold)	x_9	- 9	- 84.15
Used chemicals (disposal cost)	x_{10}	0.053	0.4956
Phosphating mixture	x_{11}	0.08	0.748

Table 7.2: Values of Technological Coefficients

k_1=1.1	k_7=0.1
k_2=1.0	k_8=0.05
k_3=1.05	k_9=0.25
k_4=1.25	k_{10}=5.30 cu.ft/kg
k_5=1.0	k_{11}=5.30 cu.ft/kg

Table 7.3: Values of Processing Cost Coefficients

f_2=	₹ 123.45/kg/lot
f_3=	₹ 49.38/kg/lot
f_4=	₹ 131.9/kg/lot
f_5=	₹ 10.92/kg/lot
f_6=	₹ 32.78/kg/lot

Systems Analysis

Solution of system equations

In the present case, the specifications of the simplified system model are:

Number of edges e = 11

Number of vertices v = 12

Number of specified 'across' variables n_x = 6

(unit cost of exogenous input/waste)

Number of specified 'through' variables n_y =1 (flow of output)

There are a total of '2e', i.e. 44 variables in the system graph, out of which ($n_x + n_y$), i.e. seven variables are specified. Therefore, to determine the values of all the variables 37, equations are to be solved simultaneously. If the branch-chord formulation is used then ($e - n_x - n_y$), i.e. 15 simultaneous equations will have to be solved. The solution procedure can be considerably simplified, as this is a pure forward flow system. By using the proposed simplified solution procedure, the values of all the unknown variables are determined by solving only one equation at a time. In the simplified model there are only 11 edges, i.e. 22 variables.

The values of all the flow variables are determined by following the steps given in chapter three for the 'through' variables. For the output of one lot (y_6 = 9.35 kg) the values of different flow variables are calculated as shown in Table 7.4.

After calculating the values of all the flow variables the values of cost variables are determined by following the steps given in chapter three for the 'across' variables, as shown in Table 7.5.

Table 7.4: Values of 'Through' Variables

Sr. No.	Material	Variable	Value
1.	Spring steel wire	y_1	13.4991 kg
2.	Coiled springs	y_2	12.2719 kg
3.	Tempered spring	y_3	12.2719 kg
4.	Ground springs	y_4	11.6875 kg
5.	Tested springs	y_5	9.35 kg
6.	Finished springs	y_6	9.35kg
7.	Wasted spring wire	y_7	1.2272 kg
8.	Scrap	y_8	0.5844 kg
9.	Rejected springs	y_9	2.3375 kg
10.	Used chemicals	y_{10}	49.5 cu.ft.
11.	Phosphating chemical	y_{11}	49.5 cu.ft.

Table 7.5: Values of 'Across' Variables

Sr. No.	Material	Cost Variable	Value (per kg) (₹)	Value (for a lot of 20000 i.e. 9.35 kg) (₹)	Value (for a single spring) (₹)
1.	Springs steel wire	x_1	18.0	168.3	0.0084
2.	Wasted springs wire	x_7	– 0.5	– 4.675	– 0.00023
3.	Scrap	x_8	– 0.5	– 4.675	– 0.00023
4.	Rejected springs	x_9	– 9.0	– 84.15	– 0.0042
5.	Used chemicals	x_{10}	0.053	0.4956	0.000025
6.	Phospating chemical	x_{11}	0.08	0.748	0.000037
7.	Coiled springs	x_2	181.7785	1699.6286	0.08498
8.	Tempered springs	x_3	246.5898	2305.6150	0.11528
9.	Ground springs	x_4	423.7943	3962.477	0.19812
10.	Tested springs	x_5	538.4129	5034.1608	0.2517
11.	Finished springs	x_d	571.8978	5347.2449	0.26736

Sensitivity Analysis

Starting from the end variable, i.e. x_d the substitutions are made in a backward manner. On making further substitutions the unit cost of output x_d is determined in terms of exogenous variables.

The technological coefficients of inputs are replaced by the technological coefficients of wastes by using the mass balance equations, e.g. $k_1 = 1 + k_7$. This modified equation has been partially differentiated with respect to various technological coefficients to carry out the sensitivity analysis of the spring manufacturing process.

The values of all partial differentials are determined and they are arranged in a descending order of their values as shown in Table 7.6.

The results of sensitivity analysis indicate that the unit cost of the final product is most sensitive to the phosphating process (k_5) then to load testing (k_9), then to end grinding (k_8), then to heat treatment (k_2), and then to spring coiling (k_{12}). The sensitivity with respect to the flow of phosphating chemical is negligible.

Table 7.6: Values of Sensitivity Analysis

Sr. No.	Process	Partial differential	Value
1.	Phospating process	$\frac{\partial x_d}{\partial k_5}$	9847.2703
2.	Load testing	$\frac{\partial x_d}{\partial k_9}$	7646.6655
3.	End grinding	$\frac{\partial x_d}{\partial k_8}$	5527.3548
4.	Heat treatment	$\frac{\partial x_d}{\partial k_2}$	4219.1473
5.	Spring coiling	$\frac{\partial x_d}{\partial k_7}$	214.7578
6.	Phosphating chemical	$\frac{\partial x_d}{\partial k_{11}}$	0.4956

7.4 DISCUSSION ON SIMPLIFICATION

It may be observed that the simplified model obtained in the above manner will be similar to a digraph model of the manufacturing systems. The digraph theory has been applied to a limited extent to bill-of-materials, cost calculations and scheduling, particularly in the context of network flows[2].

Though, in peculiar cases like that discussed in this chapter, simplified PST models will look like digraph models, the PST approach to model manufacturing systems caters for a wide variety of situations and it is not possible to reduce the number of variables in all cases by the proposed simplification method. For example, if a transportation and/or storage process is introduced between two transformation stages the proposed simplification rules will not be operational. Further, if the manufacturing system is not a pure forward flow system, but has a large number of feedbacks, it will be possible to effectively solve the PST models by making use of the specialized technique of branch, chord and branch-chord formulation and the dynamic models can be suitably converted into state-space formulation[3].

REFLECTIONS

The application of physical system theory to multi-stage forward flow manufacturing systems has been considerably simplified. The case study (manufacturing of compression springs) has made it amply clear that the physical system theory and proposed simplification can be fruitfully utilized for the purposes of costing and production planning of multi-stage manufacturing systems. It is hoped that this simplified approach will be helpful in the practical implementation of this methodology to the modelling and analysis of manufacturing systems.

Notes

1. Sushil, (1992) Costing and Production Planning Using Physical System Theory-A Case Study of Manufacturing of Comprehension Springs, *Production Planning and Control,* 3(2), 141–150.
2. Aneja A.P. and Nair K.P.K., (1982) Multicommodity Network Flows with Probabilistic Losses, *Management Science,* 28(9), 1080–1086; Assad A.A., (1978) Multicommodity Network Flows-A Survey, *Network,* 8, 37–91; Battersby A., (1970) *Network Analysis for Planning and Scheduling,* Macmillan, London; Davis R.P., and Miller D.M., (1979), Analysis of the Machine Requirements Problem Through a Parametric Markovian Model, *SME Technical Paper* MS79-96; Deganzo C.F., and Newell G.F., (1986) Configuration of Physical Distribution Networks, *Networks,* 16, 113–132; Dorsey R.C., Hodgson T.J., and Ratliff H.D., (1975) A Network Approach to a Multi-facility, Multi-product Production Scheduling Problem without Backordering, *Management Science,* 21(7), 813–822; Evans J.R., (1977) Some Network Flow Models and Heuristics for Multi-product Production and Inventory Planning, *AIIE Transactions,* 9(1), 75–81; Ford L.R., and Fulkerson D.R., (1962) *Flows in Networks* (Princeton University Press, Princeton, NJ); Jensen P.A. and Barnes W.J. (1980) *Network Flow Programming* (John Wiley, New York); Kanda A., and Rao U.R.K., (1984) A Network Flow Procedure for Project Crashing with Penalty Nodes, *European Journal of Operational Research,* 16, 174-182.
3. Koenig H.E., Tokad Y., and Kesavan H.K., (1967) *Analysis of Discrete Physical Systems,* McGraw-Hill, New York.

APPENDIX 7A

SIMPLIFIED SYSTEM MODEL

Spring Coiling

Flow equations:

$$y_1 = k_1 y_2 \tag{7.1}$$

$$y_7 = k_7 y_2 \tag{7.2}$$

Cost equation:

$$x_2 = k_1 x_1 + k_7 x_7 + f_2(y_2) \tag{7.3}$$

Heat Treatment

Flow equation:

$$y_2 = k_2 y_3 \tag{7.4}$$

Cost equation:

$$x_3 = k_2 x_2 + f_3(y_3) \tag{7.5}$$

End Grinding

Flow equations:

$$y_3 = k_3 y_4 \tag{7.6}$$

$$y_8 = k_8 y_4 \tag{7.7}$$

Cost equation:

$$x_4 = k_3 x_3 + k_8 x_8 + f_4(y_4) \tag{7.8}$$

Load Testing

Flow equations:

$$y_4 = k_4 y_5 \tag{7.9}$$

$$y_9 = k_9 y_5 \tag{7.10}$$

Cost equation:

$$x_5 = k_4 x_4 + k_9 x_9 + f_5(y_5) \tag{7.11}$$

Phosphating

Flow equations:

$$y_5 = k_5 y_6 \tag{7.12}$$

$$y_{11} = k_{11} y_6 \tag{7.13}$$

$$y_{10} = k_{10} y_6 \tag{7.14}$$

Cost equation:

$$x_6 = k_5 x_5 + k_{11} x_{11} + k_{10} x_{10} + f_6(y_6) \tag{7.15}$$

Mass Balance Equations

Spring coiling	$y_2 = y_1 - y_7$	(7.16)
Heat treatment	$y_3 = y_2$	(7.17)
End grinding	$y_4 = y_3 - y_8$	(7.18)
Load testing	$y_5 = y_4 - y_9$	(7.19)
Phosphating	$y_6 = y_5$	(7.20)

8 APPLICATION TO A LARGE SYSTEM—MANUFACTURING OF TURBOGENERATORS

8.1 INTRODUCTION

In the previous chapter, a case study of manufacturing of a small product was discussed. In order to assess the applicability of PST modelling of manufacturing systems, in this chapter a case study of a large product is presented, which indicates the difficulty of handling with large number of variables and equations.

Turbogenerator is a complex system involving a large number of components. Manufacturing of such large systems involves different kinds of materials, multiple manufacturing stages and wastes. Estimation of material flow requirements and generation of waste is very important for materials requirement planning and loading of different stages. At the same time, costing and productivity analysis is a cumbersome and involved exercise. Traditional costing methods fail to provide insights about the points of cost control which have maximum potential for cost reduction and productivity improvement. A real life case of manufacturing of turbogenerator is taken and an overview of the PST model and significant results are discussed to appreciate the applicability of PST modelling for manufacturing planning and control of large scale systems.

In the case of manufacturing of turbogenerators presented here, there are 278 edges in the simplified system graph which could have been over 500 in the conventional PST modelling. This clearly indicates the utility of the simplified framework in the context of such systems. Moreover, the step-by-step solution procedure helps in the analysis of the model considerably. By conventional approaches the matrix size would have been unduly large, whereas in the simplified procedure all the equations are to be simply listed in order and could be solved either manually or using any programming language. If the spreadsheet software is used, then even the sequencing is not required. Pin pointing the sensitive areas and their proportionate impact in such large systems is very difficult, rather impossible, by conventional means, which has been the focus kept for this study. Sensitivity analysis has facilitated the classification of stages with different levels of sensitivity: very high, high, medium, low, and very low. With the help of such an analysis selective controls can be imposed in the system, as it is difficult rather undesired to put same degree of control at all the stages. As model is large and the reader would be interested only in the process rather than the details, only relevant parts of the model and sample parameters and results are shown.

The study was carried out at Bharat Heavy Electricals Limited (BHEL), Equipment Plant at Hardwar for identification of different parts of turbogenerator, manufacturing process involved, collection of flow data, cost of exogenous items, and processing cost at different stages.[1]

8.2 SYSTEM DESCRIPTION

The BHEL's Equipment Plant manufactures Turbogenerator sets, A.C. machines, D.C. machines, and Apparatus and control gear. The major demand was estimated to be for 210 MW and 235 MW turbogenerator sets and comparatively one third for the 500 MW sets. Thus, the study was confined to only 210 MW and 235MW sets. However, the process would have been similar for the other categories also. Here, the results with respect to only 210 MW sets are provided as the main purpose is to highlight the process of modelling and its utility.

The main components of a turbogenerator are as follows:

i. Stator	**ii.** Rotor
iii. Bearing with Brush Gear	**iv.** Shaft Seals
v. Gas Coolers	**vi.** End Shields
vii. End Terminals	

i. **Stator:** The stator body with core and stator winding form the heaviest component of the entire turbogenerator. It has pipe connections to gas coolers, terminal boxes, jacking and dragging provision for handling, stator core, and stator winding. It has a three phase, double layer short pitched and bar type of winding having two parallel paths. Important considerations in winding are conductor construction, method of insulation, corona prevention, laying of stator winding, end winding, electrical and water connection, and headers.

ii. **Rotor:** This is another major component having rotor shaft, rotor winding, rotor wedges, retaining ring, fans, and field lead connections. The rotor shaft is a long forging measuring more than 9 meters in length and slightly more than one meter in diameter in case of 210 MW sets. The rotor windings are made of copper conductor with proper insulation and cooling arrangements.

iii. **Bearing with Brush Gear:** The rotor shaft is supported on pedestal type of bearings which have spherical seating to allow self-alignment. The field current to the rotor winding is provided through the brush gear, which is rigidly fixed on the extended part of the bearing pedestal on the exciter side.

iv. **Shaft Seals:** The escape of hydrogen along rotor shaft is prevented by employing shaft seals of radial flow of thrust type at either end of the generator. These are mounted between the end shields and the bearings.

v. **Gas Cooler:** Hot hydrogen is cooled by four gas coolers mounted longitudinally inside the generator stator body. Gas coolers consist of cooling tubes made out of brass with coiled copper wire wound on them to increase the surface area of cooling. Cooling water flows through the tubes.

vi. **End Shields:** To make the stator body gas tight at the two ends, two end shields are fitted with the help of bolts. Gas tightness is achieved by putting a rubber sealing chord.

vii. **End Terminals:** Three phases and six neutral terminals are brought out from the stator frame through bushings which are capable of withstanding high voltage and provided with gas tight joints.

8.3 SIMPLIFIED SYSTEM MODEL

Component Terminal Graphs

The modelling of manufacturing of turbogenerator starts with the identification of different components of turbogenerator and its manufacturing process. The number of vital components of a turbogenerator are thirteen. The number of production stages required for manufacturing those components are one hundred and two. The component terminal graphs for manufacturing turbogenerator have been developed. The response-stimulus equations for different components can be written by using the generalized component model in chapter five with appropriate values of j and i.

System Graph

According to the system postulate of PST the component terminal graphs are joined in 1 to 1 correspondence of how the components are joined in reality to obtain the system graph. The system graph thus obtained has been simplified and is shown in Figure 8.1. Since the number of edges are very large the simplified system graph is shown sub-system wise. The simplified system graphs of following sub assemblies and sub-subassemblies are shown in Figures as indicated below:

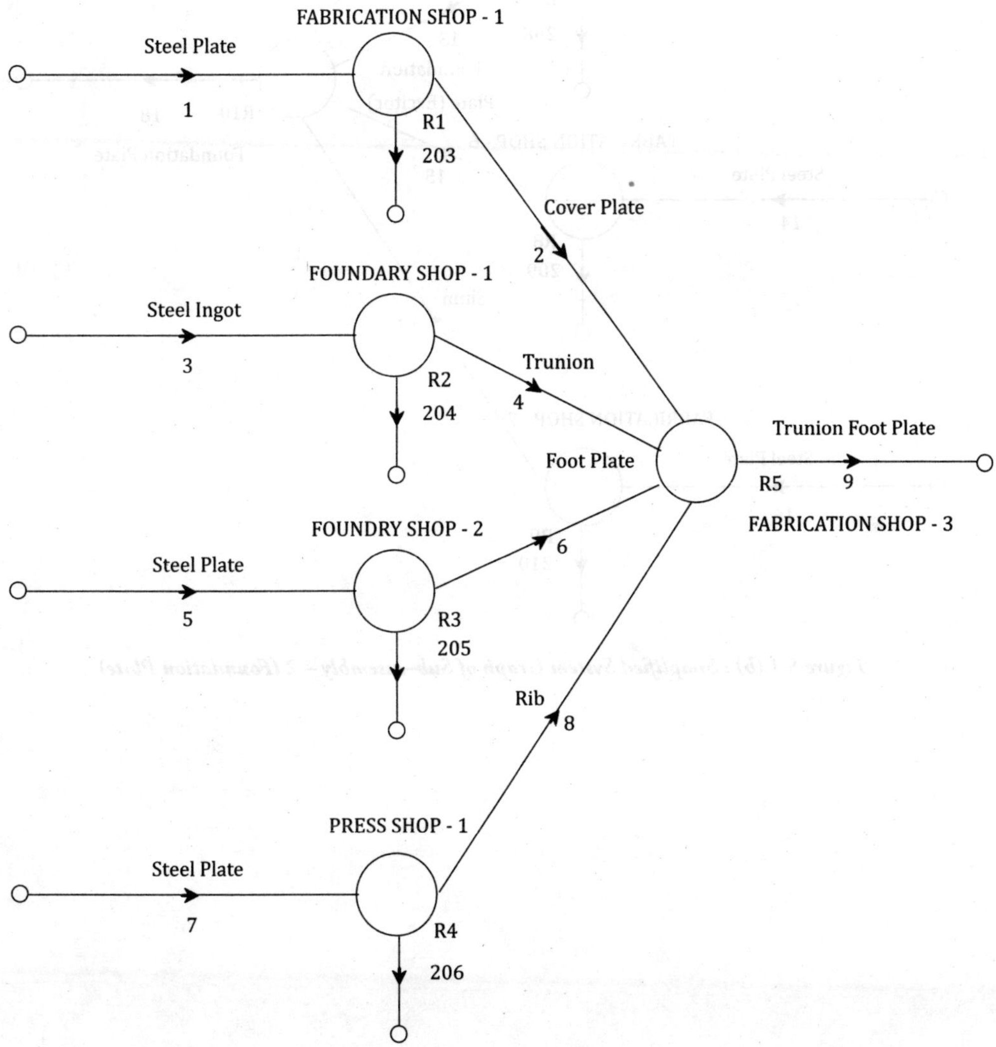

Figure 8.1 (a) : Simplified System Graph of Sub-assembly-1 (Trunion Foot Plate)

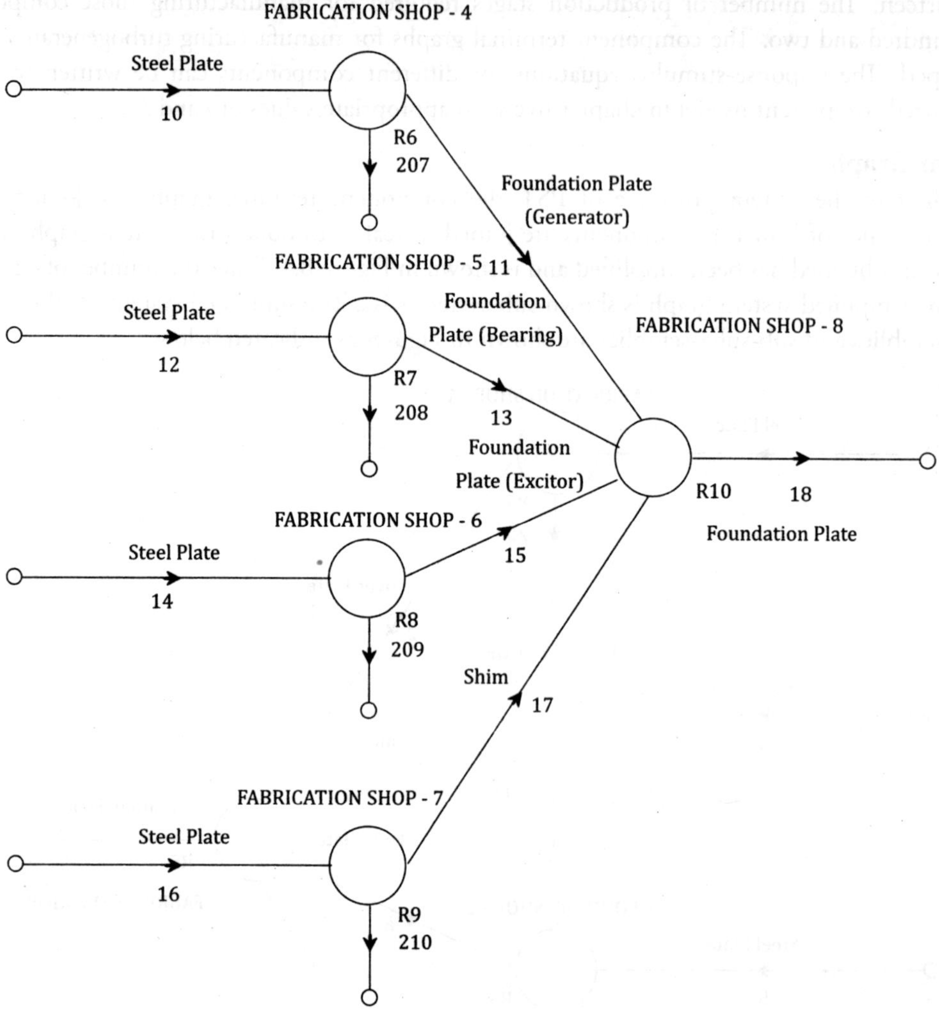

Figure 8.1 (b) : Simplified System Graph of Sub–assembly – 2 (Foundation Plate)

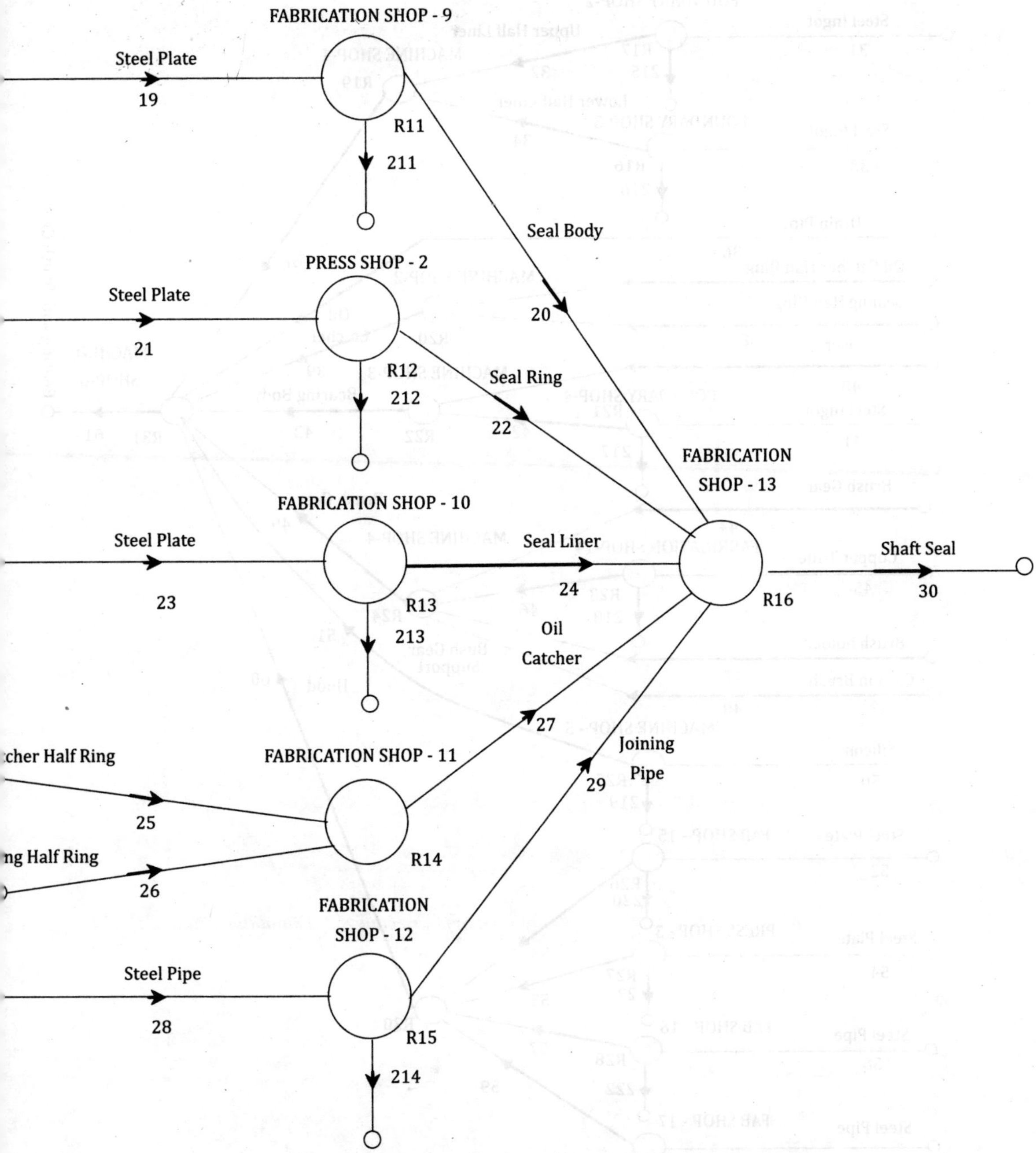

Figure 8.1 (c) : Simplified System Graph of Sub–assembly – 3 (Shaft Seal)

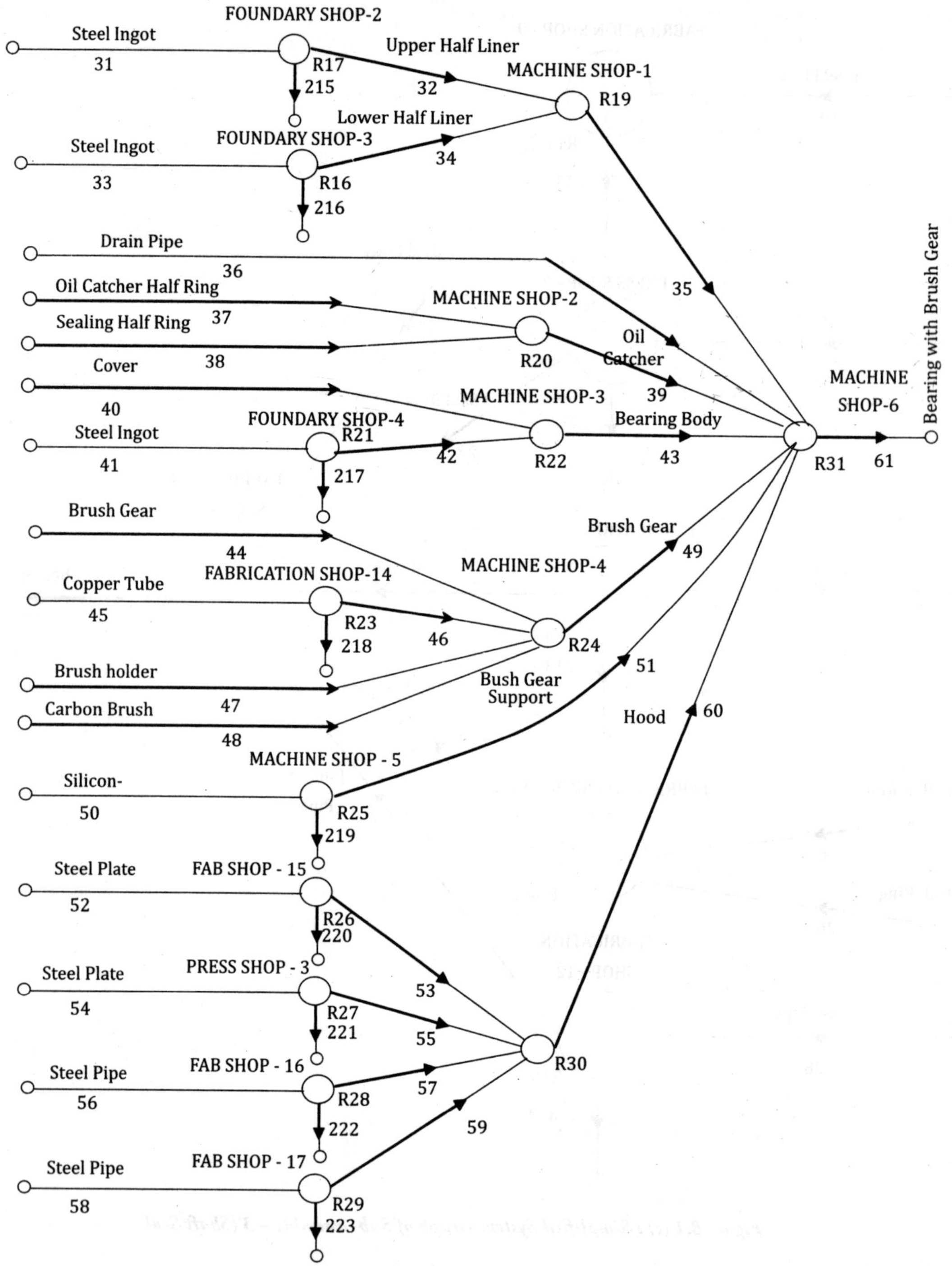

Figure 8.1 (d) : Simplified System Graph of Sub–assembly – 4 (Bearing with Brush Gear)

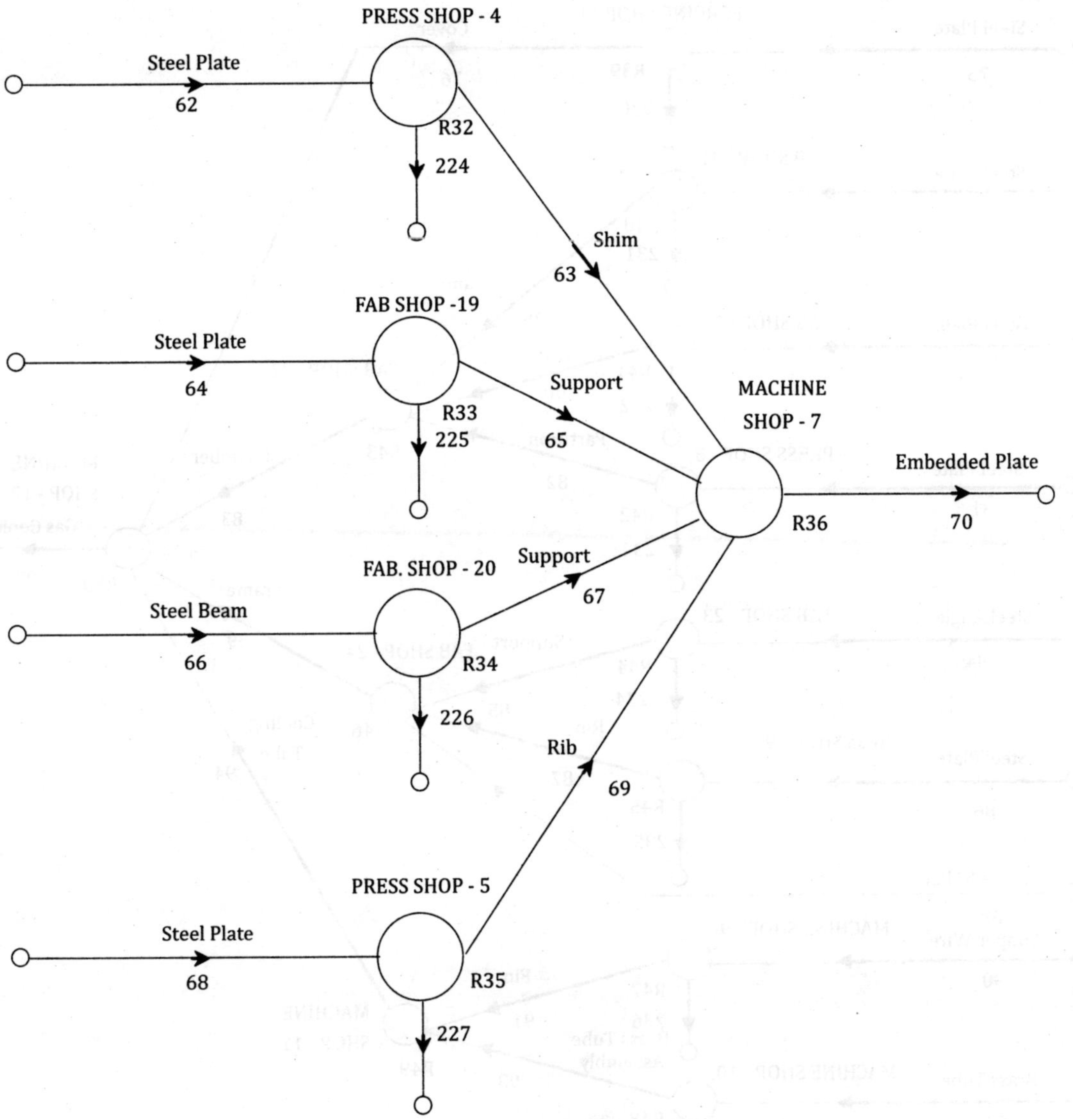

Figure 8.1 (e) : Simplified System Graph of Sub-assembly – 5 (Embedded Plate)

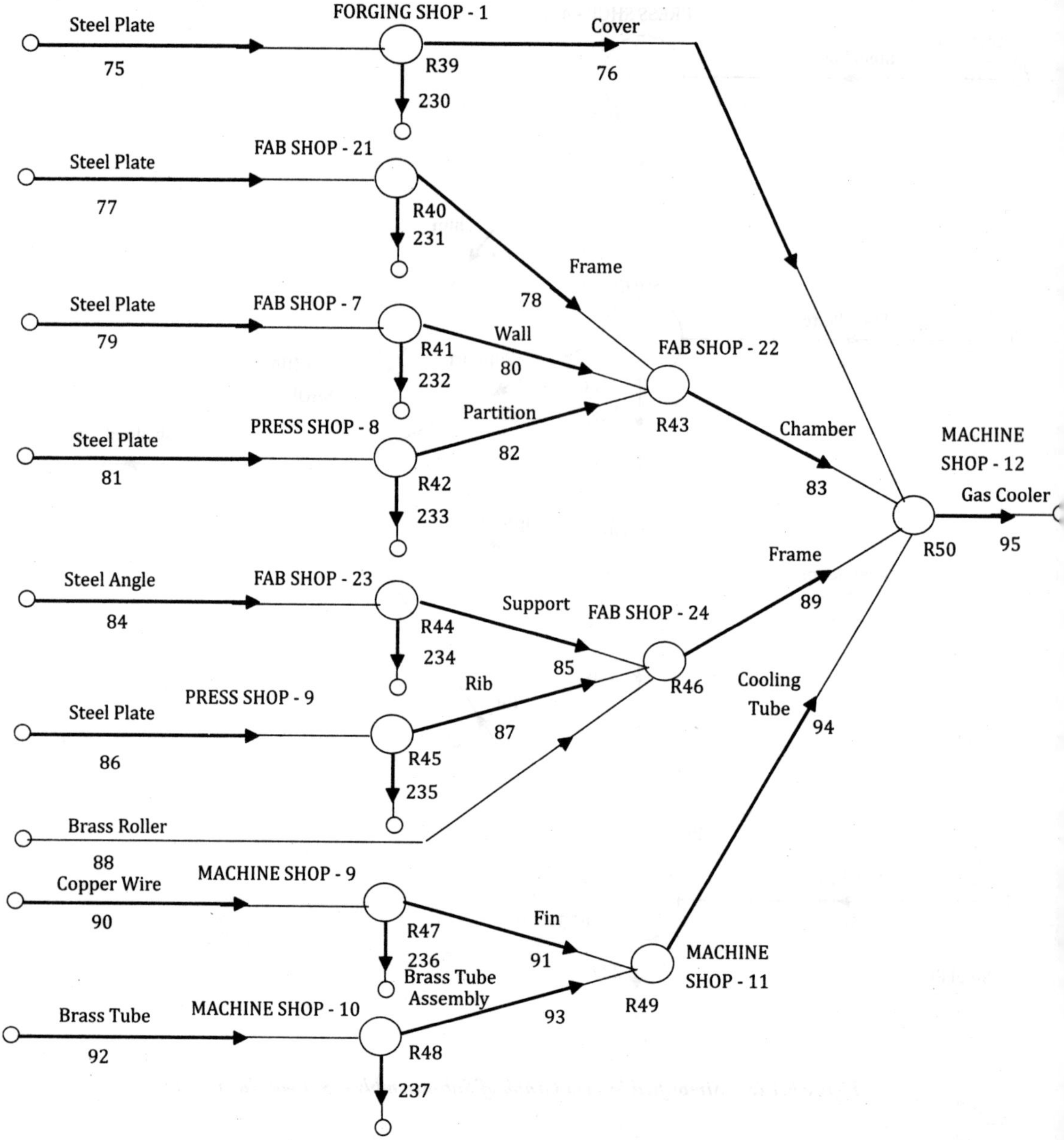

Figure 8.1 (f) : Simplified System Graph of Sub-sub-assembly – 7a (Gas Cooler)

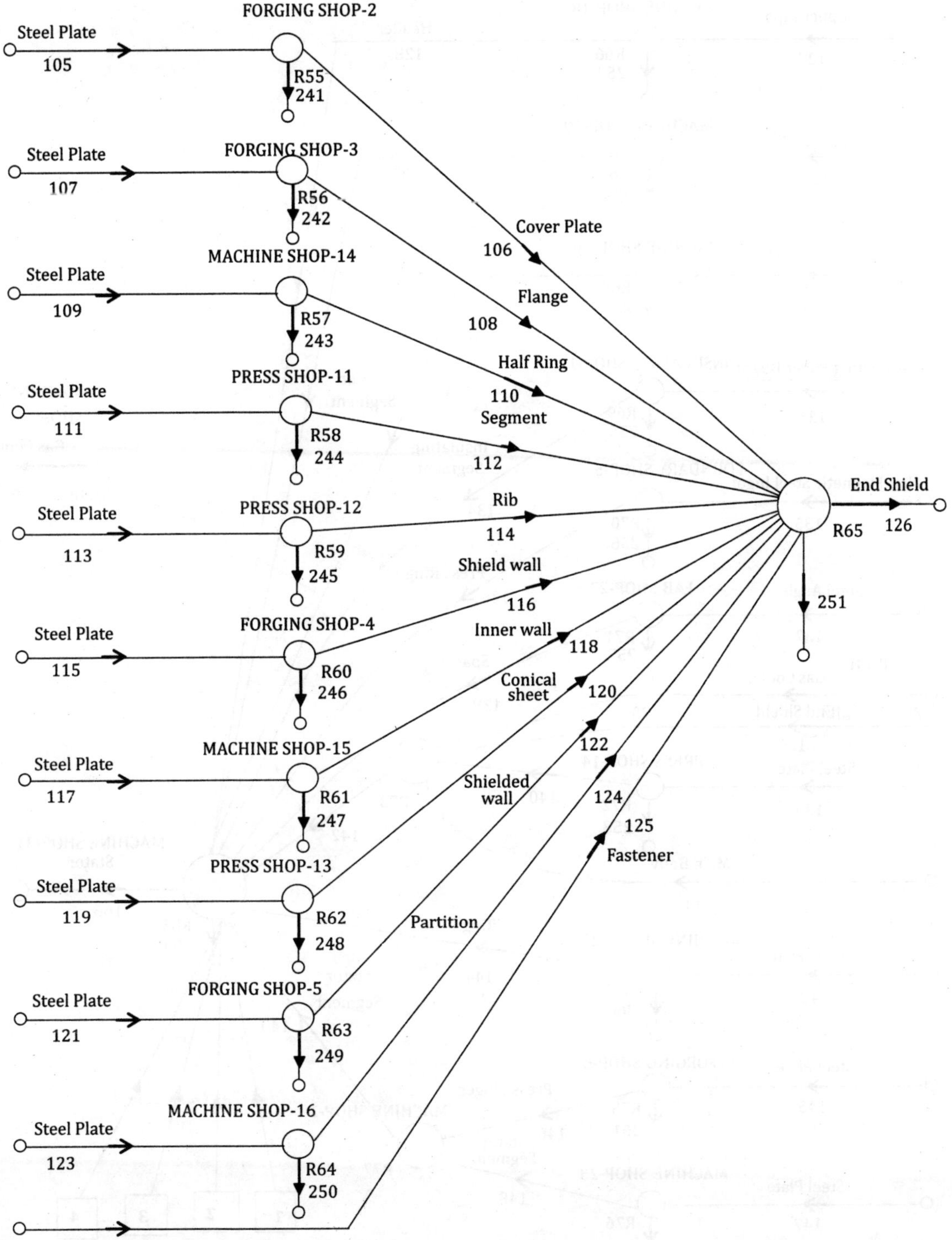

Figure 8.1 (g) : Simplified System Graph of Sub-sub-assembly – 7b (End Shield)

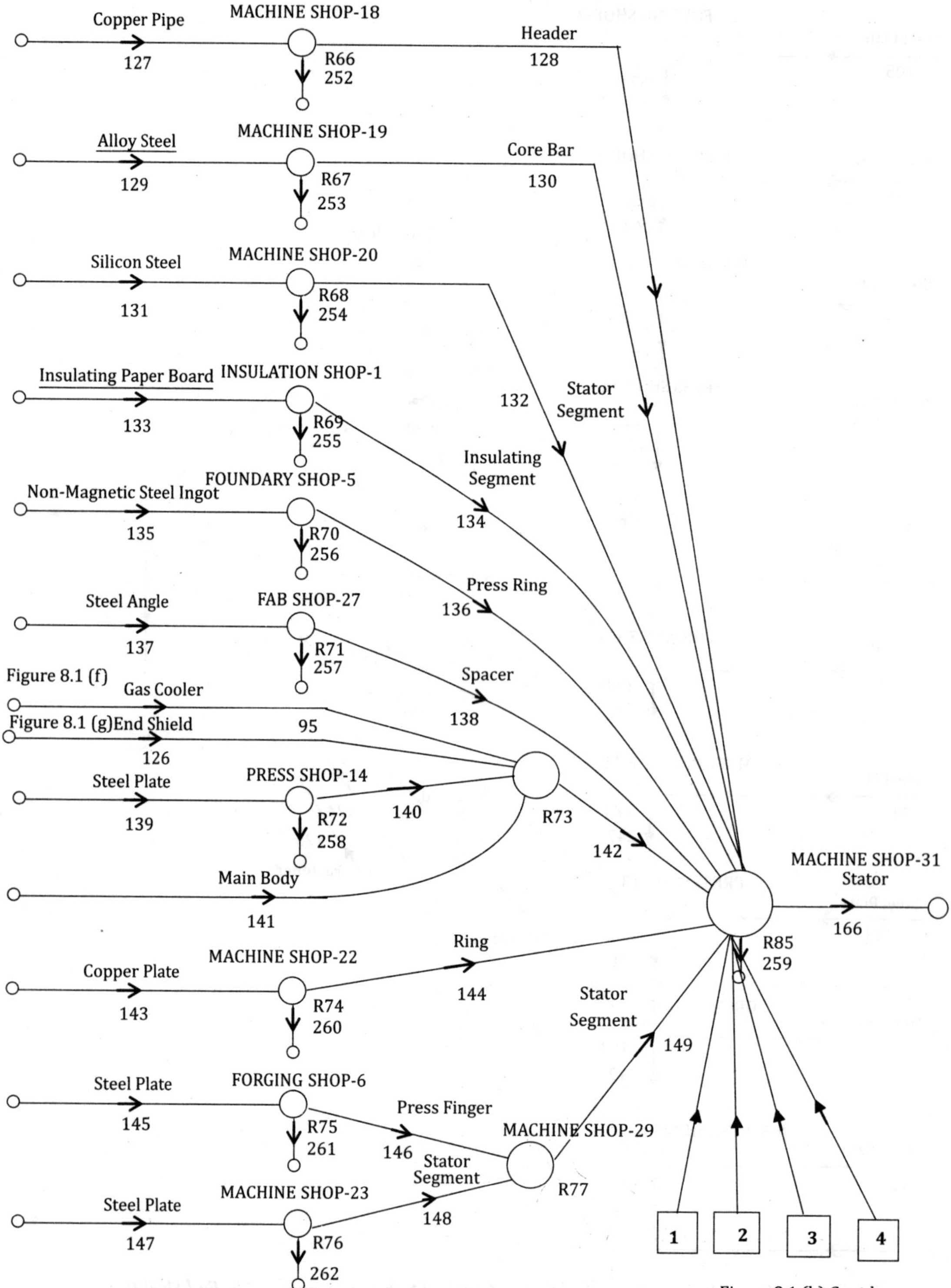

Figure 8.1 (h) Contd....

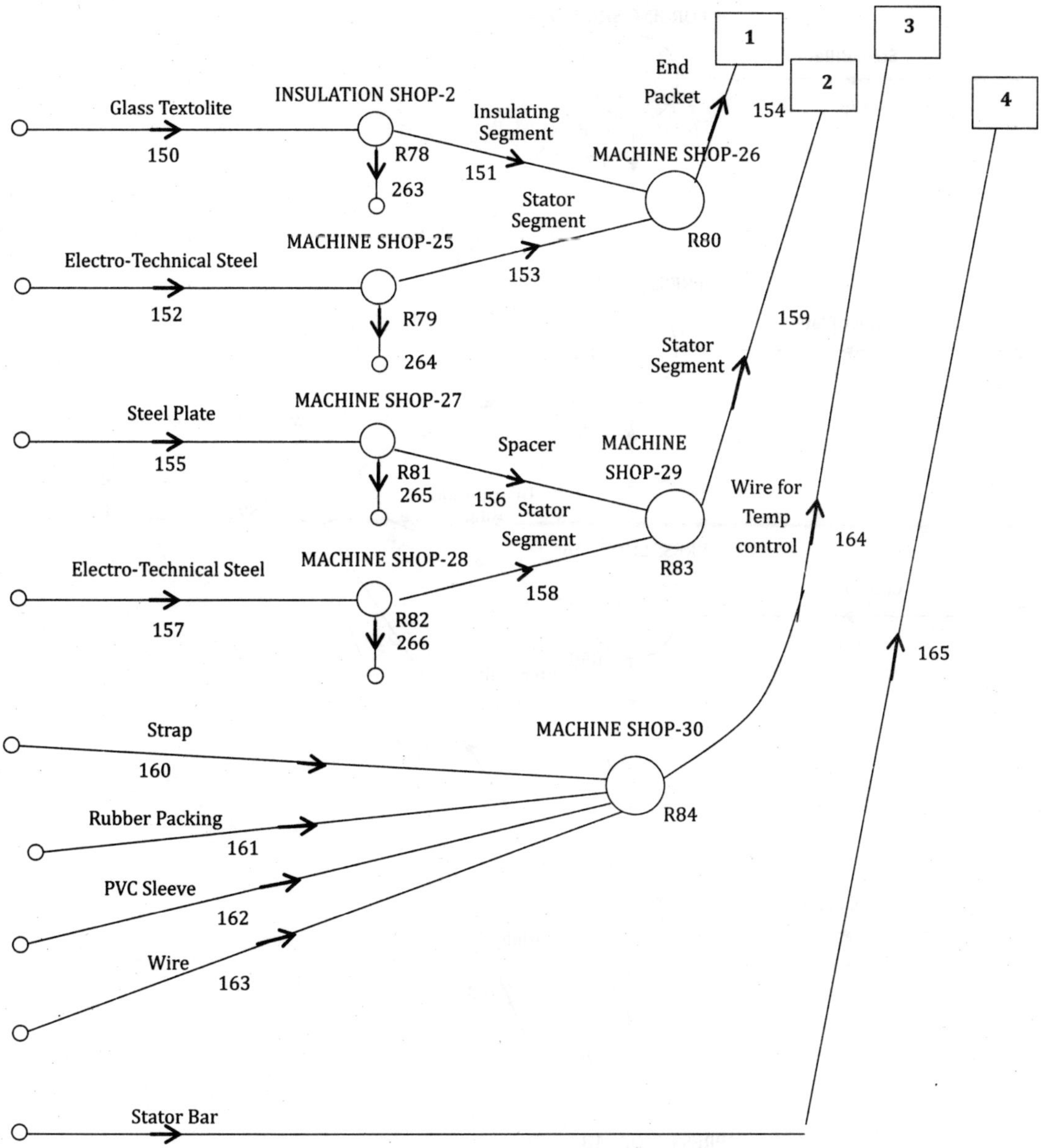

Figure 8.1 (b) : Simplified System Graph of Sub–assembly – 7 (Stator)

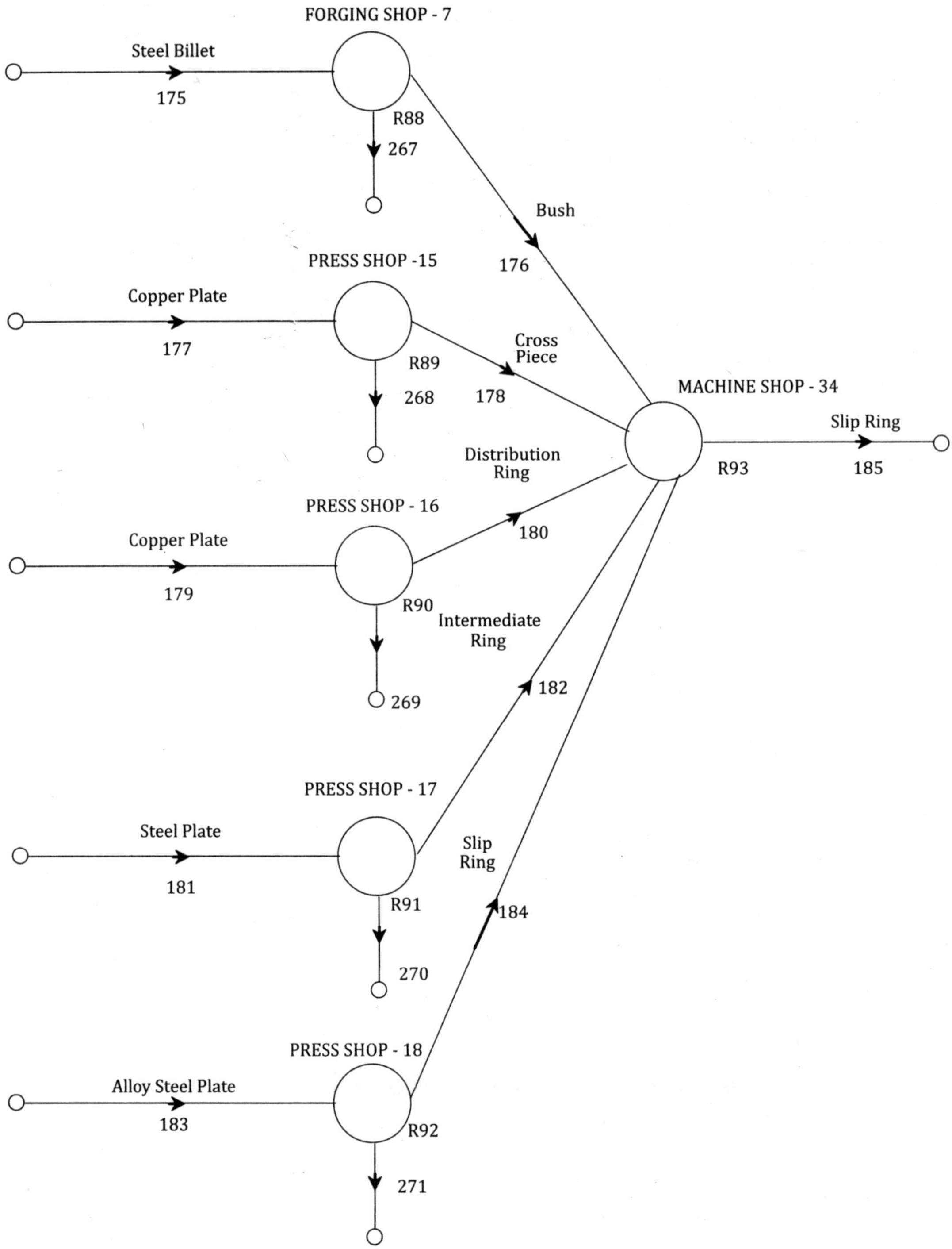

Figure 8.1 (i) : Simplified System Graph of Sub-sub-assembly – 8a (Slip Ring)

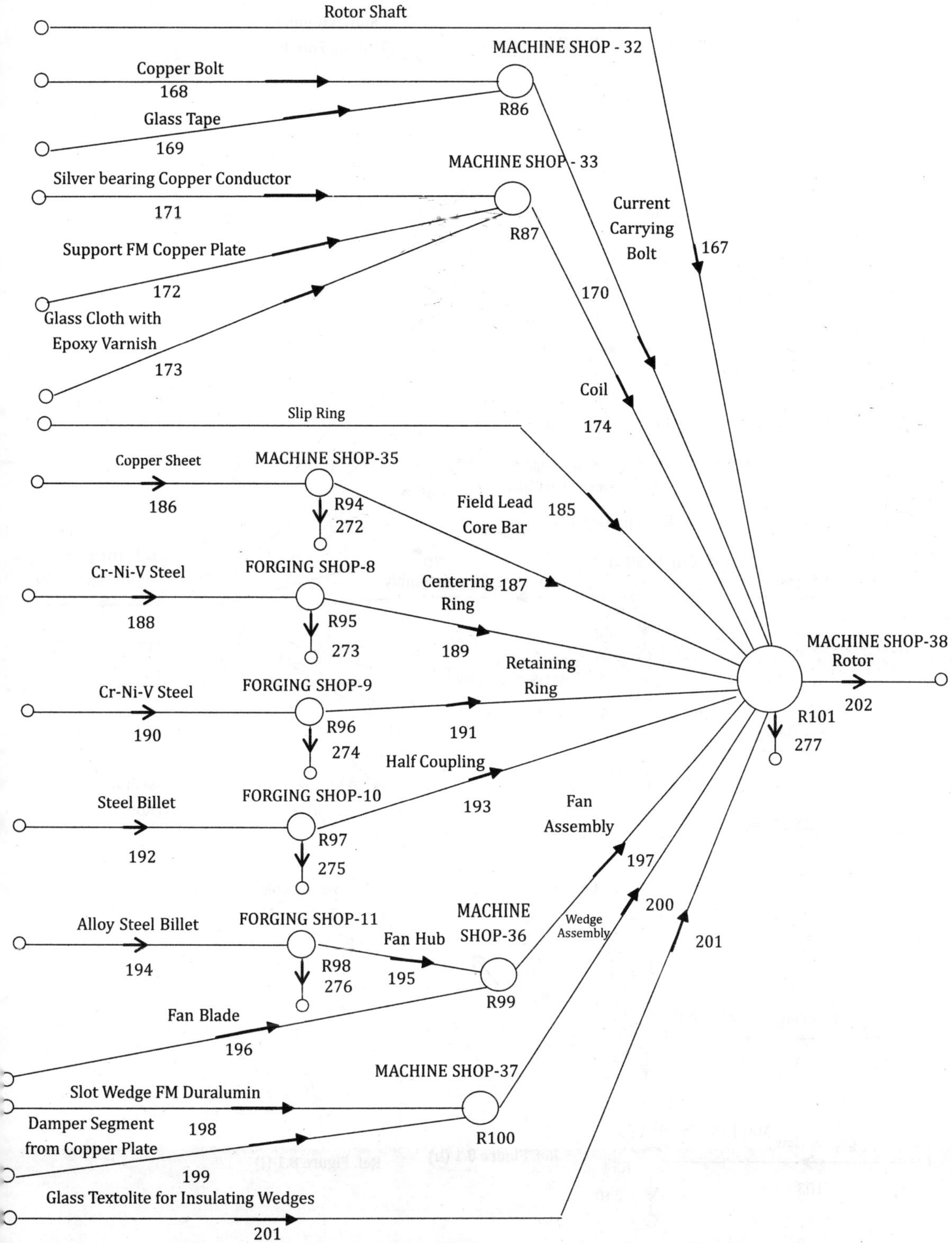

Figure 8.1 (j) : Simplified System Graph of Sub-assembly – 8 (Rotor)

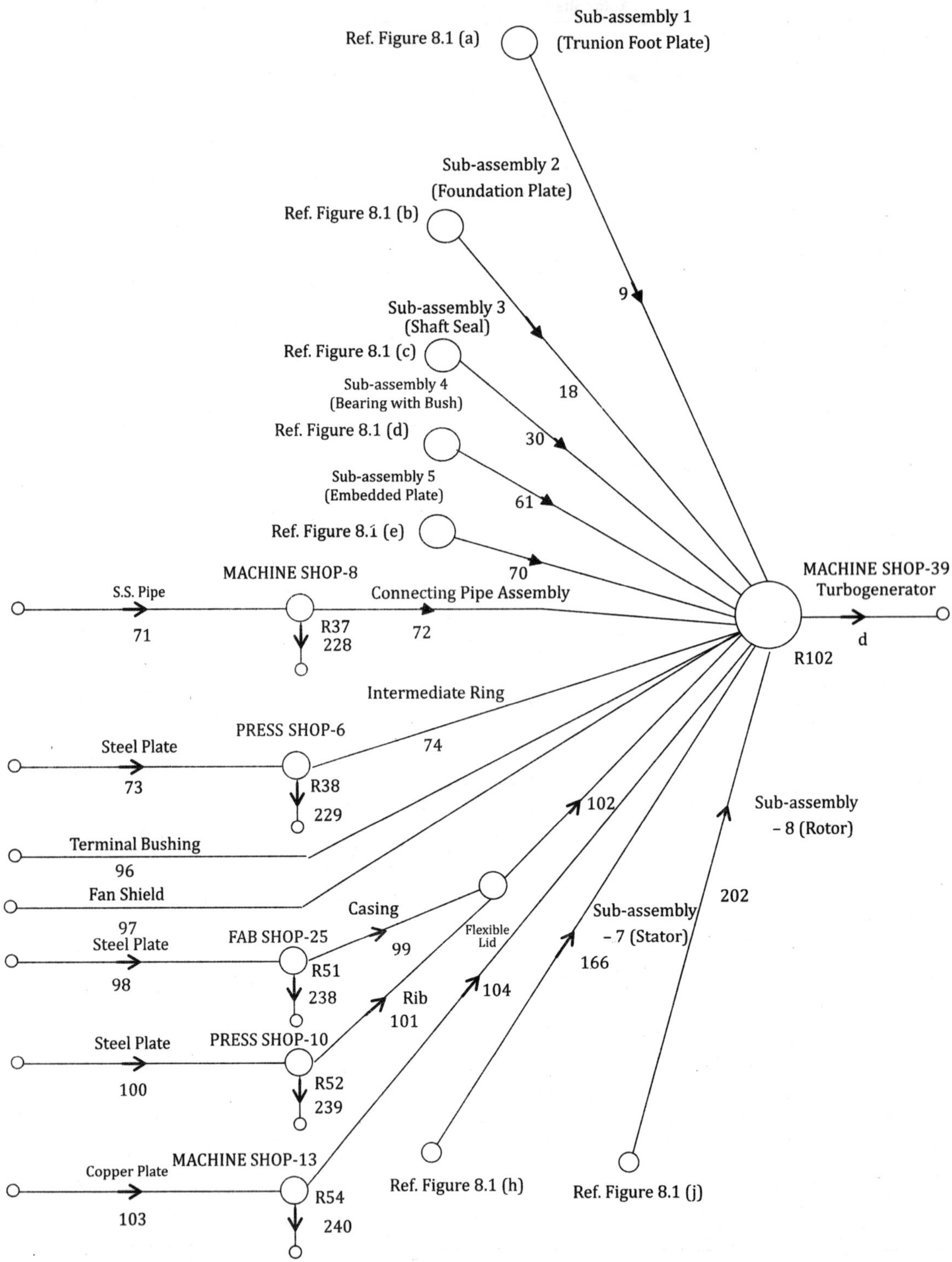

Figure 8.1 (k) : Simplified System Graph of main Assembly with Sub-assembly-6

The different manufacturing stages are marked on the processing nodes with node number given as *Ri*. All the edges are sequentially numbered and the components are written on the top of the edges for easy readability of the graph.

It appears from the system graph, that there are 278 number of edges and 103 number of vertices. From the study of the system it can be stated that the weight of different components of turbogenerator which by virtue of design consideration are constant and flows from one production stage to next stage to complete the sub-assemblies can be used to calculate linear technological coefficients.

8.4 DATA COLLECTION AND PARAMETER ESTIMATION

The BHEL's Heavy Electrical Equipment Plant at Hardwar has been used as source of practical data. The data collection mainly consists of the material, labour, capital and overhead costs incurred as also the quantity of material waste generated at various stages[2].

Flow Parameters

The flow data have been collected from BHEL's Heavy Electrical Equipment Plant, Hardwar. The waste at different processing centres has also been collected from time series data kept by production planning department. The system has been solved backward with the help of mass balance equations to compute flows at all processing centres for both 210 MW and 235 MW Turbogenerators.

Technological Coefficients

Considering the flow of material in a processing centre, the input requirement per unit of output is called technological coefficient and is denoted by '*k*'.

The values of flow variables have been used for computing the technological coefficients at different processing centres. The spreadsheet software has been used to determine the values of technological coefficients. Sample technological coefficients thus found have been tabulated in Table 8.1 for a 210 MW Turbogenerator.

It was informed by BHEL, Hardwar that neither the machine nor the process manufacturer prescribed minimum level of waste at various stages. But BHEL, Hardwar has maintained time series data of minimum average waste generated by the system which has been collected for estimating the value of technological coefficients.

Cost Parameters/Variables

Cost of Exogenous Items

For manufacturing Turbogenerator, BHEL, Hardwar procures number of fabricated parts of Turbogenerator. The cost of raw materials and exogenous items for 210 MW and 235 MW Turbogenerator have been collected from BHEL's Price Bulletin, sample costs are given in Table 8.1.

Processing Cost

The processing cost functions have been considered linear which may be represented by the following equation for the *i*th process.

$$f_i = c_i - m_i y_i$$

where, c_i and m_i are intercept and slope respectively. If scale economy is note considered, then m_i is equal to zero and thus f_i is only a constant parameter which can be written as

$$f_i(y_i) = f_i$$

The processing costs of different components and sub-assemblies have been collected from Accounts Department and sample costs are shown in Table 8.2.

8.5 ANALYSIS, RESULTS AND DISCUSSION

Flow and Cost Analysis

The weight of turbogenerator y_d was specified as 239000.00 Kgs/TG Set for 210MW sets and the flow equations were solved backwards to calculate flow of material and waste at all the stages to validate the model. Sample values of flow variables are shown in Table 8.1.

The cost parameters regarding exogenous costs of input material and waste, and processing costs were used and the cost equations were solved in a forward manner giving costs at different stages of manufacturing. Sample calculated costs are shown in Table 8.1.

Sensitivity Analysis

For the purpose of sensitivity analysis, the system equations were solved to derive the output cost per Turbogenerator set, i.e. x_d in terms of various technological coefficients, processing cost coefficients and exogenous costs of raw materials and wastes in a similar manner as given in chapter four. The sensitivity analysis has been carried out by partially differentiating the output cost (x_d) function so derived with respect to various technological coefficients of waste so as to assess the relative impact of waste at different stages on the final cost of the turbogenerator. Different stages have been clubbed into sensitivity of very high, high, medium, low and very low nature depending upon the value of partial derivatives, as shown in Table 8.3. Only the values of partial derivatives for very high, and high sensitivity stages are shown. All the stages are arranged in order of their sensitivity in different categories.

It has been found that the unit cost of the output is most sensitive to the waste generated at stage 68 (Figure 8.1 (h)) in case of both 210 MW and 235 MW Turbogenerators where stator segment is fabricated from silicon steel, while we find that the stage 29 is least sensitive in case of both 210 MW and 235 MW Turbogenerators.

From the time series data, it is observed that the maximum waste in case of both 210 MW and 235 MW Turbogenerators is generated at stage 68. The waste at stage 68 is 45% of total waste generated by the system.

From the processing cost data, it is found that the processing cost at stage 85 (Figure 8.1 (h)) is maximum.

Table 8.1: Values of Sample Technological Coefficients and Flow and Cost Variables

Sr. No.	Sub-subassembly Component	Technological Coefficient k		Flow Variable kg/TG Set (y)		Cost Variable (Rs/TG Set) (x)	
		Resource	Waste	Resource	Waste	Resource	Waste
(1)	(2)	(3)	(4)	(5)	(6)	(7)	(8)
A. TRUNION FOOTPLATE (Figure 8.1 (a))							
1.	Steel Plate	$k_1 = 1.136$	$k_{203} = 0.136$	$y_1 = 681.81$	$y_{203} = 81.81$	$x^*_1 = 7793$	$x^*_{203} = 935$

2.	Cover Plate	k_2=0.1	-	y_2=600	-	x_2	-
3.	Steel Ingot	k_3=1.176	k_{204} =0.176	y_3=5647.05	y_{204}=847.05	x^*_3	x^*_{204}=10588
4.	Trunion	k_4=0.8	-	y_4=4800	-	x_4	-
5.	Steel Plate	k_5=1.149	k_{205} =0.149	y_5=551.72	y_{205}=71.72	x^*_5=6306	x^*_{205}=819
6.	Foot Plate	k_6=0.08	-	y_6=480	-	x_6	-
7.	Steel Plate	k_7=1.087	k_{206} =0.087	y_7=130.43	y_{206}=10.43	x^*_7=1490	x^*_{206}=119
8.	Rib	k_8=0.02	-	y_8=120	-	x_8	-
9.	Trunion Foot Plate	k_9=0.025	-	y_9=6000	-	x_9	-
B. FOUNDATION PLATE (Figure 8.1 (b))							
10.	Steel Plate	k_{10}=1.176	k_{207} =0.176	y_{10}=952.94	y_{207}=142.94	x^*_{10}=10892	x^*_{207}=1633
11.	Foundation Plate (Generator)	k_{11}=0.129	-	y_{11}=810	-	x_{11}	-
12.	Steel Plate	k_{12}=1.149	k_{208} =0.149	y_{12} =2586.20	y_{208}=336.20	x^*_{12}=29560	x^*_{208}=3842
13.	Foundation Plate (Bearing)	k_{13}=0.359	-	y_{13}=2250	-	x_{13}	-
14.	Steel Plate	k_{14}=1.136	k_{209} =0.136	y_{14} =1084.09	y_{209}=130.09	x^*_{14}=12391	x^*_{209}=1486
15.	Foundation Plate (Exercitor)	k_{15}=0.152	-	y_{15}=954	-	x_{15}	-
16.	Steel Plate	k_{16}=1.087	k_{210} =0.087	y_{16} =2458.69	y_{210}=196.69	x^*_{16}=28102	x^*_{210}=2248
17.	Shim	k_{17}=0.360	-	y_{17}=2262	-	x_{17}	-
18.	Foundation Plate	k_{18}=0.026	-	y_{18}=6267	-	x_{18}	-

* Exogenously specified costs.

Discussion

This analysis, from the cost control point of view, is used as a tool for ranking the importance of different stages which will subsequently increase the productivity. Greater the sensitivity value, more is the attention required for that particular stage.

Table 8.2: Processing Cost of Sample Components of 210 MW Turbogenerator

Sr. No.	Sub-assembly Component	Processing Cost	₹/Component/ TG Set
1.	Cover Plate	f_2	1,468
2.	Trunion	f_4	11,748
3.	Foot Plate	f_6	1,142
4.	Rib	f_8	327
5.	Trunion Foot Plate	f_9	16,319
6.	Foundation plate (Generator)	f_{11}	1,948
7.	Foundation plate (Bearing)	f_{13}	5,421
8.	Foundation plate (Exciter)	f_{15}	2,287
9.	Shim	f_{17}	5,590
10.	Foundation Plate	f_{18}	16,940

The model developed indicates very high wastivity at the production stage of cutting silicon steel in stator section. This implies that the quality of raw material should be looked into very carefully.

The analysis reveals that the stage 68 where stator segment is fabricated from silicon steel is most sensitive. It has been found from Accounting Data that the silicon steel used for manufacturing stator segment is most costly material among other raw-materials used for manufacturing different components of Turbogenerator. The silicon steel is imported from abroad and is available in coils with different width sizes. Since this material is very costly, the same needs close attention during inspection and transportation. Moreover, it should be decided on the basis of sheet cutting diagram as to what exact width of the coil is to be procured which would lessen the waste.

It has been learnt that the waste generated from cutting of silicon steel are sold to an agency at considerable good price as per agreed yearly contract while the waste generated at other processing centres from steel plate and structural steel are utilized by BHEL in their Central Foundry Forge Plant at Hardwar.

REFLECTIONS

For the simplicity of the study, all the components which are available in a Turbogenerator have not been included. The small components such as fasteners, washers, gaskets, wedge, shims, small structural sections etc., which do not have much bearing as far as overall cost of Turbogenerator have been ignored.

In the absence of physical data, the model developed has been tested only for two types of Turbogenerators, i.e. 210 MW and 235 MW. It is suggested that the study should be conducted for at least five similar types of Turbogenerators where the design and technology involved is same in order to judge the validity of the model.

The flow data collected from BHEL's Time series data are old and have not been updated the same for quite a long time since they feel that their production system has been stabilized and the flow data recorded earlier would vary to a nominal extent.

The model includes only transformation processes but a production unit entails material transportation and storage processes as well. It is, therefore, suggested that the model be extended to include material transportation and storage processes. The model did not consider recycling and to make the model more effective the recycling of rejects should be considered.

Table 8.3: Results of Sensitivity Analysis for 210 MW Turbogenerator Set

(a) VERY HIGH SENSITIVITY

Sr. No.	Stage No.	Stage Description	Partial Derivative	Value of Partial Derivative
1.	R68	Machine shop-20	$\frac{\partial x_d}{\partial k_{254}}$	228100.93
2.	R85	Machine shop - 31 (Stator)	$\frac{\partial x_d}{\partial k_{259}}$	120243.86

(B) HIGH SENSITIVITY

Sr. No.	Stage No.	Stage Description	Partial Derivative	Value of Partial Derivative
3.	R101	Machine shop – 38 (Rotor)	$\frac{\partial x_d}{\partial k_{277}}$	12078.55
4.	R21	Foundary shop –4	$\frac{\partial x_d}{\partial k_{217}}$	3131.60
5.	R96	Forging shop – 9	$\frac{\partial x_d}{\partial k_{274}}$	3121.91

6.	R79	Machine shop – 25	$\frac{\partial x_d}{\partial k_{264}}$	2911.02
7.	R2	Foundary shop – 1	$\frac{\partial x_d}{\partial k_{204}}$	1630.32
8.	R70	Foundary shop – 5	$\frac{\partial x_d}{\partial k_{256}}$	1329.03
9.	R78	Insulation shop – 2	$\frac{\partial x_d}{\partial k_{263}}$	1002.15
10.	R67	Machine shop – 19	$\frac{\partial x_d}{\partial k_{253}}$	809.43
11.	R72	Press shop – 14	$\frac{\partial x_d}{\partial k_{258}}$	679.31
12.	R48	Machine shop – 10	$\frac{\partial x_d}{\partial k_{237}}$	563.98
13.	R47	Machine shop – 9	$\frac{\partial x_d}{\partial k_{236}}$	529.89
14.	R82	Machine shop – 28	$\frac{\partial x_d}{\partial k_{266}}$	316.66
15.	R7	Fabrication shop - 5	$\frac{\partial x_d}{\partial k_{208}}$	314.46

Notes

1. Choudhury D.K. (1990), Systems Modelling of Manufacturing of Turbogenerator, *M.Tech. Dissertation*, under the supervision of Sushil, IIT Delhi.
2. *Ibid.*

9 MASS-ENERGY FLOW IN A CLOSED CHAIN PRODUCTION SYSTEM WITH STORAGE

9.1 INTRODUCTION

Modelling of production systems helps in understanding, and thus aids, various planning and operating decisions. The production systems have been modelled in the literature using a number of approaches such as linear programming, simulation models, queuing models, petrinets, markov models, etc. Hirata and Fukao[1] have presented a simple macroscopic model of mass and energy flow in production processes which is based on a model of mass and energy flow in ecosystems[2]. This model has been further extended and explored[3]. In this model[4] the mass flow is considered as a variable, whereas the unit cost of energy associated with the material is treated as a parameter, which in real life is a variable.

It is proposed in this chapter that the PST framework will be more suitable for the modelling of a production system as it can handle the flow as a 'through' variable and energy cost as an 'across' variable, as shown in Figure 9.1.

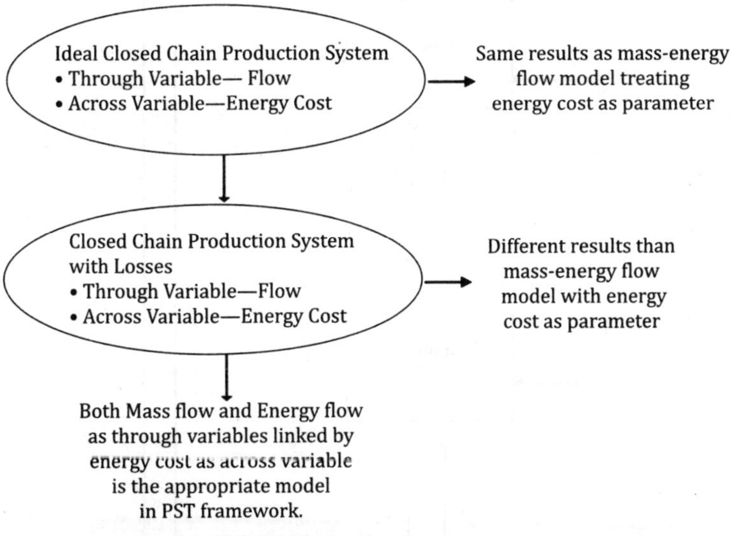

Figure 9.1: Flexible System Methodology for Closed Chain Production System

9.2 PHYSICAL SYSTEM THEORY MODEL OF A CLOSED CHAIN PRODUCTION SYSTEM

Ideal Closed Chain System

In the ideal closed chain system, complete mass recycling takes place without any loss at any of the production levels. An ideal closed chain system as shown in Figure 9.2 is modelled using concepts and principles of PST in a simplified framework. The flow of mass and energy are treated as 'through' variables (*y*), whereas the unit energy value is treated as 'across' variables (*x*). The component models, system graph and structural constraints are developed and subsequently the simplified PST model is developed by following the simplified framework of PST[5].

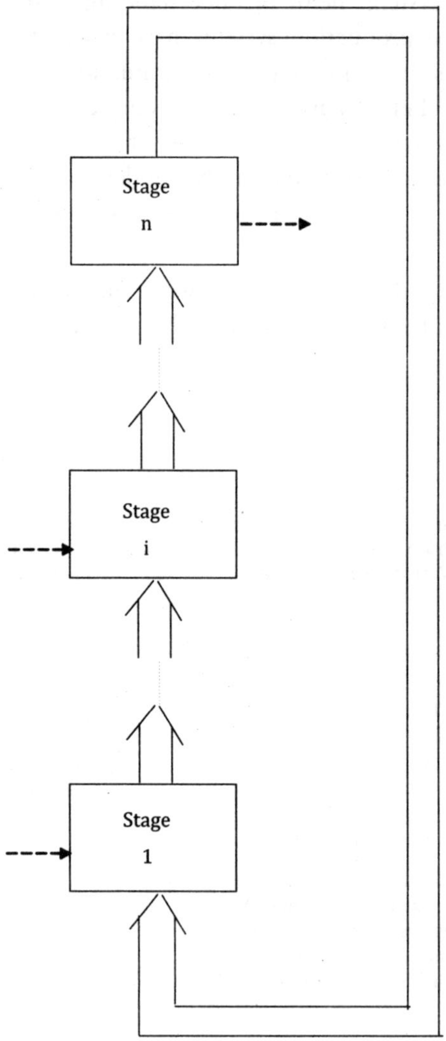

Figure 9.2: Ideal Closed Chain Model with Complete Mass Recycling—mass; ---, energy.

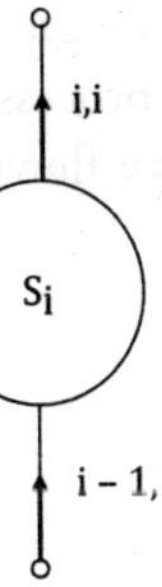

Figure 9.3 : Free-body Diagram of ith Storage Process

Component Models

The micro-feature of the *i*th production stage is conceptualized to be composed of one storage process and one transformation process.

Let y^m and y^e represent the material and energy flows and let the unit energy value be represented by x.

a. ***Storage Process:*** The free body diagram of the *i*th storage process is shown in Figure 9.3. The component terminal equations for the *i*th storage are as follows:

$$\frac{dM_i}{dt} = y^m_{i-1,\,i} - y^m_{i,\,i} \tag{9.1}$$

$$\frac{dE_1}{dt} = y^e_{\;i-1,i} - y^e_{i,i} \text{ (assuming no maintenance energy)} \tag{9.2}$$

or

$$\frac{dE_1}{dt} = y^e_{\;i-1,i} - y^m_{i-1,\,i} - y^e_{i,\,i} \tag{9.3}$$

$$x_{i-1} = E_i / M_i \tag{9.4}$$

that is

$$E_i = x_{i,i} M_i \tag{9.5}$$

In the case of the unit energy value remaining constant over time,

$$x_{i-1,\,i} = x_{i,i} \tag{9.6}$$

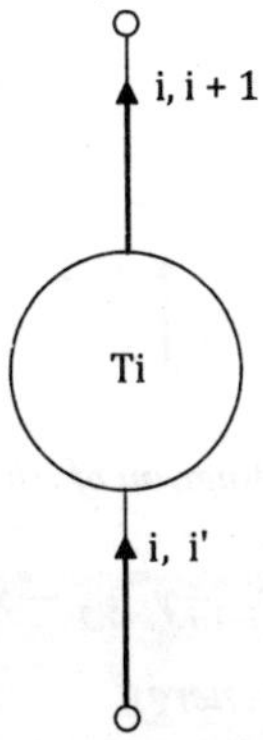

Figure 9.4: Free-body Diagram of ith Transformation Process

b. ***Transformation Process*:** The free-body diagram of the *i*th transformation process is shown in Figure 9.4. The energy balance for this process is that the total energy flowing out of the *i*th process is equal to the sum of total energy flowing in and the processing energy used at the *i*th process, i.e.

$$y^m_{i,i+1}\, x_{i,i+1} = y^m_{i,i}\, x'_{i,i} + r_i M_i \tag{9.7}$$

Dividing both sides by the material output ($y^m_{i,i+1}$), the equation for the unit energy value is as follows:

$$x_{i,i+1} = k^{m'}_{i,i}\, x'_{i,i'} + r_i / a_i \tag{9.8}$$

where $k^m_{i,i'}$ is the technological coefficient (defining the input required per unit of output) given by the flow equation as follows:

$$y^m_{i,i} = k^m_{i,i}\, y^m_{i,i+1} \tag{9.9}$$

In this case with no wastage $y^m_{i,i} = y^m_{i,i+1}$, that is $k^m_{i,i} = 1$ and

$$a_i = y^m_{i,i+1} / M_i \tag{9.10}$$

On uniting the component models of the storage process and transformation process in the micro-feature of the *i*th stage, the black-box component model is obtained taking strict continuity into account; i.e.

$$y^m_{i,i'} = y^m_{i,i} \tag{9.11}$$

The free-body diagram of the *i*th black-box production stage is shown in Figure 9.5. The response-stimulus equations are derived from previous equations:

$$\frac{dM_i}{dt} = \underset{\textit{Material input}}{y^m_{i-1,i}} - \underset{\textit{Material output}}{y^m_{i,i+1}} \tag{9.12}$$

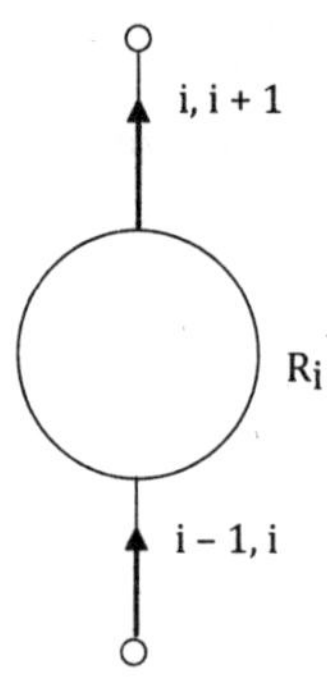

Figure 9.5: Free-body Diagram of ith Black-box Production Stage

$$\frac{dM_i}{dt} = (\underset{\textit{energy input}}{x_{i-1,i}\, y^m_{i-1,2}} - \underset{\textit{energy output}}{x_{i,i+1}\, y^m_{i,i+1}}) + (\underset{\textit{processing energy}}{r_i M_i}) \tag{9.13}$$

$$y^m_{i,i+1} = a_i M_i \tag{9.14}$$

$$x_{i,i+1} = x_{i-1,i} + r_i / a_i \tag{9.15}$$

$$x_{i-1,i} = x_i \tag{9.16}$$

$$E_i = x_i M_i \tag{9.17}$$

In case the cost is varying with time, i.e. either r_i or a_i is changing, then all the equations of the storage process and transformation process are to be solved jointly. Whereas, if the cost is not affected by time, the model can be split into two parts, as follows:

a. Solve only the transformation process model to obtain the value of x using (9.15);

b. Solve only the storage process model, i.e. using (9.12) and (9.14) we obtain

$$\frac{dM_i}{dt} = a_{i-1} M_{i-1} - a_i M_i \tag{9.18}$$

System Graph

On joining the component models of various production stages in one-to-one correspondence of how the production stages are related in real life, the system graph is obtained which is a closed graph. This system graph is used to write the continuity and compatibility equations with respect to 'through' and 'across' variables, respectively.

Simplification

As all the interconnections in this system graph are of strict one-to-one correspondence type and there is no branching, all the continuity and compatibility equations can be merged with the component terminal equations and a simplified model can be obtained.

Closed Chain System with Losses

In the closed chain system, if some losses of mass/energy take place at different production stages, then the model of the ideal closed chain system is to be modified to take care of these losses. The free body diagram of the ith production stage with loss w_i is shown if Figure 9.6. The derivation of equations for this case is given in Appendix 9 A.

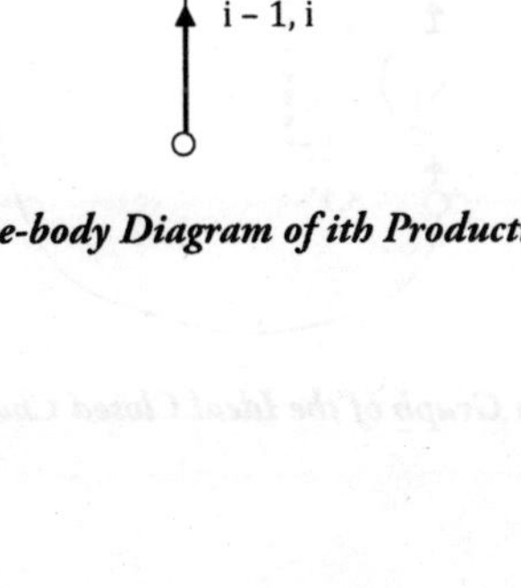

Figure 9.6: Free-body Diagram of ith Production Stage with Losses

9.3 ILLUSTRATIVE EXAMPLE

In order to facilitate the comparison the example taken for illustration is same as that given by Hirata and Fukao[4]. This consists of a simple production process with three production levels: raw materials, middle products and final products. An example is the production of iron products from raw iron. The mass recycling is represented by the feedback from the final products to the raw materials. The same system is considered with and without material waste.

Ideal Closed Chain System

This ideal production system has no material waste at any of the stages, i.e. the total mass is conserved. The processing energy enters at each level.

The PST model of this ideal closed chain system is developed using the standard procedure. The component terminal graphs of all three-production stages (divided into transformation process and storage process) are united to form the system graph as shown in Figure 9.7. The system graph is simplified by following the standard rules and the simplified system graph is obtained as shown in Figure 9.8. The system equations of the simplified PST model are as follows:

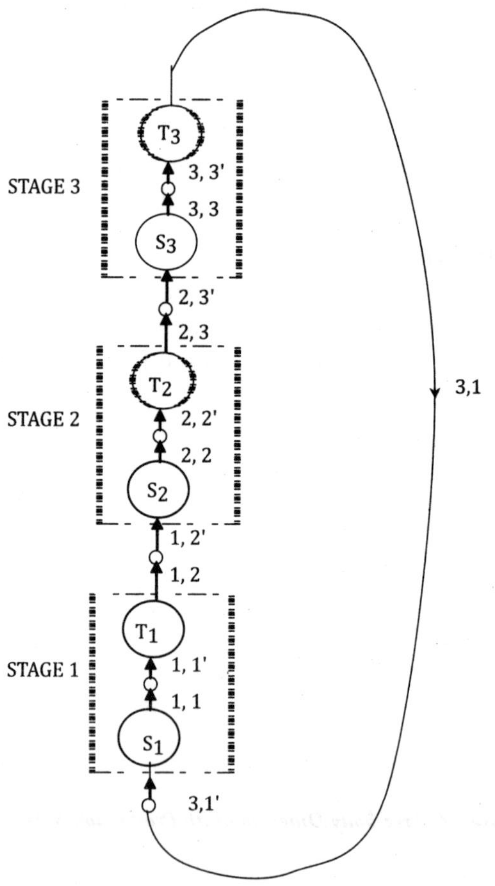

Figure 9.7: System Graph of the Ideal Closed Chain Production System

Cost Equations

$$x_{1,2} = x_{3,1} + r_1 / a_1 \tag{9.19}$$

$$x_{2,3} = x_{1,2} + r_2 / a_2 \tag{9.20}$$

$$x_{3,1} = x_{2,3} - r_3 / a_3 \tag{9.21}$$

$$y_{1,1} = y_{1,2} \tag{9.22}$$

$$y_{2,2} = y_{2,3} \tag{9.23}$$

$$y_{3,3} = y_{3,1} \tag{9.24}$$

Material Accumulation Equations

$$\frac{d}{dt}\begin{bmatrix} M_1 \\ M_2 \\ M_3 \end{bmatrix} = \begin{bmatrix} 1 & -1 & 0 \\ 0 & 1 & -1 \\ -1 & 0 & 1 \end{bmatrix} \begin{bmatrix} y_{31} \\ y_{12} \\ y_{23} \end{bmatrix}$$

or

$$\frac{d}{dt}\begin{bmatrix} M_1 \\ M_2 \\ M_3 \end{bmatrix} = \begin{bmatrix} -a_1 & 0 & a_3 \\ a_1 & -a_2 & 0 \\ 0 & a_2 & -a_3 \end{bmatrix} \begin{bmatrix} M_1 \\ M_2 \\ M_3 \end{bmatrix} \tag{9.25}$$

Energy Equations

$$\begin{bmatrix} E_1 \\ E_2 \\ E_3 \end{bmatrix} = \begin{bmatrix} x_{3,1} & 0 & 0 \\ 0 & x_{1,2} & 0 \\ 0 & 0 & x_{2,3} \end{bmatrix} \begin{bmatrix} M_1 \\ M_2 \\ M_3 \end{bmatrix} \tag{9.26}$$

In this example, if r_1=1, r_2=2, r_3= –2 and a_1=1, a_2=2, a_3=1 and $x_1 = x_{3,1}$=0, then from (9.15) and (9.16), $x_2 = x_{1,2}$ = 1 and $x_3 = x_{2,3}$= 2.

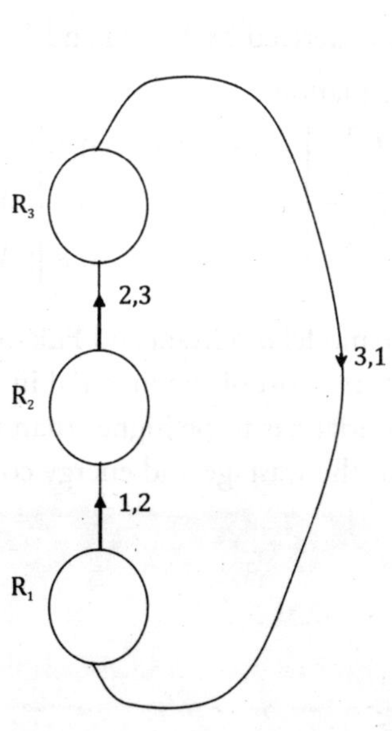

Figure 9.8: Simplified System Graph of the Ideal Closed Chain Production System

The values of a_1, a_2 and a_3 can be substituted in (9.25) which is the same as that of Hirata and Fukao[6] and thus leads to the same conclusion that the production process is stably persistent. In Hirata and Fukao's (1978) model, unit energy cost λ_i was specified at all levels and the flows per unit of material levels (i.e. a_1) were calculated, whereas in the PST formulation, unit energy cost is considered as a variable and its value is calculated by knowing the value of output flow per unit of material level a_i at all the stages. It has already been discussed that it is difficult to specify r_i and x_i independently at all the stages in view of the fact that the x_i are dependent on r_i. The utility of the approach will become more evident in the case of the closed chain system with losses.

Closed Chain System with Losses

The same production system is now considered with material waste $\rho_i M_i$ at each stage, which leads to the fact that the total mass is not conserved. The material input per unit of output at stage 1 is ϕ_1. No direct waste of energy is considered. The parameters are assumed to be the same as those of Hirata and Fukao[7], i.e.

$$r_1 = 1,\ r_2 = 2,\ r_3 = -2$$
$$\rho_1 = 0.5,\ \rho_2 = \rho_3 = 0.9$$
$$\phi_1 = H' \text{ (positive constant)}$$
$$\phi_1 = \phi_3 = \varepsilon_1 = \varepsilon_2 = \varepsilon_3 = 0$$
$$a_1 = 1,\ a_2 = 2,\ a_3 = 1$$
$$x_1 = 0$$

Then using (9.35), $x_2 = 1$ and $x_3 = 2.45$. However, if we take the values of x_1, x_2 and x_3 as given by Hirata and Fukao[8], i.e. x_1=0, x_2 =1 and x_3= 2, and calculate the values of a_i and a'_i, using (9.41), then we find that the values are very much different as shown below:

$$a_1 = 1\ (1),\ a_2 = 1.1\ (2),\ a_3 = 0.1\ (1)$$
$$a'_1 = 0.5\ (0.5),\ a'_2 = 0.2\ (1.1), a'_3 = -0.8\ (0.1) \tag{9.27}$$

The values in parentheses are those derived by Hirata and Fukao[9].

Hence, in this case the material equation is

$$\frac{d}{dt}\begin{bmatrix} M_1 \\ M_2 \\ M_3 \end{bmatrix} = \begin{bmatrix} -0.5 & 0 & 0.1 \\ 1.0 & -0.2 & 0 \\ 0 & 0.2 & 0.8 \end{bmatrix}\begin{bmatrix} M_1 \\ M_2 \\ M_3 \end{bmatrix} + \begin{bmatrix} H \\ 0 \\ 0 \end{bmatrix} \tag{9.28}$$

This is because of the fact that the model of Hirata and Fukao[10] is not able to take care of the energy cost as a variable. It takes the same energy cost of the material in both cases, i.e. when there is no waste and when there are losses at each material has to be higher than if there is not wastage. Thus, the PST model provides a better accounting of the wastage and energy cost in the system.

REFLECTIONS

The concepts of PST in a simplified framework have been used to develop a simple macro-model of a closed chain production system. The component models are developed using the ecosystem concept. The flow of material and energy are considered as 'through' variables whereas the unit energy costs are considered as 'across' variables. The modelling and analysis of a simple three stage production system indicates that the process of modelling is simple and structured, and it is able to account for the waste and energy cost in a better manner rather than treating the energy costs as parameters. Though the example presented in this paper is very simple and various assumptions about linearity are taken, if the utility of the model in costing and production planning is appreciated, it may prove to be useful in deriving the system behaviour of more complex systems. Further, the waste generated at each stage may be recycled and its effect on the energy cost can be estimated from such a model.

Moreover, the PST framework has been extended to incorporate the quality aspects into the PST framework. The proposed model can also be extended to analyse the impact of the quality of material on the energy cost. It is hoped that the PST model may prove to be a more versatile modelling approach for analysis of mass-energy flows in production systems.

Notes

1. Hirata H. and Fukao T., (1977) *Math. Biosc.,* 33, 321; 1978, *IEEE Trans. Systems Man. Cybern.,* 8, 432.
2. May R.M. (1973) *J. Theory, Biol.,* 39, 155; Ulanowicz R.E., (1972), *J. Theory. Biol.,* 34, 239.
3. Hirata H. (1979) *IEEE Trans. Systems Man. Cybern.,*9, 296; 1985, *ibid,* 15, 807; Hirata H., and Kogo H., (1978) *Int. J. Systems Sci.,* 9, 1207; Hirata H., and Sugai Y., (1988) *Int. J. Systems Sci.,* 19, 351.
4. Hirata and Fukao (1978) *op. cit.*
5. Sushil (1991) Simplified Physical System Theory Model of a Closed Chain Production Systems, *Int. J. Systems Sci.,* 22 (6), 987-999.
6. Hirata and Fukao (1978) *op. cit.*
7. *Ibid.*
8. *Ibid.*
9. *Ibid.*
10. *Ibid.*
11. *Ibid.*

APPENDIX 9 A

THE CLOSED CHAIN PRODUCTION SYSTEM WITH LOSSES

The loss of mass at stage i is explained as follows:

$$yw^m_i = \rho_i M_i + \theta_i \tag{9.29}$$

where, ρ_i and θ_i are non-positive constants for i≠1; and ρ_I and θ_I are constants of any sign. The positive ρ_I and θ_I show the mass which applies to the production process. The loss of energy at stage i is expressed as follows:

$$yw^e_i = x_i \rho_i M_I + x_i\theta_i + \varepsilon_i \tag{9.30}$$

where ε_i is a non-positive constant for $i \neq I$ and ε_i is a constant of any sign.

The response-stimulus equations for the ith stage with losses are as follows:

State equations

$$\frac{dM_i}{dt} = y^m_{i-1,i} - y^m_{I,i-1} + \rho_i M_i + \theta_i \tag{9.31}$$

material input — *material output* — *material waste or loss*

Using (9.10)

$$\frac{dM_i}{dt} = a_{i-1,}M_{i-1-\alpha_i}M_i + \theta_i \tag{9.32}$$

where

$$a'_i = a_i - \rho_i \tag{9.33}$$

The rate of change of energy,

$$\frac{dE_i}{dt} = (x_{i-1}y^m_{i-1,i} - x_{i,i+1}y^m_{i,i+1}) + (r_i M_i) + (x_i\rho_i M_i + x_i\theta_i) + \varepsilon_i \tag{9.34}$$

energy input — *energy output* — *processing energy* — *energy loss associated with material loss* — *energy loss*

Cost Equation

$$x_{i,I+1} = k_{i-1,i}x_{I-1,i} + r_i / a_i \tag{9.35}$$

(assuming the cost of waste disposal to be zero)

Flow Equations

$$y^m_{i-1,i} = k_{i-1,i}\, y^m_{i,i+1} \tag{9.36}$$

$$yw^m_i = (k_{i-1,i} - 1)y^m_{i,i+1} \tag{9.37}$$

Mass Balance Equation

$$\begin{aligned} y^m_{i-1,i} &= y^m_{i,i+1} + yw^m_i \\ &= y^m_{i,i+1} + \rho_i M_i + \theta_i \end{aligned} \tag{9.38}$$

from (9.29).

Dividing both sides of (9.38) by $y^m_{i,i+1}$, we obtain

$$k_{i-1,i} = 1 + \rho_i / a_i + \phi_i \tag{9.39}$$

where

$$\phi_i = \theta_i / y^m_{i,i+1} \tag{9.40}$$

(i.e. constant material loss per unit output).

Substituting (9.39) into (9.35) and rearranging, we obtain

$$a_i = \frac{\rho_i x_{i-1,i} + r_i}{x_{i,i+1} - (1 + \phi_i) x_{i-1,i}} \tag{9.41}$$

whereas for a closed chain model with losses, Hirata and Fukao[11] gave the same equation for a_i as that for the ideal closed chain model.

The derivation given above for a_i indicates that the value of a_i is dependent on the coefficient of the material loss. In (9.41) if we put $\rho_i = 0$ and $\phi_i = 0$, i.e. no material loss, then we obtain the equation for the ideal closed chain system, which indicates that the ideal closed chain system is a special case of the closed chain system with losses.

10 TECHNOLOGY SELECTION FOR MULTI-STAGE MANUFACTURING SYSTEMS

10.1 INTRODUCTION

Manufacturing is one of the most important issues in the world today. The modelling of manufacturing systems helps in analyzing various planning and operating decisions. Most manufacturing activities invariably involve multiple stages. And each stage constitutes a technology. For example, items requiring turning operations can be manufactured on any of the following machines: engine lathe, turret lathe, single spindle automatic lathe, multiple spindle automatic lathe, numerically controlled turning machine, etc. Each machine represents a technology with its distinct characteristics in terms of quality, cost of production, purchasing cost, etc. Since each stage in a multi-stage manufacturing situation has alternative technologies to choose from, the problem of selecting a technology at each stage becomes complex because of interactions between stages and budgetary constraints.

Some attempts have been made in the literature to address the problem of technology selection. In the case of manufacturing, Sarin and Chen[1] developed a mixed integer programming model for technology selection considering technology on an aggregate basis. Singh and Rajoria[2] addressed the problem of technology selection in the case of the Marble Industry. However, these models do not capture the realities of the multi-stage production where each stage poses a technology selection problem and where there is interaction between stages as output from the previous stage becomes input to the successive stage. Also each technology at every stage has different scrap rates, different compatibilities with other technologies and certainly different budget requirements besides continuity constraints.

In this chapter, simple models based on the concept of physical system theory and mathematical programming are developed that may be used for technology selection in multi-stage manufacturing systems. A non-linear 0-1 programming model to minimize the cost of production treating the technology as 0-1 variable is developed subject to the constraints on budget and continuity. The 0-1 non-linear model is linearized and illustrated by a simple three stage serial manufacturing system having three alternative technologies at the first stage, two at the second and two at the third using hypothetical data.

10.2 TECHNOLOGY SELECTION PROBLEM

In this section, the technology selection problem in the context of a multi-stage manufacturing system is discussed by considering a small but representative example. A generalized model is then developed in subsequent sections using the concepts of physical system theory and mathematical programming.

Consider, for example, a three-stage serial manufacturing system designed to manufacture a part requiring three operations, that is, turning, milling and drilling at the first, second and third stage respectively. At each stage, the parts not meeting the specifications are scrapped. Suppose, three alternate technologies were available for turning at the first stage: Engine Lathe (EL), Turret Lathe (TL), and Single Spindle Automatic Lathe (SSAL); two for milling at the second stage: Universal Milling Machine-Ordinary Grade I (UMOG), and Universal Milling Machine-Supergrade II (UMSG); two for drilling at the third stage: Pillar Drilling Machine (PDM), and Radial Drilling Machine (RDM).

The process capabilities, processing costs and budget requirements for machines at each stage are different. Moreover, in a serial manufacturing system the output from a stage becomes input to a subsequent stage. Accordingly, the parts manufactured in sequence on EL, UMOG and PDM will have different production output, rejects and unit output cost than being manufactured on TL, UMOG and PDM; or SSAL, UMSG and RDM. There are twelve such possibilities for this problem. Further, there are constraints on the total investment that can be made to acquire these technologies as well as on the raw material availability. Therefore, the management is faced with the problem of selecting an appropriate set of technologies. The selection criteria could either be based on unit cost minimization or a multi-objective approach.

For this purpose, a generalized non-linear 0-1 programming model is developed using the concepts of physical system theory (PST). PST allows to model elemental components separately and computes the interaction which logically results from network theoretic considerations[3]. Thus, a system theory model is custom tailored to fit each individual system modelled. The PST modelling approach can, therefore, be usefully applied to model multi-stage production systems since each production stage can be conceived as a component. Upon joining the components (stages) in a one-to-one correspondence to form a system graph, the production system characteristics can be studied considering the interaction between stages. The flexible system methodology for technology selection is shown in Figure 10.1.

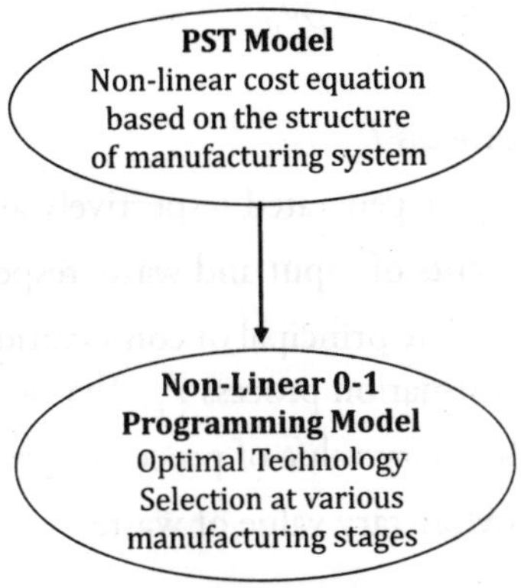

Figure 10.1: Flexible System Methodology for Technology Selection in a Multi-stage Manufacturing System

10.3 GENERALIZED TECHNOLOGY SELECTION MODEL

We develop a technology selection model for multi-stage manufacturing systems. Consider a serial production system with i stages, $i = 1, 2,..., m$, and each stage having j number of alternate technologies, $j = 1, 2,... n_i$, for the ith stage. First the component postulate the PST is used to model the transformation process, P_{ij}, corresponding the j-th alternative at the i-th stage. The technological coefficients as defined to represent the input requirements (or waste generated) per unit of output. The units not meeting the specifications are scrapped and considered as waste. It is assumed that there is one type of input, one type of output and one type of waste generated at the transformation process P_{ij}. Accordingly:

Technological coefficient of input at P_{ij}.

$$k_{ij} = (\text{Input at } P_{ij})/(\text{Output from } P_{ij}) \tag{10.1}$$

Technologies coefficient of waste at P_{ij}:

$$kw_{ij} = (\text{Waste generated at } P_{ij})/(\text{Output from } P_{ij}) \tag{10.2}$$

The free-body diagram of the transformation process P_{Ij} is shown in Figure 10.2.

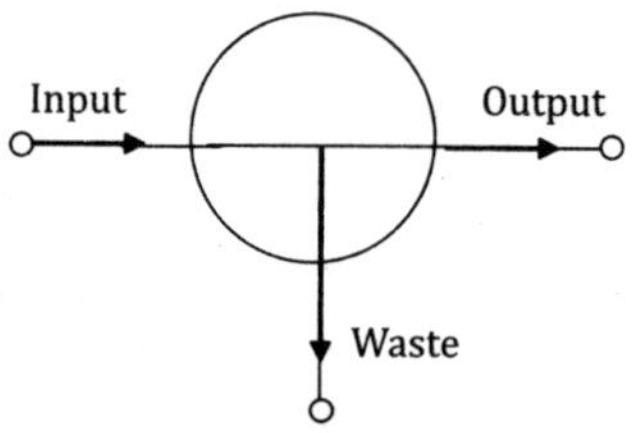

Figure 10.2: Free-body Diagram of the Transformation Process

Accordingly, the component terminal equations for P_{Ij} are:

i. ***Material flow equations***: These equations follow from the very definition of technological coefficients.

$$y_{ij} = k_{ij}.yo_{ij} \tag{10.3}$$

$$yw_{ij} = kw_{ij}.yo_{ij} \tag{10.4}$$

where

yo_{ij} : material output of process P_{ij}

y_{ij}, yw_{ij} : input required and waste generated respectively at P_{ij},

k_{ij}, kw_{ij} : technological coefficients of input and waste respectively at P_{ij}.

ii. ***Cost equations***: This follows from the principal of conservation that the money inflow rate equals money outflow rate at the transformation process P_{ij}. That is.

(Monetary value of inputs) + (Monetary value of processing costs)

= (Monetary value of output) + (Monetary value of waste).

Therefore,

$$y_{ij}\,x_{ij} + yo_{ij}\,f_{ij}\,(yo_{ij}) = yo_{ij}\,xo_{ij} + yw_{ij}\,xw_{ij}$$

or

$$xo_{ij} = k_{ij}\,x_{ij} - kw_{ij}\,xw_{ij} + f_{ij}\,(yo_{ij}), \tag{10.5}$$

where

xo_{ij} : unit cost of output of process P_{ij}

x_{ij}, : cost per unit material input at P_{ij},

xw_{ij} : salvage value per unit waste at P_{ij}

$f_{ij}(yo_{ij})$: processing cost per unit output as a function of output for process P_{ij}

$f_{ij}(yo_{ij})$: accounts for labour, capital, overheads and also takes care of scale of economies, f_{ij} can, however, be assumed to be constant at P_{ij} and independent of yo_{ij} if the scale of economies is not considered. We make this assumption in this chapter.

Using the second and third postulate of PST, the component models of various transformation processes, P_{ij} (i=1, 2, ...m; j = 1, 2...n_i) can be united to obtain the system graph which will be governed by the constraints following the principle of conservation with respect to unit flow rates and dollar flow rates. However, to simplify the treatment, we start with a three-stage system having only one technology at each stage. The component models for this system are given in Appendix 10 A. Note that the edges in the component models are assigned unique numbers for ease of representation.

These component models are then united in a one-to-one correspondence to obtain a system graph given in Appendix 10 A with accompanying governing equations. The generalized equations for unit output cost and input material requirements as a function of technology variable A_{ij} are developed in Appendix 10 B and Appendix 10 C using the results obtained in Appendix 10 A. The technology variable A_{ij} (i = 1, 2, ...m; j = 1, 2........,n_i) is defined as a 0-1 variable to represent the j-th technology at the i-th stage of process P_{ij}. Its value is one if j-th technology is selected, zero otherwise.

In the technology selection model, our objective is to select only one technology at each stage such that the unit output cost is minimized. Accordingly, the unit output cost as a function of A_{ij} is given (for details, see Appendix 10 A and 10 B) below:

$$z = \prod_{i=1}^{m}\sum_{j=1}^{n_1} a_{ij}k_{ij}x_i + \sum_{i=2}^{m}\prod_{i=1}^{m}\sum_{j=1}^{n_i} a_{ij}k_{ij}\sum_{j=1}^{n_{i-1}} a_{i-1,j}(f_{e-1,i} - k_{wi-1,1}x_{wi-1,j})$$

$$+\sum_{j=1}^{n_m} a_{mj}(f_{mj} - k_{w_{mj}}x_{w_{mj}}), \tag{10.6}$$

where

x_1 is the incoming material cost per unit at the first stage ($x_1 = x_{11} = x_{12} = \ldots = x_{1n}$).

The incoming material requirement for given finished product is given by 'through' variables at different stages solving one equation at a time and moving backwards (for derivation, see Appendix 10.C).

Accordingly, we have,

$$0 \le \prod_{i=1}^{m}\sum_{j=1}^{n_i} a_{ij}k_{ij}y_d \le y_1, \tag{10.7}$$

where

y_1 is input material available at the first stage,

y_d is desired number of units of finished output from the last stage.

The requirement that only one technology should be selected at each stage is satisfied by the following equations:

$$\sum_{j=1}^{n_i} a_{ij} = 1 \text{ for all } i. \tag{10.8}$$

Also the technologies should satisfy the budget constraints. Accordingly

$$\sum_{i=1}^{m} \sum_{j=1}^{n_i} a_{ij} b_{ij} \leq B, \tag{10.9}$$

where b_{ij} is the budget required for the i-th technology at the j-th stage; B is total budget.

a_{ij} are zero or one type variables. Therefore,

$$a_{ij} = \begin{cases} 1 \text{ if } j\text{-th type technology is selected at i-th stage for all } (i, j), \\ 0 \text{ otherwise} \end{cases} \tag{10.10}$$

Therefore, the 0-1 non-linear programming technology selection model for multi-stage manufacturing system is:

P1: minimize z (given by equation (10.6))

subject to constraints given by equations (10.7), (10.8), (10.9) and (10.10)

10.4 ESTIMATION OF PARAMETERS

The parameters in the technology selection model are: technological coefficients, unit cost of raw material input, salvage values of wastes, and the processing costs.

The raw material cost and salvage values of the waste generated are known for the type of material to be processed. The processing costs are dependent on the type of technology and the material to be processed; its elements are labour requirements, capital and overheads. These, however, constitute the standard data variable from the suppliers-production standards.

Estimation of Technology Coefficients

Each production technology basically represents a type of machine tool. The process capabilities of machine tools together with operating environment are the major determinants of the scrap or waste generated (units not meeting the specifications are scrapped). It is quite synonymous to the fraction defective in quality control theory. Once the percent defectives (rejects) are known, the technological coefficients can be determined as follows: Consider a transformation process P_{ij} shown in Figure 10.1. Assume that c_{ij} represents the percentage rejects due to technology A_{ij}. Accordingly,

$$c_{ij} = (yw_{ij}/y_{ij})\ 100 = (yw_{ij}/(yw_{ij} + yo_{ij}))\ 100$$

or

$$k_{w_{ij}} = \frac{c_{ij}}{100 - c_{ij}} \tag{10.11}$$

since $k_{ij} = y_{ij}/yo_{ij} = (yo_{ij} + y_{ij})/yo_{ij}$,

$$k_{ij} = 1 + kw_{ij} \tag{11.12}$$

10.5 SOLUTION PROCEDURE

Since a_{ij} is binary, the linear transformation of their product form in equation (10.6) and (10.7) can be easily achieved by using the strategy due to Balas[4]. Once the linear 0 –1 transformation is made, the problem can be solved by any existing package.

After linear transformation, the model will have $\sum^{m}_{i=1}\prod^{m}_{j=i}n_j$ 1-0 variables and $m+2+2(\sum^{m}_{i=1}\prod^{m}_{j=i}n_j - n_m)$ constraints.

It is worth mentioning that if the objective function has all the positive terms, then the resulting constraints would be $m+2(\Sigma^{m}_{i=1}\Pi^{m}_{j=i}n_j - n_m)$.

10.6 ILLUSTRATIVE EXAMPLE

The technology selection model is illustrated considering a three-stage serial manufacturing system discussed in the section on technology selection problem. Specifically, the system is designed to produce a part requiring three operations, that is, turning, milling and drilling each having 3, 2 and 2 sets of alternate technologies respectively. The input data regarding scrap rates, costs and budget requirements are given in Table 10.1. On the basis of scrap rates, the different technological coefficients are computed using equations 10.11 and 10.12. The computed values are given in Table 10.2.

For the example problem m = 3, n_1 = 3, n_2 = 2, n_3 = 2, the corresponding cost and technological coefficient are given in Tables 10.1 and 10.2.

Accordingly,

$$z=(100)\prod_{i=1}^{3}\sum_{j=1}^{n_j}a_{ij}k_{ij}+\sum_{l=2}^{3}\prod_{l=j}^{3}\sum_{j=1}^{n_j}a_{j-1,j}\sum_{j=1}^{i-1}a_{j-1,j}(f_{j-1,j}-k_{wl-1}x_{wl-1,j})$$
$$+\sum_{j=1}^{n^3}a_{3j}(f_{3j}-k_{w3j}x_{w3j}). \qquad (10.13)$$

Table 10.1: Input Data [a]

Technology	Scrap rate (%)	Processing cost per unit ($)	Budget required ($)
Turning			
1. Engine lathe	10	10	7000
2. Turret lathe	3	7	30000

3. Single spindle automatic lathe	1	5	40000
Milling			
1. Universal milling machine ordinary grade I	6	10	25000
2. Universal milling machine super grade II	5	15	35000
Drilling			
1. Pillar drilling machine	2	5	2000
2. Radial drilling machine	1	3	30000

[a] Unit cost of incoming material = \$ 100.00

Unit cost of scrap = \$ 10.00

Assume that 500 units of finished parts are required to be manufactured. Suppose raw material available is 550 units.

Accordingly,

$$0 \le (500)\prod_{l=1}^{3}\sum_{j=1}^{n_i} a_{ij}k_{ij} \le 550. \tag{10.14}$$

Assuming that a total budget of \$ 75000.00 is available, the budget constraint is then given by

$$\prod_{l=1}^{3}\sum_{j=1}^{n_i} a_{ij}b_{ij} \le 75000 \tag{10.15}$$

Table 10.2: Value of Technological Coefficients Calculated using Scrap Rate Values from Table 10.1 and Mass Balance Equations

Technology	Technological Coefficient of Input and Its Value		Technological Coefficient of Waste and Its Value	
Turning				
1. Engine lathe	k_{11}	1.111	kw_{11}	0.111
2. Turret lathe	k_{12}	1.031	kw_{12}	0.031
3. Single spindle automatic lathe	k_{13}	1.010	kw_{13}	0.010
Milling				
1. Universal milling machine ordinary grade I	k_{21}	1.064	kw_{21}	0.064

2. Universal milling machine super grade II	k_{22}	1.053	kw_{22}	0.053
Drilling				
1. Pillar drilling machine	k_{31}	1.020	kw_{31}	0.020
2. Radial drilling machine	k_{32}	1.010	kw_{32}	0.010

The fact that only one technology should be selected at each stage is ensured by the following equations:

$$\sum_{j=1}^{n_i} a_{ij} = 1 \text{ for } i = 1, 2, 3 \tag{10.16}$$

$$a_{ij} = \begin{cases} 1 & \text{if technology } j \text{ is selected at the } i\text{-th stage for all } (i, j), \\ 0 & \text{otherwise.} \end{cases} \tag{10.17}$$

The 0-1 linearization method due to Balas[5] introduces 16 new 0-1 integer variables and 16 additional constraints for the present problem. The optimal solution is given in Table 10.3.

Table 10.3: Results

Stage	Technology selected
Turning	Single Spindle automatic lathe
Milling	Universal milling machine ordinary grade I
Drilling	Pillar drilling machine

The minimum unit cost is \$ 129.2784 and 548.066 units worth of raw material is required to produce 500 finished shafts. The budget required is \$ 67000.00.

REFLECTIONS

The principles of physical system theory together with mathematical programming have been successfully applied in this chapter to address the strategic issues of technology selection for multi-stage manufacturing systems. Because of the explosion in the technology innovation in the manufacturing sector of world economies, the decision-maker is always faced with the dilemma of selecting an appropriate mix of technologies. This chapter provides an analytical framework to resolve such a predicament especially in a multi-stage production context considering interaction between stages. The minimization of cost models developed using the principles of physical system theory together with the mathematical programming models provide a number of alternative scenarios to the decision maker to help select appropriate technology mix. Although the model developed has been illustrated by considering an example from metal cutting technologies, many useful applications will be in metal working industries as well as in iron, copper, aluminum, gold, etc. manufacturing industries where the multi-stage production system is common and alternative technologies produce varying degrees of waste.

Notes

1. Sarin S.C. and Chen C.S., (1986), A Mathematical Model for Manufacturing System Selection, in: *Flexible Management Systems: Methods and Studies*, A. Kusiak (ed.), North-Holland, Amsterdam, 1986.
2. Singh N. and Rajoria M., (1987), Modelling and Analysis of the Marble Industry of Rajasthan, Part I: Technology Selection and Goal Setting, *International Journal of Systems Science,* 18 141-155.
3. Singh N. and Sushil (1990), Technology Selection Models for Multistage Production Systems: Joint Application of Physical Systems Theory and Mathematical Programming, *European Journal of Operational Research*, 47, 248-261.
4. Balas E., (1964), Extension de l'algorithme additif a la programmation en nombres entiers et a la programmation non lineaire, *Comptes Rendus des Seanes de l'Academie des Sciences* 258, 5136-5139.
5. *Ibid.*

APPENDIX 10A

MODEL FOR THE ILLUSTRATIVE EXAMPLE

A.1 Component Models

The following equations refer to Figure 10.3.

Note that: (i) All the scrap is sold. (ii) Scale economies are not considered.

$$\text{Cost equation: } x_{o2} = k_1 x_1 - k_{w1} x_{w1} + f_{o1}$$
$$\text{Flow equations: } y_1 = k_1 y_{o1},\ y_{w1} = k_{w1} y_{o1}$$
$$\text{Cost equation: } x_{o2} = k_2 x_2 - k_{w2} x_{w2} + f_{o2}$$
$$\text{Flow equations: } y_2 = k_2 y_{o2},\ y_{w2} = k_{w2} y_{o2}$$
$$\text{Cost equation: } x_{o3} = k_3 x_3 - k_{w3} x_{w3} + f_{o3}$$
$$\text{Flow equations: } y_3 = k_3 y_{o3},\ y_{w3} = k_{w3} y_{o3}$$

A.2 System Graph

See Figure 10.4 for the system graph and the following equations:

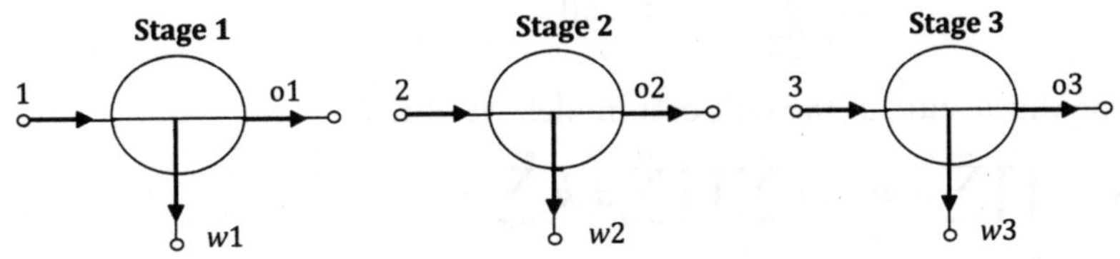

Figure 10.3: Component Terminal Graphs

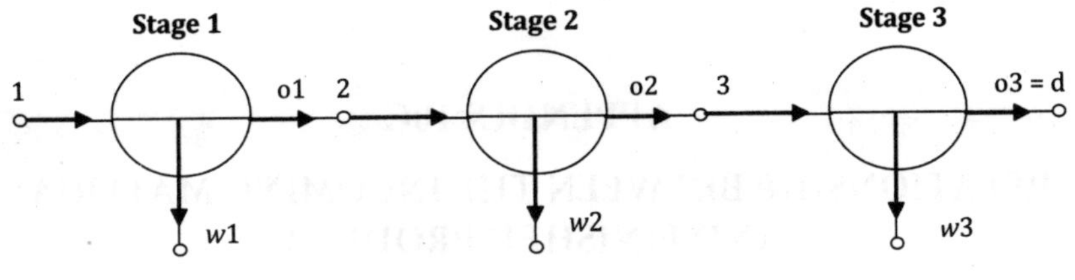

Figure 10.4: System Graph

Continuity equations:

$$y_2 - y_{o1} = 0, \qquad y_3 - y_{o2} = 0$$

Equations following the principle of conservation (dollar flow rates):

$$x_{o1} y_{o1} - x_2 y_2 = 0, \qquad x_{o2} y_{o2} - x_3 y_3 = 0$$

Mass balance equations:

$$y_1 = y_{o1} + y_{w1}, \qquad y_2 = y_{o2} + y_{w2}, \qquad y_3 = y_{o3} + y_{w3}$$

APPENDIX 10B

DERIVATION OF THE UNIT COST EQUATION

Upon successively solving the system of cost equations developed in Appendix A, we obtain the unit cost equation as follows:

$$x_d = k_3 k_2 k_1 x_1 - k_3 k_2 kw_1 xw_1 + k_3 k_2 f_1 - k_3 kw_2 xw_2 + k_2 f_2 - kw_3 xw_3 + f_3 \tag{10.18}$$

Equation (10.19) is for a three stage system. Similarly, on further generalization for the m stage systems $(i = 1, 2, \ldots\ldots, m)$ the unit cost is given by

$$x_d = \prod_{j=1}^{m} k_i x_i + \sum_{l=2}^{m} \prod_{i=l}^{m} k_i (f_{l-1} - k_{w_{l-1}} x_{w_{l-1}}) + (f_m - k_{wm} x_{wm}). \tag{10.19}$$

For the technology selection model under consideration we have i stages $(i = 1, 2, \ldots, m)$ and at the i-th stage we have j technologies $(j = 1, 2, \ldots\ldots, n_i)$. Accordingly, the technological coefficients of input and waste are defined as k_{ij} and $k^{w}{}_{ij}$ respectively. We associate a 0-1 variable A_{ij} to represent the j-th technology at the i-th stage and its value is one if the jth technology is selected, otherwise zero. Since we have to select only one technology at each stage, therefore, we have

$$\sum_{j=1}^{n_i} a_{ij} = 1 \text{for all}\, i. \tag{10.20}$$

Incorporating this feature in the unit cost model, we obtain:

$$x_d = \prod_{i=1}^{m} \sum_{j=1}^{mi} a_{ij} k_{ij} x_i + \sum_{i=2}^{m} \prod_{i=l}^{m} \sum_{j=1}^{ni} a_{ij} k_{ij} \sum_{j=1}^{ni-1} a_{l-1,j} (f_{l-1\,j - kw_{l-1} xw_{l-1,j}})$$

$$+ \sum_{j=1}^{n_m} a_{mj} (f_{mj} - k_{w_{mj}} x_{w_{mj}}) \tag{10.21}$$

APPENDIX 10C

RELATIONSHIP BETWEEN THE INCOMING MATERIAL AND FINISHED PRODUCT

We can now determine the relationship between the finished product (y_d) and the incoming material requirements (y_i) for the three-stage system by successively solving the continuity and mass balance equations as follows:

$$y_i = k_1 k_2 k_3 y_d, \tag{10.22}$$

where k_1, k_2 and k_3 are the technological coefficients identified with the stages 1, 2 and 3 respectively (This is after renumbering similar to Appendix B.)

For the i-th stage system $(i = 1, 2, \ldots m)$ we, therefore, have

$$y_i = \prod_{i=1}^{m} k_i y_d, \tag{10.23}$$

There are alternative technologies at each stage and only one technology at each stage is to be selected. Therefore, defining k_{ij} as the technological coefficient at the i-th stage for the j-th technology (i = 1, 2,..., m; and j = 1, 2, ...n,) and associating A_{ij} as a 0-1 variable with each k_{ij}, we obtain the following relationship:

$$y_i = \prod_{i=1}^{m} \sum_{i=1}^{n_i} a_{ij} k_{ij} y_d \tag{10.24}$$

Sets of combination of technologies with lowest and highest k_{ij} values at each stage would give the lower bound and upper bound on y_i. Designating these values by y_i^{min} and y_i^{max} we have the following to help set goals on y_i:

$$y_i^{\min} \le y_i \le y_i^{\max} \tag{10.25}$$

11 COST AND QUALITY ANALYSIS IN A SPINNING UNIT

11.1 INTRODUCTION

For any project there is a limited supply of resources-such as capital, materials, energy, and labour. Further, competitive environments demand a better quality product or service at the existing price or at a lower price. Survival through acceptable levels of profitability requires maintaining the current market share or improving it as much as possible. In order to attain the end goal of profit, a business unit typically has multiple objectives. Problems arise in allocating scarce resources to the variety of alternative purposes competing for their use. Matching the objectives with the resources to attain end results is not an easy task. Only those organizations that manage productivity and quality as an ongoing activity will be able to deal with these problems effectively[1].

In light of the fact that organizations are constantly involved in quality-related decisions, Duncalf and Dale[2] have come out with an analytical method for the assessment of an organization's approach to quality management. Similarly, a number of developments[3] have been reported in the literature, but the aspects of quality planning and analysis are rarely seen, as also pointed out in Juran, and Juran and Gryna[4].

It is here at this front of quality planning and analysis, that the systems approach can serve as a guide. The systems approach helps in the analysis of such complex systems by focussing attention on the interaction between different components and considering its effect on the overall functioning of the system.

11.2 QUALITY PLANNING AND ANALYSIS

Systems Thinking and Quality

The concept of a system provides logical basis for the planning and analysis of quality. Systems thinking has been used very widely for rational intervention in human activity systems. The importance of quality in these systems can hardly be overemphasized in view of its far-reaching consequences on various fronts. In order to deal effectively with the concept of quality, which has been very widely used with different connotations according to situation and convenience, a rational basis has to be created, and systems thinking can offer rich dividends in this direction. It would be pertinent to define the concept of quality before going into a design or framework for quality planning and analysis. The term quality has been defined in a number of ways by various authors. Some important definitions of quality are: "Quality is fitness for use."[5] "Quality consists of the capacity to satisfy wants."[6] "Quality (means) conformance to requirements"[7], and "Quality is the degree of excellence at an acceptable price and the control of variability at an acceptable cost."[8] Quality when viewed as conformance to specifications has great potential for being a very effective strategy. However, this definition may result in a lack of commitment to quality by members of an organization who do not acknowledge the prescribed specifications. Whereas by defining quality as fitness for use, it is viewed as something that requires every member of the organization to provide the next person in the process with an acceptable product or service.

Such a definition of quality links it with important principles in systems thinking, such as hierarchy, communication, control, and feedback. The quality of output at intermediate stages, as well as that of the final product, is greatly influenced by the structure of the system in terms of its hierarchical structure, control mechanisms, and feed-back. Effective communication about the attributes of the product between different components as well as between the system and the environment will lead to a reduction of perception gaps so that quality can be planned and controlled in a more effective manner. In order to operationalize these basic principles of systems thinking for quality planning and analysis, a framework has been presented in this chapter using the concepts of physical systems theory in conjunction with simulation.

Background Problem

To date a lot of work has been done on production systems, based on PST which takes into account the costs and production rates. But the cost of the product is also related to the quality and is not independent of it, as is assumed by the conventional PST modelling framework.

Second, the quality of output is a function not only of the quality of inputs but also of the various conversion processes that take place. Thus, a need is felt to see the impact of a particular conversion process on the quality of the product. Therefore, an attempt has been made to incorporate all these factors into the PST modelling framework and make it more meaningful and relevant in this context.

Issues

In order to attain the objective, i.e. to build up an integrated approach towards cost and quality, concepts of physical system theory and operations research have been utilized. The specific issues being dealt with are:

- **i.** Use of extended physical system theory modelling framework (chapter five).
- **ii.** Identification of relevant attributes which contribute towards the quality of the product.
- **iii.** Quantification of these attributes in a such a way as to arrive at a relevant quality index.
- **iv.** Prediction of the quality of output on the basis of the quality of the inputs and impact of various conversion processes.
- **v.** Impact of the quality on the cost.

In order to address the above issues the flexible system methodology for cost and quality analysis is shown in Figure 11.1.

11.3 CASE STUDY OF SPINNING UNIT

This section describes the particular system taken for the study, the basic assumptions and the model of the case problem.

System Description

The real life practical of a spinning unit has been chosen as a case study. The operations comprising the spinning unit are shown in Figure 11.2.

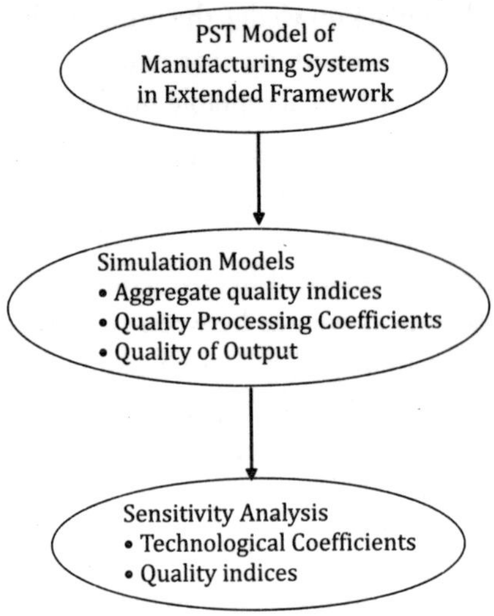

Figure 11.1: Flexible System Methodology for Cost and Quality Analysis

Cotton enters the system in the form of raw cotton, which is converted into lap in the blow room. The lap goes through the carding operation and is converted to the sliver form. This sliver undergoes two drawing operations before it is converted into roving form by the simplex. The ring frame finally converts the roving into yarn.

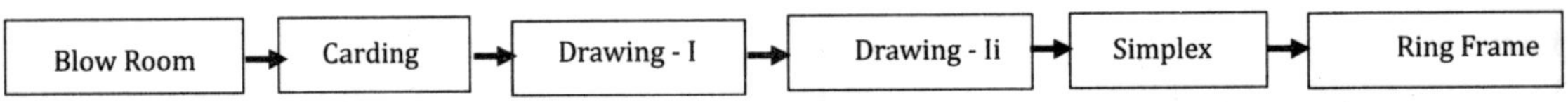

Figure 11.2: Sequence of Operations

Assumptions

While developing the model, some assumptions have been made regarding various characteristics of the system as enumerated below[9]:

- **i.** The machines which carry out particular operations of the processes are treated as indivisible elements-of the system-this means that the processes (usually continuous) occurring inside particular machines are not considered.
- **ii.** The process considered may be divided in a discrete way into particular operations without any harm to a physical sense of the model.
- **iii.** The model has been constructed considering only the material waste.
- **iv.** The flow rates have been assumed to be deterministic for the sake of simplification, though slight variations of these standard values are possible in practice.
- **v.** The model assumes the same set of conditions, e.g. machine-settings, as remaining constant for a particular set of operations for the same product.
- **vi.** The variation in the cost of the product with the change in the quality of the production has been assumed to be linear for the sake of simplification.
- **vii.** The probability distribution of the various characteristics (attributes of quality) have been assumed to be normal.

viii. All the quality transfer coefficients have been assumed to be unity, because of the fact that in practice all the processes are often looked after by a single person.

ix. The model assumes the same proportion of the processing cost per unit index of quality charged for all the processes.

x. Two different methods have been followed to determine the aggregate quality index of the product. In the first case, the weighted average of all the values of the ratings given to the attributes has been assumed to be the aggregate quality index, while in the other case the minimum rating given to a particular attribute amongst all the attributes has been assumed to be the aggregate quality index of that product.

Model Description

The model is constructed considering only the material waste. The three stages involved are:

Component Models

At this stage, each and every subsystem for the spinning unit is considered separately. Its input, output and waste are identified. The flow rate, cost and quality equations are written considering the individual component terminal graphs, as shown in Figure 11.3. All the equations written are listed in Appendix 11 A.

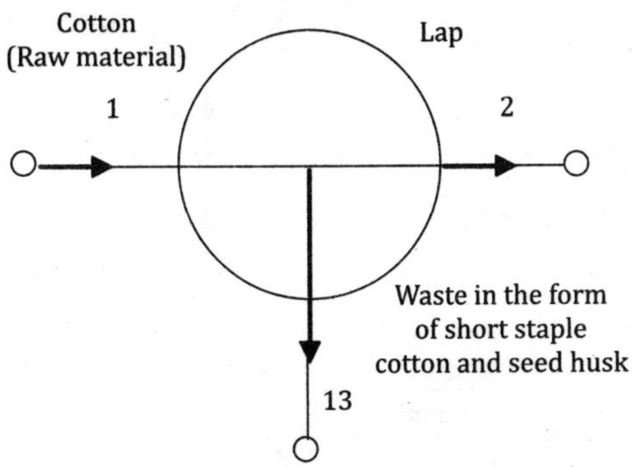

(i) Blow Room

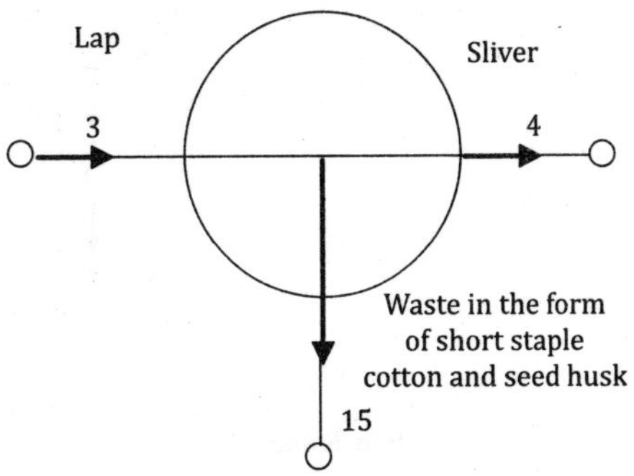

(ii) Card

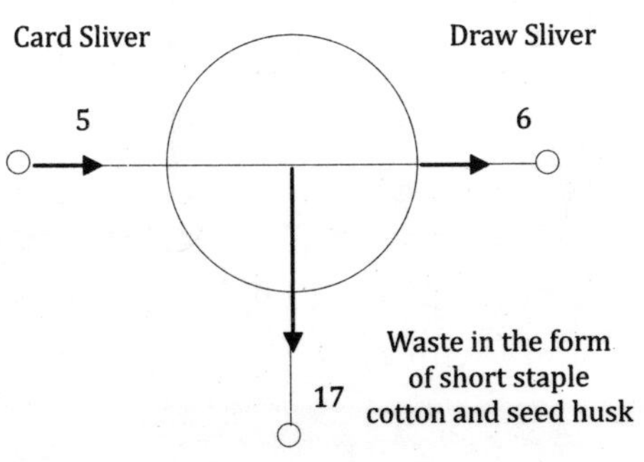

***(iii) Draw Frame* - I**

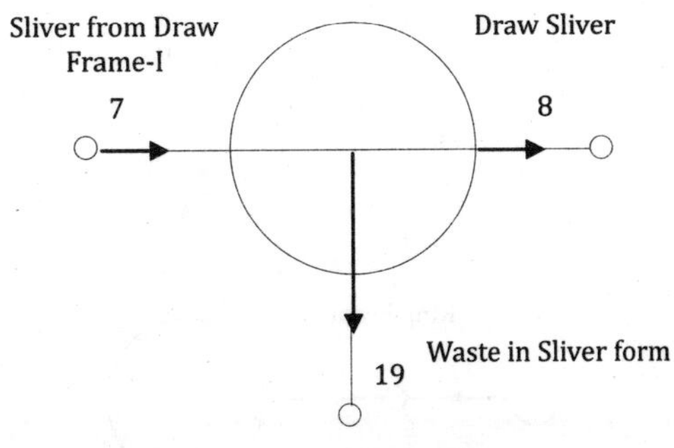

***(iv) Draw Frame* - II**

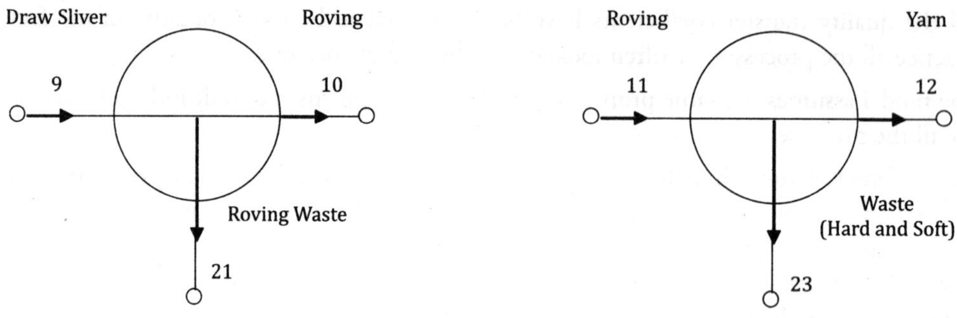

(v) Simplex *(vi) Ring Frame*

Figure 11.3: Component Models

System Graph

The system graph is constructed by joining all the component models in one-to-one correspondence and considering all the datum nodes as one point. The system graph is shown in Figure 11.4.

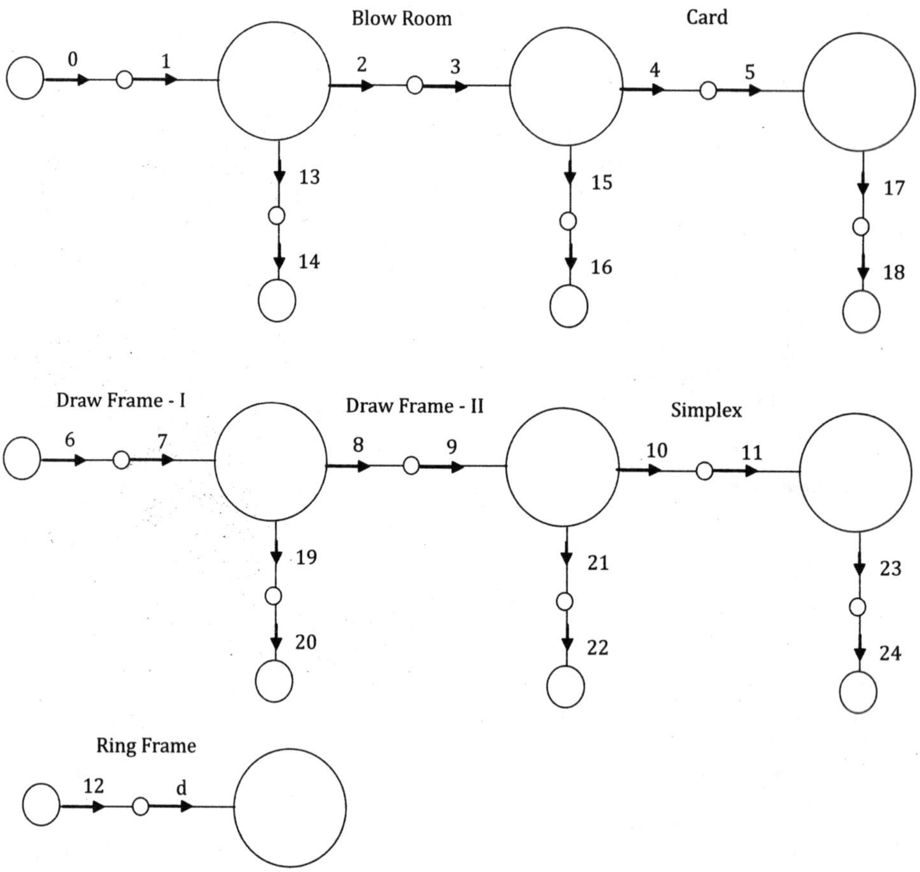

Figure 11.4: System Graph

Structural Constraints

The continuity, compatibility and mass balance equations are written down for the costs and flows. Similarly, the equations for the quality processing and the quality transfer are also written down as shown in Appendix 11 A.

Simulation

Need

The quality processing coefficients are not deterministic in actual practical situations because of variations in the quality of the inputs and the processing itself due to variation in machine conditions etc.

Moreover, there are a large number of characteristics (attributes) that contribute towards the quality of the products which themselves are stochastic in nature.

Since a mathematical analysis would tend to be complex and cumbersome, simulation is the best solution. The model could be tailor made to suit the specific requirements of the system.

Logic

Broadly, the logic of the simulation can be split into three parts.

i. ***Determination of Aggregate Quality Indices:*** Before embarking on the determination of aggregate quality indices, we should find a suitable measure of the quality of the product. This could be done in a simple manner as discussed here.

   The first step is to identify all the attributes of the quality of the product. After listing them, the attributes should be given a rating on a scale of 1–100 or 1–10 so as to suit the requirements. Weightages could also be assigned to the attributes wherever necessary. The ratings and weightages could be assigned by the person in charge of that process, depending upon his subjective judgement or the standard norms followed in the industry. The next step is aggregation by either of the two following methods:

   a. *Weighted Average Quality Index*: In this case the weighted average of the ratings given to the attributes is taken as the aggregate quality index. This is a fairly good measure of the quality as it takes into account all the relevant attributes.

   b. *Minimum Quality Index*: It is felt that the averaging method given a slightly optimistic value of the quality index which may not reflect the reality. In this case, the minimum rating of all the values of the ratings given to the attributes is considered to be the aggregate quality index of the product. This is based on the analogy from Peirce's weak link theory, that the strength of the weakest segment of the yarn is the average strength of that yarn. This analogy fits in well here since the quality of the product will be considered poor if any one of the attributes is poor, even though the other attributes have been given very high ratings. However good the other attributes may be, the functional value of the product cannot be high.

   These methods are shown in Appendix 11 B(a) and 11B (b) respectively, in the form of flow charts for simulation.

ii. ***Determination of Quality Processing Coefficients:*** The logic for determination of the quality processing coefficients of the different processes is shown in Appendix 11 B (c). These are determined on the basis of past data collected over a period of time for a particular set of operations for the same product, with the same conditions in terms of machine conditions, settings, etc.

iii. ***Determination of the Quality of the Output*:** The last phase of the simulation is comprised of the determination of the quality of the output on the basis of the quality of the inputs and the quality processing coefficients determined in the previous phase. The logic for this phase is shown in the form of the flow chart in Appendix 11 B (d).

Analysis and Results

Computer programs based on the logic described above were written. The data for the computer programs is shown in Appendix 11 C. For analysis and logical validation of the model, the programs were run for different sets of data with different quality of input. Sensitivity analysis was also carried out at this point to study the effect of different processes on the quality and the cost. Both the programs, i.e. one for a weighted average quality index method and the other for a minimum quality index method, were run for four sets of quality indices of the input in terms of the mean and standard deviation, i.e.,

i. High mean and high standard deviation

ii. High mean and low standard deviation

iii. Low mean and high standard deviation

iv. Low mean and low standard deviation.

Similarly, the programs for determination of the cost of output was run twice, once using the quality processing coefficients based on the weighted average quality index method and next with those obtained from the minimum quality index method. Again these were run twice each, with high and low input quality indices. The results obtained have been tabulated in comprehensive form in Table 11. 1 (a, b and c) and Table 11.2.

Table 11.1a: Quality Processing Coefficients

Process	Notation	Weighted Average Method	Minimum Quality Index Method
Blow Room	q_1	0.93140	0.90501
Card	q_3	1.02014	1.03217
Draw Frame – I	q_5	1.20456	1.06850
Draw Frame – II	q_7	0.84636	0.85207
Simplex	q_9	0.95595	0.94057
Ring Frame	q_{11}	0.98971	1.36565

Table 11.1b: Results for Weighted Average Method

	SET-1	SET-2	SET-3	SET-4
Mean Input Quality	8.5	8.5	4.5	4.5
Standard Deviation	1.8	0.4	1.8	0.4
Output Quality Index	8.5158	8.5744	4.4743	4.5329

Table 11.1c: Results for Minimum Quality Index Method

	SET-1	SET-2	SET-3	SET-4
Mean Input Quality	8.5	8.5	4.5	4.5
Standard Deviation	1.8	0.4	1.8	0.4
Output Quality Index	6.1854	6.2616	3.2564	3.3326

Table 11.2: Cost of Output

	Weighted average quality index method with		Minimum quality index method with	
	high mean	low mean	high mean	low mean
Mean Input Quality	8.5	4.5	8.5	4.5
Mean Output Quality	9.275	4.910	7.781	4.119
Cost of 100 Kg. Yarn (Rs)	2860.08	2521.45	2799.36	2526.91

Model Validation

Because of the absence of actual data from a particular mill only logical validation has been done. From Tables 11.1 and 11.2 the following points are clear:

a. For a particular method, the better the quality of the input, the better the output quality.

b. For the same average input quality, the lower the standard deviation, the better the output quality.

c. For the same sets of data, the minimum quality index method gives an inferior quality of output as compared to the weighted average method.

d. In both methods, the cost of superior quality output is higher than the cost of inferior quality output.

All the four observations made above are logical and expected, and thus strengthen the logical correctness and validity of the model.

Sensitivity Analysis

We can now analyse the model by deriving useful relationships and conducting the sensitivity analysis, on the basis of which we can get exact quantified information that will help in decision-making. The system equations are solved to derive the relationships in terms of exogenous variables as follows:

a. Relationship between raw material requirements and spun yarn:

$$y_1 = k_1 . k_3 . k_5 . k_7 . k_9 . k_{11} . y_{12} . \tag{11.1}$$

b. Relationship between yarn quality and quality of raw material:

$$z_1 = q_1 . q_3 . q_5 . q_7 . q_9 . q_{11} . z_{12} . \tag{11.2}$$

c. Generation of the cost formula:

$$\begin{aligned} x_d = \; & k_{11} . k_9 . k_7 . k_5 . k_3 . k_1 . x_1 . + k_{11} . k_9 . k_7 . k_5 . k_{15} . k_{16} . \\ & + k_{11} . k_9 . k_7 . k_5 . f_4 .(y_4) + k_{11} . k_9 . k_7 . k_5 . g_4 .(z_4) \\ & + k_{11} . k_9 . k_7 . k_{17} . x_{18} . + k_{11} . k_9 . k_7 . f_6 .(y_6) \\ & + k_{11} . k_9 . k_7 . g_6 .(z_6) + k_{11} . k_9 . k_{19} . x_{20} . + k_{11} . k_9 . f_8 .(y_8) \\ & + k_{11} . k_9 . g_6 .(z_6) + k_{11} . k_{21} . x_{22} \\ & + k_{11} . f_{10} .(y_{10}) + k_{11} . g_{10} .(z_{10}) + k_{23} . x_{24} . + f_{12} .(y_{12}) \\ & + g_{12} .(z_{12}) \end{aligned} \tag{11.3}$$

Carrying out the partial differentiation of x_d with respect to technological coefficients at various stages, we obtain the following results as shown in Table 11.3.

Table 11.3: Sensitivity Analysis with Respect to Technological Coefficients

Sr. No.	Stage	Partial Differential	Value
1.	Blow Room	$\delta x_d / \delta k_{13}$	0.9841046
2.	Card	$\delta x_d / \delta k_{15}$	2.5718
3.	Draw Frame - I	$\delta x_d / \delta k_{17}$	3.08499
4.	Draw Frame – II	$\delta x_d / \delta k_{19}$	3.0837
5.	Simplex	$\delta x_d / \delta k_{21}$	3.0675
6.	Ring Frame	$\delta x_d / \delta k_{23}$	2.20

By carrying out partial differentiation with respect to the quality indices we obtain the following results as shown in Table 11.4.

Table 11.4: Sensitivity Analysis with Respect to Quality Indices

Sr. No.	Stage	Partial Differential	Value
1.	Blow Room	$\delta x_d / \delta g_2(z_2)$	1.09345

2.	Card	$\delta x_d/\delta g_4(z_4)$	1.0287
3.	Draw Frame - I	$\delta x_d/\delta g_6(z_6)$	1.0283
4.	Draw Frame – II	$\delta x_d/\delta g_8(z_8)$	1.0279
5.	Simplex	$\delta x_d/\delta g_{10}(z_{10})$	1.0225
6.	Ring Frame	$\delta x_d/\delta g_{12}(z_{12})$	1.0

Interpretation of Results

The sensitivity of cost with respect to the waste generated at each stage has been determined. This can broadly be classified into three categories.

a. Almost insensitive
b. Moderately sensitive
c. Highly sensitive

The flow line is least sensitive to changes in the blow room, moderately sensitive to carding and ring frames and highly sensitive to changes in simplex and draw frames in this particular case. This may also change, depending upon the market price of the waste generated.

Similarly, when sensitivity analysis is carried out with respect to the quality indices, it is seen that it is most sensitive to the changes in the blow room and least sensitive to changes in the ring frame.

11.4 APPLICABILITY OF THE MODEL

The validated simulation model is an effective vehicle for the managerial evaluation of practical alternatives in the form of sensitivity analysis and the evaluation of alternative control policies. Some of the possible applications of the model are highlighted as follows:

i. ***Waste Aspect***

- Provides a proper accounting system for waste.
- Provides an infrastructure for collection, processing and information flow to various levels.
- Aids development of composite waste-management indices.
- Effective control of waste and effective utilization of system resources.

ii. ***Quality Aspect***

- It is an effective tool for predicting the quality of output on the basis of system performance and the quality of inputs.
- The quality of input required to produce the desired quality of output, can be deduced from retrospective calculations.
- Provides a good measure of the process capability in terms of quality processing coefficients.
- Through sensitivity analysis, shows the impact of variations in the system parameters on the quality of output.

- Indicates the need for changes in the system, such as machine settings, maintenance, etc., by pinpointing the source of trouble.
- Justifies the need for and value of modernization, if requirements are beyond the process capability.
- It would appeal to the persons in the system, as the user is the main focus of this aspect, and it takes into account his subjective judgement of the capability.

iii. ***Cost and Profitability Aspect***

- Provides a database for differential pricing according to the quality.
- Through sensitivity analysis, shows the effect of variations in changes in the flow line and the quality of the product on the cost of the product.
- Provides support for deciding the optimal quality to be produced for maximizing the profits, as it is well known that too high a quality may not always be useful to the consumer or profitable for the product.

REFLECTIONS

It is evident from the results that the model has successfully captured the aspect of quality in relation to costs. The analysis necessitates good quality input in terms of high average and low variability in order to produce good output and thus charge the right price for the product. It gives a database for charging differential prices of product, which is otherwise a highly subjective decision. Moreover, it provides a good measure of the process capability in terms of the quality processing coefficients. One may choose the minimum quality index method if the management follows a conservative policy due to stringent requirements of the customers. On the other hand, the weighted average quality index method may be used if the customers are less demanding in their requirements.

A generalized framework for quality planning and analysis has been proposed which is by no means perfect and comprehensive. It offers a lot of potential and flexibility and can be further refined and augmented according to the requirements of the users.

Some of the assumptions themselves are the limitations of the study, e.g. it has not been possible to collect data on quality aspects from a mill, hence it has not been possible to carry out the retrospective testing of the model. Similarly, a linear relationship has been assumed between the cost and the quality, which is not altogether accurate. The flow rates and the technological coefficients have been assumed to be deterministic though in practice there are slight variations. Further improvements can be brought about in this work; a few of them are suggested as follows:

i. The technological coefficients can also be regarded as stochastic.

ii. Subjective attributes of quality could also be considered in the model.

iii. Standard quality indices can also be used instead of determining them by the weighted average quality index or the minimum quality index methods, e.g., in spinning, the Fibre Quality Index (FQI) and Yarn Quality Index (YQI) may be used for this purpose.

iv. This particular model could be extended to a composite textile mill so as to take into account the quality transfer coefficients at the interfaces.

v. A careful analysis can be made to find the exact nature of the relationship between the quality and the cost of each process.

vi. The concept of quality used in this chapter is "fitness for use", which is primarily restricted to the analysis of production. The utility of the model can be enhanced by extending the quality chain beyond production by dealing with the extended definition. "Quality" is fitness for use and desirability for possession".

Notes

1. Edosomwan J.A. (1987) *Integrating Productivity and Quality Management,* New York, Marcel Dekker Inc.
2. Duncalf A.J. and Dale B.G. (1988) Quality Management Effectiveness-An Analytical Approach, *International Journal of Operations and Production Management,* 8(5), 2-45.
3. Abbott R.A. and Leaman D.C. (1982) Quality Control and Quality Assurance, In Heyel C. (Ed.) *The Encyclopaedia of Management,* New York, Van Nostrand Reinhold, 998-1009; Grant E.L. and Leavenworth R.S. (1980) *Statistical Quality Control,* New York McGraw Hill; Juran J.M. (Ed.) (1974) *Quality Control Handbook,* New York, McGraw Hill.
4. Juran J.M. (1988) *Juran on Planning for Quality,* New York, The Free Press; Juran J.M. and Gryna F.M. (1980) *Quality Planning and Analysis,* New York, McGraw Hill;
5. Juran (Ed) (1974) *op cit.*
6. Edwards C.W. (1968) The Meaning of Quality, *Quality Progress,* 36-39.
7. Crosby P.B. (1979) *Quality is Free,* New York, New American Library.
8. Broh R.A. (1982) *Managing Quality for Higher Profits,* New York, McGraw Hill; Ishtiaque S.M. (1988) *Cost Control in Spinning,* Workshop on Production Planning and Cost Control, Ghaziabad, NITRA.
9. Sushil and Pankaj (1992) System Modelling of Cost and Quality: An Application of Extended Physical System Theory and Simulation, *System Research,* 9(1), 43-66.

APPENDIX 11A

SYSTEM EQUATIONS FOR THE MODEL OF THE SPINNING UNIT

i. Component Terminal Equations

a. Blow Room Process

$x_2 \quad = k_1 . x_1 . + k_{13} . x_{13} . + f_2(y_2) + g_2(z_2)$

$y_1 \quad = k_1 . y_2$

$y_{13} \quad = k_{13} . y_2$

$z_1 \quad = q_1 . z_2$

where

y_1 - Raw material flow rate

y_{13} - Waste generation flow rate

$k_1 . k_{13}$ - Technological coefficients

x_2 - Cost associated with unit flow rate of lap

x_1 - Cost of unit flow of raw material

$f_2(y_2)$ - Processing cost of unit flow of material

z_2 - Aggregate quality index of the lap

z_1 - Aggregate quality index of the raw material

q_1 - Quality processing coefficient of the blow room

$g_2(z_2)$ - Proportion of the processing cost charged per unit of quality index of lap if it is better than average quality

Similarly for,

b. Card

$x_4 \quad = k_3 . x_3 . + k_{15} . x_{15} . + f_4(y_4) + g_4(z_4)$

$y_3 \quad = k_3 . y_4$

$y_{15} \quad = k_{15} . y_4$

$z_3 \quad = q_3 . z_4$

c. Draw Frame – I

$x_6 \quad = k_5 . x_5 . + k_{17} . x_{17} . + f_6(y_6) + g_6(z_6)$

$x_5 \quad = k_5 . y_6$

$y_{17} \quad = k_{17} . y_6$

$z_5 \quad = q_5 . z_6$

d. Draw Frame – II

$x_8 \quad = k_7 . x_7 . + k_{19} . x_{19} . + f_8(y_8) + g_8(z_8)$

$y_7 = k_7 \cdot y_8$

$y_{19} = k_{19} \cdot y_8$

$z_7 = q_7 \cdot z_8$

e. Simplex

$x_{10} = k_9 \cdot x_9 \cdot + k_{21} \cdot x_{21} \cdot + f_{10}(y_{10}) + g_{10}(z_{10})$

$y_9 - k_9 \cdot y_{10}$

$y_{21} = k_{21} \cdot x_{10}$

$z_9 = q_9 \cdot z_{10}$

f. Ring Frame

$x_{12} = k_{11} \cdot x_{11} \cdot + k_{23} \cdot x_{23} \cdot + f_{12}(y_{12}) + g_{12}(z_{12})$

$y_{11} = k_{11} \cdot y_{12}$

$y_{23} = k_{23} \cdot y_{12}$

$z_{11} = q_{11} \cdot z_{12}$

ii. Structural Constraints

a. Continuity Equations

y_0	$= y_1$	y_2	=	y_3
y_4	$= y_5$	y_6	=	y_7
y_8	$= y_9$	y_{10}	=	y_{11}
y_{12}	$= y_d$	y_{13}	=	y_{14}
y_{15}	$= y_{16}$	y_{17}	=	y_{18}
y_{19}	$= y_{20}$	y_{21}	=	y_{22}
y_{23}	$= y_{24}$			

b. Compatibility Equations

x_0	$= x_1$	x_2	=	x_3
x_4	$= x_5$	x_6	=	x_7
x_8	$= x_9$	x_{10}	=	x_{11}
x_{12}	$= x_d$	x_{13}	=	x_{14}
x_{15}	$= x_{16}$	x_{17}	=	x_{18}
x_{19}	$= x_{20}$	x_{21}	=	x_{22}
x_{23}	$= x_{24}$			

c. Mass Balance Equations

$y_1 = y_2 + y_{13}$

or $y_1/y_2 = 1 + y_{13}/y_2$

or $k_1 = 1 + k_{13}$

Similarly,

$$k_3 = 1 + k_{15}$$
$$k_5 = 1 + k_{17}$$
$$k_7 = 1 + k_{19}$$
$$k_9 = 1 + k_{21}$$
$$k_{11} = 1 + k_{23}$$

d. Quality Transfer Equations

$$z_o = t_1 z_1 \quad z_2 = t_3 z_3$$
$$z_4 = t_5 z_5 \quad z_6 = t_7 z_7$$
$$z_8 = t_9 z_9 \quad z_{10} = t_{11} z_{11}$$
$$z_{12} = t_{13} z_{13}$$

APPENDIX 11B: SIMULATION FLOW CHARTS

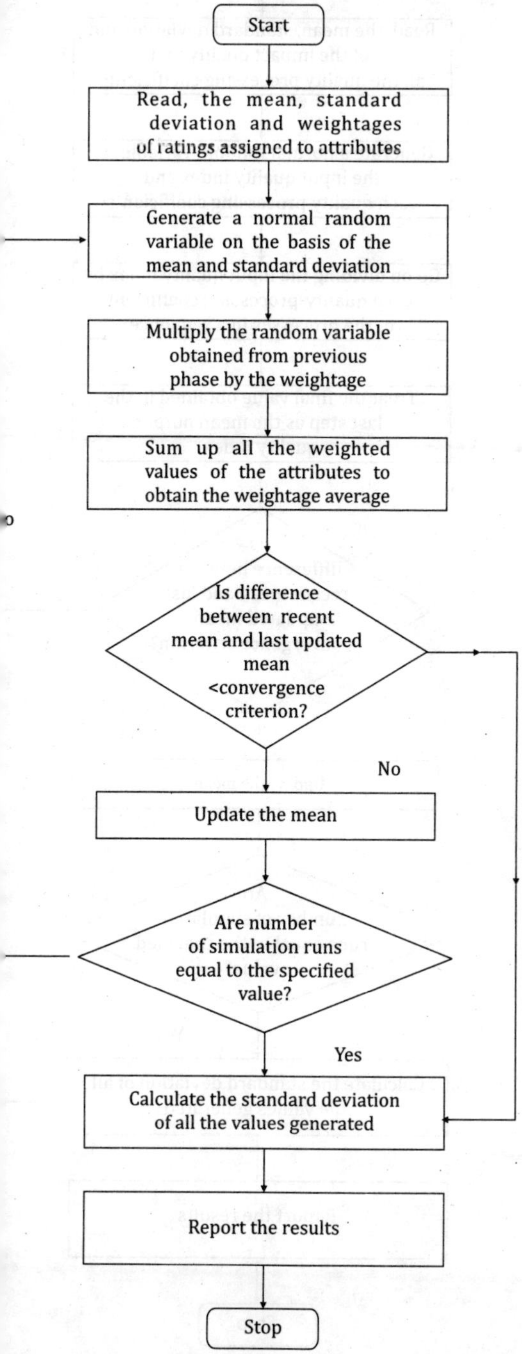

(a) Summary Flow Chart (Weighted Average Quality Index Method)

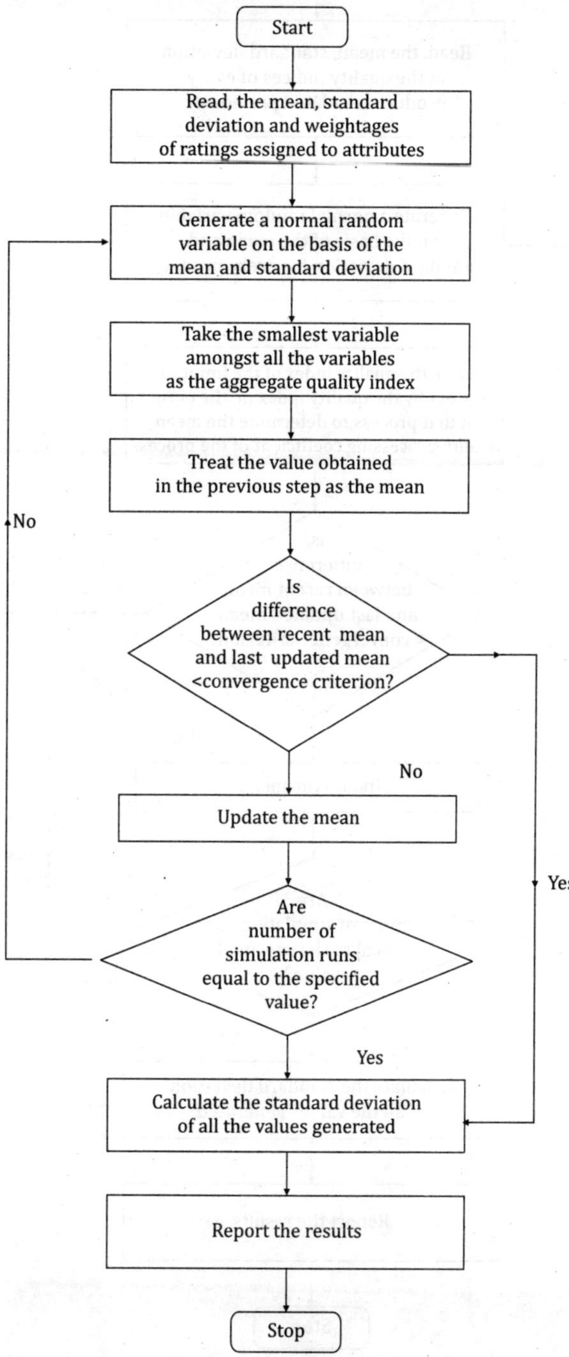

(b) Summary Flow Chart (Minimum Quality Index Method)

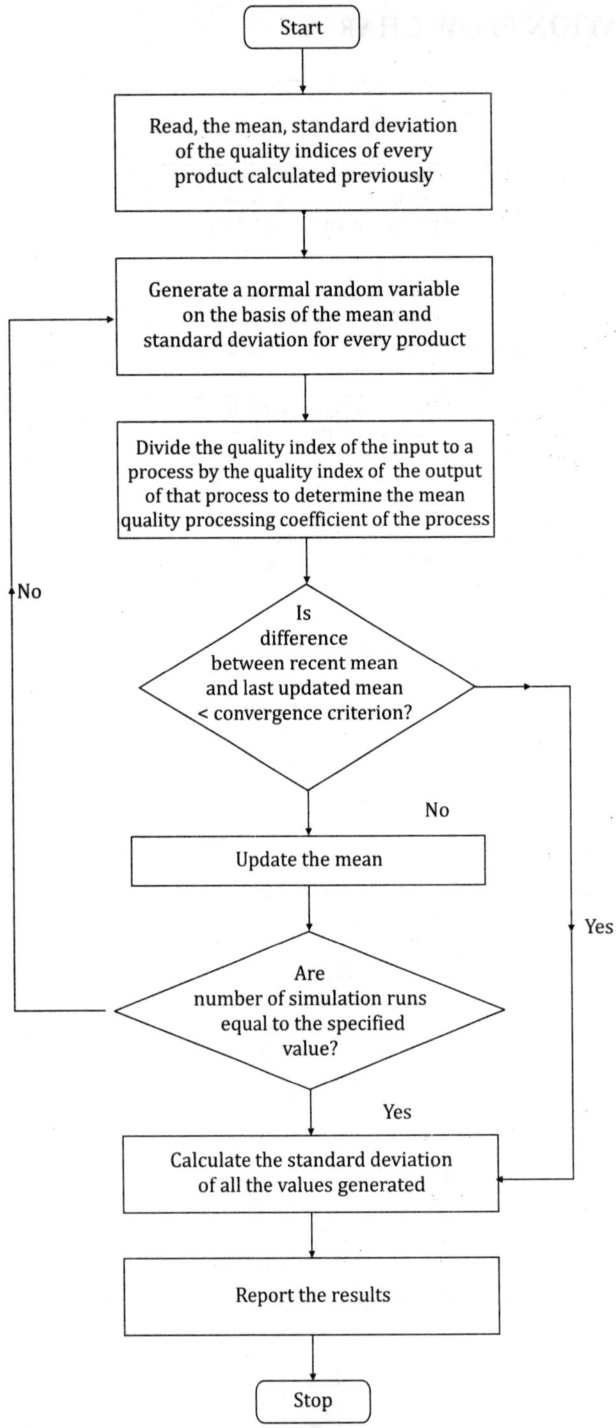

(c) *Summary Flow Chart (Determination of Quality Processing Coefficients)*

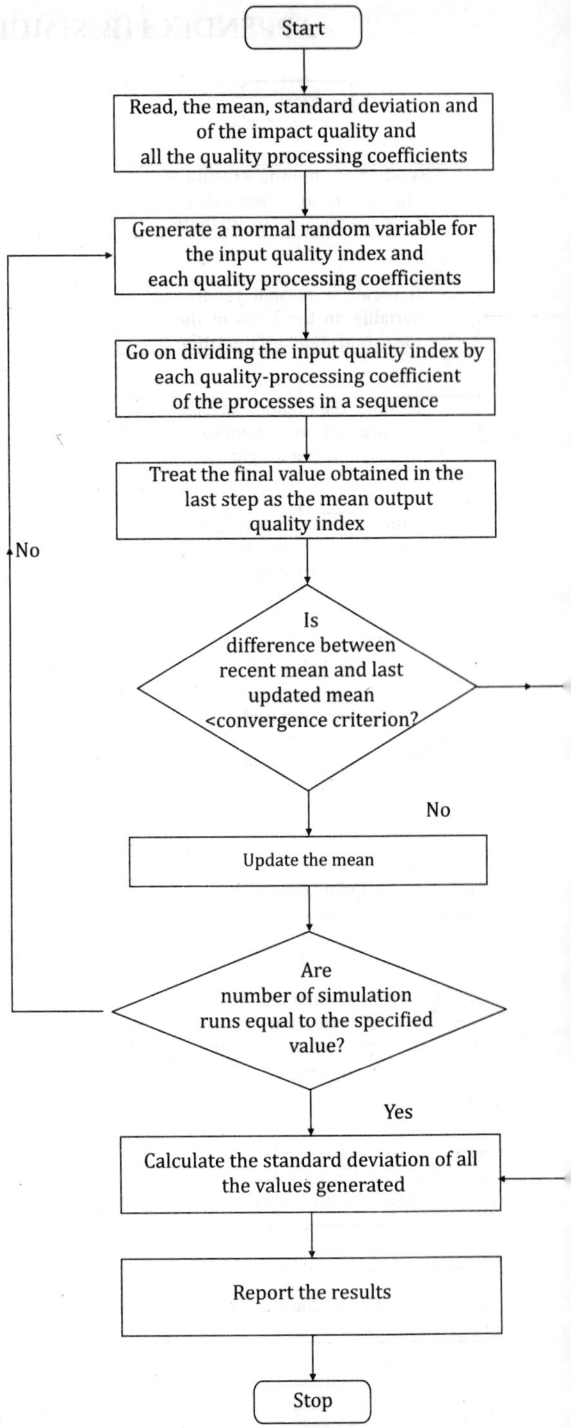

(d) *Summary Flow Chart (Determination of the Output Quality Index)*

APPENDIX 11C: DATA

a. Data for Determination of Quality Indices and Quality Processing Coefficients

Attribute	Mean	Standard Deviation	Weight
Raw material (Cotton)			
Trash percentage	8.2155	1.14	0.167
2.5% Span length	7.6127	0.9954	0.167
Uniformity Ratio	6.7584	1.2106	0.167
Maturity Ratio	7.2215	0.8788	0.167
Micronaire Value	7.3391	1.0077	0.167
Bundle Strength	7.9421	1.13	0.167
Lap			
Cleaning Efficiency %	8.8873	0.987	0.25
Trash in Lap	7.93	1.123	0.25
Full Lap CV%	8.2	1.012	0.25
1 metre Lap CV%	7.33	1.12	0.25
Card Sliver			
Trash in sliver	8.84	1.23	0.2
Cleaning Efficiency %	8.23	0.93	0.2
Draw Frame-I Sliver			
Uster %	6.35	1.089	0.5
5 metre CV %	6.76	1.121	0.5
Draw Frame-II Sliver			
Uster %	8.23	0.94	0.5
5 metre CV%	7.88	1.12	0.5
Roving			
Uster %	7.34	1.05	0.5
15 meter CV%	8.21	1.31	0.5

Yarn			
Actual Count	8.93	0.987	0.3
Actual Strength	8.97	0.664	0.05
Corrected CSP	6.42	1.04	0.08
Count CV%	6.57	1.15	0.08
Strength CV%	9.01	0.98	0.2
Yarn Uster %	8.73	1.14	0.08
Thick places and Thin Places/100 metre	6.54	1.21	0.08
Neps	6.89	1.01	0.08
TPI CV %	8.12	0.964	0.05

b. Data for Determination of Cost of Output

i. Processing Costs and Waste Extracted at Various Stages of the Spinning Process (For 100 Kgs)

Process	Processing Cost (₹)	Waste Extracted (Kg.)
Blow Room	60	5.39
Card	66	5.60
Draw Frame	33	0.03428
Simplex	70	0.471
Ring Frame	547	1.945

ii. Prices Prevailing in the Market (₹)

Waste Material	Price (₹)
Raw Material (Cotton)	1497.0
Waste from Blow Room	90.0
Waste from Card	250.0
Waste.from Draw Frame	300.0
Waste from Simplex	300.0
Ring Waste from Frame	220.0

12 INSPECTION POLICY

12.1 INTRODUCTION

In any manufacturing system, a primary consideration is proper allocation of inspection effort for quality control (QC). As the item moves through the system, it may incur defects or non-conformities as a result of processing. If a defective item is not removed/corrected, it will continue to be processed and will incur unnecessary processing costs, which are the internal failure costs. Further, if non-conformity is not detected, the item will be supplied to a customer, thus, incurring additional costs in terms of external failure costs.

It is neither feasible nor practical to inspect each and every item at every stage. Besides this the costs involved are prohibitive. A defective item supplied to a customer, besides causing a failure cost, is also going to cause loss of reputation, which is not easily quantifiable. Thus, every management attempts to determine appropriate inspection levels to avoid such situations and to minimize the quality costs.

In present day scenario of extremely intense level of competition, where the products are assumed to be nearly perfect in their performance, the subject requires greater attention. There have been many studies on the subject mainly confined to mathematical models. A majority of these models have followed dynamic programming and 0-1 linear programming approaches.

12.2 PROBLEM DEFINITION

As stated earlier, many mathematical models have been developed in the past. As an extreme, a perfect inspection accuracy can be assumed scrapping all rejected items. It was proved an extreme point solution (either 0 or 100% inspection) is optimal unless there is a constraint on final product quality level[1]. This result was further verified where rejected items are replaced by conforming items. As per another view, while working on an adaptive model, optimal inspection policies depend on previous inspection history. The effects of inspection error in a multi-stage manufacturing system with scrapping costs are found to be substantial[2].

All these models while adopting dynamic programming, failed to incorporate few essential costs like rework, repair, etc. Yun and McDowel[3] formulated a 0-1 mixed integer linear programming approach to resolve the inspection effort problem for a serial system. This approach has advantages over dynamic programming as it can be easily modified to include both production and material constraints. The results obtained by them are however only extreme point solution (0% or 100% inspection) and do not yield optimal inspection percentages. There is also a requirement to subject this model to more rigorous tests to validate its soundness.

12.3 MODEL OF PRODUCTION STAGE WITH INSPECTION

The proposed model has following features:

i. Modelling is more explicit.

ii. Material cost is included.

iii. The scrap of two types is treated separately, as the fraction to be recovered/ repaired will be entirely different and the cost of same will also be different.

iv. It makes use of multi-level analysis.

v. It can be easily simulated.

vi. It may incorporate scale economies also.

The flexible system methodology for inspection policy analysis is shown in Figure 12.1.

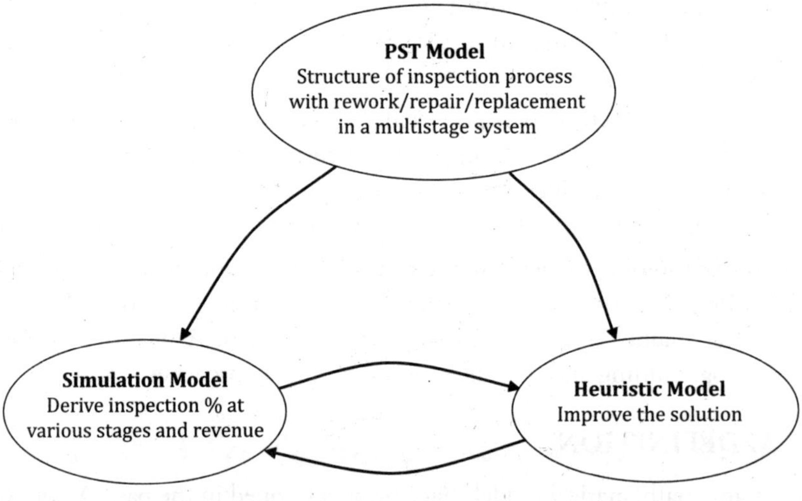

Figure 12.1: Flexible System Methodology for Inspection Policy Analysis

Component Models

The component models are developed for various processes in a manufacturing stage. The component terminal graphs are shown in Figure 12.2 and the cost and flow equations are given in Appendix 12A.

i. j^{th} operation: The process involved here is a combined transformation process as explained earlier in chapter two. The process is as per the diagram at Figure 12.2 (a).

ii. Classification after j^{th} Operation: This process is based on the classification process as discussed in chapter two. No costs are involved.

a. Conforming Units: The output of conforming units is divided into two parts i.e. v_{j1} is sent for inspection and t_{j1} is not inspected and sent to next process. The same is shown in Figure 12.2 (b).

b. Non-Conforming Units: The component terminal graph for classification of non-conforming units after jth operation is shown in Figure 12.2 (c).

iii. j^{th} Inspection: This is modelled as a combined transformation process as depicted in the diagram at Figure 12.2 (d).

Inspection Errors

Type I error: Probability of conforming units rejected (αj) = yu_{j1}/yv_{j1}

Type II error: Probability of non-conforming units accepted (βj) = $yw_{j+1,2}/yv_{j2}$

iv. Classification of Rejected Conforming Units: The rejected conforming units are divided into units for rework, repair/replace, and scrap (Figure 12.2 (e)). For rework these are sent back to jth operation. As the quality of rejected conforming and non-confirming units will be different, these functions will be different for both the categories. Similarly, the costs will also be different.

v. Classification of Rejected Non-Conforming Units: The rejected non-conforming units are classified into repair, rework and scrap as shown in Figure 12.2 (f).

vi. j^{th} Repair of Conforming Units: It is assumed that during the repair no scrap is being generated, hence $ku_{j2}1 = 1$. However, if some scrap is generated then $ku_{j2}1 > 1$. In the Figure 12.2 (g) $f_{11}1$ (Yu_{j1}) indicates processing cost for repair.

vii. j^{th} Repair of Non-Conforming Units: It is assumed that during the repair no scrap is being generated, hence $ku_{j2}1 = 1$. However, if some scrap is generated then $ku_{j2}1 > 1$.

viii. Aggregation of Conforming Units at Output of jth Stage: This stage involves recombining of inspected units, repaired units and un-inspected units at the output. See Figure 12.2 (i).

ix. Aggregation of Non-Conforming Units at Output of jth Stage: This stage involves recombining of inspected units, repaired units and un-inspected units at the output. See figure 12.2 (j).

x. Aggregation of Conforming Units at Input to j^{th} Stage: This stage describes aggregation of input to jth stage. At this stage besides input from previous stage, there would be additional units for rework. See Figure 12.2 (k).

xi. Aggregation of Non-Conforming Units at Input to j^{th} Stage: This stage describes aggregation of input to jth stage. At this stage besides input from previous stage, there would be additional units for rework. See Figure 12.2 (i).

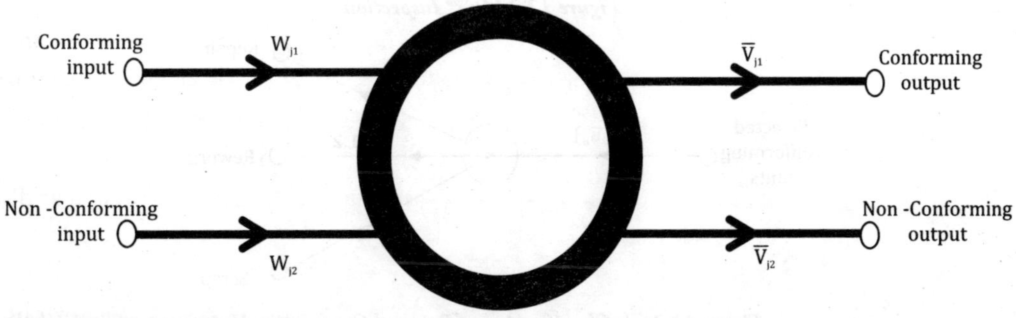

Figure 12.2 (a): j^{th} Operation

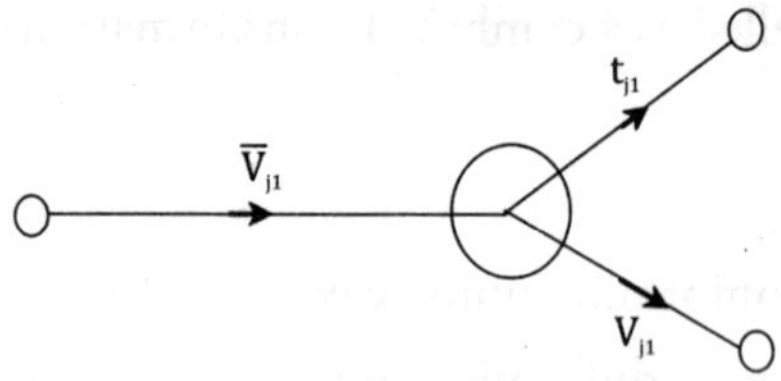

Figure 12.2 (b): Classification of Conforming Units after j^{th} operation

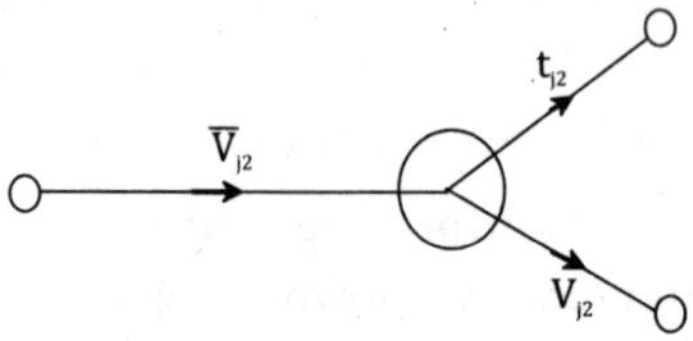

Figure 12.2 (c): Classification of Non-Conforming Units after j^{th} operation

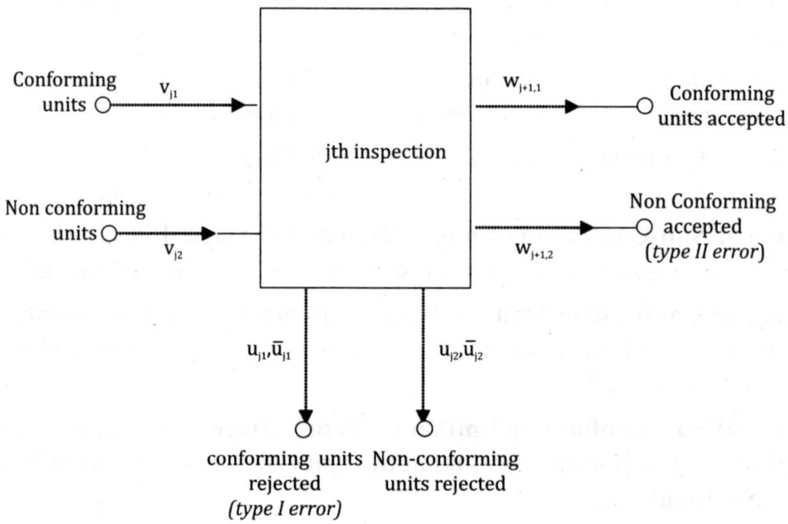

Figure 12.2 (d): j^{th} Inspection

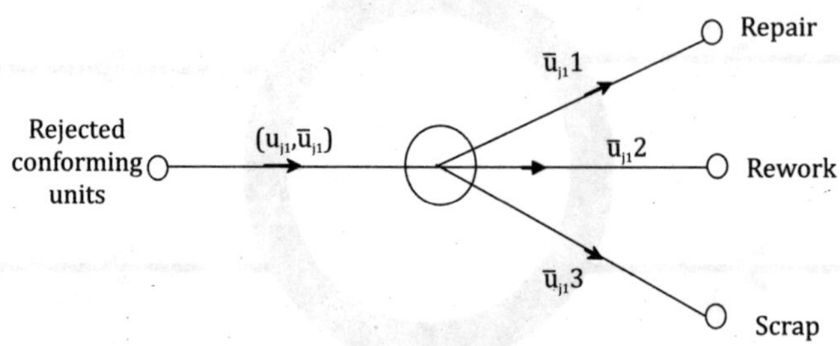

Figure 12.2 (e): Classification of Rejected Conforming Units

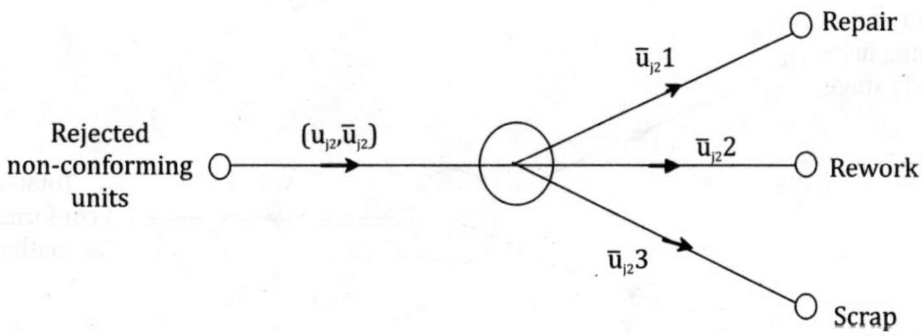

Figure 12.2 (f): Classification of Rejected Non-Conforming Units

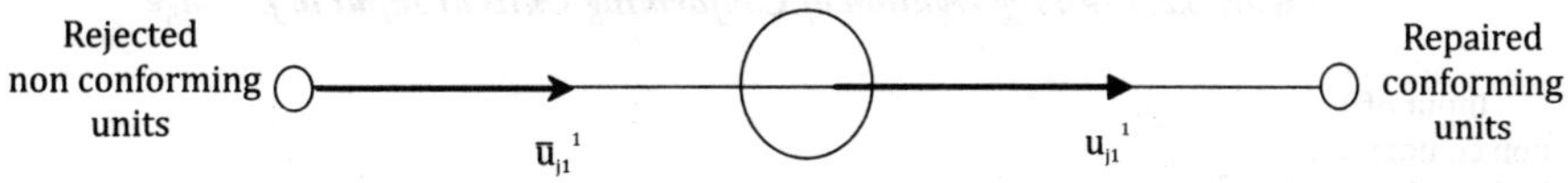

Figure 12.2 (g): Repair Conforming Units

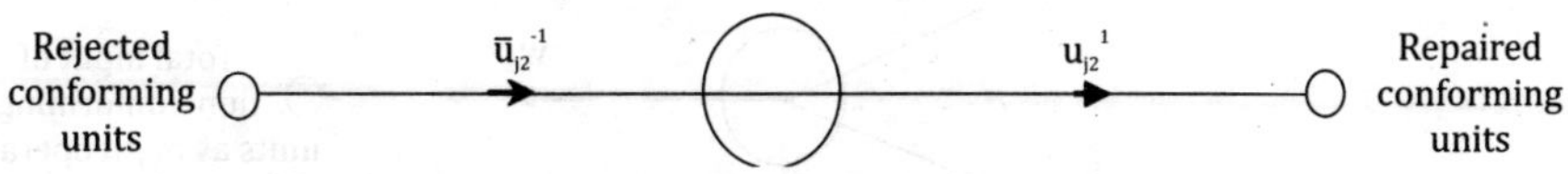

Figure 12.2 (h): Repair Non-Conforming Units

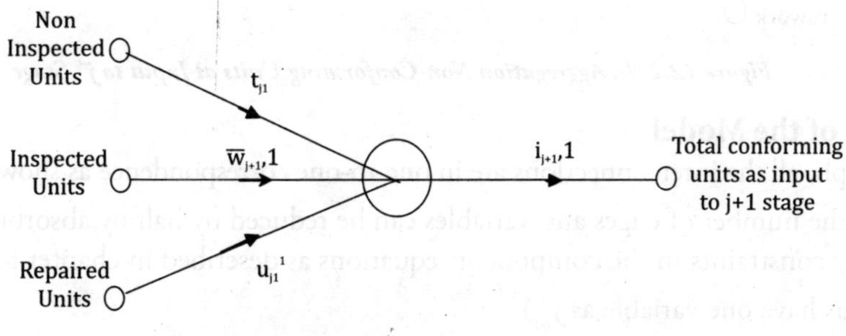

Figure 12.2 (i): Aggregation of Conforming Units at Output

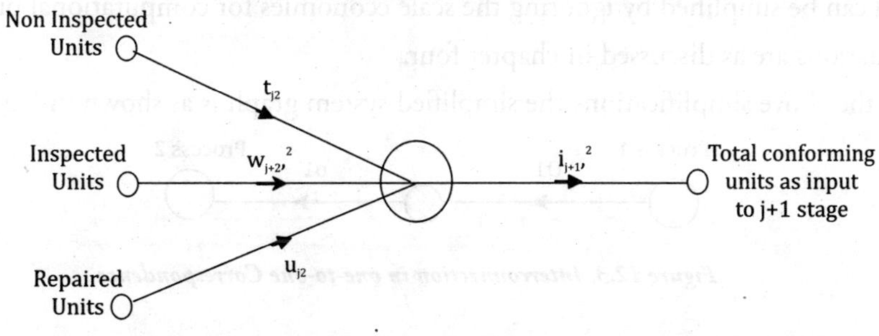

Figure 12.2 (j): Aggregation of Non-Conforming Units at Output

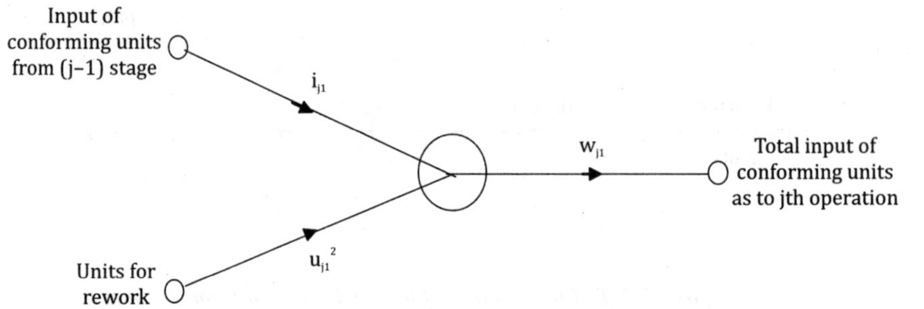

Figure 12.2 (k): Aggregation of Conforming Units at Input to j^{th} Stage

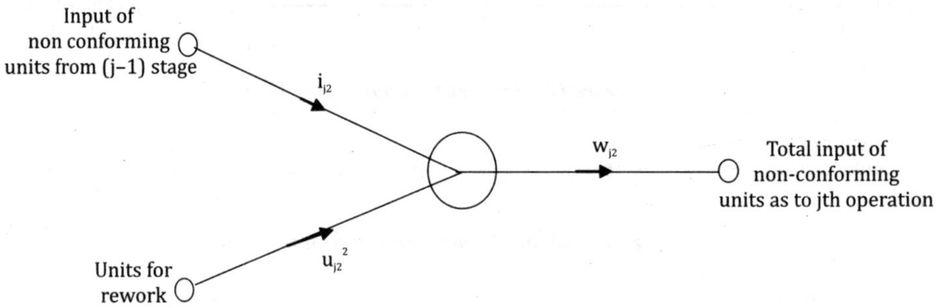

Figure 12.2 (l): Aggregation Non-Conforming Units at Input to j^{th} Stage

Simplification of the Model

In the system graph, all the interconnections are in one-to-one correspondence as shown in Figure 12.3.

In this case, the number of edges and variables can be reduced by half by absorbing the continuity and compatibility constraints in the components equations as described in chapter four. Hence:

$y_{o1} = y_{o1}'$ (thus have one variable as y_{o1})

$x_{o1} = - x_{o1}'$ (have one variable as x_{o1} but – ve sign removed)

The model can be simplified by ignoring the scale economies for computational purposes, i.e. all cost equations are as discussed in chapter four.

In view of the above simplifications the simplified system graph is as shown in Figure 12.4.

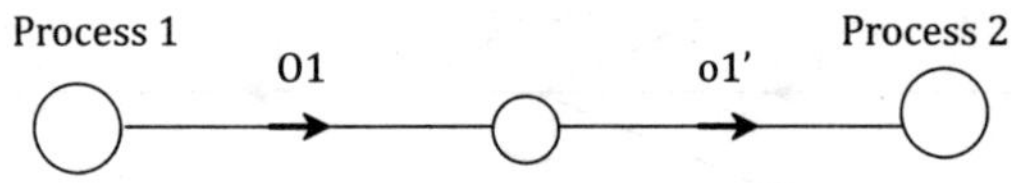

Figure 12.3: Interconnection in one-to-one Correspondence

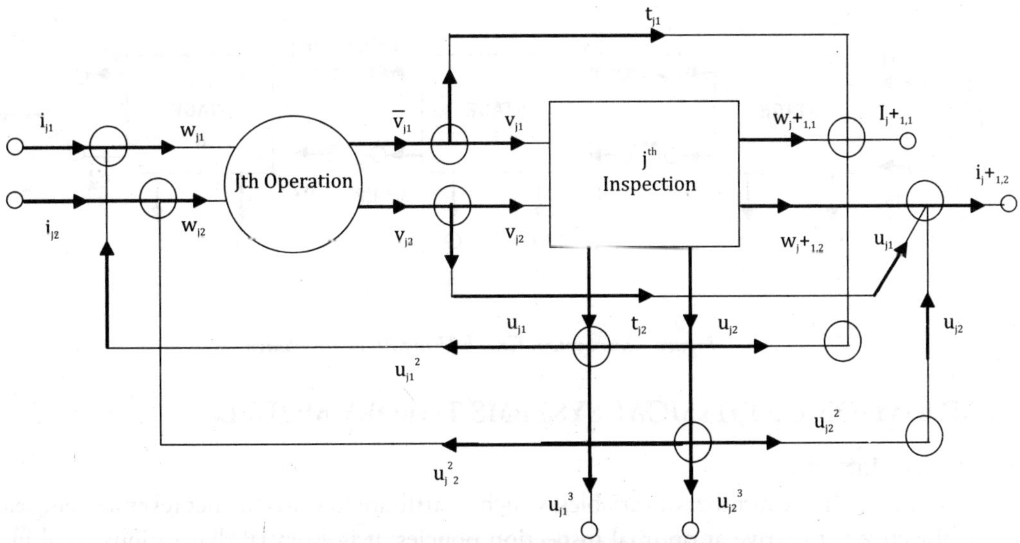

Figure 12.4: Simplified System Graph – j^th^ Stage

12.4 MULTI-STAGE ANALYSIS

The entire stage *j* is treated as a black box and relationships can be derived in terms of basic parameters. The next stage is taken as a component and simplified system graph is developed for jth stage (Figure 12.4). The black box concept of stage j can be seen in Figure 12.5 and system graph for multi stage system in Figure 12.6.

The derivation of the equations for the black box representatives of jth stage is shown in Appendix 12 B.

The model discussed above is a non-linear model, which can be simulated with ease. Using d_j (Inspection percentages) as a decision variable, (being generated randomly), one can simulate any number of stages. For the purpose of this study, only three stages are simulated.

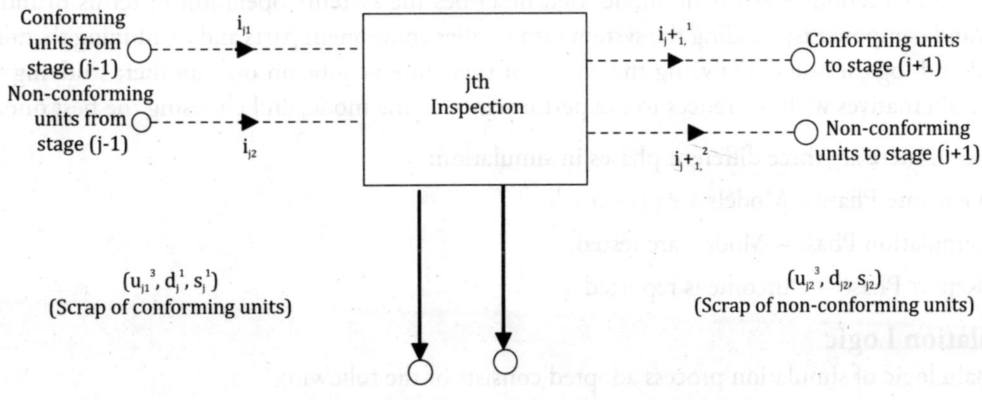

Figure 12.5: Black Box Representation of Stage j

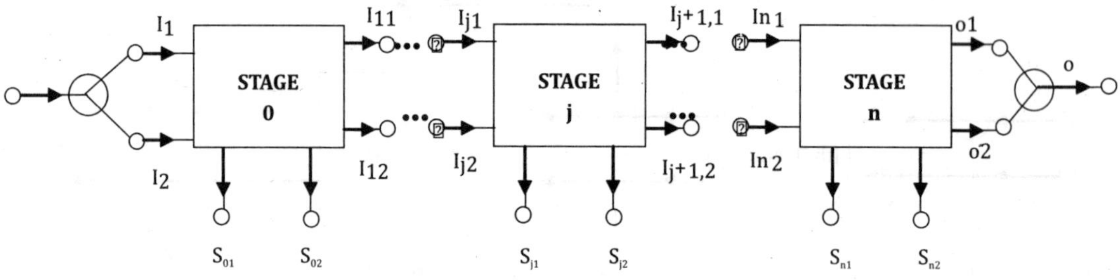

Figure 12.6: System Graph Using Black Box Stage

12.5 SIMULATION OF PHYSICAL SYSTEMS THEORY MODEL

Need for Simulation

The model deals with large number of variables, which contribute towards the net revenue being earned. Since the objective is to arrive at optimal inspection policies, it is essential that various combinations are tried out. Mathematical analysis would tend to be complex and tedious. The situation is further complicated if one wishes to do multistage analysis. In these circumstances, simulation approach is perhaps most suited.

Simulation – An Introduction

Simulation of a system is the representation of a problem in terms of its logical and mathematical rules. This has been used for years as a technique of evaluating system performance. This technique is appropriate to situations where size and/or complexity of the problem makes use of optimizing technique either difficult or impossible.

Simulation is a method of solving decision-making problems by designing, constructing and manipulating a model of real system. It is defined to be an action of performing experiments on a model of a given system. It duplicates the essence of a system or activity without actually obtaining the reality. Here a system is defined as a collection of entities or components which act and interact together towards the accomplishment of some logical goal and model as representation of the system. This involves construction of symbolic model that describes the system's operation in terms of individual events and components; dividing the system into smaller component parts and combining them in their natural and logical order; analyzing the effects of their interactions on one another; studying various specific alternatives with references to the performance of the model and choosing the best one.

In general, there are three different phases in simulation:

- **i.** Generate Phase – Models are generated.
- **ii.** Simulation Phase – Models are tested.
- **iii.** Report Phase – Outcome is reported.

Simulation Logic

The main logic of simulation process adopted consists of the following:

- **i.** Selection of parameter values

ii. Application of Inspection Policy Variables (d_j) generated randomly.

iii. Estimation of various flow and cost variables in a three stage model.

Selection of Parameter Values

The parameter values selected are same as used by Yum and Mcdowell[4]. These values have been selected primarily for ease of comparison of results.

Only one set of equations are being used in the program. Depending on number of stages being simulated, the program is made to loop back. In this case, a three-stage model is being depicted.

Inspection Policy Variable Generated Randomly

The program has been kept flexible, catering for following:

i. Number of simulation runs is flexible, subject to maximum of 1000 runs. This can be easily increased. The program generating random numbers is a subroutine of the main program. At the time of execution, the number of simulation runs desired has to be specified.

ii. The program asks for specifying seed values for the random number series. It is recommended that three prime numbers (two digits each) be specified.

Calculation of Flow and Cost Variables

As stated earlier, there is a requirement to impose two constraints, i.e. production, and material. These constraints require different approaches to simulation.

i. Material Constraint: This constraint applies to input, i.e. the input is not unlimited. For this model, input is being fixed at 1.0 units. Here after, calculation of various parameters is a forward flow, proceeding to next stage, one after another. The flow chart is at Appendix 12 C.

For this study, five simulations, each consisting of 1000 runs has been used. For each of these five simulations, fresh random number series were generated, based on new values of seed.

ii. Production Constraint: This constraint imposes limitations on final output, for both conforming and non-conforming units. Once again final output assumed is 1.0 units (combined for both conforming and non-confirming). Because of this constraint, the processing has to start backward, assuming the final output values. Having calculated the flow values from backward flow, one has to proceed forward for cost variables. This approach was used for a simulation of 1000 runs.

Results and Conclusions

Results obtained have been tabulated at Table 12.1.

Simulation techniques adopted here, involves use of three separate random number series. Each series provides an inspection value (d_j) for the stage. By doing a large number of simulation runs, behaviour of model can be better predicted and understood. It may also identify areas requiring further work and insights.

Table 12.1: Results of Simulation
Material Constraint

Seed Values	Average Value	Highest Value	Lowest Value	Inspection Percentages		
				Stage I	**Stage II**	**Stage III**
87,7,23	-35.20	-39.80	-30.75	72	03	97
11,13,7	-35.19	-39.98	-29.84	35	04	95
47,31,59	-35.20	-40.04	-31.02	06	13	98
61,87,73	-35.18	-40.52	-30.71	17	05	99
23,71,57	-35.20	-40.03	-30.74	18	11	98
		Production Constraint				
87,7,23	-22.84	-28.56	-20.78	87	07	23

12.6 HEURISTIC APPROACH

The results obtained from simulation have limitation to the extent of accuracy of random numbers being generated and probability of getting the right combination of inspection values. For a model having more than three stages, probability of getting the right combination is even more remote. Hence, there is a need to fine tune the values obtained from the simulation model to get optimal and better results.

Logic of Approach

There can be many approaches in any heuristic method. It is likely that one approach may lead to another and better approach. For the purpose of this study following approaches have been used:

i. Phase I. Uniform inspection policies at all stages.

ii. Phase II. Modification of inspection policies based on generation of non- conforming units and cost of rework at that particular stage. This has further been subjected to following modifications in parameter values:

   a. Same cost of scrap at all stages.

   b. Changes in rework cost at all stages.

   c. Changes in fraction of non-conforming units being generated at each stages

   d. Changes in inspection errors at each stage.

   e. Changes in fraction of units being sent for repair, rework, and scrap.

iii. Phase III. Further modification of inspection policies based on relative value of scrap.

iv. Phase IV. Comparison with simulation incorporating above stated parameter changes.

Uniform Inspection Policies (Phase I)

In this approach, value of inspection variable (d_j) for each stage is same. The values were varied from 0.1 to 0.99 in steps of 0.1 initially and subsequently in steps of 0.01. Results achieved have been tabulated in Table 12.2 and plotted on a graph at Figure 12.7. From the data it is evident that maximization of revenue is around values of 61 per cent, suggesting further study on the subject around these values.

Inspection Policy based on generation of Non-Conforming Units and incorporation of Cost of Rework (Phase III).

Table 12.2: Uniform Inspection Policies

Conforming input	Non-Conforming input	Revenue (monetary units)	Inspection % (d_j)
0.8117	0.1835	29.71	01
0.8064	0.1440	31.51	11
0.8011	0.1102	32.93	21
0.7959	0.0819	34.01	31
0.7906	0.0585	34.76	41
0.7854	0.0398	35.22	51
0.7801	0.0251	35.42	61
0.7749	0.0143	35.37	71
0.7697	0.0067	35.12	81
0.7645	0.0021	34.69	91
0.7603	0.0002	34.23	99

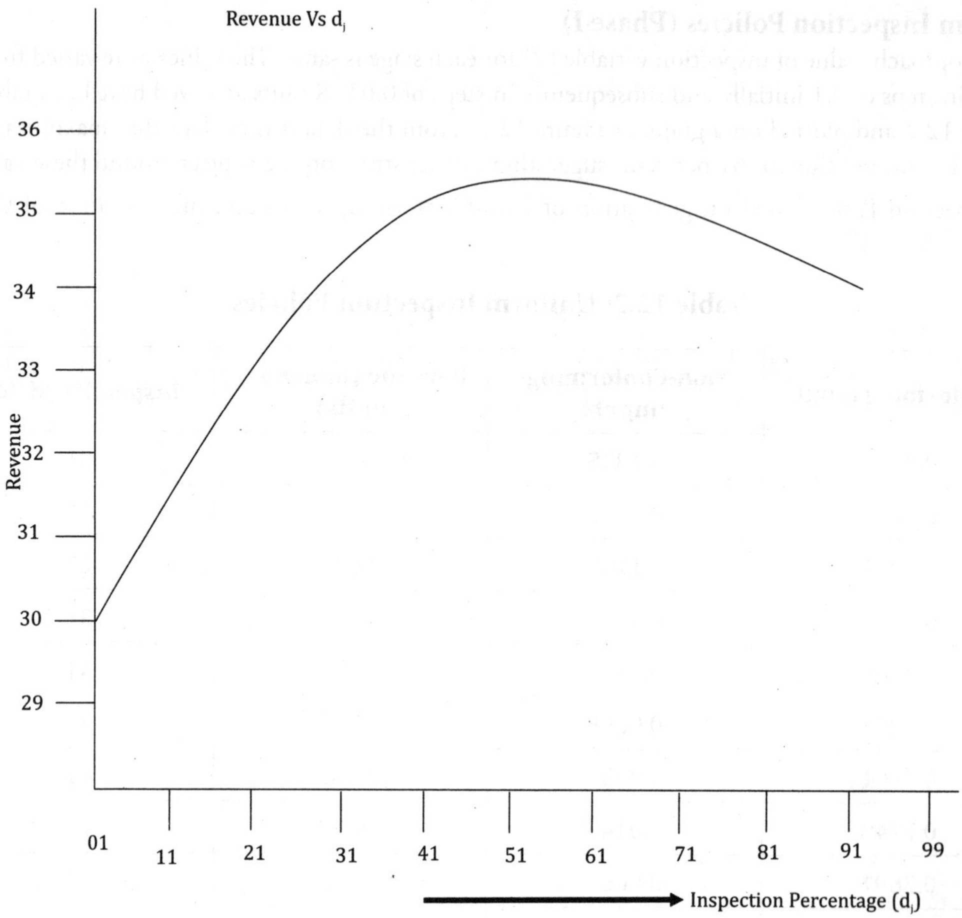

Figure 12.7: Result of Uniform Inspection Policies

Modification of Inspection Policies (Phase II)

Non-Conforming Aspect

Generation of non-conforming units at any stage is due to the following:

- **i.** A percentage of initial input is non-conforming
- **ii.** A percentage of conforming units acquire defects during the processing and thus add to the non-conforming units.

Based on above, total non-conforming units attributable to a stage can be formulated as:

$$pi = ni + ci * ei$$

where

ni = Input of non-conforming units from previous stage or initial input.

ci = Input of conforming units from previous stage or input initially.

ei = Probability of conforming units becoming defective during the process.

pi = Total non conforming units at *i*th stage

The logic followed here is that should the quantum of non-conforming units being generated at stage be less than a reference stage (stage 1 in this case), there should be proportional reduction of inspection effort at that stage. This is broken down to following logical steps:

i. Assume *dj* value for stage 1 as 100% (In actual $dj1 = 0.99$, as 100% would lead to indeterminable solution). Similarly, for 0% inspection, 0.01 is assumed, and calculate $p1$ (total non-conforming units generated at stage 1)

ii. Assume $dj2 = -0.01$ and calculate $p2$.

iii. Let $r21 = p2/p1$

If $r21 > 1$ than $dj2 = dj1$ (for $dj1 = 0.99$)

or $dj2 = r21 * dj1$

If $r21 < 1$ than $dj2 = r21 * dj1$

iv. Repeat above process for stage 3.

Cost of Rework Aspect

This approach attempts to further improve the solution incorporating cost of rework alongwith generation of non-conforming units. The *dj* values obtained in previous approach are further modified by ratios of cost of rework.

The logic followed here is, that if cost of rework at a stage is excessive, inspection at that stage should be more in order to minimize costs. This has been formulated as under:

Let $c21 = fr2/fr1$ and $c31 = fr3/fr1$

where *frn* indicates the rework cost of stage *n*

than $djn = djn' * cn1$

where *djn'* is inspection variable calculated earlier and *djn* is modified variable incorporating the cost of rework. Once again the reference is from stage 1.

Changing of Parameter Values

The values of various parameters have been changed (from those originally suggested by Yum and Mcdowel) in above stated approaches and also for simulation in Phase IV. The aim was to attempt to improve the optimal value. Flow chart for this is at Appendix 12 D.

Cost of Scrap (Phase III)

The approach followed here is similar to one followed above and is a subsequent step in optimizing to earlier phases. The stage incorporating least value of scrap should be inspected least. Thus, selecting the most profitable inspection policy in Phase II output, the result is further improved by reducing inspecting effort. In this model, cost of scrap is negligible (zero) at stage 1, hence this is subjected to gradual modification.

Simulation (Phase IV)

This phase uses the simulation logic of material constraint discussed earlier. The program is modified, incorporating parameter changes discussed above, one at a time. This has been done primarily to validate the results of previous phases.

Results and Conclusion

Results of various changes incorporated are shown in Table 12.3.

In this section, an attempt has been made to expose the model to various heuristic approaches, in order to get better results. As stated earlier, heuristic approach can seldom be complete. There is always one untried better heuristic approach, as approaches are directly a function of analytic qualities. Various methods adopted therefore should be accepted with this limiting factor.

Table 12.3: Summary of Results: Heuristic

Parameter	Original Value	Changed Value	Run	Phase II	Phase III	Simulation Phase (comparison)
Scrap value	0,-2,40	10,10,10	Run 1	38.55 (91,01,99)	-	38.35 (90,01,95)
Rework cost	2,0,3	1,2,1	Run 2	35.15 (99,99,01)	40.80 (99,01,99)	41.74 (30,01,99)
ϵj value	.05,1,0	.1,0,.05	Run 3	37.86 (91,01,.61)	38.01 (99,01,81)	40.16 (30,01,99)
αj βj	αj = .02,.05,.01 βj = .01,.1,.01	.05,.01, .02 .1,.01,.01	Run 4	39.57 (51,01,99)	40.75 (0,0,99)	40.08 (30,01,99)
Fraction of NC	.2,0,0 .3,0,.5 .5,1.0,.05	.4,.3,.33 .3,.6,.33 .3,.1,.34	Run 5	36.69 (51,01,99)	39.12 (0,0,99)	36.88 (30,01,99)

(d_{j1}, d_{j2}, d_{j3})–inspection percentage at there stages

12.7 FINDINGS

Simulation Approach

i. As per this approach five separate simulations, using different random number series, have resulted in values which are very close to each other. (Highest value ranges from 39.80 to 40.52) and Lowest value from 29.84 to 31.02).

ii. Production constraint has resulted in much lower values.

Heuristics Approach

i. Uniform inspection policies suggest, that application of uniform policies is not cost effective, as at some stage costs of inspection, overtake other benefits obtained.

ii. In all cases, Phase III has yielded better results. This validates the concept of cost of scrap associated with each stage.

iii. Simulation runs incorporating changed results have yielded better results, as compared to straight simulations.

REFLECTIONS

Based on findings in previous section, following conclusions can be arrived at:

i. Physical Systems Theory can be applied with reasonable accuracy and degree of certainty to a serial manufacturing system.

ii. The model as proposed here is sound and can be adopted for a multi-stage serial manufacturing system.

iii. It is not cost-effective to have uniform inspection policies at all stages. The policies must be modified based on few of the following constraints:

- Cost of scrap
- Generation of non-conforming units at each stage.
- Probability of Type I and II errors.
- Cost of rework.
- Handling of rejected units.

Notes

1. Yum B.J. and McDowell E.D., (1986) Optimal Inspection Policies in a Serial Manufacturing System Including Scrap Rework and Repair: A MILP Approach, *Int. J. of Pro. Res.*, 10, 1451–64.
2. Lehman, (1977) *Computer Simulation and Modelling*, Lawrence Erlbaum Associates Publishers, New York.
3. Yum and McDowell (1986) *op. cit.*
4. *Ibid.*
5. Sushil, Simplification of Physical Systems Theory in Modelling of Manufacturing, Organizational and other Socio Economic Systems, *Department of Management Studies, IIT Delhi.*
6. Case and Benett, (1977), The Economic Effect of Measurement Error on Variables Acceptance Sampling, *Int. J. Prod. Res*, 15.
7. Hanock W., Lunderberg D., Kelton D. Bischak D., (1987), Economically Optimal Sample Size, Frequency and Capability Indices Using X Charts, *Int. J. Prod. Res.*, 25, 967–978.

APPENDIX 12A

jth Operation

Assuming *dj* as the fraction of units passing through *j*th inspection, the various equations are:

Conforming Units

i. Cost Equation

$$x\bar{v}_{ji} = xt_{j1} = xv_{j1}$$

ii. Flow Equations

$$yt_{j1} = (1-d_j) * yv_{j1}$$
$$yv_{j1} = f_j * y\bar{v}_{j1}$$

Non-Conforming Units

i. Cost Equation

$$x\bar{v}_{j2} = xv_{j2} = xt_{j2}$$

ii. Flow Equations

$$yv_{j2} = d_j * y\bar{v}_{j2}$$
$$yt_{j2} = (1-d_j) * y\bar{v}_{j2}$$

jth Inspection

i. Cost Equations

$$xw_{j+1}, 1 = kv_{j1}.\ xv_{j1} - ku_{j1}.\ Xu_{j1} + fw_{j+1}, 1(yw_{j+1}, _1)$$
$$xw_{j+1}, 2 = kv_{j2}.\ XV_{j2} - ku_{j2}.\ Xu_{j2} + fw_{j+1,\,2}\ (yw_{j+1,\,2})$$

where xu_1 - denotes cost of dividing Yu_{j1} into rework/repair/replace/scrap and handling.

xu_{j2} - denotes cost of dividing Yu_{j2} into rework/repair/replace/scrap and handling.

$x\bar{u}\ j1$ - Net value gained by selling scrap of conforming units.

$x\bar{u}\ j2$ - Net value gained by selling scrap of non-conforming units.

ii. Flow Equations

$$yv_{j1} = kv_{j1} * yw_{j+1}, 1$$
$$yu_{j1} = ku_{j1} * yw_{j+1}, 1$$
$$yv_{j2} = kv_{j2} * yw_{j+1}, 2$$
$$yu_{j2} = ku_{j2} * yw_{j+1}, 2$$

iii. Derivations

$$\alpha 1 = \frac{yu_{ji}}{yu_{j1}} = \frac{yv_{j1} - yw_{j+1,}1}{yv_{j1}} = \frac{1 - yw_{j+1,}1}{yv_{j1}}$$

$$\therefore yv_{j1} = \frac{1}{1-\alpha_i}$$

$$\text{Similarly } ku_{j1} = \frac{1_j}{1-\alpha_j}$$

Also $kv_{j2} = \dfrac{1}{\beta_j}$

and $ku_{j2} = \dfrac{1-\beta_j}{\beta_j}$

Classification of Rejected Conforming Units

i. Cost Equation

$$xu_{j1} = x\bar{u}_{j1}1 = x\bar{u}_{j1}2 = xd_{j1} = x\bar{u}_{j1}$$

xu_{j1} = is cost of dividing (administrative cost)

xd_{j1} = Net value gained by selling scrap

ii. Flow Equations

$$y\bar{u}_{j1}1 = k\bar{u}_{j1}1.\, yu_{j1}$$
$$y\bar{u}_{j1}2 = k\bar{u}_{j1}2.\, yu_{j1}$$
$$y\bar{u}_{j1}3 = k\bar{u}_{j1}3.\, yu_{j1}$$

Classification of Rejected Non-Conforming Units

i. Cost Equations

$$xu_{j2} = x\bar{u}_{j2}1 = x\bar{u}_{j2}2 = xd_{j2} = x\bar{u}_{j2}$$

xu_{j2} – is cost of dividing (administrative cost)

xd_{j2} – is net value gained by selling scrap

ii. Flow Equations

$$y\bar{u}_{j2}1 = k\bar{u}_{j2}1.\, yu_{j2}$$
$$y\bar{u}_{j2}2 = k\bar{u}_{j2}2.\, yu_{j2}$$
$$y\bar{u}_{j2}3 = k\bar{u}_{j2}3.\, yu_{j2}$$

Repair of Conforming Units

$$xu_{j1}1 = ku_{j1}1 * xu_{j1}1 + f^{j1}(yu_{j1})$$
$$yu_{j1}1 = ku_{j1}1 * yu_{j1}1$$

Repair of Non-Conforming Units

$f_{j2}\,(yu_{j2}1)$ indicates processing cost for repair.

$$xu_{j2} = k\bar{u}_{j2}1 * x\bar{u}_{j2}1 + f_{j2}\,(yu_{j2})$$
$$y\bar{u}_{j2}1 = k\bar{u}_{2}1 * yu_{j2}$$

Aggregation of Conforming Units of Output of jth Stage

i. Cost Equations

$$xi_{j+1'}\,1 = kt_{j1} * xt_{j1} + kw_{j+1'1} * xw_{j+1'1} + ku_{j1} * xu_{j1}1$$

ii. Flow Equations

$$yt_{j1} = kt_{j1} * yi_{j+1}, 1$$
$$yw_{j+1,1} = * kw_{j+1}, ^1 yi_{j+1}, 1$$
$$y\bar{u}_{j1} 1 = k\bar{u}_{j,1,} * yi_{j+1,} 1$$

iii. Derivations

$$yt_{j1} = (l - \alpha_j)\bullet(y\bar{v}_{j1}) = \frac{1-\alpha}{x_j}\bullet yv_{j1}$$

$$= \frac{1-\alpha_j}{\alpha_j} . kv_{j1} . kw_{j+1,}^{1}$$

$$\boxed{kt_{j1} = \frac{1-\alpha}{\alpha_j} kv_{j1} . kw^1_{j+1,}}$$

$$yu_{j1}^{1} = yu_{j1}^{1}, (as\ k\bar{u}_{j1} = 1)$$
$$= k\bar{u}_{j1}\, yu_{j1} = ku^{j1}.\ ku_{j1} .\ yw_{j+1,} 1$$

$$\boxed{kt^1_{j1} = k\overline{u}_{j1} . k\overline{w}^1_{j+1,}}$$

Aggregation of Non-conforming Units at Output of jth Stage

i. Cost Equations

$$xi_{j+1}, 2 = kt_{j2} . xt_{j2} + kw_{j+1,} 2.\ xw_{j+1,2} + ku_{j2}^{1} .\ xu_{j2} 1$$

ii. Flow Equations

$$yt_{j2} = kt_{j2} . yi_{j+1}, 2$$
$$yw_{j+1,2} = kw_{j+1,} 2.\ yi_{j+1,} 2$$
$$yu_{j2}^{1} = ku_{j2}^{1} . yi_{j+1}, 2$$
$$kt_{j2} = \frac{1-\alpha_j}{\alpha_j} . kv_{j2} . k\overline{w}_{j+1,} 2$$

iii. Derivations

$$yu_{j2}^{1} = ku_{j2}^{1} .\ ku_{j2.}\ kw_{j+1,} 2$$

$$kw_{j+1,} 2 = \frac{1}{\frac{1-\alpha}{\alpha_j} . kv_{j2} + 1 + ku_{j2} . ku_{j2}}$$

Aggregation of Conforming Units at Input of jth Stage

i. Cost Equations

$$xw_{j1} = ki_{j1} \cdot xi_{j1} + ku_{j,2,} .\ xu_{j} 2$$

ii. Flow Equations

$$yi_{j1} = ki_{j1} \cdot yw_{j1}$$
$$yu_{j,2} = ku_{j2} \cdot yw_{j1}$$

iii. Derivations

$$ku_{j1}^{2'} = \frac{ku_{ji}^{2} . ku_{j1} . a_{j}}{kv_{j1} . kw_{j1}}$$

$$ki_{j1} = 1 - kui_{1}^{2'}$$

Aggregation of Non-conforming Units at Input of jth Stage

i. Cost Equations

$$xw_{j2} = ki_{j2} \cdot xi_{j2} + kuj^{2'}_{2} \cdot xu_{j\,2}^{2}$$

ii. Flow Equations

$$yi_{j2} = ki_{j2} \cdot yw_{j2}$$
$$yu_{j\,2}^{2} = ku_{j\,2}^{2'} \cdot yw_{j2}$$

iii. Derivations

$$ku_{j\,2}^{2'} = \frac{ku_{j\,2}^{2} . ku_{j2} . a_{j} . kv_{j2}}{kv_{j2} . kw_{j2}}$$

$$ki_{j}^{2} = 1 - ku_{j\,2}^{2'}$$

APPENDIX 12B

Derivation for Black Box Representation

$$xi_{j+1}, 1 = \Phi i_{j1}.\ xi_{j2} + \Phi d_{j1}\ xd_{j1} + \Phi u_{j1}^{\ 3},\ xu_{j1}^{\ 3} + fj_{j+1,1}\ (yi_{j+1,1})$$

$$xi_{j+1,}\ 2 = \Phi i_{j2}.\ xi_{j2} + \Phi i_{j1}\ xi_{j1} + \Phi d_{j1}^{\ 2} \cdot xd_{j2} + \Phi u_{j2}^{\ 3} xu_{j2}^{\ 3} + \Phi u_{j1}^{\ '3} + f_{zi+1,2}(yi_{\ j+1,1})$$

$$yi_{j1} = \Phi i_{j1} \cdot yi_{\ j+1,1}$$

$$yd_{j1} = \Phi d_{j1} \cdot yi_{\ j+1,\ 1}$$

$$yu_{j1}^{\ 3} = \Phi u_{j1}^{\ 3} \cdot yi_{\ j+1,\ 1}$$

$$yi_{j2} = \Phi i_{j2} \cdot yi_{\ j+1,\ 2}$$

$$yi_{j1} = \Phi i_{j1}^{\ 1} \cdot yi_{\ j+1,\ 2}$$

$$yd_{j2} = \Phi d_{j2} \cdot yi_{\ j+1,\ 2}$$

$$yu_{j2}^{\ 3} = \Phi u_{j2}^{\ 3} \cdot yi_{\ j+1,\ 2}$$

$$yu_{j1}^{\ 3} = \Phi u_{j2}^{\ 31} \cdot yi_{\ j+1,\ 2}$$

APPENDIX 12C

SIMULATION FLOW CHART

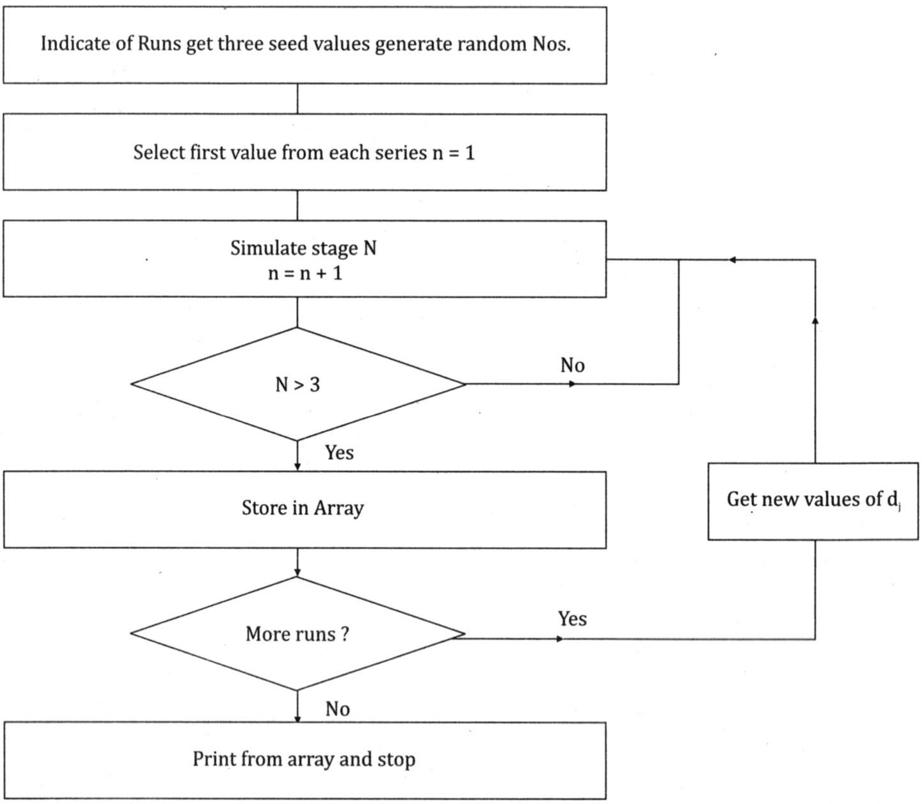

APPENDIX 12D

Flow Chart: Heuristic

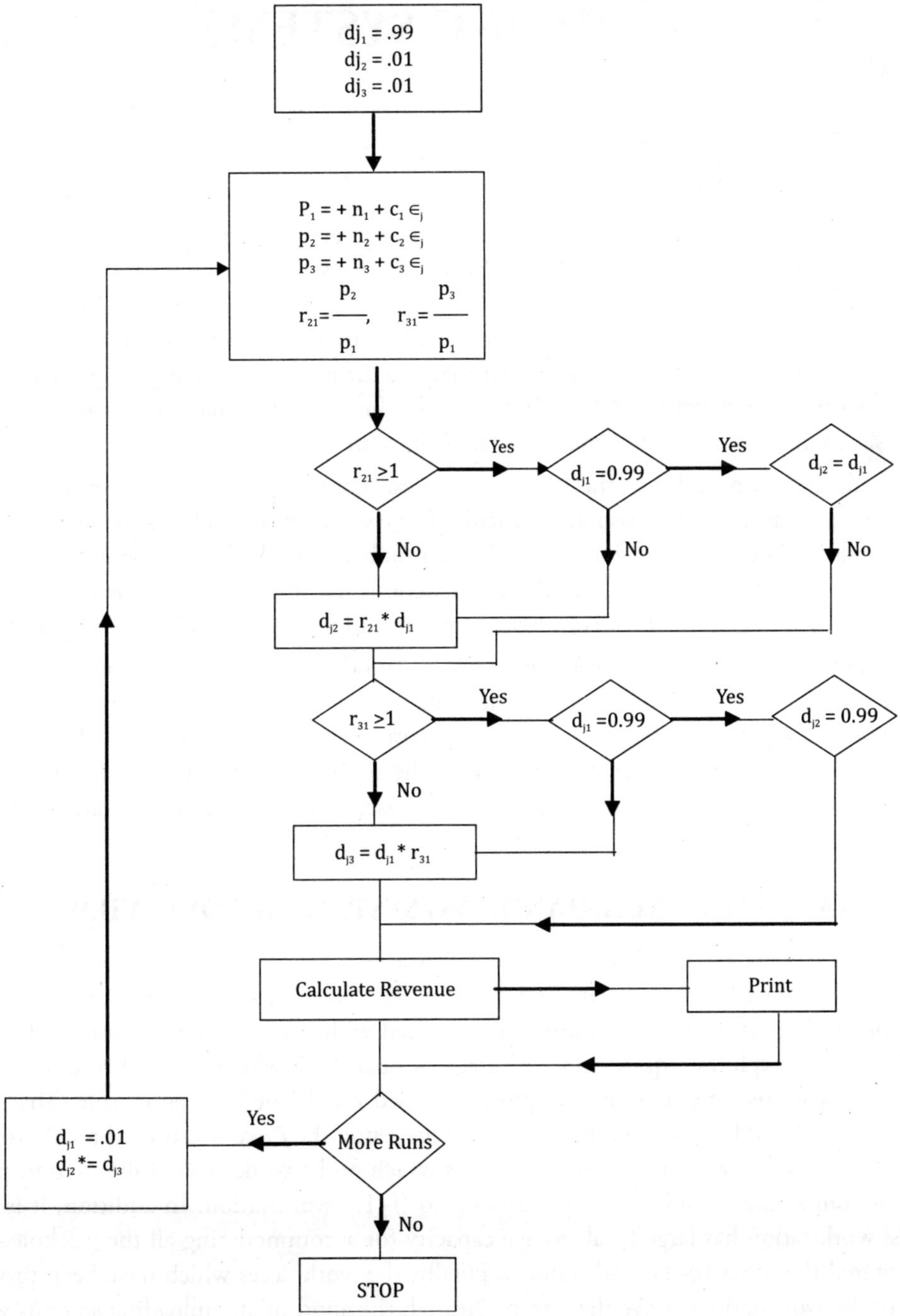

13 INTEGRATED MATERIAL HANDLING SYSTEMS

13.1 INTRODUCTION

Material handling is an overall job that directly affects each area in a plant and as such requires careful planning. The structure of material handling system varies with the industry, the type of manufacturing, its bulk and its value. An integrated manufacturing system can be defined as manufacturing facility consisting of a set of workstations with an inventory system linked by a material handling system. The performance of integrated manufacturing systems is largely affected by various operational factors. The operational factors include loading, material handling, storage, processing operations, scheduling, and layout of the manufacturing system.

As the scale and complexity of the system increases, many more factors and relationships come into picture which makes the analysis more difficult. The systems approach helps in the analysis of such complex systems by focussing attention on the interaction between different components and its effect on the overall functioning of the system. These interactions, in physical system theory framework are represented by a system graph, which is governed by continuity and compatibility constraints.

The focus of this chapter is on developing a systems model of a specific integrated manufacturing system[1] consisting of a set of workstations, a loading station and unloading stations, an inventory system, linked by an integrated material handling system. The aim of this model will be to study the effect of an operational plan on the performance of the system. The system taken in this chapter is illustrative and small in size, but similar concepts can be used to model and analyze large-scale material handling systems.

13.2 INTEGRATED MATERIAL HANDLING SYSTEM – AN OVERVIEW

The system which is taken here consists of a n workstations connected by a set of conveyors. Each conveyor has a specified length, velocity and capacity. Each workstation consists of a machine with a specified processing rate. Each workstation is associated with a local storage of specified value. It is assumed that the workpieces require only one stage of operation, which it could receive from either one of the workstations. The flow of workpieces is based on the ordered entry rule. That is, if the workstations are indexed from 1 to n, the workpieces first enter the jth workstation, if the jth workstation is full, the blocked workpieces (i.e. those workpieces which find a workstation full) are transported by the conveyor connecting j^{th} and $(j+1)^{th}$ workstation to $(j+1)^{th}$ workstation. In addition, it is assumed that the last workstation has large local storage capacity for accommodating all the workpieces which have not entered the other $(n-1)$ workstations. Finally, the workpieces which have been processed at either one of the workstations, leave the system through the appropriate unloading stations which are located at each workstation.

In the above system, two types of blocking could occur. Firstly, an arbitrary workpiece may be blocked if it finds one of the workstations full. Secondly, an arbitrary workpiece may be blocked, if it finds a conveyor full. Since overflowing from the first (n–1) workstation is allowed, the first type of blocking will not occur. For the second type of blocking, the velocity parameter of the conveyors can be varied to control the congestion along it. This procedure enables us to properly design the integrated material handling system such that the average congestion inside the system will not exceed the allowable limits.

13.3 GENERALIZED MODEL

The scheme of modelling is shown in Figure 13.1, which shows the integration Physical System theory Model with simulation model.

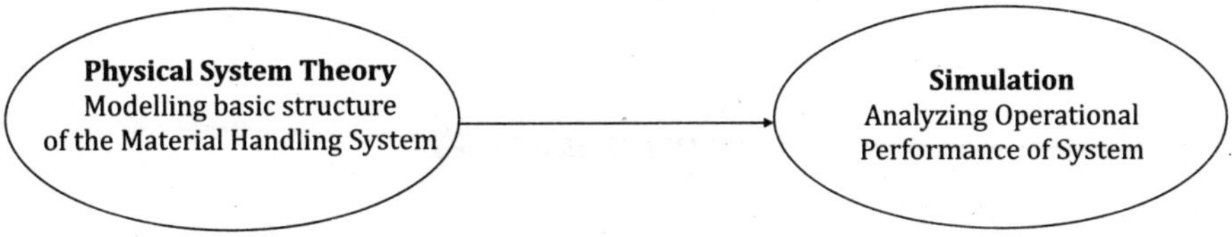

Figure 13.1: Flexible System Methodology Design for Modelling and Analysis

Physical System Theory Model

The generalized model of the n stage integrated material handling system has been developed in the PST framework. The component models for generalized transportation process are developed. These components can be combined to form the system graph and the structural constraints of continuity and compatibility can be written which are then used to develop the simplified system graph as per the rules given in chapter four. The development of the simplified system graph has been illustrated by an example. The system equations of the simplified PST model are then used in the simulation model to derive meaningful results about the system behaviour.

(i) ***Transportation process*:** The component terminal graph of jth transportation process is shown in Figure 13.2 (a). The variables for this process are defined as follows:

Across variable = b_i (Congestion factor)

= Average number of workpieces travelling along the jth conveyor line.

Through variable = y_i

= Number of workpieces flowing through per unit time.

The response-stimulus equations for the process are as follow:

$$b_i = (l_i/v_i)*y_I \qquad (13.1)$$

where, l_i = length of the conveyor

v_i = Velocity of the conveyor

as, $(l_i/v_i) = c_i$ a constant, the equation 13.1 is reduced

$$\text{to, } b_i = c_i*y_i \qquad (13.2)$$

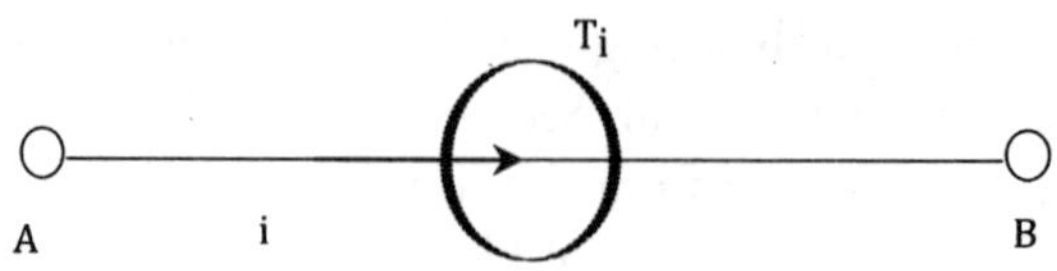

(a) Transportation Process

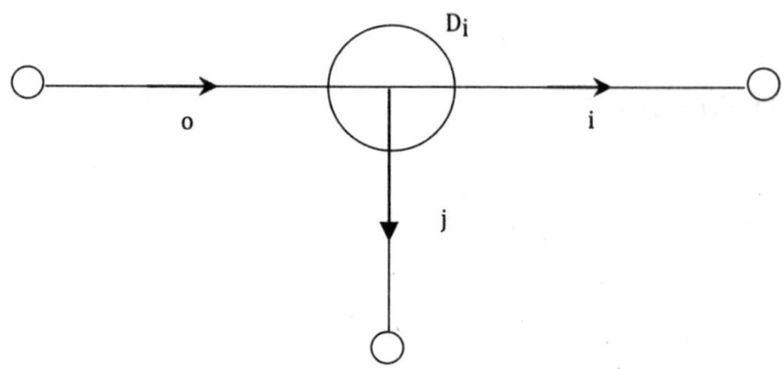

(b) Distribution Process

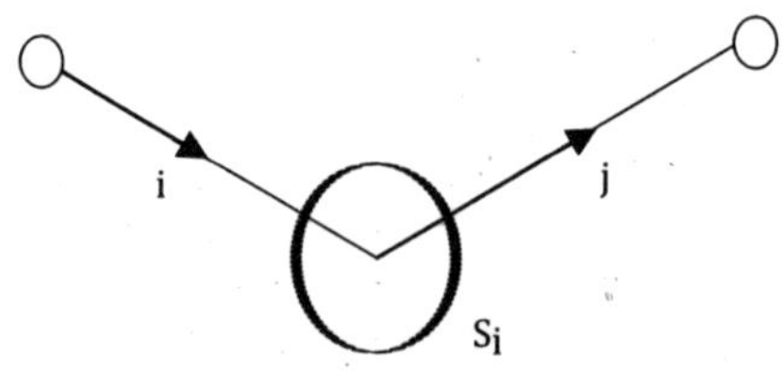

(c) Storage Process

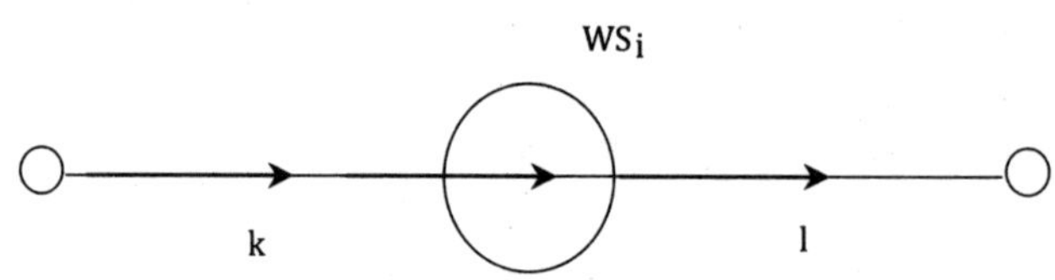

(d) Transformation Process (Work Station)

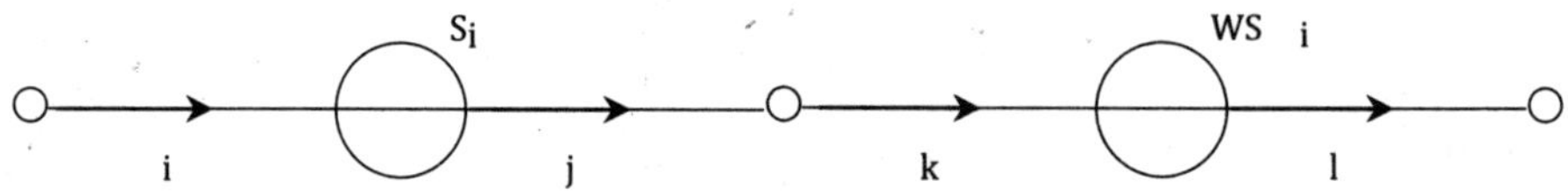

(e) A Storage Process and Work Station

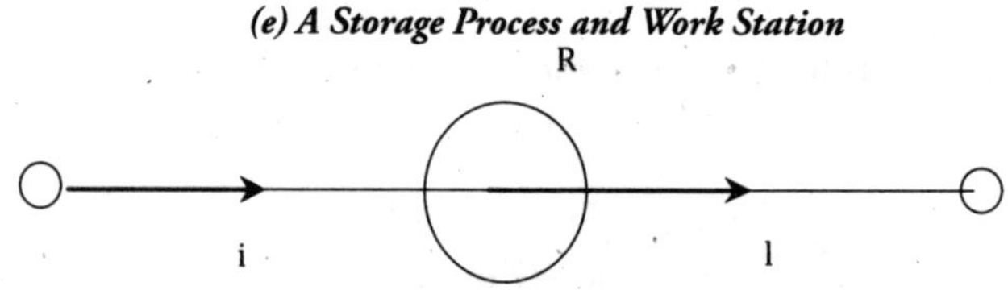

(f) A Combined Process

Figure 13.2: Component Terminal Graphs of Elements of Integrated Material Handling System

(ii) *Distribution process:* The component terminal graph of $(i+j)^{th}$ distribution process is given in Figure 13.2(b). The response-stimulus equations for this process are as follows:

$$y_i = k_1 * y_o + k_o * (MG_i - ME_i\,(t-1)) \tag{13.3}$$

$$y_j = k_2 * (y_o - MG_i - ME_i\,(t-1)) \tag{13.4}$$

where, y_o, y_j, y_j = the flows in number of workpieces in a specified time interval on links o, i and j respectively.

MG_i = Capacity of storage of workstation i.

$ME_i\,(t-1)$ = Inventory level of workstation i in previous time period.

k_o, k_1, k_2 = Constants that take on values 0 and 1 depending on values of y_o, MG_i and $ME_i\,(t-1)$

If $y^o \leq (MG_i - ME_i\,(t-1))$

then $k_o = 0,\ k_1 = 1\ k_2 = 0$ (13.5)

else $k_o = 1,\ k_1 = 0,\ k_2 = 1$

(iii) *Storage process:* The component terminal graph of jth storage process is shown in Figure 13.2 (c). The response-stimulus equation for this process is as follows:

$$ME_i\,(t) = ME_i\,(t-1) + (y_j\,(t) - y_j\,(t)) \tag{13.6}$$

where, ME_i = Inventory level at workstation i.

y_i, y_j = Flow in number of workpieces in a specified time interval (through variable) on links i and j respectively.

(iv) *Transformation process:* The free body diagram of jth transformation process is shown in Figure 13.2(d).

Since no wastage has been considered, $y_j\,(t) = y_j\,(t)$. The Utilization Factor is considered as an across variable. Utilization factor is defined as:

$uf_i = e_I/p_i$ $\qquad i = 1$ to n

where, $e_I = y_o - y_j$ $\qquad i = 1$ to n

Since the Utilization Factor depends on the flow input to storage and since flow output of storage is equal to flow input to workstation, these two can be combined to form one component as shown in Figure 13.2(e). The component terminal graph of the reduced component is shown in Figure 13.2(f).

$$uf_i = y_I(t)/p_i \tag{13.7}$$

where, p_i = Processing rate of workstation i

The components of the integrated material handling system and the associated variables considered in the model are summarized in Appendix 13 A.

Simulation

The system behavior can be studied by simulating the model. The basic flows are generated as per the data given and the various system equations are solved dynamically overtime.

13.4 ILLUSTRATIVE EXAMPLE

System Description

The system taken here consists of three workstations, having local storage, connected by a set of five conveyors as shown in Figure 13.3. The various components of the system can be identified as follows:

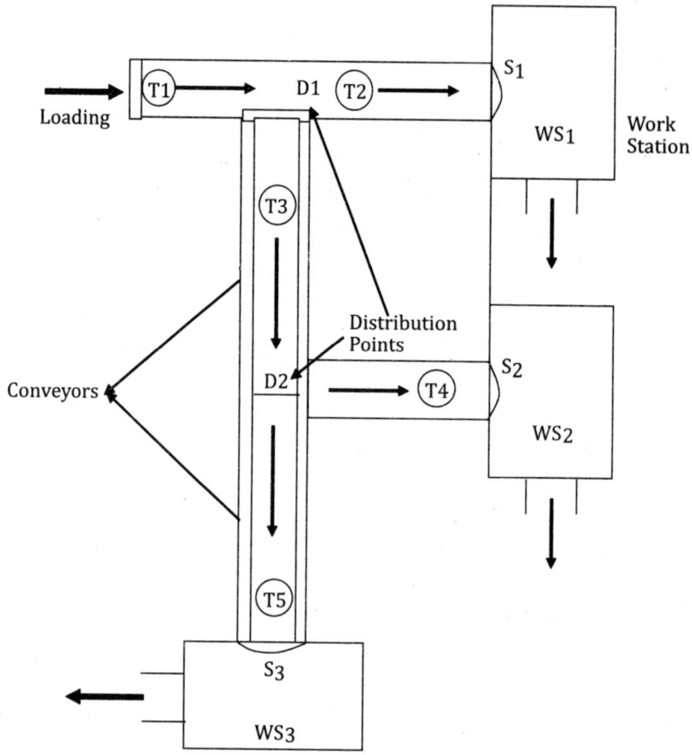

Figure 13.3: System Considered for Analysis

a. Production Center:

#1 Workstation 1 (WS1)

#2 Workstation 2 (WS2)

#3 Workstation 3 (WS3)

b. Storage:

#4 Storage 1 (S1)

#5 Storage 2 (S2)

#6 Storage 3 (S3)

c. Transportation Links (conveyors):

#7 Link AB (T1)

#8 Link BC (T2)

#9 Link BD (T3)

#10 Link DE (T4)
#11 Link DF (T5)

d. Distribution Process:

#12 Distribution Process 1 (D1)
#13 Distribution Process 2 (D2)

Model

The response-stimulus equations for different components are written by using the generalized component models. The components are taken in order in which they will be connected in the system graph. The component models, and the structural constraints for the example problem are given in Appendix 13A. By following the rules given in chapter the PST model is simplified and the simplified system graph is developed. The total number of equations in the simplified model is 14.

Simulation and Results

Based on the model, following studies were made to analyze the behaviour of the integrated material handling system. For each study 100 simulation runs were made. The results are the average of these 100 simulation runs.

i. In the first study, the length and velocity of each conveyor was taken to be equal to l_i = 20 length units and v_i = 10 length unit/5 min. The arrivals were assumed to be Markovian with rate y_i = 2.6 per min. The processing rate of each workstation was taken as p_i = 1.0 per min. The storage capacity at workstation 1 and 2 was taken as 10 and 20 units respectively. The maximum capacity of each conveyor was taken as 20 length units (40 workpieces in number).

The results showed that although the average congestion factor was lower than the maximum permissible one but there were some values on conveyor 1 which were exceeding the maximum, therefore its velocity was increased to 15 length unit/5 min. Also, the velocity of conveyor 2 was reduced to 5 length unit/5 min. Since the maximum requirement from this conveyor is only 10 units which is satisfied by this speed.

Using the above modifications, this study was done to validate the model with the original model of Pourbabai[2]. Here processing rates of workstation 1 and 2 were varied with their total sum equal to 2.0 constant and inventory level of 1 unit at each workstation. The results of this model compares favourably with the original model.

ii. In the second study, the effect of arrival rate on congestion factor, inventory level and utilization factor was found. Figure 13.4 shows that with an increase in the arrival rate, the congestion factor of line conveyor 3 and conveyor 5 increases continuously, whereas for the conveyors 2 and 4 it stabilizes to a value after reaching to its maximum capacity. The curve in Figure 13.5 shows that both the workstations 1 and 2 reach their capacity as arrival rate is increased.

iii. In the third study, the storage capacity at the workstation 1 and 2 was changed but keeping their total equal to 30 constant. Figure 13.6 shows that the utilization factor of their workstation first decreased and then increased whereas for the workstation 1 it is continuously increasing and for the workstation 2 it is continuously decreasing. This behaviour is also seen for inventory level in Figure 13.7.

iv. In the fourth study, the processing rates for the workstation 1 and 2 were varied keeping their sum equal to 2.0 constant. Figure 13.8 shows that the congestion factor on the line conveyor 3 continuously decreases whereas on the line conveyor 5 it first increases, then decreases and then again increases. In Figure 13.9 the inventory level of workstations 2 and 3 also reflect this behaviour.

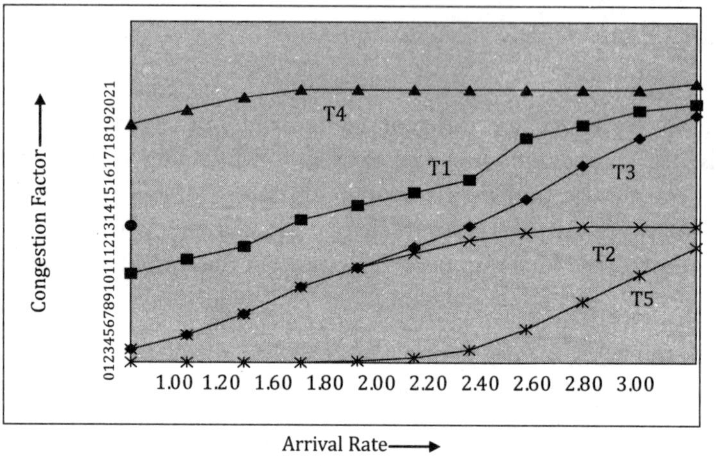

Figure 13.4: Effect of Arrival Rate on Congestion Factor

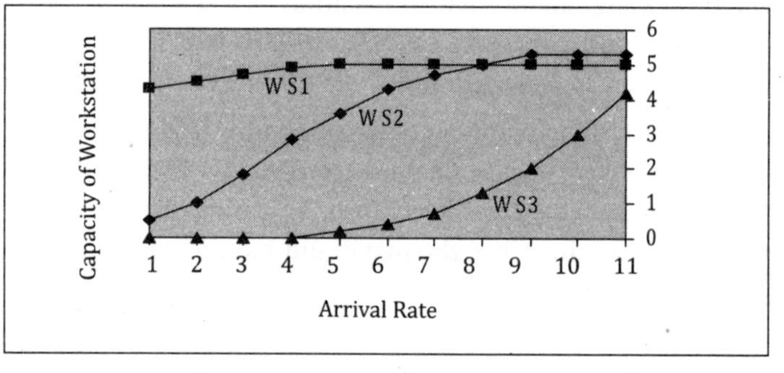

Figure 13.5: Effect of Arrival Rate on Capacity

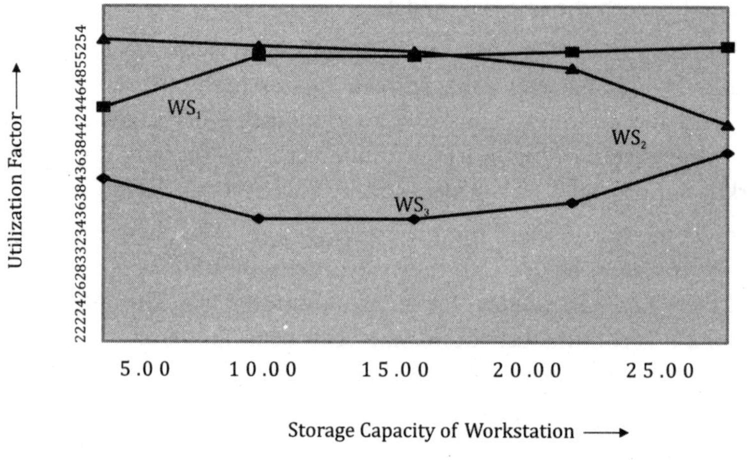

Figure 13.6: Effect of Storage Capacity on Utilization Factor

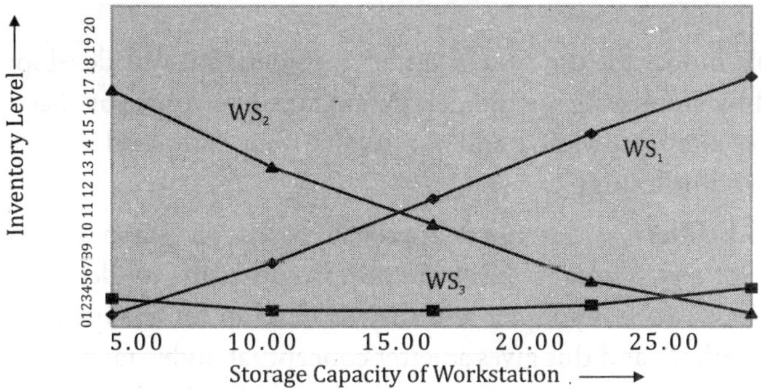

Figure 13.7: Effect of Storage Capacity on Inventory Level

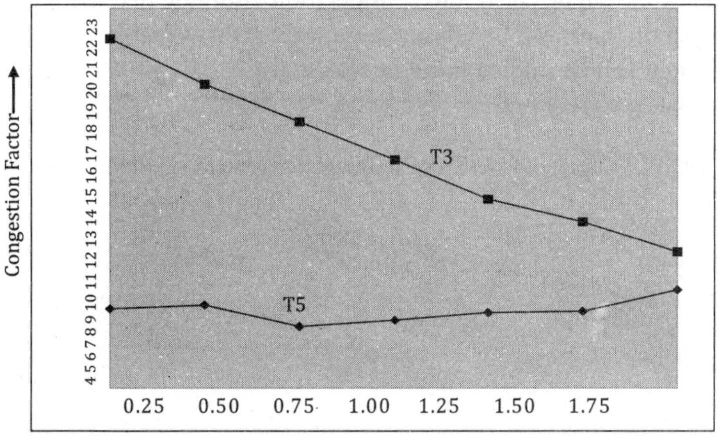

Figure 13.8: Effect of Processing Rate on Congestion Factor

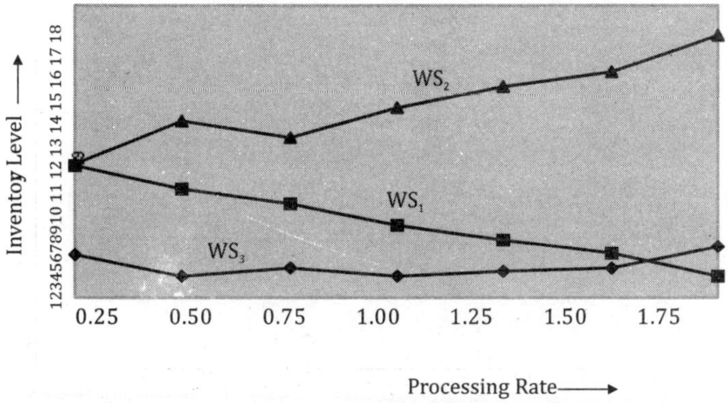

Figure 13.9: Effect of Processing Rate on Inventory Level

REFLECTIONS

In this chapter, a simple model for the system under consideration was developed. The behaviour of the system was studied by simulating the model. Parameters like conveyor speed, arrival rate, storage capacity, and processing rate were varied and the results were compared in terms of the congestion factor, inventory level, and utilization factor.

The PST framework offers a structured approach based on graph theory that facilitates the development of model for any system. It has been observed that the model developed in this chapter in the simplified PST framework is simpler and easy to develop for the integrated material handling system than that of Pourbabai[3], and this gives a better conceptual understanding of the system structure and behaviour.

Further, work can be carried out in the areas of waste recycling and cost estimation. For waste recycling, scrap and rework can be taken into consideration assuming that the reworked pieces are sent back to the loading station. Cost can be taken as a uniform across variable for all types of components and the system behaviour can be studied in terms of cost also.

Notes

1. Pourbabai B., (1988), Analysis and Design of an Integrated Material Handling System, *Int. J. Prod. Res.*, 26 (7), 1225–1236.
2. *Ibid.*
3. *Ibid.*

APPENDIX 13A

A. Physical System Theory Model of the Illustrative Integrated Material Handling System

Component number	Component terminal graph	Response-stimulus equations
# 7	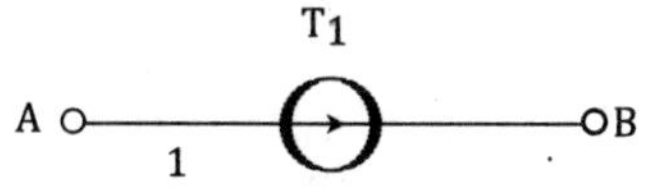	$b_1(t) = y_1(t) * c_1$ where $c_1 = i_1/v_1$
# 12	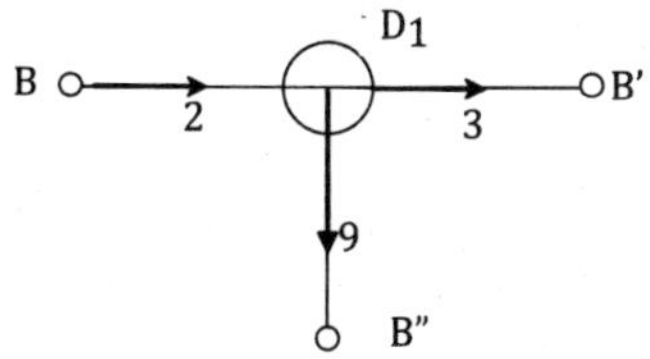	$y_3(t) = k_1 * y_2(t) + * k_o * (\text{MG1–ME1}(t–1))$ $y_9(t) = k_2 * (y_2(t)–(\text{MG1-ME1}(t–1))$ If y_2 $(\text{MG1–ME1}(t–1))$ then $k_o = 0,\ k_1 = 1,\ k_2 = 0$ else $k_o = 1,\ k_1 = 0,\ k_2 = 1$
# 8	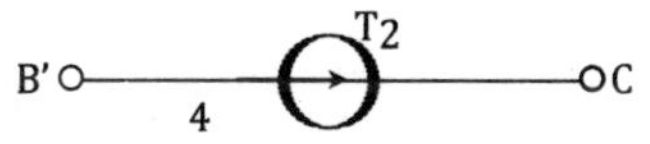	$b_4(t) = y_4(t) * c_4$ where $c_4 = l_4/v_4$
# 4	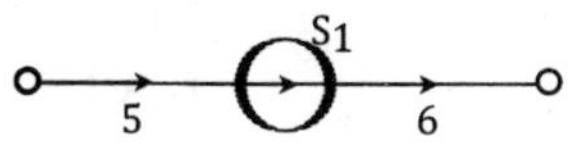	$\text{ME1}(t) = \text{ME1}(t–1) + y_5(t) – y_6(t)$ or $y_6(t) = \text{ME1}(t–1) – \text{ME1}(t) + y_5(t)$
# 1	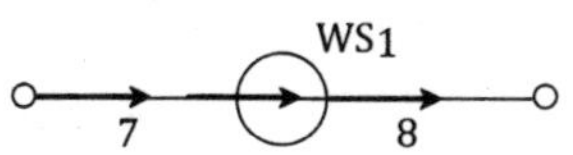	$y_6(t) = y_7(t) = y_8(t)$ $UF_1 = y_5(t)/P_1$
# 9	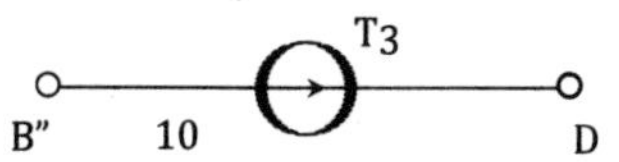	$b_{10}(t) = y_{10}(t) * c_{10}$ where $c_{10} = l_{10}/n_{10}$
# 13	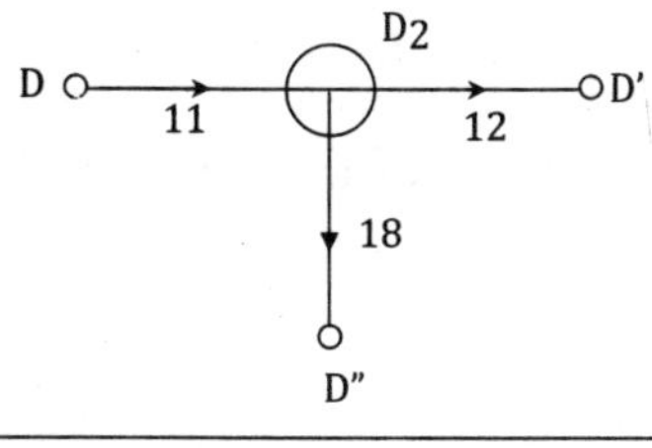	$y_{12}(t) = k_1 * y_{11}(t) + k_o * (\text{MG2-ME2}(t–1)$ $y_{18}(t) = k_2 * (y_{11}(t)–(\text{MG2-ME2}(t–1))$ If $y_{11} = (\text{MG2–ME2}(t–1)$ then $k_o = 0,\ k1 = 1,\ k2 = 0$ else $k_o = 1,\ k1 = 0,\ k2 = 1$
# 10	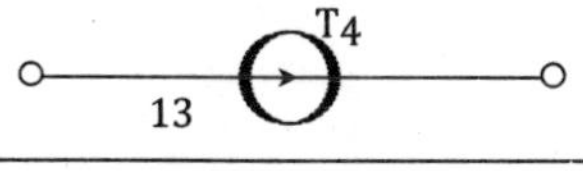	$b_{13}(t) = y_{13}(t) * c_{13}$ where $c_{13} = l_{13}/v_{13}$
# 5	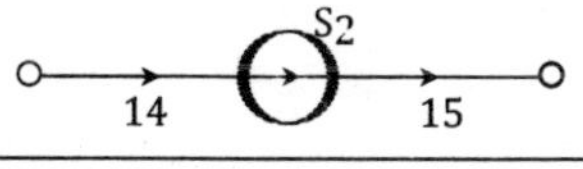	$\text{ME2}(t) = \text{ME2}(t–1) + y_{14}(t) – y_{15}(t)$ or $y_{15}(t) = \text{ME2}(t–1) – \text{ME2}(t) + y_{14}(t)$
# 5	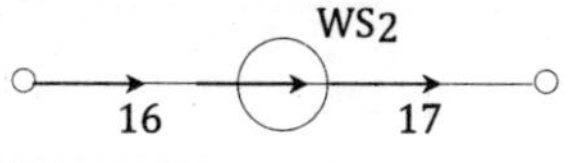	$y_{15}(t) = y_{16}(t) = y_{17}(t)$ $UF_2 = y_{14}(t)/p_2$

# 11	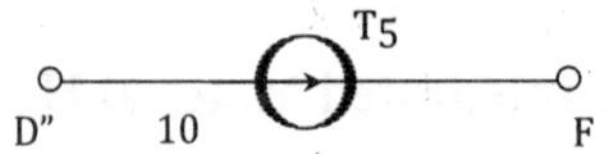	$b_{19}(t) = y_{19}(t) * c_{19}$ where $c_{19} = l_{19}/v_{19}$
# 6	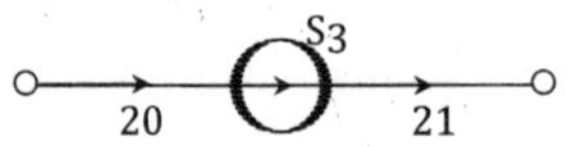	ME3 (t)= ME3 $(t-1) + y_{20}(t) - y_{21}(t)$ or $y_{21}(t)$ = ME3 $(t-1)$ – ME3 $(t) + y_{20}(t)$
# 3	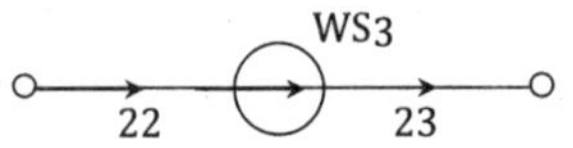	If $y_{20}(t)$ + ME3 $(t-1) = y_{21}(t)$ then $UF_3(t) = (y_{20}(t) + ME3(t\text{-}1))/p_3$ else $UF_3 = y_{21}(t)/p_3$ $y_{21}(t) = y_{22}(t) = y_{23}(t)$

B. System Graph

The component models in the section A are combined to form the system graph.

C. Structural Constraints

a. Continuity Equations

$y_1(t) = y_2(t)$

$y_3(t) = y_4(t)$

$y_4(t) = y_5(t)$

$y_6(t) = y_7(t)$

$y_8(t) = y_{01}(t)$

$y_9(t) = y_{10}(t)$

$y_{10}(t) = y_{11}(t)$

$y_{12}(t) = y_{13}(t)$

$y_{13}(t) = y_{14}(t)$

$y_{15}(t) = y_{16}(t)$

$y_{17}(t) = y_{o2}(t)$

$y_{18}(t) = y_{19}(t)$

$y_{19}(t) = y_{20}(t)$

$y_{21}(t) = y_{22}(t)$

$y_{23}(t) = y_{o3}(t)$

b. Mass Balance Equations

$y_2(t) = y_3(t) + y_9(t)$

$y_{11}(t) = y_{12}(t) + y_{18}(t)$

c. Compatibility Equations

Since across variables are different for different types of the components, the concept of compatibility equations does not come into picture.

PART IV

APPLICATIONS IN ORGANIZATIONAL AND TACTICAL MANAGEMENT

14 INTEGRATED FINANCIAL DECISION-MAKING

14.1 INTRODUCTION

Financial planning and investment decisions have traditionally been at the centre of attention in any firm doing business. First, it is the allocation of capital that starts the process which eventually leads to the commitment of labour and materials. Second, once the capital has been allocated, the investment is irreversible; and third, the usually large magnitude of the capital outlay makes the investment decision a major influence on the future profitability of the firm.

14.2 FINANCIAL DECISION-MAKING–AN OVERVIEW

Capital Structure Decisions

Financial planning, also known as capital structure, refers to composition of long-term sources of funds. A company, in addition to exploring and increasing its demand for capital funds, must face the problem of determining the sources of money.

A useful distinction can be made between internal and external sources of funds. The chief internal sources are depreciation charges, reserves and surpluses (*i.e.* retained earnings). External sources are principally debentures, long-term debt, preference share capital, and equity share capital.

The financing decision of a firm relates to the choice of the proportion of above sources to finance the investment requirements. The theory of capital structure shows the theoretical relationship between the employment of debt and the return to shareholders. The use of debt implies a higher return to the shareholders as also the financial risk. A proper balance between debt and equity to ensure a trade-off between risk and return to the shareholders is necessary.

The financial manager must therefore, plan an optimum capital structure for the firm. The optimum capital structure is obtained when the market value per share is maximum or the average cost of capital is minimum.

Leverage

The term leverage may be defined as the employment of an asset or sources of funds for which the firm has to pay a fixed cost or fixed return[1]. Consequently, the earnings available to the shareholders as also the risk are affected. There are two types of leverage— "operating" and "financial". The leverage associated with investment activities is referred to as operating leverage, while leverage associated with financing activities is called financial leverage for purposes of the financing decision of the firm.

Financial leverage not only intends to magnify shareholder's returns under favourable conditions, but also exposes them to financial risks. Financial risk is associated with the financing decision or capital mix of the firm. A totally equity financed firm will not have any financial risk. But when debt is used in the capital structure of the firm, financial risk is involved. Financial risk results because when the debt is used it increases:

(a) The variability of shareholders' return

(b) The probability of insolvency.

Cost of Capital

When a company considers using outside sources to finance investment, a basic factor is the cost of capital. The firm's cost of acquiring funds, or its cost of capital, is that rate of return on the project that will not change the market price of the firm's equity share. In the selection of an investment, it therefore becomes the minimum rate of return allowed on a prospective investment.

The cost of each component of capital, *i.e.* equity shares, debt and retained earnings is different. The firm has to maintain a balance between debt and the equity component of capital. The actual cost of capital which is of significance to the firm is the composite or combined cost of capital. The combined cost of capital is the weighted cost of capital. The weighted average cost of capital may be expressed in a single formula that shows cost of capital as the sum of the weighted individual costs of each component of the capital structure[2]. The formula may be expressed as follows:

$$k_o = \%\,D_{mkt}\,(k_i) + \%\,ES_{mkt}\,(k_e) + \%\,RE_{mkt}\,(k_{re})$$

where

k_o = Overall cost of capital

k_i = Cost of debt

k_e = Cost of equity

k_{re} = Cost of retained earnings

$\%\,D_{mkt}$ = Percentage of debt in capital structure

$\%\,ES_{mkt}$ = Percentage of equity share in capital structure

$\%\,RE_{mkt}$ = Percentage of retained earnings in capital structure

Cost of Debt: The cost of debt is generally considered to be the most reliable cost to calculate because interest charges are known and fixed by agreement between the firm and its creditors. The interest paid on debt is tax deductable. The higher the interest charges the lower will be amount of tax, so the effective cost of debt, *i.e.* the after tax cost of debt is substantially less than the before tax cost. The before tax cost should be adjusted for tax effects. Cost of debt can be mathematically expressed as follows:

$$k_i\,(AT) = k_i\,(BT)\,(1-TR)$$

where

$k_i\,(AT)$ = After-tax cost of debt

$k_i\,(BT)$ = Before -tax cost of debt

TR = Corporate tax rate

Cost of Equity Capital: Determining the cost of common stock presents greater difficulties than the known costs associated with preferential shares or debt. The shareholder does not expect to receive any fixed, predetermined return on his purchase of equity shares. Rather, he receives the right to participate in sharing future earnings of the firm and the right to receive future dividends. Therefore, the required rate of return by the shareholder is the cost of equity capital. The cost of the equity can be measured by the dividend model, which states:

$$k_e = \frac{D_1}{P_0} + g$$

where

k_e = Cost of equity

D_1 = Dividend per share at the end of 1st year

P_0 = Price per share today

g = The rate at which the expected stream of dividend is to grow.

The above relationship is based on the following assumptions:

a. The market value of share is a function of expected dividends.

b. The initial dividend D_1>0.

c. The dividend grows at a constant rate 'g'.

d. The dividend pay out ratio (i.e. dividend as a percentage of earnings) is constant.

Realizing the above practical problems in calculation of cost of equity, the capital asset pricing model[3] is mostly used.

Cost of Retained Earnings: Actually retained earnings are nothing but unpaid dividends and involve an opportunity cost. The opportunity cost of retained earnings is the dividend foregone by shareholders. The cost of retained earnings is the return expected by the common shareholder on their investment.

Capital Budgeting–Investment Decisions

Strategic investment decisions involve large sums of money and may also result in a major departure from what the company has been doing in the past. Acceptance of a strategic investment will involve a significant change in the company's expected profits and the risks to which these profits will be subject. The future success of a business depends on the investment decisions made today. These decisions once made are not easily reversible without much financial loss to the firm[4].

Thus, efficient allocation of capital is one of the most important function of financial management. Financial management not only deals with the procurement of capital but its efficient use with the objective of maximizing the owner's wealth.

Capital budgeting decision can be defined as the firm's decision to invest its funds most efficiently in long-term activities, in anticipation of an expected flow of future benefits over a period of time.

There are three major steps in capital budgeting process. These are listed below:

a. The generation of investment alternatives.

b. The estimation of benefits and costs.

c. The selection of the alternatives.

Methods of Capital Budgeting

The methods for measuring worth of investment proposals range from highly subjective-intuitive judgement approaches to objective-quantified approaches. The methods of appraising capital expenditure proposals can be classified into following categories:

a. Traditional

b. Time adjusted.

The first category includes the following:

a. Average rate of return method.

b. Pay back method.

The second category is more popularly known as Discounted Cash Flow Techniques as they take the time factor into account. It is characteristic of most investments that benefits will flow over several years. Since money received today is not the same as money received one year from now, some adjustment must be made to the cash flows. This category includes the following methods:

a. Net Present Value method (NPV).

b. Internal Rate of Return method (IRR).

The main objective of this chapter is to make use of physical system model for financial decision-making. We will confine to use NPV method to show the relevance of the model.

14.3 MODEL FOR INTEGRATED FINANCIAL DECISION-MAKING

The systems approach helps in the analysis of such a system by focussing attention on the interaction between different components. Physical system theory and Goal Programming have been successfully used in this regard. The overall integrated model is shown in Figure 14.1. The inputs and outputs of the physical system model are shown in Figure 14.2.

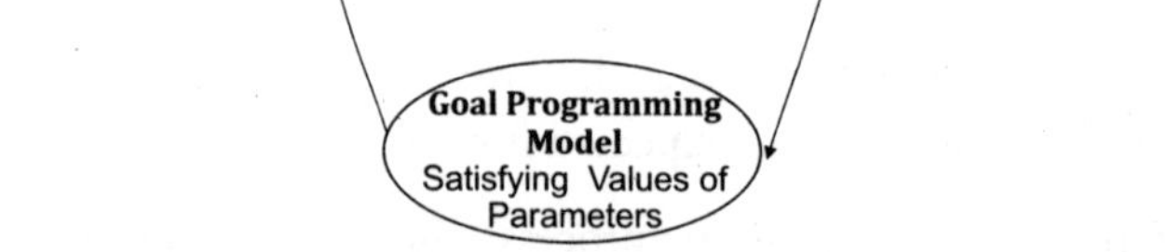

Figure 14.1: Flexible System Methodology for Integrated Financial Decision Model

Physical system theory is a powerful tool available for considering the impact of dependable variables by analyzing the whole process as a system on the whole.

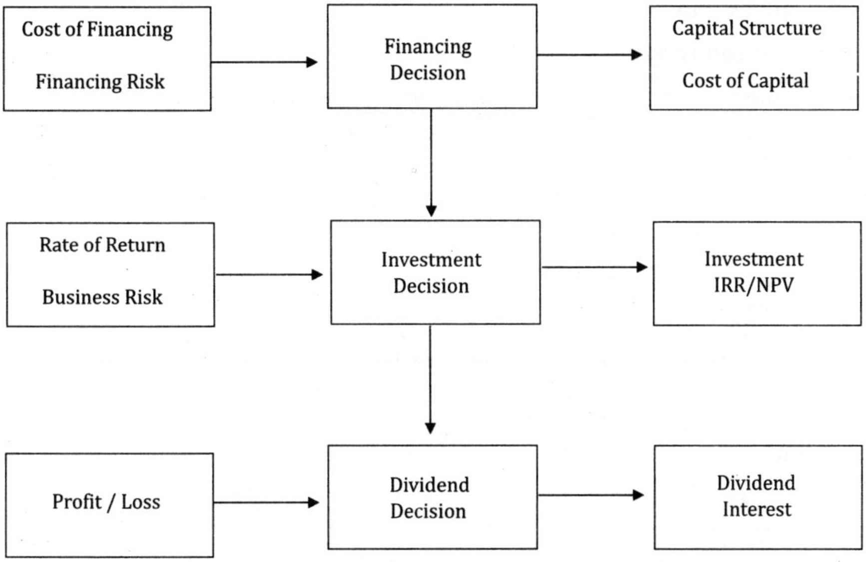

Figure 14.2: Inputs and Outputs of Physical Systems Model

In this model, the following variables have been used:

Across Variable	:	Cost of capital
		Rate of return on capital
Through Variable	:	Flow of capital

Five stages of a typical financial cycle have been identified to evaluate the characteristics of the model. These stages are enumerated below:

- **i.** The capital structure stage
- **ii.** Investment decision stage
- **iii.** Return on investment stage
- **iv.** The aggregation of returns stage
- **v.** The dividend stage

Assumptions

To develop this model, only three types of finances are considered. The firm can invest these finances in two different projects. In actual practice, the firm may have a number of avenues for generating finances and investing them in more number of projects.

Component Models

The component models for an illustrative system with three sources of finance and two-investment options are presented, which are then generalized. The generalized model is given in Appendix 14 A.

Stage I – The Capital Structure Stage

For developing this model, we will assume three sources of finance (Figure 14.3).

1. Debt: at a specified rate of interest.
2. Equity: issues of equity shares.
3. Retained Earning: returns obtained by firm due to its activities.

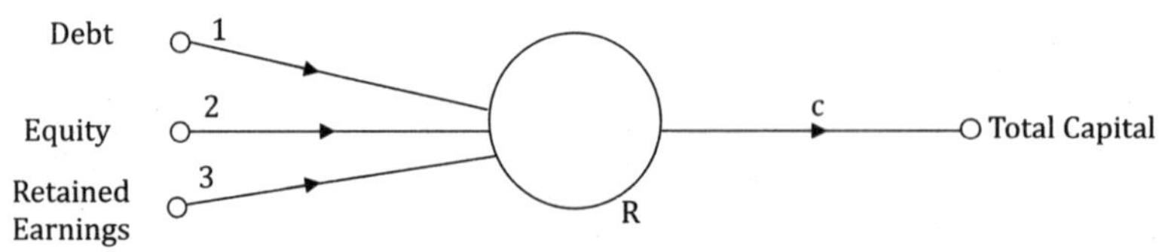

Figure 14.3: Component Terminal Graph of Capital Structure Stage

The different equations can be written as follows:

a. Cost equation

$$x_c = k_1 x_1 + k_2 x_2 + k_3 x_3 + f_c$$

where x_1, x_2, x_3 are relevant costs of capital

k_1, k_2 k_3 are weights of sources of finance

f_c – Transaction cost to raise the capital (e.g. Floatation cost of issues)

b. Flow equation

$$y_1 = k_1 y_c$$
$$y_2 = k_2 y_c$$
$$y_3 = k_3 y_c$$

where y_1, y_2 and y_3 are the flow of capital from sources 1, 2, and 3.

Stage II - Investment Decision Stage

The proportion of investment in the investment projects is different and may not be equal or same. The component terminal graph is shown in Figure 14.4.

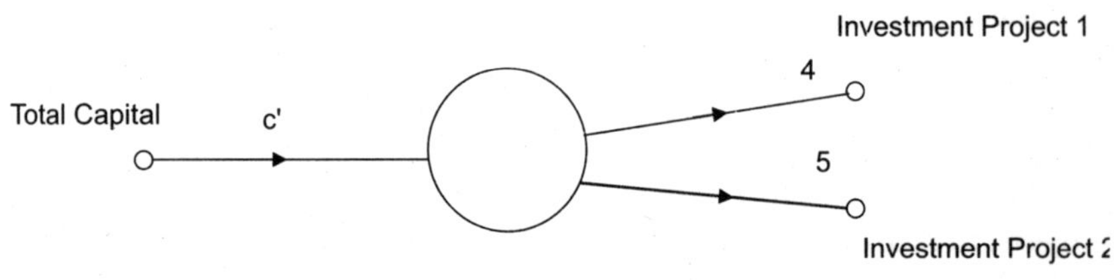

Figure 14.4: Component Terminal Graph of Investment Stage

The equations can be written as follows:

a. Cost equation

$$x_c' = x_4 = x_5$$

Since, the funds generated are utilized for investing in different projects the cost of capital of each is same.

b. Flow equations

$$y_4 = k_4 y_c'$$

$$y_5 = k_5 y_c'$$

where $$k_4, k_5 = \frac{\text{Investment in particular project}}{\text{The total funds available for investment}}$$

x_c', x_4, x_5 = Cost of capital

x_c', y_4, y_5 = Flow of capital

The mass balance equation is given by:

$$k_4 + k_5 = 1$$

Stage III – Returns on Investment Stage

The component terminal graph for ith project and specific projects in the illustrative example are shown in Figure 14.5.

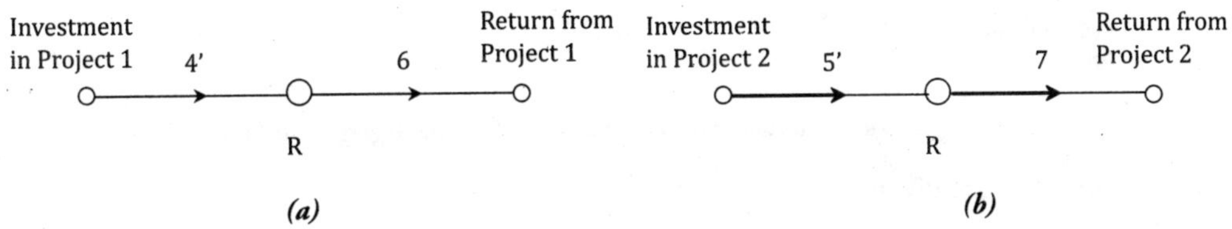

Figure 14.5: Component Terminal Graphs of Returns on Investment Stages

Any investment done in a project yields some return from the project. The return from a project is required to be greater or equal to the minimum required rate of return (MRRR) for the project to be financially viable. If the return is less than MRRR then the project is not profitable.

The minimum required rate of return would be a function of the cost of capital, *i.e.* at what cost one is able to generate capital and at what business risks one is going to invest in particular project. Thus, minimum required rate of return is a function of cost of capital, *i.e.* x_c'. In mathematical terms it can be represented as follows:

MRRR = Cost of Capital + $fn\ (x_c')$ of project x

where f_c = Risk factor

x_c' = Cost of capital

The equation of the project can be written as follows:

The cost equation

$$x_6 = x_4' + f_6\ (y_6)$$

where x_4 = Total return from project 1

x_6 = MRRR of project 1.

Similarly, we can have for 2nd project.

$$x_7 = x_5' + f_7(y_7)$$

The flow equations are as follows:

$$y_4' = k_4' y_6$$
$$y_5' = k_5' y_7$$

It is evident that the amount of return (y_6) is related to the capital invested in the project (y_4')

$f_6(y_6)$ and $f_7(y_7)$ represent the risk factor of each business venture. They are made up of two components. One, the business risk, which varies from project to project, and second, financial risk which is independent of the project.

Stage IV – The Aggregation of Returns Stage

The aggregation of returns is nothing but the summation of returns from each project (Figure 14.6).

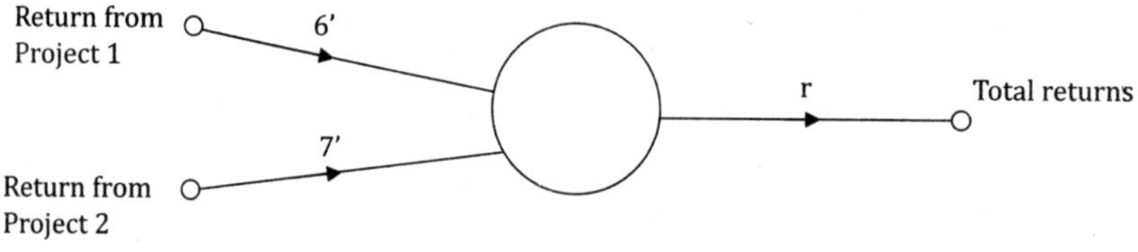

Figure 14.6: Component Terminal Graph of the Aggregation Stage

The different equations are given below:

Cost Equation:

$$x_r = k_6' x_6' + k_7' x_7' + f_r(y_r)$$

where $k_6' = y_6'/y_r'$ and $k_7' = y_7'/y_r$

x_6', x_7' = MRRR of project 1, 2 respectively

$$= \frac{(1+r_1)k_4}{(1+r_1)k_4 + (1+r_2)k_5} \text{ and}$$

$$= \frac{(1+r_2)k_5}{(1+r_1)k_4 + (1+r_2)k_5} \text{ respectively}$$

where r_1, r_2 are return from different projects.

x_r = MRRR of firm

$f_r(y_r)$ = Cost of collecting different components of total aggregate flow.

Flow Equations:

$$y_6' = k_6' y_r$$
$$y_7' = k_7' y_r$$

The flow equations represent the flow from each project.

Mass Balance Equation:

$$y_r = y_6' + y_7'$$

where

y_6' and y_7' = returns from project 1, 2 respectively.

y_r = total returns of firm

Stage V – The Distribution of Returns Stage

In this stage, the total returns from investment decisions/projects are distributed to various heads. Some of the common heads where these returns are distributed. (Figure 14.7):

a. Dividend on equity/preference shares

b. Interest payment on debt.

c. Retained earnings

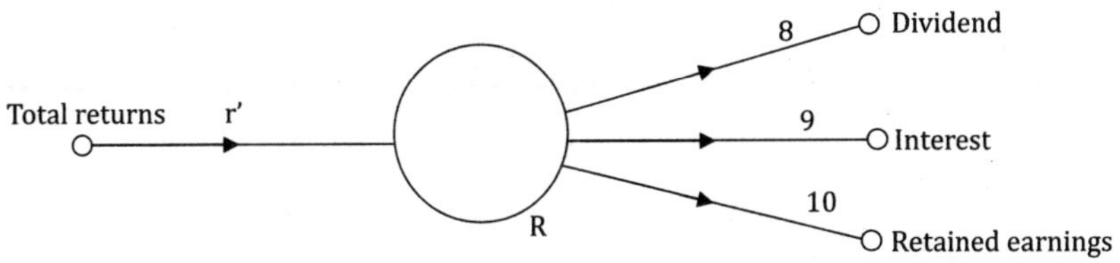

Figure 14.7: Component Terminal Graph of the Distribution of Returns Stage

Cost Equation:

$$x_r' = k_8 x_8 + k_9 x_9 + k_{10} x_{10} + f_r (y_r')$$

Subject to Following Constraints:

$$x_8 < x_r'$$

$$x_9 < x_r'$$

$$x_{10} = x_8$$

where $k_8 = \dfrac{y_8}{y_r} \quad k_9 = \dfrac{y_9}{y_r}, k_{10} = \dfrac{y_{10}}{y_r}$

The term $f_r'(y_r')$ represents the distribution charges in terms of postage etc.

where

x_8 = Rate of dividend payment

x_9 = Rate of interest payment

x_{10} = Cost of capital of retained earnings, this is same as rate of dividend payment

y_8 = Amount paid as dividend

y_9 = Amount paid out as interest on debt

y_{10} = Amount held as retained earnings.

x_8 i.e. rate of dividend, is an exogenous variable and consists of:

a. Market rate of returns on riskless investment (I)

b. Business Risk (B_r – Measured from operating leverage)

c. Financial Risk (F_r – Measured from financial leverage)

$$x_8 = I + B_r + F_r$$

Flow Equations:

$$y_8 = k_8 y_r'$$
$$y_9 = k_9 y_r'$$
$$y_{10} = k_{10} y_r'$$

Structural Constraints:

The compatibility and continuity equations of this system are:

Compatibility Equations:

$$x_c = - x_c'$$
$$x_4 = - x_4'$$
$$x_5 = - x_5'$$
$$x_6 = - x_6'$$
$$x_7 = - x_7'$$
$$x_r = - x_r'$$

Continuity Equations:

$$y_c = y_c'$$
$$y_4 = y_4'$$
$$y_5 = y_5'$$
$$y_6 = y_6'$$
$$y_7 = y_7'$$
$$y_r = y_r'$$

Variables

The list of different across and through variables in different stages of financial cycle is given in Table 14.1.

System Graph

The component terminal graphs of each stage are combined to form the system graph. The system graph is shown in Figure 14.8.

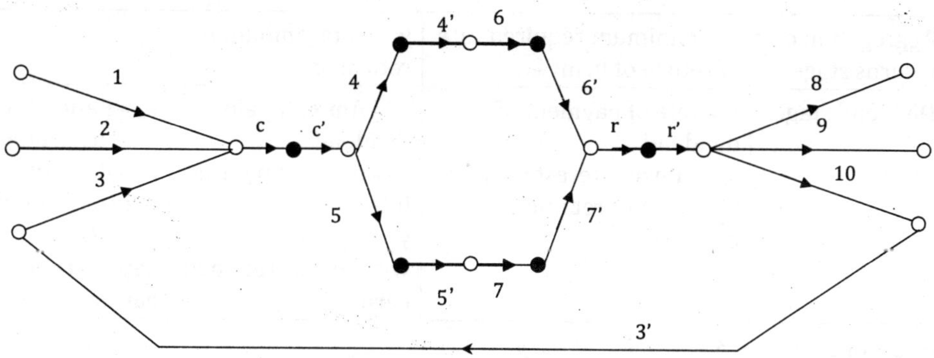

Figure 14.8: System Graph of Financial Decision Model

After integrating the above system graph the simplified system graph is as shown in Figure 14.9.

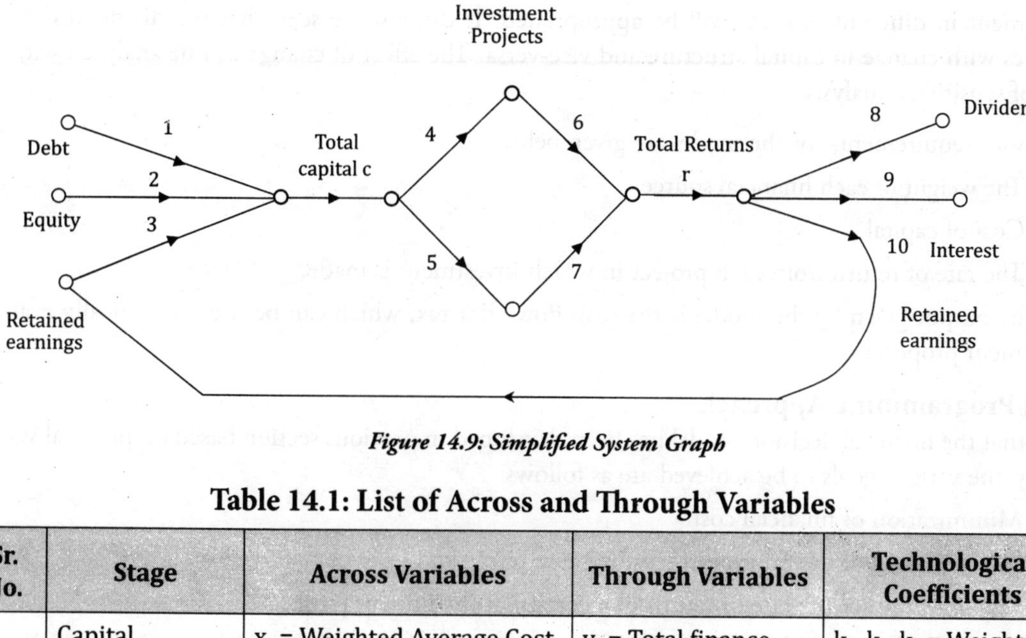

Figure 14.9: Simplified System Graph

Table 14.1: List of Across and Through Variables

Sr. No.	Stage	Across Variables	Through Variables	Technological Coefficients
1.	Capital Structure Stage	x_c = Weighted Average Cost of capital x_1, x_2, x_3 = Cost of respective sources of capital	y_c = Total finance y_1, y_2, y_3 = Amount from each source	k_1, k_2, k_3 = Weights of sources of Finance
2.	Investment Decision Stage	x_4, x_5 = Cost of capital of investment in different projects	y_4, y_5= Amount invested in different projects	k_4, k_5= Ratio of investment in different projects to total funds available
3.	Returns on investment stage	x_6, x_7 = Minimum required rate of return from different projects	y_6, y_7= Returns from different projects	

Contd...

4.	Aggregation of returns stage	x_r = Minimum required rate of return of firm	y_r = Total amount of return of firms	
5.	Dividend stage	x_8 = rate of payment of dividend x_9 = rate of interest payment x_{10} = Cost of retained earnings	y_8 = Amount paid as dividend y_9 = Amount paid as interest y_{10} = Amount withheld as retained earnings	k_8 = Ratio of return paid as dividend k_9 = Ratio of return paid as interest k_{10} = Ratio of return kept as retained earnings

n (sources of capital) = 3

m (investment projects) = 2

In this system graph, if the firm's capital structure is known it is possible to know the investment opportunities. The dividend decision is also known *i.e.* in what ratio the funds generated from the investment in different projects will be appropriated. It can also be seen that the dividend decision changes with change in capital structure and vice-versa. The effect of change can be analyzed with the help of sensitivity analysis.

The input requirements for the model are given below:

a. The weight of each financial source.

b. Cost of capital

c. The rate of return from each project in which investment is made.

The output given by the model is the cash flow after tax, which can be used to critically evaluate investment proposal.

Goal Programming Approach

Now that the financial decision model has been developed in previous section based on physical system theory, the various goals to be achieved are as follows:

a. Minimization of financial cost.

b. Minimization of the minimum required rate of return of the company/firm.

c. Maximization of the rate of return of a particular investment project.

The different goals and the objectives function expressed in mathematical equations are given in Appendix 12 B. The following assumptions are made while deriving the goal programming equations.

i. There are maximum four financial alternatives

ii. There are maximum three projects.

14.4 ILLUSTRATIVE EXAMPLE

The usefulness of any model can only be verified when it is applied to real life problem and the results achieved with the help of the model are compared with the results obtained through conventional methods. For solving this model some assumptions/ data has to be made available. The cost of capital and the expected return from different projects over the life of the projects have to be known. The expected return can be assumed same, throughout the life of project or it may be varied to see the effect on the performance of the project. The results of the model will reflect different returns the firm can get over and above, the minimum required rate of return by adopting different financial proposals and using the funds so generated judiciously.

For analyzing the financial decision model developed in previous section, an illustrative problem has been solved. For the purpose of solving this model the following assumptions are made:

a. There are three financing options namely debt, equity, and retained earnings.

b. Two investment projects.

c. Varying rate of return for the projects life.

The input data are given in Table 14.2. The rates of return from the two projects for the life on the projects are given in Table 14.3. The expected output of this model consists of variables enumerated in Table 14.4.

Table 14.2: Input Data

Sr. No.	Parameters	Details	Value
1.	k_1	Weight of debt	60%
2.	k_2	Weight of equity	30%
3.	k_3	Weight of retained earnings	10%
4.	y_1	Actual debt taken	₹ 15 lakhs
5.	Interest Rate	Interest at which debt is taken	12%
6.	Tax Rate	Corporate Tax	50%
7.	FR	Financial Risk	02%
8.	BR	Business Risk Project I Project II	 03% 05%
9.	k_4	Investment in Project I	60%
10.	k_5	Investment in Project II	40%
11.	Life of Project		10 years

Table 14.3: Rate of Return from Projects

Year	Rate of Return	
	Project I	Project II
1.	12%	12%
2.	15%	12%

Contd...

3.	20%	15%
4.	22%	18%
5.	25%	20%
6.	30%	25%
7.	30%	28%
8.	25%	25%
9.	20%	18%
10.	15%	14%

Table 14.4: Output Results

Sr. No.	Parameter	Details
1.	ROR	Overall Rate of Return expected by firm from the two investment proposals
2.	MRRR	Minimum Required Rate of Return expected by firm from each project
3.	Tax	Payment of Tax
4.	Depreciation	Depreciation on projects
5.	Interest Payment	Payment of Interest on debt
6.	CFAT	Cash flow after tax
7.	NPV	Net Present Value

Methodology

Cash flows are calculated using this model. These cash flows depend on the returns on investments which are different for different years. Based on the cash flows the net present value is calculated discounting at the cost of capital. The decision to invest or not in the projects is based on the fact, if net present value is positive or negative respectively. The decision criterion can be expressed as follows:

a. If NPV > 0 Select the Projects.

b. If NPV <=0 Reject the projects.

Flow and Cost Analysis

Cash Flow After Tax (CFAT)

a. Cost of debt = 15 (1 – 0.5) = 7.5%

b. Minimum required rate of return on equity shareholder = 18%

= 11.5

c. Cost of Retained Earnings = Cost of Equity = 11.5

$$\text{Cost of equity} = \frac{D_1}{P_0} + g$$

$$= \frac{18}{12} + 10$$

$$= 25 \text{ lakhs}$$

Equity = 25 × 0.3 = 7.5 lakhs

Retained earnings = 25 × 0.1 = 2.5 lakhs

d. Investment in different projects

Projects 1 = 25 x 0.6 = 15 lakhs

Projects 2 = 25 x 0.4 = 10 lakhs

e. The returns on the two projects for the first year are 12% each.

Yield from Project 1 = 15 × 0.12 = 1.8 lakhs

Yield from Project 2 = 10 × 0.12 = 1.2 lakhs

Total Yield 3.0 lakhs

f. Interest on debt taken = $\frac{15 \times 12}{100}$ = 1.8 lakhs

g. Assuming straight-line depreciation is adopted by the firm. The across variables are calculated. The value of different across variable are evaluated as given below:

a. Weighted Cost of Capital

$$x_c = k_1 x_1 + k_2 x_2 + k_3 x_3$$

$$= 0.6 \times 0.075 + 0.3 \times 0.115 + 0.1 \times 0.115$$

$$= 0.091$$

$$= 9.1\%$$

b. In the 2nd stage the capital generated in the 1st stage is invested in different projects. Hence the cost of capital would remain the same as in 1st stage.

Therefore, $x_c = x_4 = x_5 = 9.1\%$

c. The 3rd stage represents the returns stage

$$x_6 = x_4 + f_6(y_6)$$

where x_6 = required rate of return of project 1

x_4 = cost of capital for project 1

$f_6(y_6)$ = This comprises of business and financial risk.

$$x_6 = 0.091 + 0.03 + 0.02 = 0.141 = 14.1\%$$

Similarly,

$$x_7 = 0.091 + 0.03 + 0.02 = 0.141 = 14.1\%$$

d. At the final stage the overall minimum required rate of return is given by

$$x_r = k_6 x_6 + k_7 x_7$$

$$\text{where } k_6 = \frac{(1+r_1)k_4}{(1+r_1)k_4 + (1+r_2)k_5}$$

$$= \frac{(1+0.12)0.6}{(1+0.12)0.6+(1+12)}$$

$$= 0.6$$

Similarly

$$k_7 = \frac{(1+r_2)k_5}{(1+r_1)ku+(1+r_2)k_3}$$

$$= \frac{(1+0.12)0.4}{(1+0.12)0.6+(1+0.14)0.4}$$

$$= 0.4$$

$$x_r = 0.6 \times 0.141 + 0.4 \times 0.161$$

$$= 0.149$$

$$= 14.9\%$$

Depreciation on Project 1 = 15 × 1/10 = 1.5 lakhs/year

Depreciation on Project 2 = 10 × 1/10 = 1.0 lakhs/year

Total depreciation = 2.5 lakhs/year

h. Taxable Income = Earnings before tax – Interest – Depreciation

= 3.0 – 1.8 – 2.5

Since, the taxable income is negative no tax is required to be paid, since the income is negative there is no profit for the first year.

i. The cash flow of the firm is not what is shown as profits but the returns from the projects minus the tax payment.

Therefore, CFAT = 3.0 lakhs

Similar calculations can be done for finding the CFAT for the remaining duration of the project. The Net Present Value of these CFAT can be determined by discounting at the cost of capital. Similarly, the different CFAT for whole life of the project at different rates of return are computed as shown in Tables 14.5 and 14.6.

1. Interest on Debt = 15%
2. Corporate Tax Rate = 50%
3. Weighted Cost of Capital = 9.1%

Table 14.5: Cost of Capital

Capital Structure	Weight (k)	Total Finance (Lakhs)	Cost of Capital
Debt	0.6	15.0	7.5
Equity	0.3	7.5	11.5
Retained Earnings	0.1	2.5	11.5

Table 14.6: Calculation of Net Present Value

Year	Investment Cash Outflow (Lakhs)	Cash inflow (Lakhs)	PVF at Cost of Capital	Net Flow (Lakhs)	Present Value (Lakhs)
1.	- 15	-	1	- 15	- 15
2.	- 10	3.0	0.917	- 7	- 6.419
3.	- 10	3.45	0.84	- 6.55	- 5.50
4.		4.4	0.77	4.4	3.338
5.		4.7	0.71	4.7	3.337
6.		5.025	0.65	5.025	3.266
7.		5.65	0.59	5.65	3.333
8.		5.80	0.54	5.80	3.132
9.		5.225	0.50	5.275	2.6375
10.		4.55	0.46	4.55	2.093
11.		3.65	0.42	3.65	1.533

NPV = -4.25

Since NPV is less than zero, the investment proposal is rejected.

REFLECTIONS

A model for financial decision-making has been formulated. Some of the important aspects of this model have to be highlighted.

In this model the stages identified would remain same irrespective of the size of the model. The components at each stage may increase or decrease, *i.e.* the number of financial alternatives and projects, however, the number of stages will remain the same. The distribution process has also been taken as one of the stages.

The across variables change from cost of capital to returns on investment. Whereas the through variable remain same throughout the system, *i.e.* flow of capital.

This model helps the decision-maker for taking right decision in financial management. An overall view of the implication of taking a financial decision can be seen. Acceptance or rejection of projects can be arrived at once. The CFAT and NPV's are calculated over the life of project. Sensitivity analysis by verifying the input variable can be carried out. The model can also be applied to capital budgeting problems or to financial planning problems in isolation.

The model deals with deterministic data. Business risk associated with each project has to be determined earlier. The returns from projects over the life of projects have to be forecasted accurately to give a realistic decision.

Every firm has its own preferences and policies. Hence the combination of these preferences and policies cannot be reflected in one general model. Alternations and modifications may be required in the model to cater to a particular problem.

Notes

1. Khan M.Y. and Jain P.K. (2012) *Financial Management*, Tata McGraw-Hill, New Delhi.
2. Brealey R.A. and Myers S.C. (1988) *Principles of Corporate Finance*, McGraw-Hill, New Delhi.
3. Gitman L.J. (1976) *Principles of Managerial Finance*, Harper and Row, New York.
4. Bierman H. and Smidt S. (1974) *The Capital Budgeting Decision*, Macmillan, New York.

APPENDIX 14A

Generalized Component Model of Capital Structure Stage

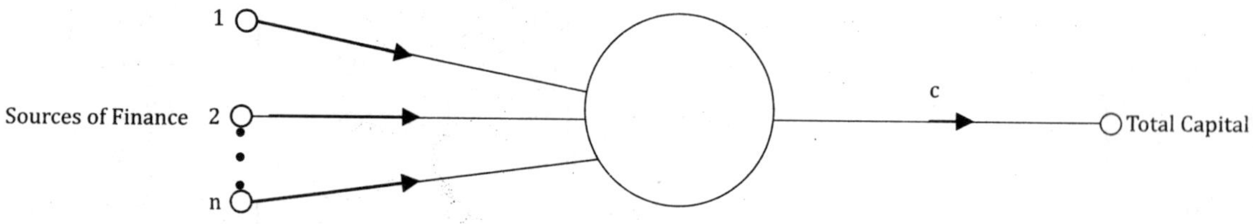

$$x_c = \sum k_j x_j = f_c(y_c)$$

$$y_j = k_j y_c \qquad \forall_j$$

Generalized Component Model of Investment Stage

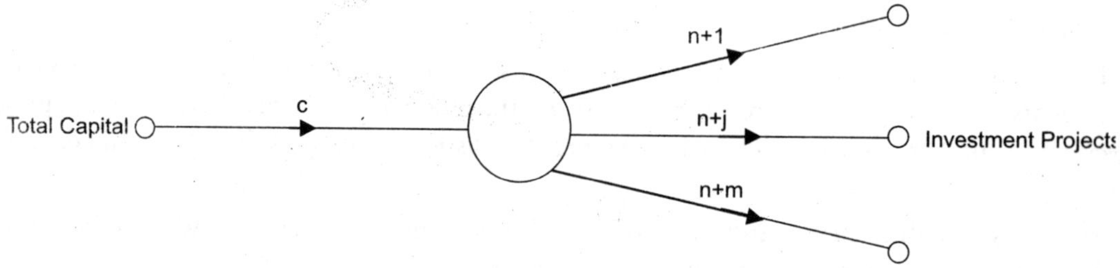

$$x_c = x_{n+j} \qquad \forall_j$$

$$y_{n+j} = k_{n+j}.y_c \qquad \forall j$$

Generalized Component Model of Returns on Investment Stage

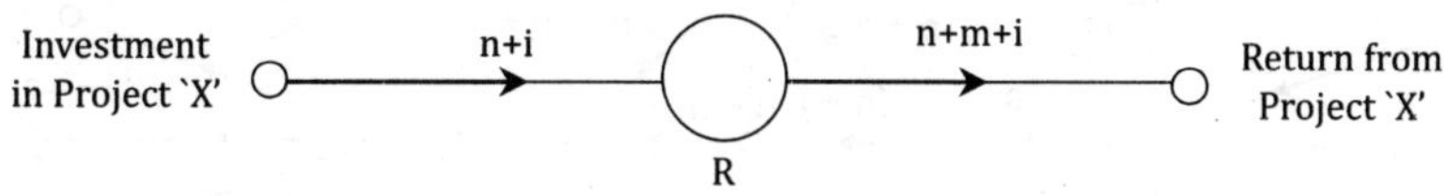

$$x_{n+i} = x_{n+m+i} + f_i(y_i)$$

where $i = n+m+1$ to $n+2m$

and $y'_m = k'_m(y'_{n+m+1})$

Generalized Component Model of Aggregation Stage

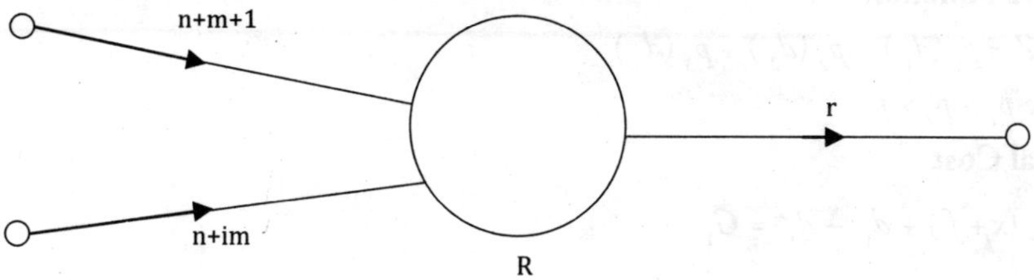

$$x_r = \sum_{i=n+m+1}^{n+2m} k_i x_i$$

Generalized Component Model of Distribution of Returns Stage

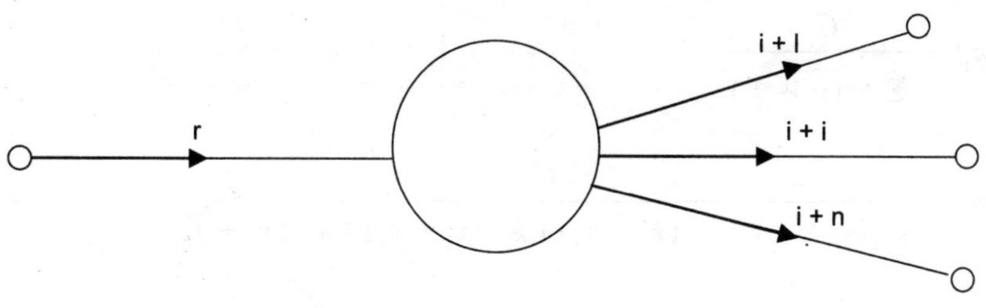

$y_{i+j} = k_j y_r$ where $j = 1 \ldots n$

Generalized Simplified System Graph

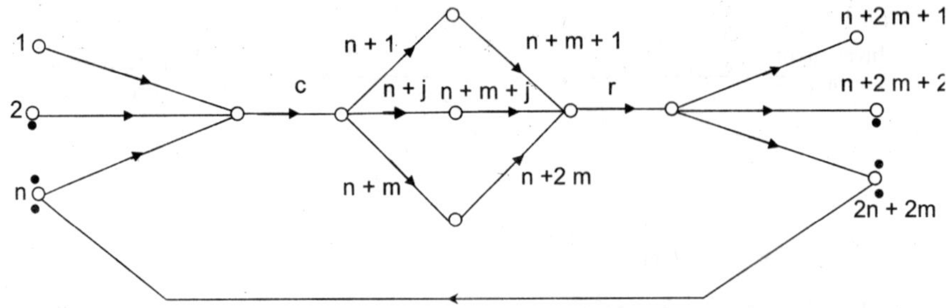

APPENDIX 14B

Goal Pogramming Model for Integrated Financial Decisions

a. Objective Function

$$\text{Min } Z = p_1 (d_1^+) + p_2 (d_2^+) + p_3 (d_3^-)$$

where $p_1 > p_2 > p_3$

b. Financial Cost

$$\sum_{i=1}^{n} k_{fi} (x_i + f_i) + d_1^- - d_1^+ = G_1$$

where

x_i = Cost of ith financial alternative

f_i = Financial risk of ith financial alternative

k_{fi} = Proportion of ith financial alternative

Solving the above equation we get

$$d_1^- - d_1^+ = \frac{G_1}{\sum_{i=j}^{n} k_{fi}(x_i + f_i)}$$

$$d_1^- - d_1^+ = \frac{G_1}{k_{f1}(x_1 + f_1) + k_{f2}(x_2 + f_2) + k_{f3}(x_3 + f_3) + k_{f4}(x_4 + f_4)}$$

d_1^+ = Positive deviation of financial cost

d_1^- = Negative deviation of financial cost

G_1 = Expected financial cost

Minimum required rate of return (MRRR)

$$\sum_{i=j}^{n} k_{fi} (x_i + f_i) + \sum_{j=1}^{m} k_{pj} b_j + d_2^- - d_2^+ = G_2 (1 + R)$$

where R = Overall ROR

$$k'_{pi} = \frac{(1+r_i)k_{pi}}{\sum_{j=1}^{m}(1+r_j)k_{pj}}$$

b_j = Business risk of jth project
d_2^+= Positive deviation of MRRR
d_2^-= Negative deviation of MRRR
r_j = ROR of jth project
k_{pj}= Investment ratio in project J
k'_{pj}= MRRR project j.
G_2 = MRRR of firm.

Solving the above equation we get

$$[\sum_{j=1}^{m} k_{pj}\, b_j] + = d_2^{\,-} - d_2^{\,+} = \frac{G_2(1+R)}{\sum_{j=1}^{n} k_{fi}(x_l + f_l)}$$

$$K_{p1}b_1 + K_{p2}b_2 + K_{p3}b_3 + d_2^- \, d_2^+ = G_2\,(1 + R) \,/\, kf_1\,(x_1 + f_1) + kf_2\,(x_2 + f_2) + kf_3\,(x_3 + f_3) + kf_4\,(x_4 + f_4)$$

We have
for project 1

$$K_{p1} = \frac{(1+r_j)K_{p1}}{(1+r_1)K_{p1} + (1+r_2)K_{p2} + (1+r_3)K_{p3}}$$

$$(1+r_1)\,k_{p1} + \frac{K_{p2}}{k'_{p2}}\,(1+r_2)\,(k'p_2 - 1) + (1+r_3)\,k_{p3} = 0$$

Similarly for project 2, 3 we get

c. ROR

where

G_3 = Overall ROR
$d_3^{\,+}$ = Positive deviation of ROR
$d_3^{\,+}$ = Negative deviation of ROR

Solving above equation we get

$$k_{p1}r_1 + k_{p2}r_2 + k_{p3}r_3 + d_3^{\,-} - d_3^{\,+} = G_3$$

We have

$$k_{p1} + k_{p2} + k_{p3} = 1$$

$$\sum_{j=1}^{m} k_{pj}\, r_j + d_3^- - d_3 + = G_3$$

15 ORGANIZATIONAL ANALYSIS FOR AUTHORITY AND RESPONSIBILITY RELATIONSHIPS

15.1 INTRODUCTION

Organizations are for the people, by the people, and of the people. In order to achieve some goals set by a group of people, the rest of them work, cooperate and perform as per certain agreed norms and standards. This leads to formation of groups. The construction of human groupings leads formation of a structure due to two reasons:

a. Creation of division of labour, i.e. placing individual at the appropriate place in order to utilize his specific skill or expertise, and

b. Hierarchy of operation, i.e. to effect decision upon its impact on achievement of goals. In addition, organizations have rules and procedures to handle specific situations.

Thus, the organizational structure is an important factor which has a great bearing in the accomplishment of the organizational goal(s). The organizational structural design is not static; it might have to undergo minor or major, changes, with changing environmental factors which affect the processes of fulfilment of its goals. The task of restructuring becomes more difficult and complicated as the size of the system grows over a period of time. Hence, after reaching every major milestone towards achievement of its goals. The organization should pause and review its structure. History is replete with many instances where large organizations have gone 'down the drain' for not reorganizing themselves to meet the changing environments.

The important steps involved in the restructuring process include the identification of goals, breaking them down into sub-goals, grouping of logically connected activities and the determination of span of control by managers. The process is governed by a whole set of formal duties, tasks, relationships, policies, procedures, standards, rules, regulations, and so on. The formal structure is presented in the form of charts, manuals, job descriptions, etc.

Till 1960s, organizational analysis and understanding was dominated by Weber's idea of ideal bureaucracy[1]. Weber's discussion provided the framework for numerous studies, which provided evidence that, in practice, most large organizations did not resemble the ideal bureaucracy. Ideals from industrial psychology, social psychology, and sociology dealing with such issues as cohesion, leadership, job satisfaction and determinants of productivity raised further issues for organizational theory. In 1960s, emphasis shifted from formulation of organizational theories to studies of organizational processes. As a result, studies dealt with smaller frameworks for studying the processes rather than comprehensive frameworks for studying the organizations as a whole. The emphasis also shifted towards obtaining practical advice for problem solving and decision-making related to actual realities existed on ground. This led to development of applied organizational science in the form of organizational design and development including studies on structure, technology, and growth.

In mid 1970s, the theory of group structures centered around study of micro processes of structural change in the small organizations. The pioneering work in this respect was carried out by Mackenzie[2] and was joined by Lippitt[3]. Mackenzie, after an extensive study and research, developed a "Theory of Group Structures" [4]. Mackenzie[5] also applied the measure of degree of hierarchy to the problems of organizational design. The concept of power in organizational design has also been discussed and analyzed by many authors. Mackenzie[6] identified a natural organizational phenomena called virtual positions to the enactment of power. A critical position is a composite position of a group of participating workers engaged as a recurring process, and it may be regulated by virtual position. While analyzing authority-task problems, Lippitt and Mackenzie[7] considered the problems of organizational analysis and design at two levels; a micro level dynamic problem solving organization based on task process; and a macro level official organization system, which is comparatively static in nature. A review of theory on the organizational analysis and design is given by Sushil [8].

While carrying out organizational analysis for structural balance and stability many issues existing in the organization need to be studied and analyzed. They include, amongst other things, hierarchical relationships, chain of command, delegation of authority, authority-responsibility relationships, centralization, and so on. Amongst many concepts, the concept of authority-responsibility relationships plays an important role in the structuring of any organization.

This chapter is an attempt to design a technique for organizational structural analysis at micro level based on task processes with special reference to authority-responsibility relationships. In doing so, this technique uses the concepts and principles of physical system theory for analyzing the authority-responsibility relationships.

15.2 ROLE OF AUTHORITY AND RESPONSIBILITY IN ORGANIZATIONS

Managers at every level, in any organization perform two important functions: planning and execution. The first step in the execution process is the establishment of organizational patterns which can contribute to the effective and efficient accomplishment of organization's objectives. Intrinsic to this function are the concepts of 'responsibility' and 'authority'.

Authority and responsibility constitute the framework of organizational management-based on which any organization is founded.

Authority

According to a dictionary definition, 'authority' is the legal or rightful power; a right to command or to act. Though the concept of authority can be viewed differently with different connotations, it can be broadly conceptualized and classified as the individual's right to pre-empt, direct, to act, decide and control. Authority is the sum of the powers and rights entrusted to make possible the performance of work delegated or assigned. Authority includes such rights of power like spending specified amount of money, allocation of resources, hiring and firing of people. It may also include such powers as to render advice and provide service.

Responsibility

Responsibility, an inescapable corollary of authority, is the obligation that one has to act in response to an order issued by higher authority. Responsibility indicates the individual's assumption of an obligation, i.e. an agreement, promise, contract or recognition of duty. Responsibility also means what one is expected to do or not to do as a responsible person. Responsibility is also considered as the work assigned to a position or function. It encourages all mental and physical activities which must be performed to carry out a task or duty.

15.3 AUTHORITY AND RESPONSIBILITY RELATIONSHIPS

Responsibility, in the popular sense, is a containing obligation which arises from superior-subordinate relationship. This interpretation of responsibility to mean the obligation to do which is expected extends the concept beyond the superior-subordinate relationship to others in a group situation, e.g. peers and associates.

Similarly, authority in organizations also revolves around its members. Authority is a form of power that does not imply force. Directives or orders are followed because it is believed that they ought to be followed. Compliance is voluntary, because all members of the organization generally have the same value system.

As discussed earlier, based on the division of labour a hierarchical structure is formed in every organization for accomplishment of its goals. All the components of the hierarchy have a specified task to perform. The members of the organization are the agents through whom all the tasks get performed, the targets and goals get accomplished. Each manager in the hierarchy had to perform certain jobs assigned to him by the organization. In an indirect sense, these jobs can be termed as his responsibility, and to accomplish these tasks he is vested with certain resources and power to utilize them, which may be termed as his authority. For ensuring effectiveness in achievement of the goal, each manager fulfils the responsibilities rested on him with judicious use of resources given to him in terms of authority.

It is a well-known and accepted feature that for a manager to be effective he must be given adequate authority commensurate with his responsibilities. An imbalance in this equation will lead to, either emergence of an undesirable power centre or under utilization of scare resources, and ultimately will lead to setting in of entropy in the organization.

A delicate balance between authority and responsibility must be maintained in the entire hierarchy so that the organization becomes effective. Though these two factors are widely discussed in thoery, very few models are available which try to capture the reality existing in an organization. The quantitative analysis of these two highly abstract terms will lead to correct measurement of these factors leading to some prescriptive actions.

15.4 ROLE IN ORGANIZATIONAL ANALYSIS AND DESIGN

According to Pfeffer [9] organizational design is essentially a political process as well as a technical and theoretical problem. Kilmann[10] has defined organizational design as "the arrangement and the process of arranging the organization's structural characteristics to attain or improve the efficiency, effectiveness, and adaptability of the organization". Organizational design has been considered as a maintenance process by Nystrom and Starbuck[11] as the organizations are subjected to internal as well as external changes. The organizational maintenance cycle has been described by Mackenzie[12]in three phases: (i) Organizational Audit–to help in identifying organizational maladaptations; (ii) Organizational Analysis–to assess the probable impact of suggested changes; and (iii) Organizational Implementation–to install the planned changes.

The proposed model will have implications at all the three phases of organization design cycle. As the development of the model considers the details of activities being carried out at each managerial position it helps in identifying inconsistencies and maladaptations during the Organizational Audit. The preparation of responsibility balance will indicate whether the managers believe in delegation or there is growing unwillingness to accept responsibility. The values of authority and responsibility at each managerial position obtained from the model compared to the actual values will indicate whether there is parity between authority and responsibility. This will help in organizational design by balancing the authority-responsibility relationships at different levels of hierarchy. The sensitivity analysis of the model will help in categorizing the positions in the organizations into highly sensitive, moderately sensitive, and less sensitive with respect to various coefficients. During organizational design the more sensitive positions should be assigned with new responsibilities more carefully.

The maladaptation being identified during the Organizational Audit lead to the development of alternatives that are analyzed in close interaction with the managers involved. During the Organizational Analysis various alternatives that can be attempted, to see the role of authority-responsibility relationships, are as follows:

- To analyze and observe that what happens if additional responsibilities are given at a managerial position, i.e. by what percentage the authority should be increased?
- In order to meet the needs of some new activities or to reduce the imbalances in the organizational structure, or as a consequence of the design of information systems if there is a need to introduce new managerial position(s), or some positions are to be merged, what impact will it have on the authority-responsibility relationships? In what way the authority and responsibility should be redistributed?
- To compare the consequences of centralization and decentralization decision on the authority-responsibility relationships. The responsibility coefficients in the proposed model can be appropriately changed and the values of authority and responsibility can be compared for different scenarios.
- To analyze the interaction at different levels in terms of authority and responsibility leading to the determination of optimal delegation of authority.

The optimization can be attempted either by integrating the proposed system model with a mathematical programming (*i.e.* goal programming) model or by simulating the system model with different combinations of the authority-responsibility relationships.

In view of the impact of various changes generated during the organizational analysis, final recommendations can be made and implemented regarding organizational restructuring during the organizational implementation phase. The flexible system methodology for organizational analysis is shown in Figure 15.1.

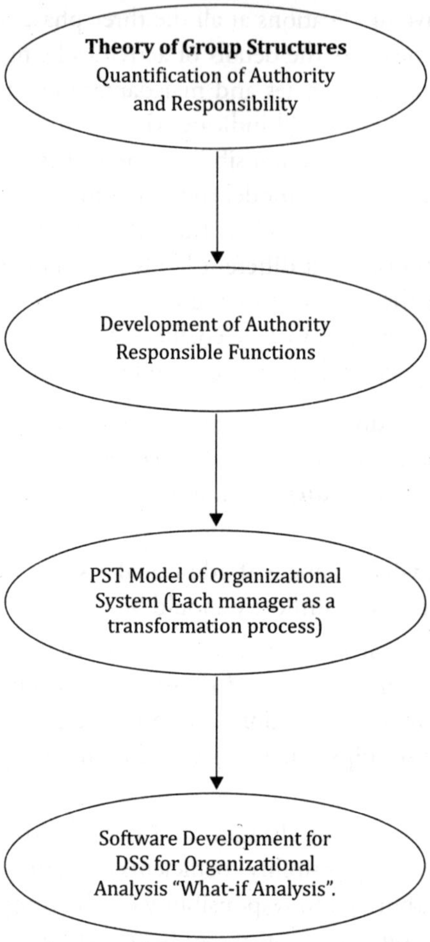

Figure 15.1: Flexible System Methodology for Organizational Analysis

15.5 QUANTIFICATION OF AUTHORITY AND RESPONSIBILITY

In reality, the concept of authority and responsibility are difficult to comprehend. However, for any meaningful analysis we need to device a model to quantify these soft variables. The operational concept of these variables is utilized in their quantification, which could be carried out at micro and macro levels.

The micro level problem solving organizational structures are derived from the task processes using the theory of group structures developed by Mackenzie[13]. In the theory of group structures, structural changes are described as occurring by means of a voting process.

Lippitt and mackenzie[14], after considerable research considered and organization giving three role systems, viz. task role system (R_T), authority role system (R_A) and official organizational chart (R_C).

The number of activities performed by a participant, indicated by '1' entries in the task role matrix is taken as a measure of responsibility of that participant. The authority has been considered as the legal right to preempt activities and is measured by the number of '1' entries for a participant in the authority role matrix. The unit authority, i.e. authority per unit of responsibility is calculated by dividing the total authority of a participant by his /her total responsibility.

The macro level measures for official organizational structure can be either fact based or opinion based.

An example for a five-person organization is shown in Tables 15.1, 15.2 and 15.3 to illustrate the measure of responsibility and authority; only six activities are considered.

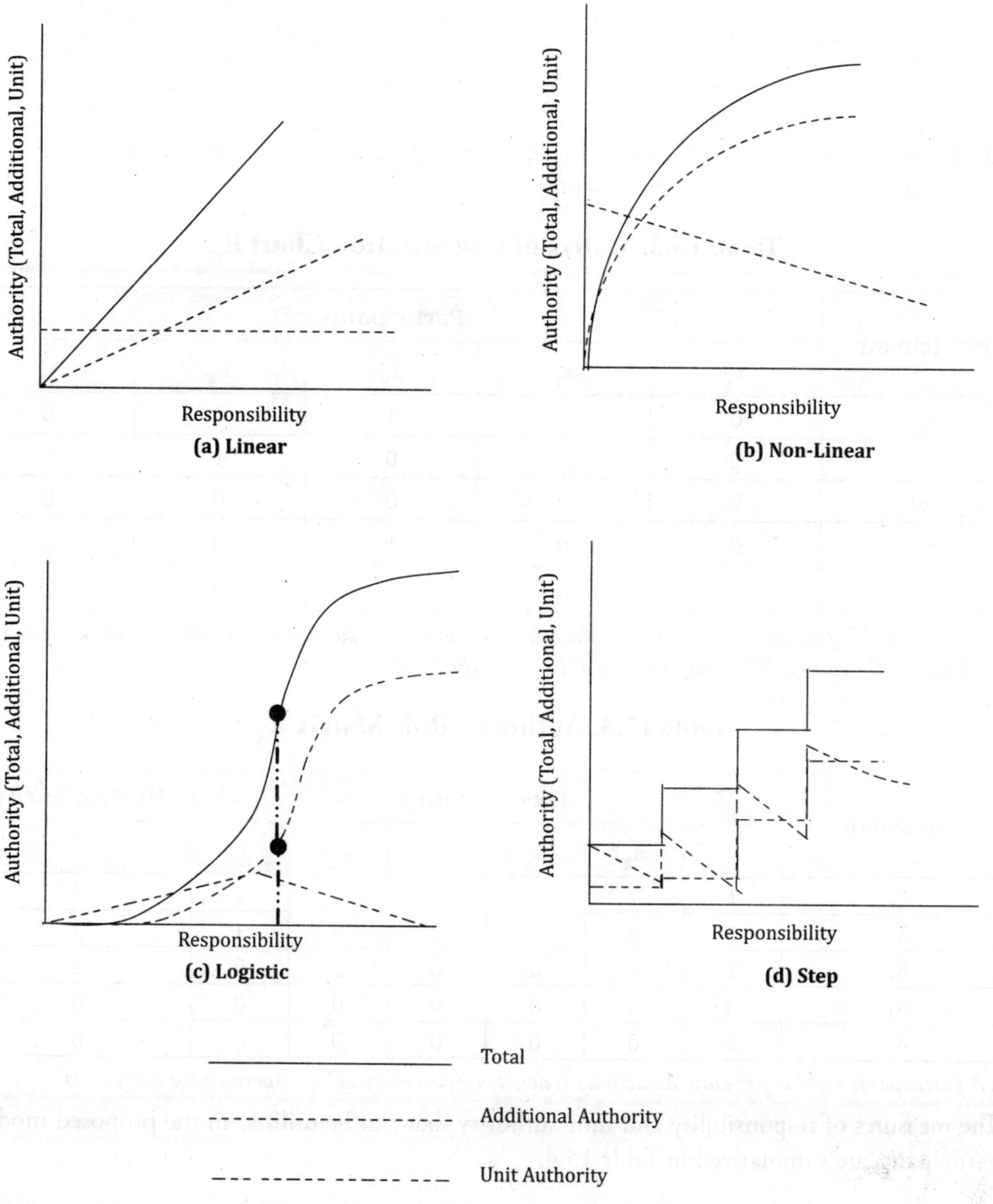

Figure 15.2: Sample Authority-Responsibility Functions

Table 15.1: Task Role Matrix R_T

Participant	Task Activities						Responsibility
	a_1	a_2	a_3	a_4	a_5	a_6	
x_1	1	1	1	1	1	1	6
x_2	1	0	1	1	0	1	4
x_3	1	0	0	1	1	1	4
x_4	1	0	0	1	0	1	3
x_5	1	0	0	1	0	0	2

Note: *Performance of activity or responsibility for performance of an activity is denoted by an entry of* "1". *Non performance and non responsibility is denoted by* "0".

Table 15.2: Matrix of Organization Chart R_C

Participant	Participants				
	x_1	x_2	x_3	x_4	x_5
x_1	0	1	1	0	0
x_2	0	0	0	1	1
x_3	0	0	0	0	0
x_4	0	0	0	0	0
x_5	0	0	0	0	0

Note: *An entry is* "1" *if the participant heading the column reports to the participant heading the row in which the cell is located. The entry is* "0" *otherwise. The diagonal is always* "0".

Table 15.3: Authority Role Matrix R_A

Participant	Task Activities						Unit Authority
	a_1	a_2	a_3	a_4	a_5	a_6	
x_1	1	1	1	0	1	1	5
x_2	1	0	1	0	0	1	3
x_3	1	0	0	0	1	0	2
x_4	0	0	0	0	0	0	0
x_5	0	0	0	0	0	0	0

Note: *A participant's right to pre-empt an activity is denoted by an entry of* "1" *otherwise the entry is* "0".

The measures of responsibility and unit authority that can be utilized in the proposed model, fo five participants are summarized in Table 15.4.

Table 15.4: Measures of Authority and Responsibility

Participant	Responsibility (1)	Authority (2)	Unit Authority = (2)/(1)
x_1	6	5	0.83
x_2	4	3	0.75
x_3	4	2	0.50
x_4	3	0	0
x_5	2	0	0

15.6 AUTHORITY - RESPONSIBILITY FUNCTION

An authority-responsibility function depicts the change in authority with respect to increase in responsibility. An authority-responsibility function is represented either as a change of total authority with total responsibility; or as change in unit authority (total or additional) with total responsibility. The additional authority at a position is the authority in addition to the authority of subordinates. The function $f(y)$ used in the model is representing the change of additional unit authority with total responsibility. The different kinds of authority-responsibility functions that can be used in the proposed model are as follows:

a. **Linear function:** If the authority is increasing linearly with the responsibility, a linear function may be used as shown in Figure 15.2 (a). The unit authority that indicates the slope of the curve is constant. The function $f_i(y_i)$ as used in the proposed model is $f_i(y_i) = c_i$.

b. **Non-linear function:** If authority increases rapidly with the additional responsibility in the beginning, but after obtaining sufficient responsibility the additional responsibility needs less additional authority, a non-linear function as shown in Figure 15.2(b) may be used. The unit authority (slope of the curve) is gradually decreasing with the increase in responsibility *i.e.*

   $f_i(y_i) = c_I - m_i y_i$.

c. **Logistic function:** A logistic function may be used if the rate of increase of authority with additional responsibility has a flex point (R*) as shown in Figure 15.2(c). If the responsibility at a managerial position is less that R* the unit authority is linearly increasing, *i.e.*

   $f_i(y_i) = m_i^1 y_i$

   (where m^1_i is slope).

   If the responsibility is more than R* then unit authority linearly decreases, *i.e.*

   $f_i(y_i) = c_i - m^2_i y_i$

   (where c_i is the intercept and m^2_i is the slope of unit authority curve).

d. **Step function:** A step function as shown in Figure 15.2 (d) is of common occurrence, which signifies that authority is not continuously increasing with increase in responsibility. For a range of responsibility the authority needed is the same, but then suddenly the authority is increased for the additional responsibility. The unit authority is represented by,

$$f_i(y_i) = c_i - m_i y_i$$

where c_i and m_i are the parameters that are different for different ranges of responsibility (y_i).

15.7 MODEL OF ORGANIZATIONAL SYSTEM

Components and Interactions

Different managers at various levels are being considered as components of the organizational system. There can be 'l' different hierarchical levels and n_l managers at each level. Then, the total number of components = $\sum_l n_l$. Each component is interacting with some component at a higher level as well as a few components at a lower level. There can be uncle and cousin relationships as well.

The system model consists of individual component models, system graph and structural constraints [16].

Component Model

The free body diagram of *i*th component (Manager), in case of pure hierarchical system is shown in Figure 15.3.

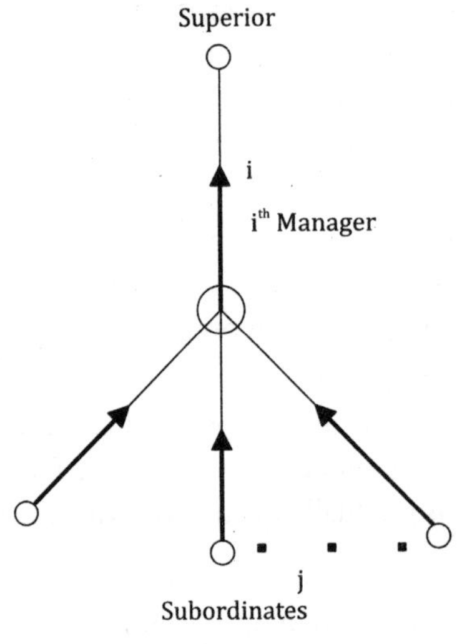

Figure 15.3: Free Body Diagram of ith Manager

Let,

x_i = Unity authority (authority per unit responsibility) of *i*th manager

x_j = Unity authority of *j*th manager (subordinate to *i*th manager) reported to *i*th manager

y_i = Responsibility of *i*th manager

y_j = Responsibility of *j*th manager (subordinate to *i*th manager) reported to*i*th manager

$$k_j = \frac{\text{Responsibility coefficient of } j\text{th subordinate}}{\text{Responsibility of } j\text{th subordinate reporded to } i\text{th manager}}$$

= Responsibility of *i*th manager under consideration

f_i = Authority-responsibility coefficient of *i*th manager

= Additional unit authority per unit responsibility of the *i*th manager under consideration.

Then, the *authority equation* is written as follows:

Total authority flowing out of *i*th level (Manager)	=	Total authority flowing into *i*th level	+	Additional authority at *i*th level (Manager) to fulfil special responsibilities:

$$y_i x_i = \sum_j y_j x_j + f_i(y_i)\, y_i \qquad (15.1)$$

Dividing both sides by y_i:

$$x_i = \sum_j k_j x_j + f_i(y_i) \qquad (15.2)$$

The *responsibility equations* define the respective responsibility coefficients,

$$y_i = k_j y_j \ \forall_i \qquad (15.3)$$

There will be, in all, $\sum_l n_l$ component models.

This component model can be generalized for a component with uncle and cousin relationships, as shown in Figure 15.4.

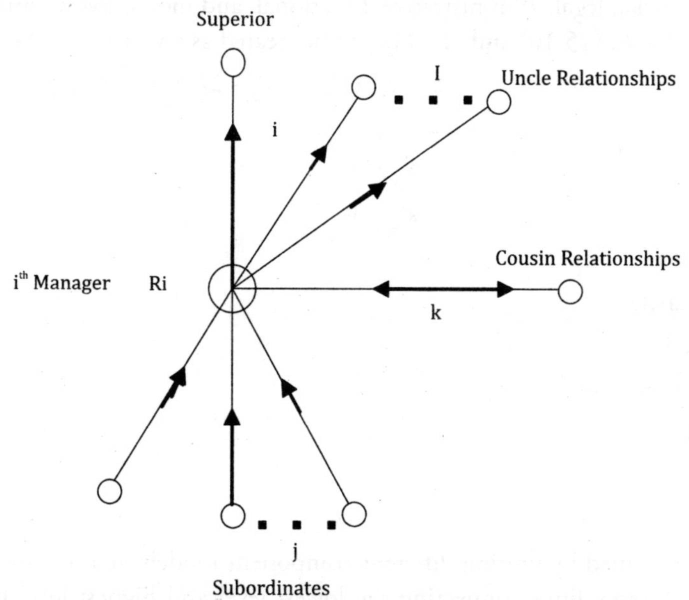

Figure 15.4: Free Body Diagram of ith Manager with Uncle and Cousin Relationships

The authority and responsibility equations are generalized as follows:

$$x_i = \sum_j k_j x_j \pm \sum_k k_k x_k - \sum_l k_l x_l + f_i(y_i) \qquad (15.4)$$

$$y_j = k_j y_i \ \forall_j \qquad (15.5)$$

$$y_k = k_k y_i \quad \forall_k \tag{15.6}$$

$$y_l = k_l y_i \quad \forall_l \tag{15.7}$$

where,

x_i = Unit authority of *i*th manager reported to immediate superior

x_k = Unit authority reported to manager *i* from *k*th peer (cousin)

x_l = Unit authority of *i*th manager reported to *l*th superior (uncle)

y_i = Responsibility of *i*th manager reported to immediate superior

y_k = Responsibility shared with *k*th peer

y_l = Responsibility reported to *l*th superior (uncle)

k_k = Responsibility coefficient of *k*th cousin relationship

k_l = Responsibility coefficient of *l*th uncle relationship

The component model for a pure hierarchical system is a special case of this model with no uncle or cousin relationships.

The component model presented above represents the authority-responsibility relationships in an aggregative form. This can be generalized to separately analyze various kinds of authority and responsibility relationships, e.g. financial, legal, administrative, functional, and moral. Every variable and co-efficient in equations (15.8), (15.9), (15.10) and (15.11) can be treated as a vector, as follows:

$$x^r_i = \sum_j k^r_j x^r_j \pm \sum_k k^r_k x^r_k - \sum_l k^r_l x^r_l + f^r_i (y^r_i) \tag{15.8}$$

$$y'^r_j = k^{r'}_j y_i \quad \forall_j \tag{15.9}$$

$$y^r_k = k^r_k y_i \quad \forall_k \tag{15.10}$$

$$y^r_l = k^r_l y_i \quad \forall_l \tag{15.11}$$

where,

r = 1 for Financial

= 2 for Legal

= 3 for Administrative

= 4 for Functional

= 5 for Moral

System Graph

The system graph is obtained by uniting different component models in a one-to-one correspondence and by introducing dummy links connecting the lowest level and highest level nodes to a reference node. The reference node is introduced to close the system graph and acts as datum node with reference to which all the authorities are measured. The authority across dummy links, connecting first level nodes, may take any value zero, positive or negative. A zero value indicates that there is no lower level existing; a positive value indicates the existence of one or more lower levels, whereas a negative entry indicates a reporting from higher level to a lower level (which is, usually, not in practice). The dummy links also help in making the organizational system compatible, which may not be so in real life.

The system graph is to be divided into tree and cotree. The system graph with '*e*' edges and '*v*' vertices has $(v-1)$ tree branches and $(e-v+1)$ cotree chords.

Continuity Constraints

This is the Generalization of the Kirchhoff's Current Law (GKCL), according to which the sum of through variables for a cut-set of edges is equal to zero. In this case for a particular relationship:

Responsibility of any manager	=	Responsibility of the same manager reported to immediate peer/superior

Compatibility Constraints

This is the Generalization of Kirchhoff's Voltage Law (GKVL), according to which the sum of across variables for a circuit of edges is zero. In this case, for a particular relationship:

Unit authority of any manager	=	Unit authority of the same manager as reported to immediate peer/superior

Responsibility Balance Equations

In case of ideal or total delegation (where the role of superior is of a coordinator), the responsibility of the superior will be equal to the sum of responsibility of all the subordinates, i.e.,

$$y_i = \sum_j y_j \tag{15.12}$$

In this case, the sum of responsibility at each level will be same. But in practice, this is not the case. The responsibility of *i*th superior, i.e. the equation (15.12) can be generalized as

$$y_i = \sum_j y_j \pm r_j \tag{15.13}$$

where r_j is the responsibility gap/surplus with *i*th level.

Simplified System Graph

After assuming strict continuity and compatibility, the simplified system graph can be obtained as discussed in chapter four. The model presented here is based on simplified system graph.

Implications of Coefficients and Constraints in the Model

The proposed model is having two types of coefficients, viz. responsibility coefficient (*k*), and authority-responsibility coefficient (*f*). The responsibility coefficients define the structure of the organizational system in terms of the ratios of responsibilities of subordinates and the manager. This implicitly assumes that in an organizational system, responsibility across two adjoining levels is maintained in a fixed ratio. If the overall responsibilities are increasing or decreasing, these are apportioned to various managers in the similar proportions without affecting their respective responsibility relationships.

The authority-responsibility coefficients (*f*) in the model are able to take care of the pattern of change in authority at a particular level with the change of responsibility.

The model incorporates two types of constraints, viz. continuity constraints and compatibility constraints, which define that the responsibility or authority of any manager is equal to the responsibility or authority reported to immediate superior. This implies that there are no perception gaps between two managers about their authority or responsibility. Such an assumption is true for a rational-legal type of organization, which is taken as a basis for analysis. This may be a somewhat strong assumption in certain cases, and the methodology can be gradually modified to suit the organizations not following the mechanistic form, using the extended framework as discussed in chapter five.

Model Validation

The model considers a rational-legal type of organization as a basis which according to Weber, with its bureaucratic organizational form, is the final stage in depersonalization, and considers organization as sets of officials whose roles are circumscribed by written definitions of their authority. The apportioning of authority and/ or assigning responsibility generated by the model respectively is to be compared with actual allocation of the authorities and/or responsibilities in the organizational system respectively. The difference, if any, between the model results and actual values of authority-responsibility for different managerial positions indicate imbalances in the allocation of authority and/or responsibility to various positions. There may be various causes for such an imbalance. One important factor is the variation in the individual capability and experience of managers in position. Hence, if there is an imbalance between two managerial positions about the allocation of authority/responsibility, the individual capabilities and expertise of the managers in position should be compared. If there is no significant difference between the two managers, then it can be concluded that the imbalance is there in the structure of the organization.

In order to effectively consider these aspects in the model, the model can be expanded to take care of a third variable, i.e. managerial effectiveness. For this purpose the extended framework of physical theory can be applied, in which the managerial effectiveness can be taken as a quality variable.

15.8 DEVELOPMENT OF SOFTWARE

By developing a user-friendly software, it is proposed to make this model more useful to the managers including those who do not possess working knowledge of computer. The proposed model is operationalized by developing an interactive decision support system for organizational analysis named OASIS (Organizational Analysis with Interactive System).[17]

Qualitative Requirements of Software

A requirement analysis was carried out to assess the required qualities of the software to be developed for organizational analysis. They are as follows:

(a) *To be based on microprocessors*: Considering the proliferation of microprocessors in the form of PCs (Personal Computers) the software should be developed based on the facilities available on microprocessors. This will enable even the medium size organizations to use PC and this software for analysis.

(b) *Data base manipulation facilities*: The strength of the software to be developed is in its ability to create, view, modify, edit the database that is used for the analysis. This will facilitate the analyst in carrying out the cause-effect analysis of changes that he would like to incorporate. Popularly known as 'what-if" analysis, on-line, gives added advantage to the analyst, to quickly see the changes that take place with respect to changes in the input data. Such a facility is possible only in database packages, and electronic spreadsheets, in the PC environment.

(c) *Interactive facility*: It has been appreciated that the organization analyst may be from any discipline of science or engineering and hence, he should face no problem in using software based and developed on Personal Computer. The programs should be user friendly and proceed logically, taking inputs through interactive messages and computer prompting. In addition, some of the software packages available on PCs lend themselves to the development of user-friendly and interactive programs. In short, the interactive facility offers the analyst a tool to carry out on-line analysis of databases.

Overview of Logic

In any interactive software, the user must be taken through logical steps for his analysis to use the software. The basic four steps that are involved in any analysis of database systems are:

(a) Creation of data bases

(b) Validation of data bases

(c) Edit / modification of data bases

Generation of Output/Analysis

Menu Assistance

To obtain a higher degree of user friendliness, the software package should take the analyst through logical steps, with each step offering all possible operations or processes. This is generally achieved through user friendly 'MENU'. The 'MENU' is a dialogue between the software system and the system analyst. This arrangement is more effective under the circumstances where the analyst is not very conversant with package or the software. Over a period of time, he may develop adequate expertise to handle the package.

Creation of Databases

Task Role Matrix

The first database to be created is the task role matrix. To create the task role matrix the following procedure is adopted:

(a) All the tasks are broken down to possible subtasks or sub-sub tasks and are numbered.

(b) With respect to each member, each sub-sub task is assessed for his responsibility (towards voting process), by taking the opinion of the member, his boss, and his subordinate (where applicable). This data is analyzed based on a logic that if all of them agree on the responsibility of a sub-sub task the member is assigned to that responsibility.

The discrepancies are resolved, in case where the boss's view differs from the member's view. Thus, an intermediate file called "Responsibility Determination File" is created.

(c) From this file, task role matrix is created giving the member-wise responsibility units.

Authority Role Matrix

The creation of the authority role matrix is similar to the task role matrix. The following steps are followed in creation of the database:

(a) All the tasks that are broken down to sub-sub tasks are considered for pre-emptively by each member.

(b) With respect to each sub unit of task, the opinion of the member, his boss and his subordinate is taken. Where all of them concur that the authority to preempt is that of the concerned member, he is assigned the authority. The task where discrepancies arise are resolved and appropriately the authorities are assigned to the member.

(c) Thus, an intermediate file giving the authority determination file is created.

(d) From this file an authority role matrix is created giving the details of participant / memberwise authority units.

Unit Authority

After creating the databases for responsibility (task role matrix) and authority (authority role matrix), a third data base is created called "unit authority matrix". This data base file indicates the memberwise responsibility, authority and the unit authority of each member.

Organization Chart

The next important database is the organization chart, which indicates the hierarchical levels existing in an organization. The data can be captured for organization chart and can be filled on-line. The database indicates, with respect to each member, the level, of the members who report to him and the total number of such members. From this database, the processing for generation of equations starts. The database before its use for generation of equations correct its level and creates dummy levels at the top (where necessary) and at the bottom.

Validation of Databases

It is very essential that all the data bases created are validated before they are used for further processing.

Task Role Matrix

The validation of the task role matrix involves the following checks on the database:

(a) Maximum responsibility units will be given on screen along with the member who has maximum responsibility and responsibility units.

(b) Tasks which have not been assigned to any member.

(c) Tasks and the number of members who have been assigned this task.

Authority Role Matrix

The validation of the authority role matrix has the following checks:

(a) The maximum authority assigned to a member and the minimum authority assigned to another member.

(b) Activity which has not been assigned to any member.

(c) Activities which have more than one/two members who have authority to preempt.

Organization Chart

In case of organization chart the validations built-in are as follows:

(a) No member can report to himself.

(b) No member can report to two persons.

(c) If top most level has more than one member, it will automatically generate a dummy member to whom others will report.

Generation of Equations and Solving of Equations

There are two types of equations that are generated based on the information contained in the databases. These are unit authority equations and responsibility equations. The responsibility equations generate the responsibility quotient of each member, given the total responsibility. Similarly, the unit authority equations generate the authority quotient of each member, given the individual's inherent additional authority function ($f_0 = c_0 - m_0 y_0$). The continuity and compatibility constraints are inbuilt in the generation of the responsibility and unit authority equations.

Responsibility Equations

For generating the responsibility (flow) equations the programme uses database of organization chart and the responsibility chart. The following logical steps are followed to generate flow equations:

(i) Organization chart database is invoked.

(ii) Member record is ready. (Say Member 1)

(iii) Each member's record is scanned to see who are the members who report to this member, i.e. members who report to Member1.

(iv) With respect to each member who reports to the member (i.e., Member1) a flow equation is generated. For example, if Member 2, and Member 3 report to Member 1 then two equations will be generated. Each equation will be stored as a record. In this example, the following equations will be generated:

$$y_2 = k_2 y_1,$$
$$y_3 = k_3 y_1.$$

(a) Thus, an intermediate file will be created, having a record for each of the equations. The fields of the intermediate files are Left side Y (LY), Right side Y, KVALUE (KVAL) and YVALUE (YVAL).

(b) This file once created, will be used to compute the K Value.

(c) The programme prompts (asks) for the initial value of Y i.e. total responsibility value. In this example, the initial value of Y will be given.

(d) Subsequently, the programme takes each equation and solves them progressively. The left side Y value is obtained from responsibility database of the corresponding member. The right side Y value is either given on line or has been computed based on the already supplied value. Based on Y VALUE of left and right side K VALUE is computed.

(e) The programme lists the equations and gives the Y value of each member.

Unit Authority Equations

The unit authority equations are generated by the programme. It makes use of the database created for organization chart and the K values computed by the flow equations. In addition, the programme also makes use of the database which gives individual members authority quotient. The following logical steps are followed in the generation of unit authority equation (x equation):

(a) The organization chart database is invoked.

(b) Each member's record is scanned (*i.e.* Member 1 record is under scan).

(c) All the members who report to the current record are identified. For example, Member 2 and Member3 report to Member 1.

(d) With respect to current member the equation generated is $x_1 = k_2x_2 + k_3x_3 + f_1$

(e) The parameters of this equation are captured in intermediate data base file.

(f) Once all the equations are generated and stored, the intermediate file is opened and scanned from bottom.

(g) The initial values of authority of dummy members will be equal to F values which are automatically obtained from the data base file FVALUE.

(h) The K values are obtained from flow equation database.

(i) The right side X value is obtained by scanning the current data base file and the left side X value is computed.

(j) The equations are printed.

(k) The X values are computed and printed.

15.9 ILLUSTRATIVE EXAMPLE

An organizational system is represented by hierarchy of managers placed at different levels, having various degree of authority and responsibility. Every manager in the organization has been conceptualized as a multi-terminal component interacting with some component at the higher, as well as lower levels of hierarchy. Each interaction has been considered as a two terminal component. The unit authority across two levels is different and hence taken as an across variable '*x*'; The responsibility which flows through the organization is considered as a through variable '*y*'. This modelling technique can be applied at any level in the organization.

The proposed model is illustrated by applying it to a illustrative organizational system with twelve managerial positions at three hierarchical levels, as shown in Figure 15.5, and thirteen activities. The input data and results for this example are shown in Appendix 15 A. The simplified system graph of the organizational model is shown in Figure 15.6.

If measures of actual authority are known, the calculated values can be compared with them to assess the variation in practice from the parity of authority and responsibility. If it is a new organization to be designed the values of authority derived from the model will act as a guideline to distribute the rights among various managerial positions.

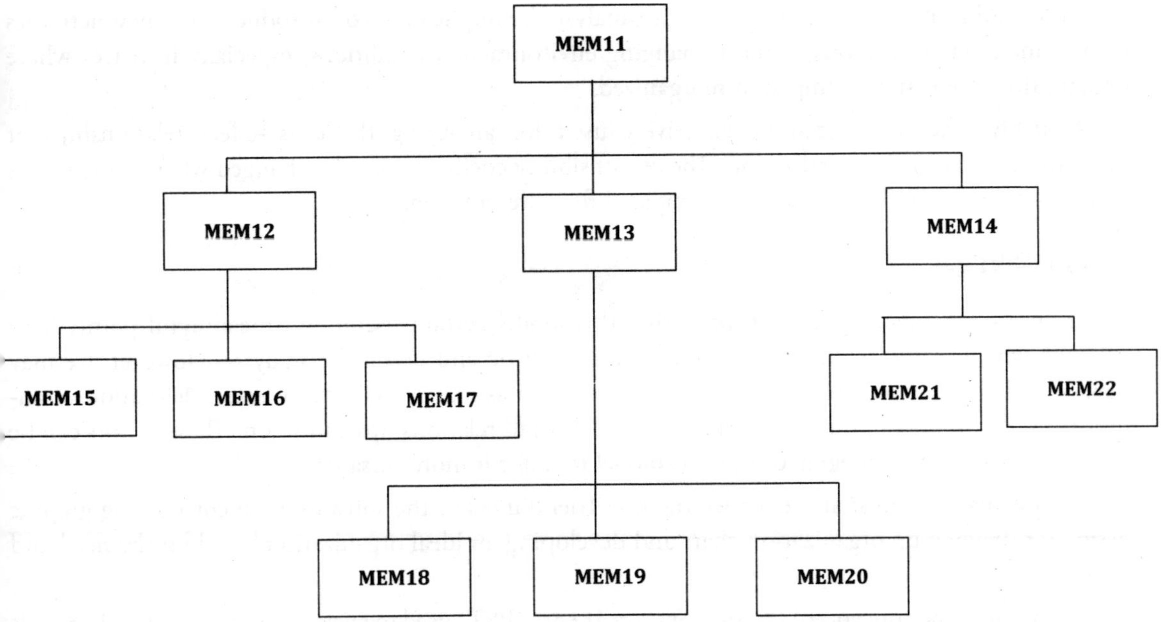

Figure 15.5: Organization Chart for the Illustrative Example

Applicability of Software

Firstly, the concepts of PST can effectively be used for analysis of authority-responsibility relationships. The analysis and measurement of responsibility alone will suffice to generate the authority, so that the disparities can be identified.

Secondly, this software package which is user-friendly and interactive can be used for 'what-if' analysis; to see the changes of responsibilities at any managerial position and the quantum of authority required to balance the structure.

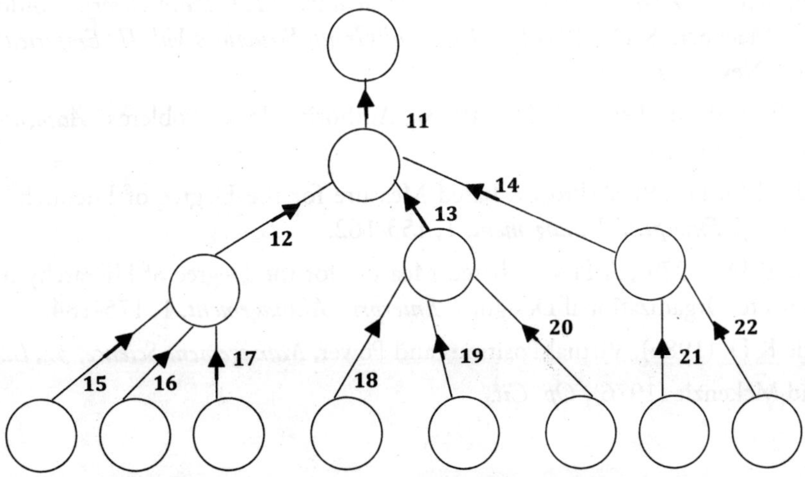

Figure 15.6: Simplified System Graph of the Illustrative Organization

Thirdly, this interactive package will also analyze the implications of introduction of new activities or new managerial positions to meet changing environmental conditions, especially in places where organizations are being revamped or reorganized.

Fourthly, The model can be effectively used for analyzing the cause-effect relationships of centralization versus decentralization. The responsibility coefficient can be changed whereas the values of authority and responsibility can be compared for different scenarios.

REFLECTIONS

Based on the experience gained in preparing this model certain useful and meaningful points have emerged which would make this technique more realistic and useful for analysts. However, we may have included and integrated certain other factors such as managerial effectiveness, delegation of authority, power in organizations, virtual position; line-staff relationships and so on. These factors can be taken one by one, and integrated into this model to make it more versatile.

Future analysts, must also enhance the user friendliness of the software by incorporating graphic design for drawing up organization chart and developing an ideal organization based on the need and norms.

In addition, the concepts of Physical System Theory (PST) used to enrich the organizational analysis may be further strengthened with other qualitative techniques for data capture and data base creation. For example, use of 'Fuzzy-Sets' for analysis will facilitate the captured data regarding authority and responsibility.

This technique can further be explored for adoption in the analysis of Project Management environment for review and allocation of managerial resources based on intensiveness of activities.

Notes

1. Weber M. (1947), *The Theory of Social and Economic Organization*, Oxford University Press, New York.
2. Mackenzie K.D. (1976), *A Theory of Group Structures, Vol. I: Basic Theory*, Gordon and Breach, New York; Mekenzie K.D. (1976) *A Theory of Group Structures Vol. II: Empirical Tests*, Gordon and Breach, New York.
3. Lippit M.E. and Mackenzie K.D. (1976), Authority Task Problems, *Administrative Science Quarterly*, 21, 643-660.
4. Mackenzie K.D. (1979), A Process based Measure for the Degree of Hierarchy in a Group I, The Measure, *J. Enterprise Management*, 1, 153-162.
5. Mackenzie K.D. (1979), A Process based Measure for the Degree of Hierarchy in a Group, III, Applications to Organizational Design, *J. Enterprise Management*, 1, 175-184.
6. Mackenzie K.D. (1986), Virtual Positions and Power, *Management Science*, 32, 622-642.
7. Lippit and Mckenzie (1976), *Op. Cit.*

8. Sushil (1990), Modelling and Analysis of Authority-Responsibility Relationships in an Organizational System–A Physical Systems Theory Approach, *Systems Research,* 7; 255-271.
9. Pfeffer J. (1978), *Organizational Design*, AHM Publishing Corporation, Arlington Heights.
10. Kilmann R.H. (1977), *Social System Design: Normative Theory and the MAPS Design Technology,* Elsevier, New York.
11. Nystrom P.C. and Starbuck W.H. (eds.) (1981), *Handbook of Organizational Design: I and II,* Oxford University Press, New York.
12. Mackenzie K.D. (1979), *Op. Cit.*
13. MeKenzie (1976) *Op. cit.*
14. Lippit and McKenzie (1976) *Op. cit.*
15. Sushil (1990) *Op. cit.*
16. *Ibid.*
17. Sushil and Raghunathan B. (1994) Interactive Decision Support System for Organizational Analysis, *Decision Support System,* 11, 283-298.

APPENDIX 15A: Parameters and Results of Illustrative Example

(a) Authority Role Matrix

Manager	ACT11	ACT12	ACT13	ACT14	ACT15	ACT16	ACT17	ACT18	ACT19	ACT20	ACT21	ACT22	ACT23	Total Author-ity
MEM11	1	1	1	1	1	1	0	0	1	0	0	1	0	8
MEM12	1	0	1	1	0	0	0	0	0	0	0	0	0	3
MEM13	0	0	0	0	1	1	1	1	0	0	0	0	0	4
MEM14	0	0	0	0	0	0	0	0	1	1	0	1	1	4
MEM15	0	1	0	0	0	0	0	0	0	0	0	0	0	1
MEM16	1	0	0	0	0	0	0	0	0	0	0	0	0	1
MEM17	0	0	0	1	0	0	0	0	0	0	0	0	0	1
MEM18	0	0	0	0	1	0	0	0	0	0	1	0	0	2
MEM19	0	0	0	0	0	1	0	0	0	0	0	0	0	1
MEM20	0	0	0	0	0	0	1	1	0	0	0	0	0	2
MEM21	0	0	0	0	0	0	0	0	1	1	0	0	0	2
MEM22	0	0	0	0	0	0	0	0	0	0	1	1	1	3

(b) Task Role Matrix

Manager	ACT11	ACT12	ACT13	ACT14	ACT15	ACT16	ACT17	ACT18	ACT19	ACT20	ACT21	ACT22	ACT23	Total Responsibility
MEM11	1	1	1	1	1	1	1	1	1	1	1	1	1	13
MEM12	1	1	1	1	0	0	0	0	0	0	0	0	0	4
MEM13	0	0	0	1	1	1	1	1	0	0	0	0	0	5
MEM14	0	0	0	0	0	0	0	0	0	1	1	1	1	4
MEM15	1	0	0	0	0	0	0	0	0	0	0	0	0	1
MEM16	0	1	1	0	0	0	0	0	0	0	0	0	0	2
MEM17	0	0	0	1	0	0	0	0	0	0	0	0	0	1
MEM18	0	0	0	0	0	1	0	0	0	0	0	0	0	1
MEM19	0	1	0	0	1	0	0	0	0	0	0	0	0	2
MEM20	0	0	0	0	0	0	1	1	0	0	0	0	0	2
MEM21	0	0	0	0	0	0	0	0	1	1	0	0	0	2
MEM22	0	0	0	0	0	0	0	0	0	0	1	1	1	3

(c) Organization Chart

Manager	MEM11	MEM 12	MEM 13	MEM 14	MEM 15	MEM 16	MEM 17	MEM 18	MEM 19	MEM 20	MEM 21	MEM 22	Level	Subordinates
MEM11	0	1	1	1	0	0	0	0	0	0	0	0	1	3
MEM12	0	0	0	0	1	1	1	0	0	0	0	0	2	3
MEM13	0	0	0	0	0	0	0	1	1	1	0	0	2	3
MEM14	0	0	0	0	0	0	0	0	0	0	1	1	2	2
MEM15	0	0	0	0	0	0	0	0	0	0	0	0	3	0
MEM16	0	0	0	0	0	0	0	0	0	0	0	0	3	0
MEM17	0	0	0	0	0	0	0	0	0	0	0	0	3	0
MEM18	0	0	0	0	0	0	0	0	0	0	0	0	3	0
MEM19	0	0	0	0	0	0	0	0	0	0	0	0	3	0
MEM20	0	0	0	0	0	0	0	0	0	0	0	0	3	0
MEM21	0	0	0	0	0	0	0	0	0	0	0	0	3	0
MEM22	0	0	0	0	0	0	0	0	0	0	0	0	3	0

(d) Responsibility Coefficients

Manager	Reporting to	Resp. Coeff.	Value
MEM12	MEM11	K12	0.31
MEM13	MEM11	K13	0.38
MEM14	MEM11	K14	0.31
MEM15	MEM12	K15	0.25
MEM16	MEM12	K16	0.50
MEM17	MEM12	K17	0.25
MEM18	MEM13	K18	0.20
MEM19	MEM13	K19	0.40
MEM20	MEM13	K20	0.40
MEM21	MEM14	K21	0.50
MEM22	MEM14	K22	0.75

(e) Values of Authority-Responsibility Functions

Manager	CVAL	MVAL	RESP	FVAL
MEM11	1.500	0.005	100.00	1.000
MEM12	0.850	0.007	31.00	0.633
MEM13	0.800	0.009	38.00	0.458
MEM14	0.925	0.012	31.00	0.553
MEM15	0.030	0.000	7.75	0.030
MEM16	0.030	0.000	15.50	0.030
MEM17	0.032	0.000	7.75	0.032
MEM18	0.030	0.000	7.60	0.030
MEM19	0.025	0.000	15.20	0.025
MEM20	0.030	0.000	15.20	0.030
MEM21	0.028	0.000	15.50	0.028
MEM22	0.025	0.000	23.25	0.025

(f) Responsibility Values

Manager	Variable	Responsibility
MEM11	Y11	200.00
MEM12	Y12	62.00
MEM13	Y13	76.00
MEM14	Y14	62.00
MEM15	Y15	15.50
MEM16	Y16	31.00
MEM17	Y17	15.50
MEM18	Y18	15.20
MEM19	Y19	30.40
MEM20	Y20	30.40
MEM21	Y21	31.00
MEM22	Y22	46.50

(g) Unit Authority of Managers

Manager	Variable	Unit Authority
MEM11	X11	0.759
MEM12	X 12	0.446
MEM13	X 13	0.144
MEM14	X 14	0.214
MEM15	X 15	0.030
MEM16	X 16	0.030
MEM17	X 17	0.032
MEM18	X 18	0.030
MEM19	X 19	0.025
MEM20	X 20	0.030
MEM21	X 21	0.028
MEM22	X 22	0.025

(h) Comparison Table

MGR	Total Auth (Model)		Total Auth (Actual)	
	Auth (Model)	Auth Normalized	Auth. (Actual)	Auth. Normalized
MEM11	152.00	100.00	8.00	100.00
MEM12	47.12	31.00	3.00	37.50
MEM13	57.76	38.00	4.00	50.00
MEM14	47.12	31.00	4.00	50.09
MEM15	11.78	7.75	1.00	12.50
MEM16	23.56	15.50	1.00	12.50
MEM17	11.78	7.75	1.00	12.50
MEM18	11.55	7.60	2.00	25.00
MEM19	23.10	15.20	1.00	12.50
MEM20	23.10	15.20	2.00	25.00
MEM21	23.56	15.20	2.00	25.00
MEM22	35.34	23.25	3.00	37.50

Auth–Authority
Resp–Responsibility

16 ANALYSIS OF FLOW, COST AND QUALITY IN PROJECT MANAGEMENT

16.1 INTRODUCTION

Network planning is an important aspect of project management. It is, usually, carried out to analyze the time and cost of various activities. Another important variable is the flow of resources for various activities. Each activity can be considered as a transformation process that takes input of various resources and gives output is terms of the progress of the project. Thus, the activity on node diagram can be converted into system graph for analysis of flow and cost variables.

A significant variable that has not received adequate attention in project planning is "quality". This can be incorporated in the planning by using the extended framework of physical system theory as given in chapter five. The system of the project network can be effectively used for quality planning. It can be used to convert the target output project quality in terms of the required quality of various inputs. Also, it can be used to estimate the output quality keeping in view the quality of available inputs. Further, it can relate quality with the cost and can help in decision-making about the required quality of project within the available budget.

The chapter illustrates an application of the extended framework in the context of a highway construction project and reports the model development, parameter estimation, and quality, cost and sensitivity analysis. The flexible system methodology for project management is shown in Figure 16.1.

16.2 PROJECT SITUATION

For applying the extended framework of physical system theory, the case of a road project has been selected. The project consists of widening existing single lane pavement and simultaneously strengthening the pavement to meet the demand of traffic. The existing road will have to be realigned at a number of locations in order to improve the geometrics and avoid congested towns and villages.

Prior to any execution of a project, sound planning is a must. Highway planning is also a basic need for highway development. The planning phase includes

(a) Economic studies

(b) Financial studies

(c) Traffic studies

(d) Engineering studies

In case, these studies establish the feasibility of a road development, detailed design work follows.

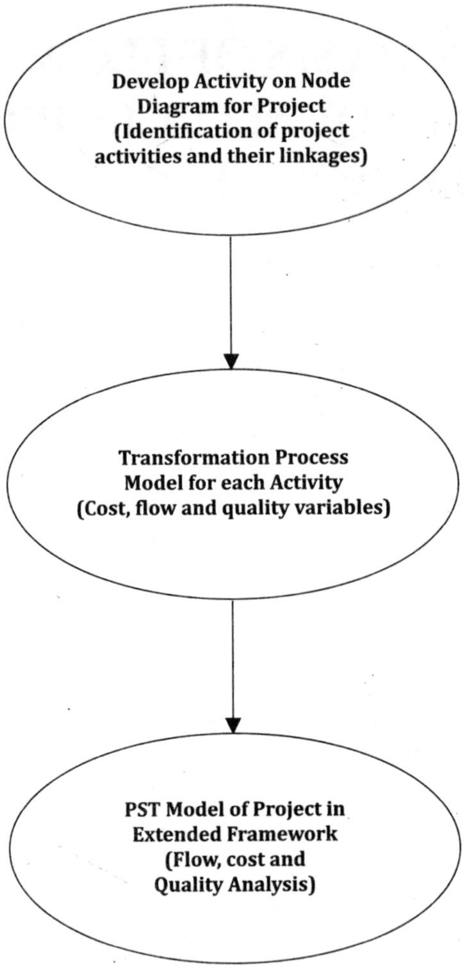

Figure 16.1: Flexible System Methodology for Project Management

In highway construction, usually 'stage construction' technique is adopted. This means improving the pavement structure in stages to match with increase in road traffic. This becomes all the more imperative for developing countries like India, because of their financial handicaps with regard to initial investment or the high type of pavement construction like cement concrete.

The highway construction project may be broadly divided in two phases:

i. Earthwork and preparation of subgrade

ii. Pavement structure

The earthwork mainly consists of preparation of the subgrade to be suitable to receive the subsequent construction of pavement structure.

The major construction activities are as follows:

- Earthwork

- Excavation
- Embankment Construction
- Preparation of Sub Grade
- Soil Compaction
- Laying Sub-base
- Laying Base Course
- Bituminous Course
- Miscellaneous

16.3 ACTIVITIES IN THE PROJECT SITUATION

In the project situation, the total length of road has been divided into four convenient sections. It is assumed that the work for all the sections can proceed independently. Since all the sections involve similar activities, it is assumed that it will be sufficient to model and analyze one section only. Other sections behave in the similar manner.

The project work has been broken down into various activities. These activities have been used to construct a network same as that used in scheduling. The network consists of 56 activities, select ones are listed in Appendix 16A. The network diagram is converted into precedence diagram or Activity on Node diagram (Figure 16.2) which can directly be used for the development of system graph.

16.4 DEVELOPMENT OF SYSTEM MODEL

Assumptions

While building up the model, assumptions have been made regarding various characteristics of the system. The assumptions made have been presented as follows:

- The model has been developed considering only the material wastes.
- With the variation in the quality indices of various inputs, their cost is assumed to vary linearly but not less than the given minimum permissible rates.
- Costs of various inputs are assumed to be standard cost.
- The processing cost per unit is assumed to be constant, i.e. economies of scale have not been considered.
- The technology employed at processing units does not change, i.e. technological coefficients do not change during this exercise.
- The processing cost does not vary with variation in quality index of output
- The flow rates assumed and quality indices developed are deterministic in nature.
- Only transformation processes have been considered.

Development of Component Models

Different components of the system under consideration are conceptualized to be involved in transformation process.

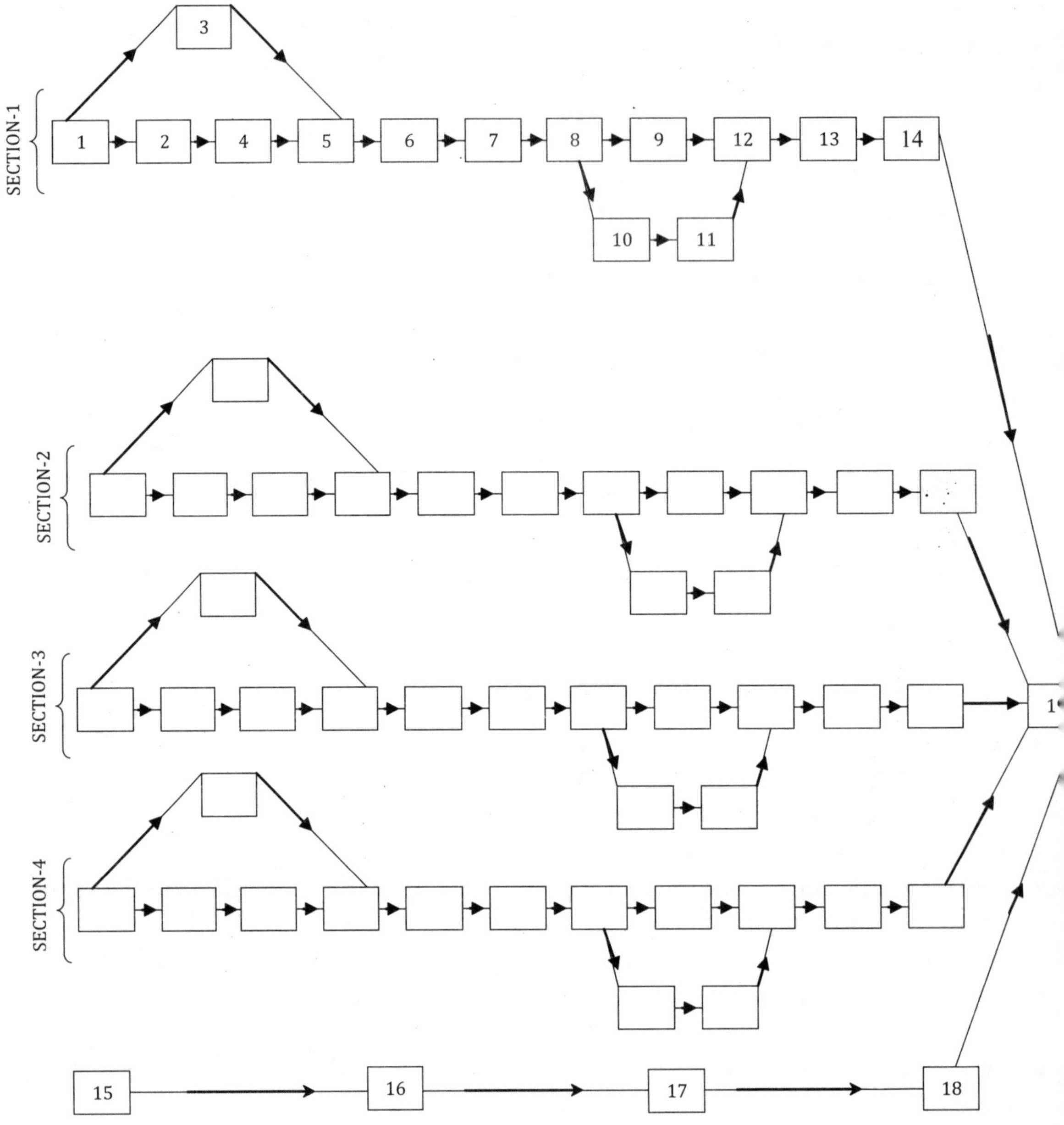

Figure 16.2: Precedence Diagram for a Typical Highway Widening and Improvement Project

(*Activity Nos. are explained in Appendix 16 A*)

Every process in the system transforms input into output and generates some waste in some cases. There are a large number of exogenous inputs to the system. But most of these can be classified in either of the following categories.

(1) Land

(2) Labour

(3) Non Bituminous material

(4) Bituminous material

(5) Manual equipment

(6) Mechanical equipment

Every process generates an output in the form of completed activity, and an exogenous waste which may or may not have salvage value.

The component models for all the sub-systems have been developed in the light of precedence diagram prepared for the system. Various component models have been presented in Figure 16.3. While developing component models, and writing system equations and quality equations, the following notation have been used:

w_{1n} = Waste from *i*th processing unit to nature

d_i = Land as input to *i*th processing unit

l_i = Labour as input to *i*th processing unit

e_I' = Manual equipment as input to *i*th processing unit

e_i = Mechanical equipment as input to *i*th processing unit

m^a_i = Non-bituminous material as input to *i*th processing unit

m^b_i = Bituminous material as input to *i*th processing unit

ij = Output from ith processing unit and input to *i*th processing unit

z'_{di} = quality index of land as input in *i*th Processing unit

z_{li} = Quality index of Labour as input to *i*th Processing

z_{ei} = Quality index of Manual equipment as input *i*th processing unit

z_{ei}' = Quality index of Mechanical equipment as input to *i*th processing unit

$z_{m\,i}^{a}$ = Quality index of type A material as input to *i*th processing unit

$z_{m\,i}^{b}$ = Quality index of type B material as input to *i*th processing unit

z_{ij} = Quality index of output from ith Processing Unit and input to *j*th processing unit.

q_{di} = Quality processing coefficient per land at *i*th processing unit

q_{li} = Quality processing coefficient per labour at *i*th processing unit

a_{ij} = Modified quality processing coefficient associated with *ij*

q_{ei} = Quality processing coefficient for manual equipment at *i*th processing unit

$q_{e\,i}'$ = Quality processing coefficient for mechanical equipment at *i*th processing unit

$q_{m\,i}^{a}$ = Quality processing coefficient for material of type A at *i*th processing unit.

qm^b_i = Quality processing coefficient for material of type B at *i*th processing unit

q_{ij} = Quality processing coefficient for output from *i*th processing unit at *j*th processing unit

t_{ij} = Quality transfer coefficient for output of ith processing unit as perceived at *j*th processing unit.

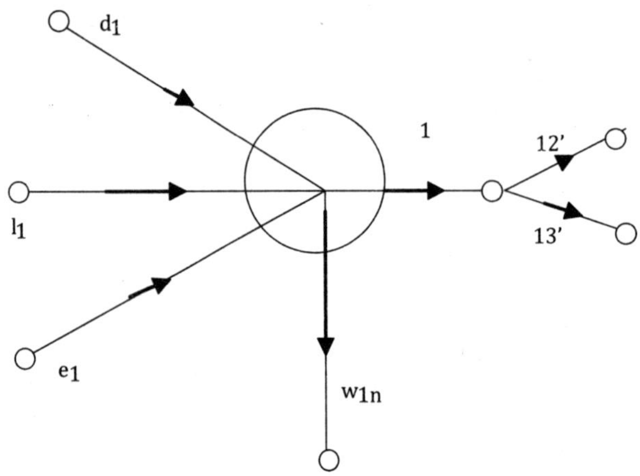

Activity No. 1

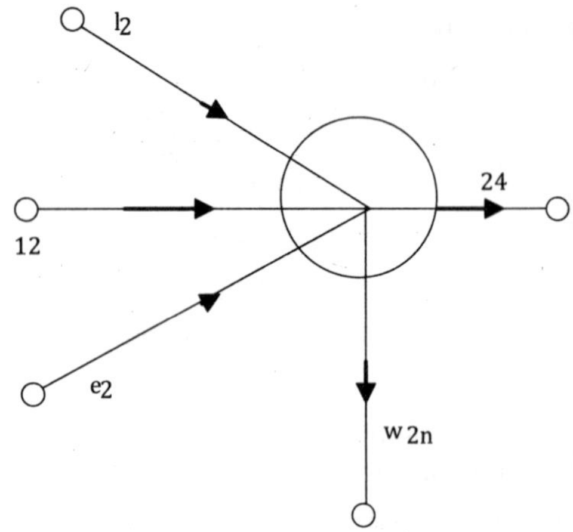

Activity No. 2

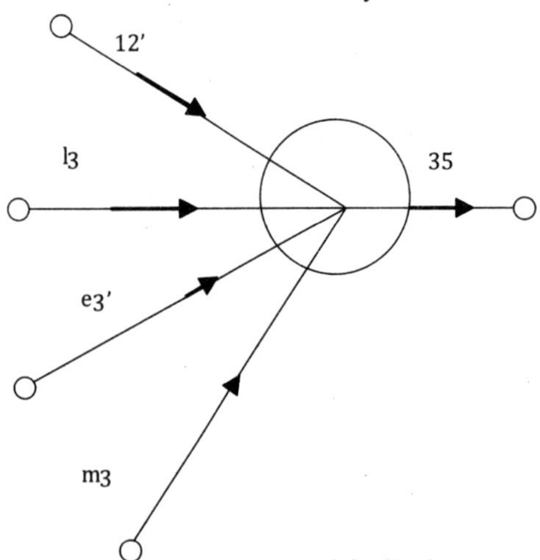

Activity No. 3

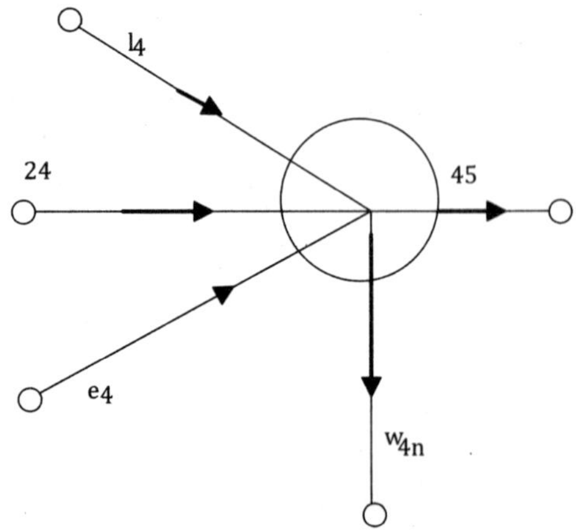

Activity No. 4

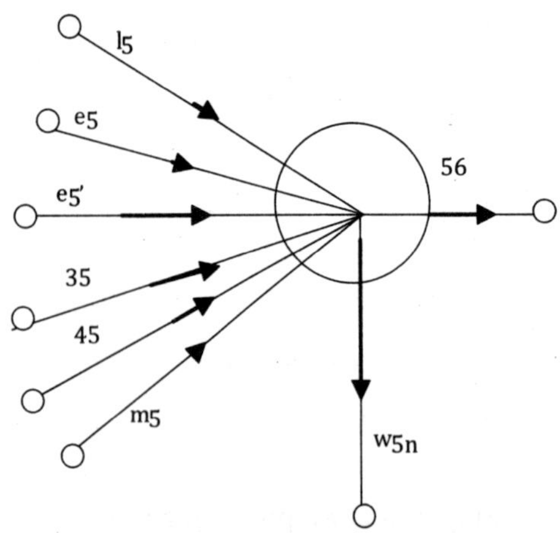

Activity No. 5

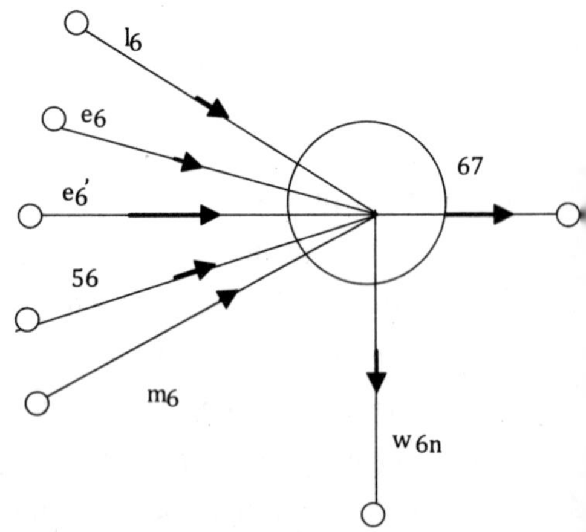

Activity No. 6

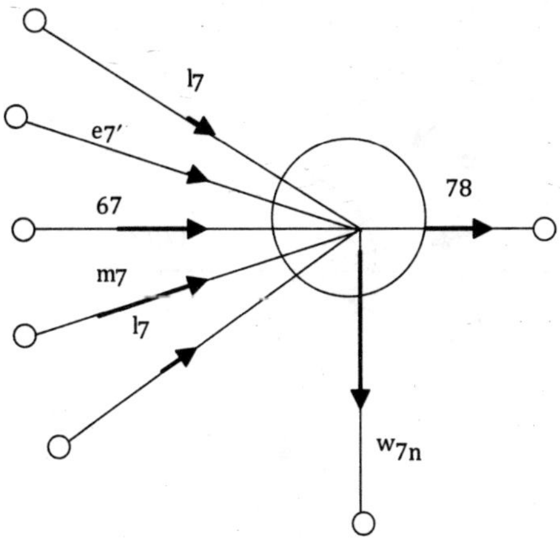

Activity No.7

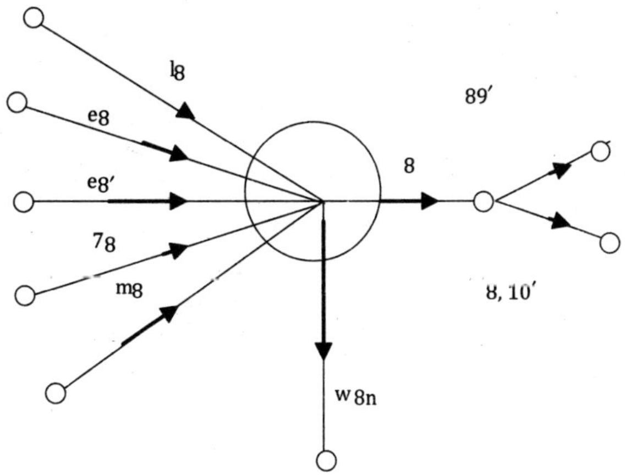

Activity No.8

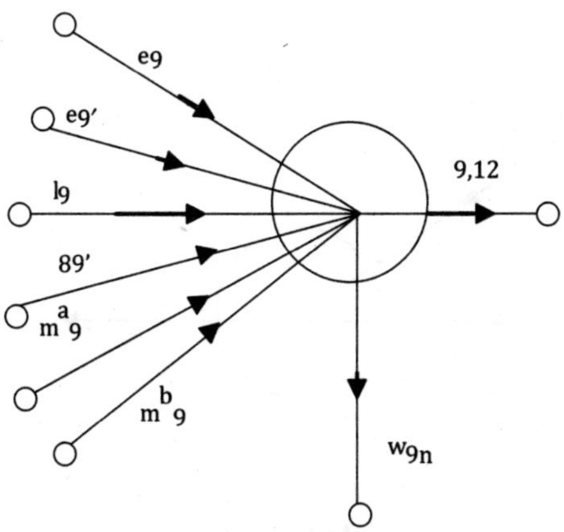

Activity No.9

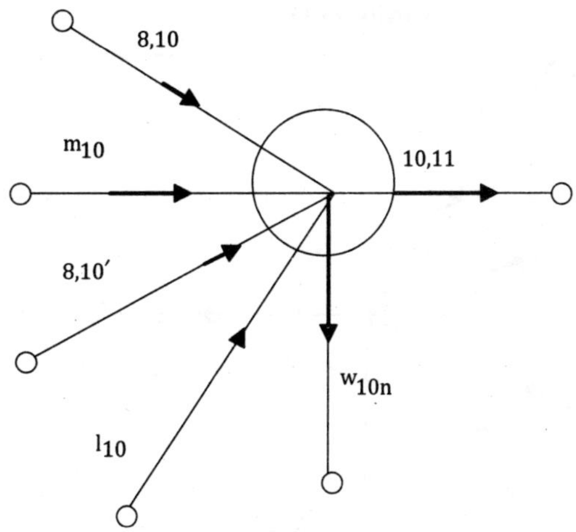

Activity No.10

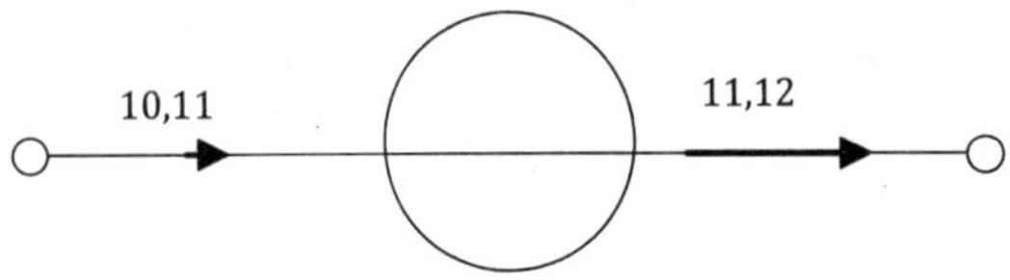

Activity No.11

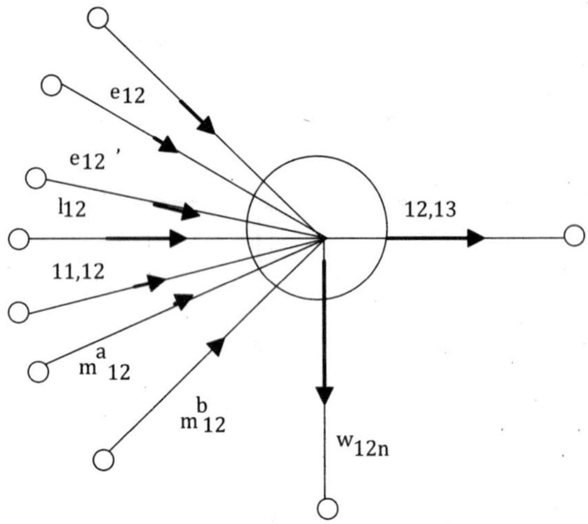

Activity No.12

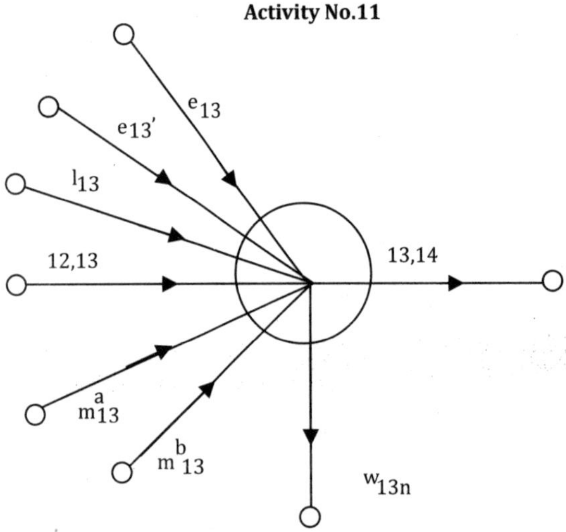

Activity No.13

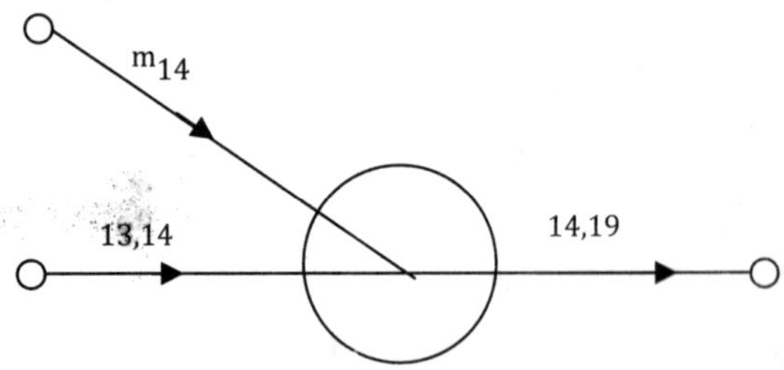

Activity No.14

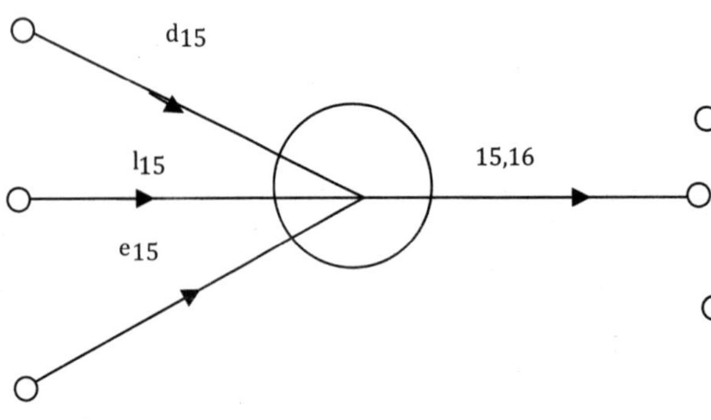

Activity No.15

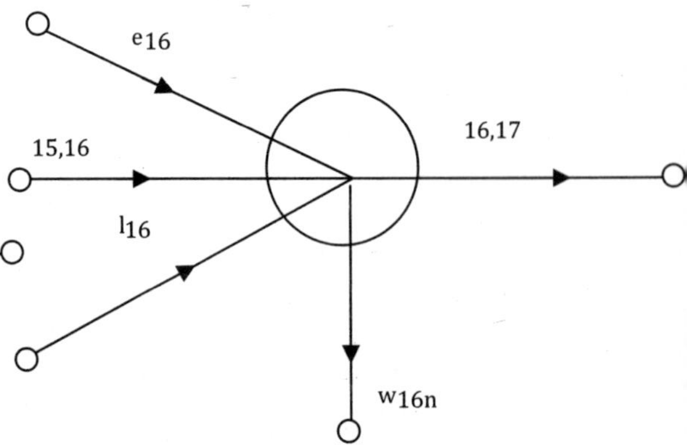

Activity No.16

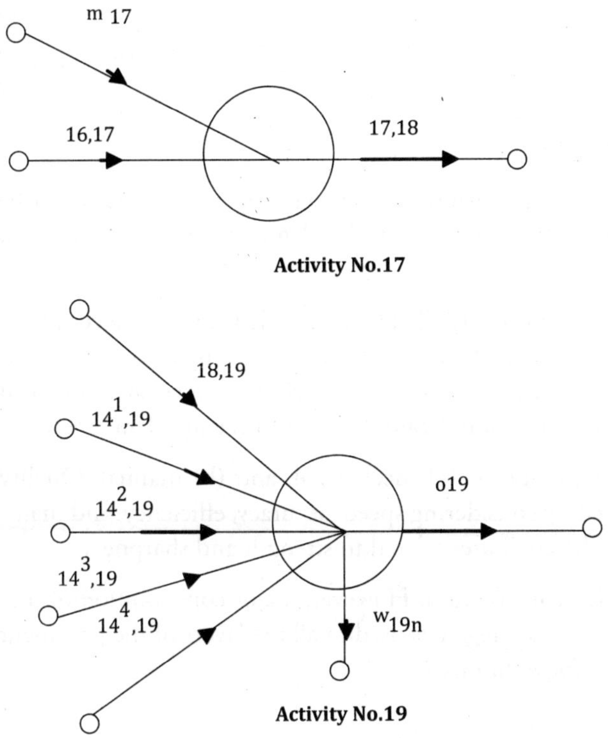

Figure 16.3: Component Models of Various Activities

Associated system equations are written as per the model of transformation process given in chapter five by substituting the appropriate notation. These equations being similar in nature are not reproduced here.

Simplified System Graph

As per systems postulate, the system graph is formed by uniting component terminal graphs in one to one correspondence to the actual interfacing of components.

Simplified system graph has been prepared. It is to be noted that since the behaviour of all the four sections are assumed to be alike, only one section has been completely enumerated. All the analysis in the later sections has been carried out for section one.

Estimation of System Parameters

Before conducting any experiment on the model, various system parameters have to be estimated. These parameters are:

- Technological coefficients
- Quality processing coefficients
- Quality transfer coefficients

Methodology

The first step is identification and classification of inputs. In a construction project, most of the inputs can be broadly classified into the following categories:

i. Land
ii. Labour
iii. Equipments
iv. Materials

Land: Quality attributes of land can be related to type of terrain and soil properties. Terrain can be plain, rolling or mountainous. Various soil properties to be considered may be moisture content, C.B.R value, plasticity index etc.

Labour: Normally labour input can be unskilled labour, semiskilled labour, skilled labour or a combination of these. Requirement of these depends on the type of work to be done. Quality attributes for labour have been subjectively selected as diligence and efficiency. Again, the weights to these attributes can be assigned on some subjective basis. In the project situation, labour referred to is equivalent labour.

Equipment: Construction equipment can be of two types: (a) mechanical, and (b) manual. Quality index of the mechanical equipment can be arrived by considering speed, accuracy, efficiency, and maintenance as attributes. Manual equipments have the attributes related to strength and sharpness.

Materials: A large variety of materials are used for construction. However, major consumption of materials is that of bitumen and stone aggregates. Stone aggregate is used in all the layers of the pavement and bitumen is used in upper layers of crust, i.e. above sub-base.

Various attributes for stone aggregates can be:

- Gradation
- Plasticity index
- Moisture content
- C.B.R. value
- Abrasion value
- Impact value
- Flakiness index, etc.

Whereas, for bitumen, its physical and chemical properties are important.

The second step is to assign quality rating to the various attributes. For our purpose, a 10 point scale has been adopted.

Scale	Quality rating
10	Perfect
9	Very good
7	Good
5	Marginal
3	Poor
1	Exposed

The third step is to arrive at quality index of inputs on the basis of quality rating of attributes and the associated weights. In the fourth step, quality processing coefficients and quality transfer coefficients can be estimated by substituting the quality indices of inputs and outputs in quality equations for various processing units.

To estimate the technological coefficients, data related to flow rate is required.

Data Collection

Data on quality costs was hardly available, for few analysts have collected information on both quality costs and quality performance. Experts in the field were consulted to identify various attributes for inputs and output for all the processing units, and to assign weight to these attributes. Only then quality parameters could be estimated.

To estimate technological coefficients, data on flow rates was required which was not readily available in standard form.

Estimation of Quality Parameters

A summary of Quality Processing Coefficients (QPC); Quality Transfer Coefficients (QTC); and Modified Quality Processing Coefficient (MQPC) is presented in Appendix 16B.

Estimation of Technological Coefficients

For all the processing units, technological coefficients were estimated using the available data from field conditions. By substituting flow rates in material flow equations, technological coefficients were arrived at. These have been presented in Appendix 16 B.

16.5 QUALITY ANALYSIS

Quality Indices of Inputs for Planned Output Quality Indices

An experiment was conducted to analyze the effect of various planned output quality indices on quality index requirement for all the exogenous inputs. In the project situation selected, there are 51 exogenous inputs at various processing units. To study the impact, three cases have been considered.

Case (A): when planned output quality index for the project on a 10 point scale is 9 i.e. $z_{19} = 9.0$

Case (B): when planned output quality index is 7 i.e. $z_{19} = 7.0$

Case (C): when planned output quality index is 5 i.e. $z_{19} = 5.0$

For all the three cases, requirement of quality index for all the 51 exogenous inputs have been worked out using response-stimulus equations and can be seen in Appendix 16 C. From the table it can be interpreted that for a higher planned output quality index, the requirement of input quality is invariably high.

Output Quality Index for Available Input Quality Indices

The experiment was conducted to analyze the impact of change in input quality indices on output quality index. The experiment was so designed that one could know the input category where maintaining higher quality indices with most favorably reflect on output quality.

To study this three cases were as follows:

Case (A): For all the processing units, land inputs, equipment inputs and material inputs were taken to be of good quality and the quality rating for labour was assumed to be very good.

$$i.e.\ z_{di} = 7 \qquad z_{li} = 9$$
$$z_{ei}^{1} = 750 \qquad z_{mi} = 7$$

Case (B): For all the processing units, land labour and material inputs was taken to be of good quality whereas equipments used were of very good quality

$$i.e.\ z_{di} = 7 \qquad z_{li} = 7$$
$$z_{ei} = z'_{ei} = 9 \quad \text{and} \quad z_{mi} = 7$$

Case (C): For all the processing units, land, labour and equipment inputs were taken to be of good quality in contrast with the material inputs of very good quality.

$$i.e.\ z_{di} = 7 \qquad z_{li} = 7$$
$$z_{ei} = z'_{ei} = 7 \quad \text{and} \quad z_{mi} = 9$$

On analyzing the results, it can be interpreted that highest output quality index is achieved for Case (C), i.e. in order to achieve any planned output quality, higher quality indices for material inputs must be emphasized first. It is also seen from the results that variation in output quality index in Case (A) and (B) is almost nil which means that equipment and labour inputs have nearly the same impact on output quality index.

16.6 COST ANALYSIS FOR PLANNED OUTPUT QUALITY INDICES

To study the cost implication of various output quality indices, this experiment was conducted. Here also, three cases were considered. These cases were same as that for experiment discussed in the previous section.

It is to be noted that for some of inputs, minimum permissible unit rates (equivalent to Government approved minimum rates, e.g. in case of labour) have been fixed. It can be seen that for output quality indices of 9.0, 7.0 and 5.0, the unit rates for output have been worked out, at assumed unit rates for inputs, as Rs 270.52/m^2, Rs 209.16/m^2, and Rs 181.39/ m^2 respectively. For another situation with output quality index of 8.49, the output cost is Rs 232.90/ m^2. Since very limited cases have been studied, a complete generalization regarding cost-quality relation can't be made. But it certainly follows that with increase in output quality index, output cost increases at a higher rate with respect to quality, i.e. the additional cost incurred for improving quality index increases with increase in quality index.

16.7 SENSITIVITY ANALYSIS

From the model, useful relationships can be derived and sensitivity analysis be conducted, on the basis of which we can extract quantified information that will help in decision-making. The system equations can be solved to derive the relationships in terms of exogenous variables.

For the project situation, sensitivity analysis of output quality with respect to quality indices of 51 exogenous inputs was done by partially differentiating the quality function w.r.t. input quality variables. The analysis revealed that the output quality index is most sensitive to type A material input at Processing Unit No. 13 (i.e. stone aggregates in surfacing course) and material input at Processing unit No. 14 (i.e. fixing road furniture) having highest values as 0.37 and 0.25 respectively. In general, it was observed that output quality was more sensitive to material inputs. It confirms the validity of result for experiment described in the previous section. Sensitivity analysis of output quality with respect to modified quality processing coefficients associated with 51 exogenous inputs was conduced. It revealed that output quality index was most sensitive to quality processing of labour. The highest value was obtained when output quality was differentiated with respect to labour associated with processing unit No. 13.

REFLECTIONS

The chapter presented a system model of a highway construction project and predominantly presented the quality analysis. This can be suitably integrated with other tools of quality management such as Quality Function Deployment (QFD) to provide quality in various phases of projects as per customer requirements. The crucial area in implementation of the proposed model is the estimation of quality parameters.

APPENDIX 16A

Activities of Highway Construction Project

Activity	Description
1	Clearing and grubbing in Section I
2	Dismantling existing pavement in section 1
3	Construction of temporary service roads for realignment in Section I
4	Dismantling houses in Section I
5	Earthwork inclusive of carriage and compaction in Section I
6	Gravel sub-base in Section I
7	Water Bound Macadam Base course in Section I
8	Water Bound Macadam Top course in Section I
9	Providing levelling course of lean bituminous Macadam in Section I
10	Providing Tack coat in Section I
11	Dummy Activity
12	Providing Bituminous Macadam in Section I
13	Providing Semi-sense carret for Section I
14	Fixing kilometre stones, road signs in Section I
15	Construction of diversion for traffic
16	Dismantling existing culverts
17	Construction of culverts
18	Dummy Activity

APPENDIX 16B

System Parameters

Processing Unit No.	Technological Coefficient		Q.P.C./Q.T.C.		Modified Q.P.C.	
	Parameter	Value	Parameter	Value	Parameter	Value
(1)	(2)	(3)	(4)	(5)	(6)	(7)
1	kd_l	1.00	q_{dl}	0.875	adI	0.742
	kl1	0.27	ql1	1.041	al1	0.168
	ke1	0.27	qe1	1.050	ae1	0.167
	kwin	0.10	t12'	1.000		
			t13'	0.800		
2	k12'	1.00	q12'	0.952	a12'	0.991
	kl2	0.03	ql2	0.992	al2	0.028
	ke2	0.03	qe2	1.000	ae2	0.028
	kw2n	0.10				
3	k13'	1.00	q13	0.771	a13	0.835
	Kl3	0.20	ql3	0.987	al3	0.013
	ke'3	.033	qe'3	0.892	ae3	0.024
	km3	0.500	qm3	1.084	am3	0.297
4	k24	1.00	q24	1.077	a24	0.774
	ke4	0.10	qe4	1.077	al4	0.077
	kl4	0.10	ql4	1.066	ae4	0.078
	kw4n	0.10				
5	k35	0.30	q35	0.965	a23	0.147
	k45	1.00	q45	0.907	a45	0.520
	kl5	0.18	ql5	0.977	ae5	0.087
	k65	0.18	qe5	0.952	al5	0.089
	ke5'	0.01	qe5'	1.000	ae5'	0.005
	km5	0.45	qm5	1.047	am5	0.230
6	ks6	1.00	q56	0.977	a56	0.781
	kl6	0.10	ql6	0.931	al6	0.082
	ke6	0.10	qe6	0.955	ae6	0.105
	ke6'	0.01	qe6'	0.977	ae6'	0.008

	km6	0.10	qm6	0.852	am6	0.090
	kw6n	0.01				
7	k67	1.00	q67	1.00	a67	0.840
	ke7	0.06	qe7	0.955	ae7	0.053
	ke7′	0.02	qe7′	0.977	ae7′	0.017
	kl7	0.06	ql7	0.931	al7	0.054
	km7	0.05	q m7	0.948	a m7	0.044
	kw7n	0.005				
8	k78	1.00	q 78	1.00	a 78	0.840
	ke8	0.06	q e8	0.931	a e8	0.054
	ke8′	0.02	q e8′	0.977	a e8′	0.017
	kl8	0.06	q l8	0.931	a l8	0.054
	km8	0.05	q m8	0.948	a m8	0.044
	kw8n	0.005	t 89′	1.000	t 89′	
			t 8,10′	1.000	t 8,10′	
9	k89	1.00	q 89	1.048	a 89	0.212
	kma9	0.05	qma9	0.976	ama9	0.011
	kmb9	3.30	qmb9	1.071	amb9	0.683
	kl9	0.07	ql9	0.975	al9	0.016
	ke9	0.07	qe9	1.000	ae9	0.016
	ke9′	0.02	qe9′	1.000	ae9′	0.004
	kw9n	0.005	qw9n		aw9n	
10	k8,10′	1.000	q8,10′	1.011	a8,10′	0.831
	kl10	0.020	ql10	0.920	al10	0.018
	ke10	0.020	qe10	1.034	ae10	0.016
	km10	0.150	qm10	1.034	am10	0.122
	kw10n	0.10	qw10n		aw10n	
11	k10,11	1.000	q10,11	1.000	a10,11	1.000
12	k11,12	0.500	q11,12	1.012	a11,12	0.070
	k9,12	0.500	q9,12	0.977	a9,12	0.073
	kl12	0.500	ql12	0.952	al12	0.075
	ke12	0.050	qe12	0.977	ae12	0.007
	ke12′	0.020	qe12′	0.977	ae12′	0.003
	kma12	5.400	qma12	0.953	ama12	0.807

	kmb12	0.050	qmb12	1.047	amb12	0.007
	kw12n	0.005	qw12n		aw12n	
13	k12,13	1.000	q12,13	1.024	a12,13	0.334
	kl13	0.040	ql13	0.975	al13	0.014
	ke13	0.040	qe13	1.000	ae13	0.014
	ke13'	0.020	qe13'	1.000	ae13'	0.007
	kma13	1.800	qma13	0.995	ama13	0.620
	kmb13	0.020	qmb13	1.071	amb13	0.006
	kw13n	0.100	qw13n		aw13n	
14	k13,14	1.000	q13,14	0.996	a13,14	0.502
	km14	1.000	qm14	1.068	am14	0.468
15	km15	0.050	qm15	1.000	am15	0.455
	ke15	0.030	qe15	0.933	ae15	0.292
	kl15	0.030	ql15	0.926	al15	0.295
16	k15,16	1.000	q15,16	1.000	a15,16	0.833
	ke16	0.100	qe16	0.9333	ae16	0.089
	kl16	0.100	ql16	0.926	al16	0.090
	kw16n	0.100	qw16n		aw16n	
17	km17	0.0002	qm17	1.000	am17	0.00019
	k16,17	1.000	q16,17	1.000	a16,17	1.000
18	k17,18	1.000	q17,18	1.000	a17,18	1.000
19	k14119	0.250	q14119	0.993	a14119	0.201
	k14219	0.250	q14219	0.993	a14219	0.201
	k18,19	0.250	q18,19	1.060	a18,19	0.189
	k14319	0.250	q14319	0.993	a14319	0.201
	k14419	0.250	q14419	0.993	a14419	0.201
	kw19n	0.250				

where,

Q.P.C. = Quality Processing Coefficient

Q.T.C. = Quality Transfer Coefficient

APPENDIX 16C

Comparative Statement of Input Quality Indices for Given Output Quality Index

Process Unit No.	Output Quality Variable	Output Quality Index for			Input Quality Variable	Input Quality Index for		
		Case (A)	Case (B)	Case (C)		Case (A)	Case (B)	Case (C)
(1)	(2)	(3)	(4)	(5)	(6)	(7)	(8)	(9)
19.	z_{19}	9.00	7.00	5.00	$z^{1}_{14\ 19}$ $z^{4}_{14\ 19}$ $z_{18,19}$	8.94 8.94 9.54	6.95 6.95 7.42	4.97 4.97 5.30
18.	$z_{18,19}$	9.54	7.42	5.30	$z_{17,18}$	9.54	7.42	5.30
17.	$z_{17,18}$	9.54	7.42	5.30	$z_{16,17}$	9.54	7.42	5.30
16.	$z_{16,17}$	9.54	7.42	5.30	$z_{15,16}$ z_{e16} z_{l16}	9.54 8.90 8.83	7.42 6.92 6.87	5.30 4.94 4.91
15.	$z_{15,16}$	9.54	7.42	5.30	z_{m15} z_{e15} z_{l15}	9.54 8.83 8.90	7.42 6.87 6.92	5.30 4.91 4.94
14.	$z_{14,19}$	8.94	6.95	4.97	$z_{13,14}$ z_{m14}	8.90 9.55	6.92 7.42	4.94 5.31
13.	$z_{13,14}$	8.90	6.92	4.94	$z_{12,13}$ z_{l13} z_{e13} z_{e13}' $z^{a}_{m\ 13}$ $z^{b}_{m\ 13}$	9.11 8.68 8.90 9.11 8.86 9.53	7.09 6.75 6.93 7.09 6.89 7.41	5.07 4.83 4.95 5.07 4.93 5.30
12.	$z_{12,13}$	9.11	7.09	5.07	$z_{11,12}$ $z_{9,12}$ z_{l12} z_{e12} $z_{e12'}$ $z^{a}_{m\ 12}$ $z^{b}_{m\ 12}$	9.22 8.90 8.67 8.90 8.90 8.68 9.54	7.18 6.93 6.75 6.93 6.93 6.76 7.42	5.13 4.95 4.83 4.95 4.95 4.83 5.31

11.	$z_{11,12}$	9.22	7.18	5.13	$z_{10,11}$	9.22	7.18	5.13
10.	$z_{10,11}$	9.22	7.18	5.13	$z_{8,10'}$ z_{l10} z_{e10} z_{m10}	9.32 8.48 9.53 9.53	7.26 6.67 7.42 7.42	5.19 4.72 5.30 5.30
9.	$z_{9,12}$	8.90	6.93	4.95	z_{89}' z_{l9} z_{e9} $z_{e9'}$ z_{m9}^{a} z_{m9}^{b}	9.33 8.68 8.90 8.90 8.69 9.53	7.26 6.76 6.93 6.93 6.76 7.42	5.19 4.83 4.95 4.95 4.83 5.30
8.	z_8'	9.33	7.26	5.19	z_{78} z_{l8} z_{e8} $z_{e8'}$ z_{m8}	9.33 8.69 8.91 9.12 8.84	7.26 6.76 6.93 7.09 6.88	5.19 4.83 4.96 5.07 4.92
7.	z_{78}	9.33	7.26	5.19	z_{67} z_{l7} z_{e7} $z_{e7'}$ z_{m7}	9.33 8.69 8.91 9.12 8.84	7.26 6.76 6.93 7.09 6.88	5.19 4.83 4.96 5.07 4.92
6.	z_{67}	9.33	7.26	5.19	z_{56} z_{l6} z_{e6} $z_{e6'}$ z_{m6}	9.12 8.69 8.91 9.12 7.95	7.09 6.76 6.93 7.09 6.19	5.07 4.83 4.96 5.07 4.42
5.	z_{56}	9.12	7.09	5.07	z_{35} z_{45} z_{e5} $z_{e5'}$ z_{l5} z_{m5}	8.80 8.27 8.91 9.12 8.68 9.55	6.84 6.43 6.93 7.09 6.75 7.42	4.89 4.60 4.95 5.07 4.83 5.31
4.	z_{45}	8.27	6.43	4.60	z_{24} z_{e4} z_{l4}	8.91 8.91 8.82	6.93 6.93 6.85	4.95 4.95 4.90
3.	z_{35}	8.80	6.84	4.89	z_{l3} z_{e3} $z_{e3'}$ z_{m3}	8.69 7.85 6.78 9.54	6.75 6.10 5.27 7.41	4.83 4.36 3.77 5.30

2.	z_{24}	8.91	6.93	4.95	z_{12} $z_{12'}$ z_{e2}	8.84 8.48 8.91	6.87 6.60 6.93	4.91 4.71 4.95
1.	$z_{1'}$	8.48	6.60	4.71	z_{d1} z_{11} z_{e1}	7.42 8.83 8.90	5.78 6.87 6.93	4.12 4.90 4.95

where

z_{19} represents the given output quality index from processing unit no. 19

17 META ANALYSIS OF MANAGEMENT INFORMATION SYSTEMS

17.1 INTRODUCTION

Rapidly changing national and international business scenario introduces a variety of options and situations thereby putting organisational information system of a large system to a test tougher than earlier. This necessitates new and innovative approaches to planning, analysis, design and implementation of Management Information System (MIS). MIS planning particularly has gained considerable interests among the researchers[1]. Early researchers confined to developing better conceptual models of IS planning[2]. Integrating IS planning with business planning was recognized as the main issue. Now business planning is advocated to be linked with information technology (IT) objectives[3]. Issues of strategic IS planning were also highlighted by many researchers[4]. Other issues found attention of the researchers are users' involvement, training of IS staff and inadequate resource commitment[5]. Despite these many efforts, there are instances of only few methodologies that could provide broad guidelines to IS planning.

MIS structure or architecture has been proposed as an appropriate basis to evolve and build upon the ingredients of a useful methodology as it describes the network of information-decision relationship for the entire organisation[6]. This, however, finds cursory treatment in the literature. Primary reason could be the failure to recognize the relationship between IS and organisational structure[7]. Large-scale systems usually have multiple organisational structures that necessitate different MIS structure to suit their requirement. A well-defined system structure makes it possible to develop a clearly articulated MIS architecture which in turn acts as a means to maintain system clarity and smooth functioning. Information linkages in the architecture represent a high level of abstraction, summarizing details of internal exchange and processing at a decision point. Information detailing is, therefore, highly dependant on the type of decision to which it is linked. Thus, the information linkage defines the key design features, viz. 'integration', 'co-ordination', and 'control'.

MIS architecture is considered as an important planning tool for MIS development and a participative approach for designing the same is suggested using Interpretive Structural Modelling[8].

The objective of the present chapter is to describe a modelling scheme useful in performing a meta analysis of the MIS architecture as shown in Fig. 17.1. This facilitates an evaluation of different types of information for their varying flow and quality against the resource investment (cost) made. An insight of the systems' capabilities is also gained by the designer. It begins with characterising a decision unit in the MIS architecture by three basic processes, viz. information processing, transmission, and storage. This chapter is limited to addressing decision units as only information processing units. Informational linkages across the decision units are analysed in terms of three unique variables (viz. information flow rates, their quality and cost) defining an abstract information linkage from all the possible design dimensions. A suitable measure of these has been suggested and used in the modelling scheme. Physical System Theory (PST) has been chosen as basic framework for modelling because the MIS configuration in the form of MIS architecture is very close and commensurate with the kind of systems most appropriately modelled in this framework to study their behaviour. PST has been ingeniously supplemented with basic Fuzzy logic[9] and Analytic Hierarchy Process[10].

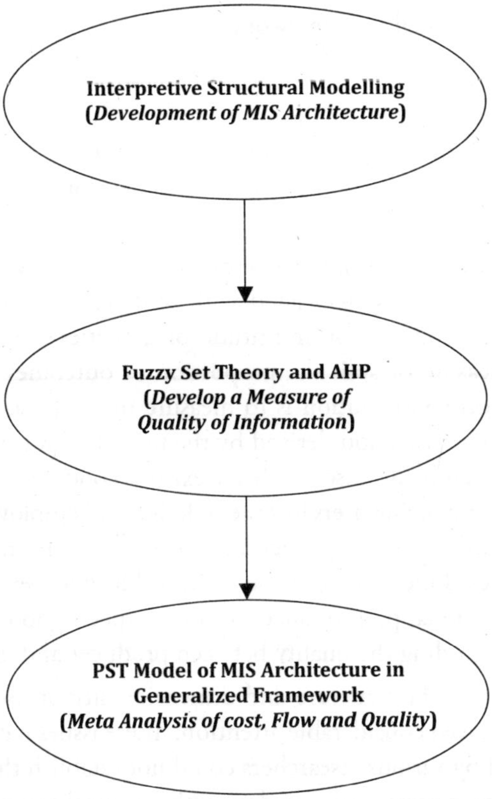

Figure 17.1: Flexible System Methodology for Meta Analysis of MIS

17.2 MODELLING AND ANALYSIS OF MIS ARCHITECTURE

Development of MIS architecture establishes the fact that information is the basic entity that holds together the entire system, and thus, gives a meaning to the system's structure. Further analysis of the architecture with modelling is aimed to arrive at the detailed specifications of individual informational interrelationships. This would lead to certain prior decisions essential for detailed design and implementation of an MIS project. These decisions relate to planning of the specific design activities. This becomes more important when the system is large and the MIS architecture is very complex, e.g. MIS of large-scale systems at national level. Modelling MIS architecture for analysis is done in many stages. It begins with identifying design variables associated with information linkages (flow, quality, and cost), decision units as information processing activities, developing relationships for information variables across a decision unit, estimation of parameters and subsequently carrying out analysis with help of the defined relationships.

Design Variables

For a meta-level analysis, three unique variables can be identified to be associated with each information linkage, viz. information flow rates, quality of the information flow, and cost associated with the flow. These variables are unique in the sense that they, most appropriately, explain an abstract information linkage from all the possible design dimensions. New approaches are being proposed to measure these variables and to be used in the modelling framework.

Quality of Information

Literature on MIS has paid more attention to the measurement of value instead of quality of information resource. Value is the economic dimension of quality. Earlier efforts to provide theoretical base to value measures have been in isolation and not in the stream of IS development procedure causing difficulty in their application.

Information is an intangible organisational resource whose utility is not direct. It is relatively difficult to quantify the cost and value associated with obtaining and using it. Also, the improved organisational performance is produced by multitude of activities taking place concurrently. It is, therefore, very difficult to measure or split the proportion of outcomes as value contributed by the information system. A practical consideration is to measure the value as relevancy of information to the user's requirements and the satisfaction derived by the user. This was first recognized by Cyert and March[11]and empirically endorsed by Swanson[12]. There exists a good amount of literature dealing with appropriateness and efficacy of involving users in system design and employing users' perception as basic mechanism (to evaluate user information satisfaction (UIS)) to ensure that IS being developed meets their requirements. It becomes difficult to test the relationship between satisfaction and bottomline indicators of user and organisational performance. Another aspect ignored in the previous schemes is the issue of perception gaps regarding the quality between producer and user.

This establishes the fact that the structure and dimensionality of user satisfaction are important theoretical issues that have received considerable attention. These issues have not been fully resolved. As a result, approaches suggested by various researchers could not establish their utility in physical system design. Information processing systems have to deal with labyrinthine paths of information flowing from varying sources to many other users. There may be possible redundancy and quality bearing some cost. What is needed is a strategy enabling the evaluation of different types of information as to their varying efficiency and effectiveness against the resource investment. This would have been of great advantage in identifying the source of trouble versus cost of improvements and deciding the detailed design strategies. The present chapter attempts to work out such a framework of systems analysis.

To use a UIS based quality measure in the proposed framework, we need a composite quality index. This could be based on the evaluation of various quality attributes on a well-defined scale and also by considering the relative weighing of the attributes. An approach based on this principle[13] is described below.

In the context of MIS, quality of an information resource may be characterised with its several attributes, viz. time-oriented (timeliness, maintainability and reliability), structural (format, sufficiency, accuracy, comparability, scope of multiple users, database, cost efficiency), sensory (presentation style), commercial (warranty), and behavioural or ethical (decision relevancy, consistency, understandability and free from bias). Parameters of fitness of use may be, quality of IS design and conformance, systems' capabilities, and organisational support.

The above attributes are sufficiently vague and unquantifiable, and can be treated most appropriately as fuzzy variables. Opinion about the degree of satisfaction may be sought on an appropriate scale (0-1 or 0-5) about each quality attribute and also their priority weight over each other for a particular information resource. This provides a normal fuzzy set in which user's response is implied as a membership function of quality variable in a fuzzy set. Weight reflects the intuitive judgement of the user about overall importance of the attribute in serving the broad purpose. Analytic Hierarchy Process (AHP) can be used in quantifying these weights. Advantage with AHP is that it takes care of maintaining the consistency in judgement by effectively handling the problem of individual bias in the situation of subjective judgement analysis. This is the reason why its application is wide. AHP determines the priority rank of attributes as well as magnitude of the priority of an attribute; this magnitude is regarded as the priority weight measuring relative importance (αl, l = 1, n) of the attributes over each other. It is obtained through a process of paired comparison and geometric mean with a check of consistency in judgement.

Weight should also reflect the 'degree of certainty' or 'certitude' present when the user decides the level of satisfaction derived from a particular quality attribute of information resource. This is a complementary concept to the 'degree of fuzziness' or incertitude, associated with a fuzzy variable. De Luca and Termini[14] have suggested a measure of the fuzziness in a system (El, l= 1,n). The complementary value of the fuzziness (dl, l= 1, n) represents the total amount of certitude associated with the variable described as fuzzy. Normalised certitude (βl, l=1, n) and priority (αl, l = 1, n) weights together account the overall weight (wl, l= 1,n) for an attribute over the other defining quality of an information resource.

Thus, an overall quality index of information should define an aggregate impact of the information quality on the decision-making. For an attribute having dominant importance over the others, any improvement or degradation in that attribute will significantly affect the overall quality measure. The weighted average measure of information quality is taken as overall quality index of an information resource.

If a^l (l = 1,2,...n) defines the quality attribute of information in a set z and μ (al) is its membership value, then Quality index z of attribute a is defined as,

$$z(a^l) = \mu(a^l).wl \tag{17.1}$$

and weighted average value as quality index is defined by the relation,

$$z = \sum_{l=1}^{n} [\mu(a^l).w^l] \tag{17.2}$$

where, w^l (≥) is normalised combined weight of priority (α^l) and normalised certitude (β^l) associated with each quality attribute so that,

$$\sum_{l=1}^{n} w^l = 1 \tag{17.3}$$

Flow of Information

It refers to the amount of information flow per unit of time. It is a quantitative measure of information. Most part of the information systems in organisations, now a days, is computerised. Determination of information flow rate in these systems is easy in the unit of Kbyte. For manual processing systems, the following relation is proposed to be useful,

$$y = b + e \sum_{l=1}^{n} r^l s^l \qquad (17.4)$$

where,

y is information flow rate in kbyte

b is the kbytes of computerised information

n is number of manual report transactions within a certain period considered as unit of time

r^l is average number of pages in each report

s^l is average number of words per page

e is the equivalence factor for converting word to the unit of kbyte.

This relation provides an approximate measure for manually operated information flows. It is suggested to break up a manual report into its significant segments. The total number of words in each segment is counted. An equivalence factor (e) is introduced to convert an average word size to the unit of kbyte.

There may be some specific practical problems like shared documents or identifying potential flow vs actual flow or finding an appropriate measure for pictures, graphs, charts, etc. These can only be addressed on a case to case basis, and relationship can suitably be tuned. Quantum of the flow of normal text is generally found high enough as compared to the other forms of information in manual system. The present formulation will, therefore, normally serve the purpose for a meta level system analysis to be carried out within the proposed modelling framework where we need an approximate measure of the flows.

Cost of Information

Cost estimation in business systems has been dealt with in detail in most of the accounting journals. Estimation of the cost of information is a complex and illusive process. Total cost may account for input cost and the cost of processing related operations. Factors like manpower, machine time, depreciation, stationary, etc. are included in costing at each stage. A linear relationship may be used to study and summarize the operating cost of MIS in ordinary situations. However, a comprehensive model becomes necessary to allow for and forecast initial cost of developing, testing, installing and adapting systems. Such a model should also incorporate the jointness of costs in an MIS. Our ability to forecast these costs remain abysmal. These may be treated appropriately as fixed development cost. A linear relationship for total cost may be expressed as,

$$x = h + f(y) + g(z) \qquad (17.5)$$

Here h is input cost and $f(y)$ and $g(z)$ are processing cost functions with respect to flow rate and quality of information resource at a point of time. These functions implicitly take into account the fixed development cost.

A more rigorous analysis involves the study of the impact of variations in specific quality and flow attributes $(g(z(a^l))$ and $f(y(a^l)))$ over the total cost. This may be significant for the designer point of view as the cost consequences of improving the quality and flow of information for an attribute is easily known. In order to identify the most appropriate factor for consideration in a multiple regression equation, one must use a theoretical base, empirical reasoning and statistical means. Users can be interviewed to identify most of the real factors that influence the dependent variable. There may be some missing factors that can be introduced based on theoretical considerations. Attributes of flow may be defined as volume, frequency of flow, level of aggregation/processing, message content, etc. The model would become,

$$x = h + f(y(a^l)) + g(z(a^l)) \qquad \forall\, l \tag{17.6}$$

Non-linear behaviour is closer to real life and can be defined suitably by a non-linear function. Box Jenkins methodology is an efficient and practical procedure for using autoregressive schemes to establish the model that best fits a set of data and for testing the adequacy of model. The same may be tested to establish the suitable one describing the real behaviour. However, a power function with multiple variables is more realistic representation and may be expressed as,

$$C = r\, f(y(a^l))^t\, g(z(a^l))^s \qquad \forall\, l \tag{17.7}$$

Here r, s and t are the parameters, values of which can be estimated with the help of regression analysis. This relationship is also useful in finding out a value of the information that is generally related to payoffs. Payoff based models, however, have not been successful so far. It is easier to consider value as the user's estimate of the overall capacity of the resource to satisfy his/her need set defined by the quality attributes. In other words, value of information would depend on how close it comes as resource to the user's level of satisfaction. Value is maximized by achieving enhanced level of satisfaction at minimum cost. Practical significance of this is highlighted in the advantages derived by an improved user designer communication, quick user requirement identification, and incremental evaluation of costs and benefits. This enables the management to have an option thereby resulting in a user tailored system.

The Generalized Model

For the sake of meta analysis of MIS architecture, every decision unit is defined for doing any one or a combination of the basic processes, i.e. information processing, transmission, and storage. Information processing may change the form of presentation or achieve a defined change in the message content. A decision unit receives information emanating from other units for its use and after utilisation and further processing, transmits new information to other decision units. Since the information-processing requirement being large enough, computer aids may be used as a supportive facility for processing and storage of data and information. Thus, at a point of time, a decision-maker acts as receiver, user, processor and sender of the information. Transmission is special type of transformation process where the resource is simply moved from one location to the other at a cost. This includes collection, translocation and distribution efforts.

The model utilizes the measurement approaches of flow, quality and cost variables suggested in the previous sections. Provision for analysis of the system is in quantitative terms except the quality variable. A quantitative analysis may be appropriate and feasible in case of review and redesign of the existing MIS.

Only information processing is being dealt in the present model. It may be further extended to incorporate the transmission and storage aspects as well. This, however, is not included here at this stage for the sake of simplicity. Information processing is an activity that adds the output information with 'purpose', 'knowledge', 'meaning', 'information', or 'decision value' after applying certain defined and classified processing methods to the inputs of data resource, facts and figures. These processes are carried out with the help of monetary support and energy. The free body diagram of *i*th or *j*th information processing activity (P_j or P_i) is identified with a set of oriented line segments and shown in Figure 17.2. The processor receives inputs of data/information resource (*ji*) and produces output information (*oj* or *oi*). These line segments are used to establish a directional reference frame for input–output flow rates. Response-stimulus relationships for various information processes are defined by following notation,

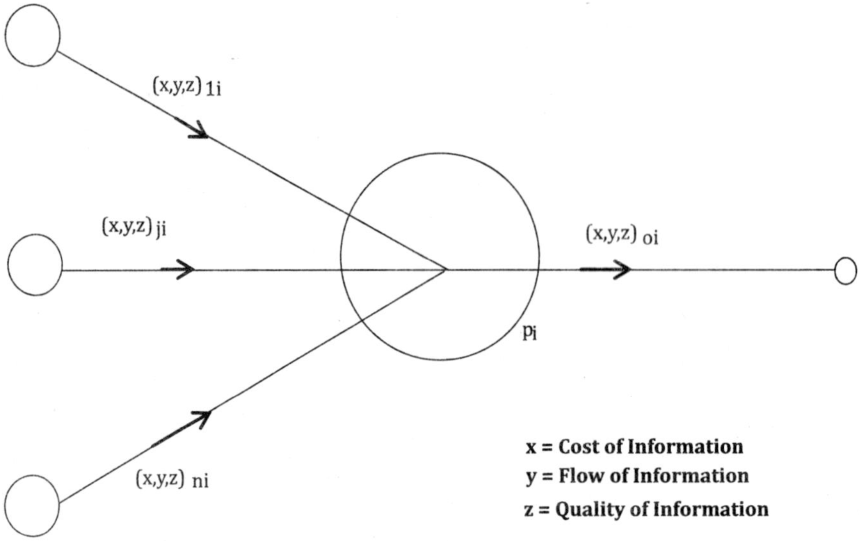

Figure 17.2: Information Flow from jth to ith Processing Unit

y_{ji} is input flow rate of information from *j*th to *i*th process.

y_{oj}/y_{oi} is called design capacity of processing unit and is specified about an information output flow.

x_{ji} is unit cost of input information (from process *j* to *i*).

x_{oi} is cost of output information from the *i*th process.

k_{ji} is Information Processing Coefficient for y_{ji}. This relates the input information flow to output information flow rate. It is a data aggregation function that defines how input informations are combined to form the output information.

$f_i(y_{oi})$ is a processing cost function to explain information-processing cost with respect to information flow rates y_{oi}. It refers to the scale economy of processing.

z_{ji} is quality index of input information.

z_{oj}/z_{oi} is quality index of output information.

q_{ji} is quality-processing coefficient for input information. It relates output information quality with the input quality.

p_{ji} is modified quality processing coefficient used for forward calculations where output quality is to be determined on the basis of quality of multiple information inputs.

$g_i(z_{oi})$ is a cost function that describes the changes in processing cost due to variation in the quality of output information.

t_{ji} is quality transfer function.

Figure 17.2 is referred to explain the response stimulus relationships (flow, quality, cost equations) and constraints for a component terminal graph as follows,

Information flow equation

$$y_{ji} = k_{ji}\, y_{oi} \tag{17.8}$$

This explains how the input information (y_{ji}) is analysed, processed, and aggregated to form the output (y_{oi}). In other words, it characterises the composition, combination and contents of the output information from input resources. This relationship, for the most part, is assumed to behave linearly. A substitution of input data or a change in their proportions to obtain essentially the same functional output is reflected in the model as a change in the coefficient k_{ji}. In this model, kbyte will be taken as basic unit of information flow.

Quality processing equation

$$z_{ji} = q_{ji}\, z_{oi} \tag{17.9}$$

Here, q_{ji} relates the quality of output information to the quality of input information resource, i.e. how the output quality is affected by the quality of input information. This relationship is useful in finding out input quality needed to produce expected output quality. In other words, it facilitates a backward analysis of the system model. q_{ji} is quality-processing coefficient which defines the quality processing capability of the decision unit.

We often encounter decision units receiving multiple information inputs. Quality of output from these will depend upon both flow and quality of input. A modified quality processing coefficient p_{ji} is introduced to account these changes caused by flow and quality of inputs, in the following relationship for quality of output information (analysing forward),

$$z_{oi} = f(z_{ji})$$

or

$$z_{oi} = \sum_j p_{ji} \cdot z_{ji} \tag{17.10}$$

where,

$$p_{ji} = \frac{k_{ji}}{q_{ji} \sum_j k_{ji}} \tag{17.11}$$

Cost equation

Working on the line of equation (17.5) as suggested earlier, unit cost of output information may be found out by the following relationship

$$x_{oi} = \sum_j k_{ji} x_{ji} + f_i(y_{oi}) + g_i(z_{oi}) \tag{17.12}$$

or,

$$x_{oi} = c_i + \sum_j k_{ji} x_{ji} - m_i y_{oi} + m'_i\, z_{oi} \tag{17.13}$$

here,

c_i refers to the fixed development cost

$\Sigma k_{ji} x_{ji}$ accounts for cost of obtaining input information

$(-m_i y_{oi} + m'_i\, z_{oi})$ accounts for the cost of processing related operations. Here m_i is decrease in the cost with the increase in volume of information, and m'_i is the increase in cost with improvement in the quality of information.

A one to one correspondence is assumed between quality index and the inherent quality of the information. A linear variation of cost function is taken for the sake of simplicity. Non-linearity in these equations can be incorporated by developing a suitable Decision Support System (DSS).

Constraints

As stated earlier, a system graph is obtained by uniting the component terminal graphs in one to one correspondence with actual interrelations of the components in the system. It is generally governed by continuity and compatibility laws. It is known as structural constraints of the system. Continuity law states that the resource is conserved in the system,

$$y_{oj} = y_{ji} \tag{17.14}$$

On the other hand, compatibility law states that the cost evaluated through whatever path followed between any two vertices will be the same, i.e.

$$x_{oj} = x_{ji} \qquad \forall j \tag{17.15}$$

Quality is a term that has valuable meaning to user-producer both. About this, the concept of perceived quality is very important as has been described in detail by Jacoby and Olson[15]. Usually producer and user differ in their perception on quality. The producer views fitness for use in terms of the ability to process and produce with less rework, less wastage of resources and high productivity. From user's viewpoint fitness for use includes effective utility of information. Apart form this, since quality of information resource is measured as users' level of satisfaction, a perceptual inconsistency between the sender and receiver is likely. Due to this gap, a relative change in quality of information during the transfer can be defined by introducing a transfer function, i.e.

$$z_{ji} = t_{ji}\, z_{oj} \tag{17.16}$$

Sensitivity Analysis

Sensitivity analysis can be done for the quality and cost of output information with respect to various system parameters. This is carried out with the help of a relationship for cost or quality of output information about all the exogenous variables and then partially differentiating them with respect to each parameter. Sensitivity analysis reveals the most sensitive system parameter that requires special attention in physical systems design. The advantage of this model is that it provides a good measure of the information processing capacity through quality and information processing coefficients. Also, it is an effective tool for predicting quality of output from performance of the processor and quality of input information. Conversely, it provides a simple and an efficient procedure to find out the quality of input information required for producing desired quality of output information. Most importantly, their cost repercussions are also known before hand through this analysis. This helps in capital allocation and investment planning on MIS project. Through sensitivity analysis, the impact of variations in the quality of output information flow on the quality and cost of output information can be studied.

Steps for Analysis

Steps of meta analysis are the following:

(i) Each decision unit of MIS architecture is represented by a component terminal graph where the orientation is used to describe information flow, its quality and associated cost and related terminal equations. Mathematical description of how these decision activities are combined to form the operational system model, is achieved by uniting vertices of component terminal graphs in one-to-one correspondence with actual informational interconnections. This results in a system graph which also helps in finding structural constraint equations.

(ii) System parameters associated with all decision units (information processing, quality processing and quality transfer coefficients) are estimated by simulating the recent past data under the same conditions. Simulation is necessary because these coefficients are not generally deterministic. Variations are possible in quality and quantum of the information flows in most of the practical situations.

(iii) Equations related to the laws of continuity, compatibility and quality transfer as well as resource balance equations related to each processing point, are written down for the information low, quality and cost variables.

(iv) Calculations are made to find out the required information quality, flow and its unit processing cost, at the subsequent downward levels (backward analysis), to meet the demand of the top-level decision-maker. These results reveal the status of existing system. A forward analysis can be done with the help of given values of exogenous input resource. This tells us about the level of output quality and flow of information that the existing system can provide to the top-level decision-makers. On the whole, these analyses give an important insight into the systems requirements, capabilities and scope of restructuring.

(v) A sensitivity analysis can also be carried out to study the effect of different processors on the quality and cost of information flows.

17.3 ILLUSTRATIVE EXAMPLE

A national level system dealing with environment control policies in Indian context is chosen to illustrate the model[16]. The MIS architecture of the same system is adapted here for meta analysis. System design effort is guided by the requirements of top-level decision units in the architecture as these remain informationally dependent on the rest of the system. A meta analysis of the architecture is required to derive necessary guidelines. For the present illustration, top four levels in the MIS architecture are considered, consisting of decision units such as Planning and co-ordination of pollution control activities (Processor 2); Environmental clearance of projects (Processor 3); Formulation of water resource development and pollution control programs (Processor 6); Formulation and enforcement of rules and guidelines for hazardous substance management (Processor 5); Formulation and enforcement of air and noise pollution control programs (Processor 4); Appraisal of the project in the regard to their environmental implications (Processor 7); Enforcement of water pollution control in co-ordination with central and state level agencies (Processor 8); Control of Hazardous and toxic chemicals and accidents (Processor 9); Review and planning at atomic energy and safety programs (Processor 10); and Control of air quality and noise pollution (Processor 11) (Figure 17.3). These decision units are treated as the processor, sender, receiver and also user of information. The rest of the MIS architecture is assumed as exogenous source of information resource to the subsystem chosen here for analysis. The top-level decision unit (1) is considered as exogenous sink of information resource coming out from the subsystem.

Component Models and System Graph

All the decision units are considered as information processing centres. By using the generalised model of *i*th processor, the component models of all the decision units are developed by constructing the component terminal graphs and the component terminal equations. For example, for the processor 2, flow, quality, cost and constraints equations would be as given in Appendix 17.A. Mathematical description of how these decision activities are combined to form the operational system model is achieved by uniting vertices of component terminal graphs in a one-to-one correspondence with actual informational linkages and obtaining the system graph as shown in Fig. 17.3. The total number of vertices and edges in the system graph is 19 and 38 respectively. Thus, for the subsystem under consideration, there will be total 128 equations defining relationships for flow, quality and cost.

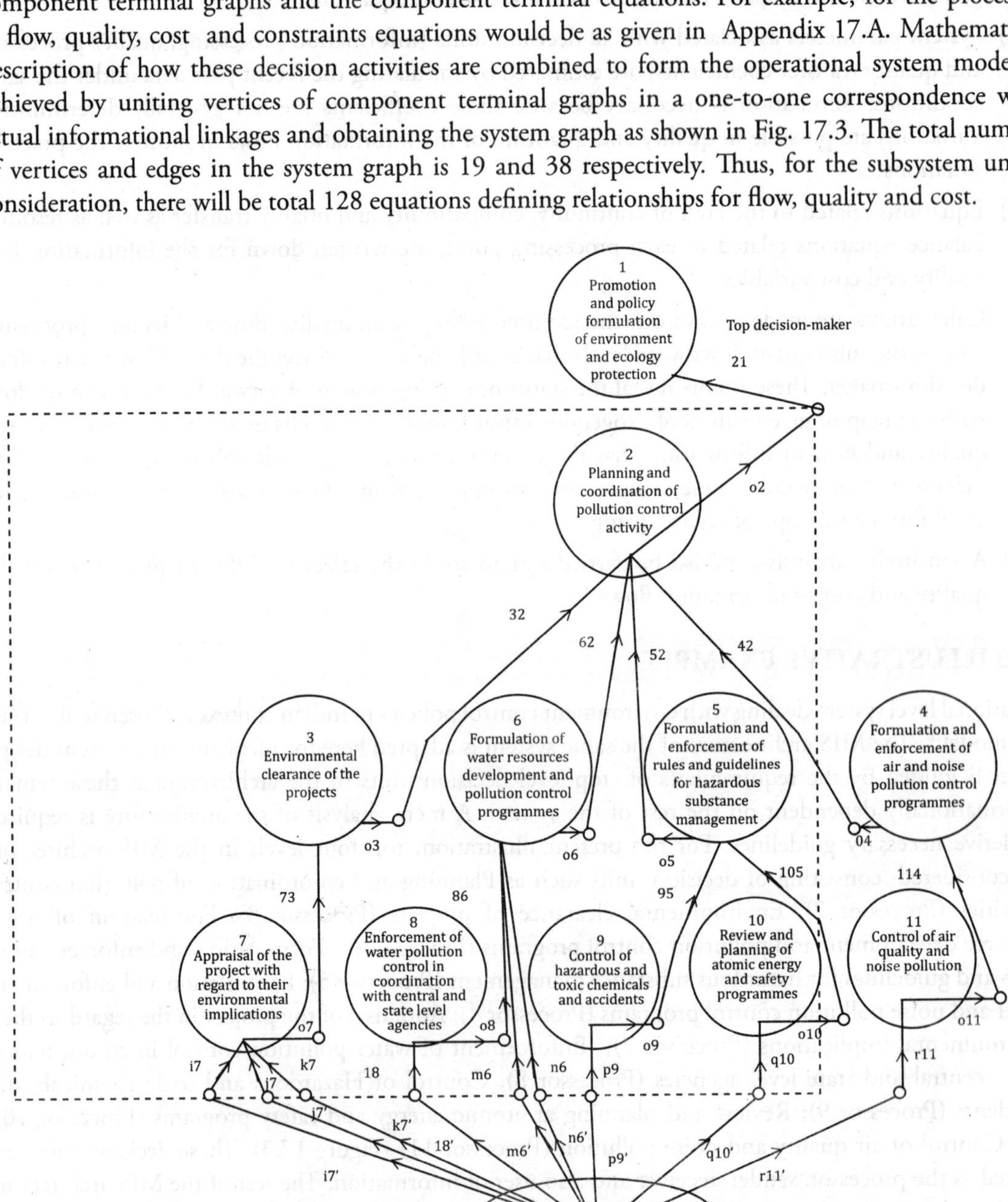

Figure 17.3: System Graph of a Sub-System of the MIS Architecture of Environment Control System

Parameter Estimation

Parameters required are, for example, information and quality processing coefficients (k_{ji} and q_{ji}), transfer coefficients (t_{ji}), minimum unit cost of information processing (assumed as a constant (c)), and slopes/ variations in unit processing cost with respect to variations in information flow rates and quality (m_i and m'_i). These are estimated with the help of data from the recent past. Real life situations are characterised by the uncertainties in the system parameters and thus parameters should represent these stochasticities. A mathematical analysis would, however, tend to be a complex and cumbersome exercise. Simulation may be resorted to deal with such situations. However, in the present case, a deterministic situation with linear variations is assumed for the sake of simplicity.

Associated with each information linkage, there will be two quality indices, one is based on the processor/sender's opinion about quality and other is based on the receiver/user's opinion. For example, z_{i7} is the quality of information resource *i7'* from the point of view of the source while z_{i7} is the quality of the same information viewed by the user. The interconnecting point denoting the transfer of the information is the point with respect to which quality transfer coefficient is estimated. A set of quality attributes of information has been defined, viz. decision relevancy (a^1), sufficiency (a^2), reliability (a^3), understandability (a^4), timeliness (a^5), and accuracy (a^6). Opinion about the level of satisfaction for these attributes, is sought from both the sender and receiver on 0-1 scale. Here, the interpretation of the scale can be done qualitatively as:

Level of Satisfaction	Nil (N)	Very very low (VVL)	Very low (VL)	Low (L)	Less than avg. (LT)	Average (A)	More than average (MTA)	High (H)	Very high (VH)	Very very high (VVH)
Values on the scale	0	0.1	0.2	0.3	0.4	0.5	0.6	0.7	0.8	0.9

Information flow rates and information processing coefficients for all the linkages have been estimated and given in Appendix 17.B. Quality indices for the information resource flows have been obtained and parameters such as quality processing, modified quality processing and transfer coefficients for these have been estimated; these are also included in Appendix 17B. Basic data regarding quality indices, entropy and eigen vector weights are not presented here.

Parameters associated with the unit cost of output information are required to be established for each processing unit. This can be done with the help of regression method. A regression equation is established which expresses the unit cost of output information in terms of information flow rates and quality; flow and quality presumably control or cause the unit cost of the processed information to increase or decrease. Data from the recent past has been used. The unit cost of output information is termed as dependent variable while variables such as flow rates and quality are called independent variables. Actual correlation should be considered between information flow rate and quality to study the corresponding variation in unit cost of output resource. However, in the present case, one variation is studied keeping the other one constant.

The estimated value of the constants associated with each processor following the above procedure have been given in the Appendix 17B.

Analysis and Discussion

Information system analysis is possible both ways, forward as well as backwards. Forward analysis is carried out by specifying the exogenous values of input information available to the system in order to find out the likely output information to the top-level decision-maker. On the other hand, backward analysis helps in finding out the required information flow and quality at other parts of the system, to meet the demand of the top-level decision-maker. Generally, forward analysis is useful in the study of quality and cost variables while backward analysis is used in the study of flow rates and its quality assessment.

Flow and continuity equations are solved backwards by specifying y_{21} (top decision unit) to analyse the variations in information flows at other places in the system. Results have been compiled in Appendix 17C for ten different output requirements at the top (y_{21}). The results show that the processors with multiple inputs such as 7, 6, 5 and 2 have maximum processing load. Also, a considerable amount of information resource comes to the processor 8 as input. In case, the present requirement is enhanced (≥1396.50 kbyte at the top), these processing units might need computerisation to cope up with the increasing processing load. A cost-benefit analysis will help in deciding the degree of automation and centralisation to be set up.

A backward analysis is also recommended to study the systems' capability in supporting the desired quality of output information at the top. For different levels of quality demand at the top (z_{21} = L, VL, VVL, LTA, A, MTA, H, VH, VVH and P), necessary input quality at the subsequent downward stages has been calculated and presented in Appendix 17B. It is important to note that, to meet a high quality expectation at the top, the basic input quality to the system has to be perfect (calculations show more than 1.0). The question arises, at the best, what quality of information output the existing system can offer to the top-level decision maker (z_{21}). A forward analysis provides the answer to this question; assuming the best input quality to the system in an extraordinary case (*i.e.* $z_{i7'} = z_{j7'} = z_{k7'} = z_{18'} = z_{m6'} = z_{n6'} = zp_{9'} = z_{q10} = z_{r11} = 1.0$), the highest possible quality of information output to the top (z_{21}) would not be more than 0.7559 (0.8 = VH). This is a significant result about strength of the system.

A forward analysis has been carried out to estimate the unit cost of output information at each processor for the given information flow rates and its quality. Details of the results have been selectively reported in the Appendix 17C; in this, impact of variations in quality and flow rates over cost has been studied, keeping one variable constant at a time. Variations are also depicted in Figure 17.3. The unit cost of output (x_{o2}) with the lowest quality of information ($z_{o2} = 0.1$), and flow (y_{o2} = 1396.5 kbyte) is Rs.2077.79 which is continuously increasing with increase in the quality of the output. Unit cost is Rs. 3389.91 for the perfect output quality with the flow requirement remaining the same (1396.50 kbyte). Similarly, with perfect quality and maximum flow (y_{o2}= 12000 kbyte), cost reduces to Rs. 2969.4. On the other hand, cost for the lowest quality ($z_{o2} = 0.1$) and maximum flow (y_{o2} = 1200 kbyte) comes out to be only Rs. 1657.39. It is obvious now that the impact of quality is more significant than flow on the total cost. Total yearly budget of MIS for the environment control system under study has been presented for these extreme cases in Table 17.1. The total budget for MIS will be 348.20 million rupees for the existing requirement at the top (y_{o2} = 1396.5 kbyte) with lowest quality. The budget for the similar requirements with highest or perfect quality would be 568.08 million rupees. These are the two extreme limits of the budget if flow rate at the top remains the same as the existing one. These variations are valid for improvement in quality within the present set up. However, a large scale restructuring or redesign will require additional resource investments on automation or further improvements in the infrastructure. The total budget may have to be adjusted based on a cost-benefit analysis and considering the variations of flow rates and quality for the future system.

The above analysis gives an insight into the system about information flow, quality, and cost. The budget allocation for the MIS development of the specific activities can be decided. Other macro level decisions taken up after this analysis are, regarding the level of automation, combination of batch and on-line processing, and additional sources of information if necessary. These decisions should have a justification with respect to likely benefits expected to be accrued; this requires a cost- benefit analysis. Autonomous decision units in the MIS architecture are designed in the end. Design of subsystem consisting of independent decision units follows the guidelines drawn upon from the analysis and design of MIS subsystem with dependent decision units (towards the top in architecture).

Table 17.1: Budget Estimates for the MIS Project of Environment Control System

Sr. No.	Quality of output information	Flow rate of output information	Unit cost of output information	Total MIS cost per month	Total yearly budget required for the MIS
	z_{02}	y_{02} (kbyte)	x_2(Rs)	$y_{02} \times x_2$(Rs)	(Rs)
	0.1	1396.5	2077.79	29,01,633.07	348,19,605.00
	0.1	12000.0	1657.39	198,88,680.00	238,64,000.00
	1.0	1396.5	3389.91	47,34,009.30	568,08,112.00
	1.0	12000.0	2964.40	365,32,800.00	4275,93,000.00

Sensitivity Analysis

The sensitivity analysis has been performed for assessing the impacts of variations in the parameters on both, the output quality and its cost. For this, quality and cost equations have been solved backwards, starting from the top and obtaining the final relationship in terms of exogenous variables.

The derived equation is partially differentiated with respect to various modified quality processing coefficients (p_{ji}) and the quality transfer coefficients (t_{ji}). The values of various partial differentials have been calculated. The results indicate that the output quality at the top is most sensitive to the processing at the processor 2 with respect to the input information from the processors 4, 5, 6 and 3. It is least sensitive to the exogenous input information resources. In terms of perception gaps, the output quality gets affected maximum by the individual differences at the input stage on the top (i.e. t_{21}). However, it is less sensitive to the perception gaps elsewhere in the system which is quite logical. Again, perception gaps at the exogenous level have least impact on the output quality of information at the top, except the input information resource to the processor 9. Also, impact of the quality processing is marginally higher than the perception gaps in this particular subsystem except at the top.

The sensitivity analysis for the cost is performed by solving the cost equation for the output information resource at the top. The equation is expanded to relate with the exogenous input variables for a backward analysis. The derived equation is partially differentiated with respect to information processing coefficients (k_{ji}), quality processing and transfer coefficients (q_{ij} and t_{ji}) and also with respect to the constants and slopes associated with the cost (c_i, m_i, and m'_i). The results indicate that the information-processing coefficient for the input resources at the processor 2 has significant impact over the unit cost of the output information. At the lower level in the system, only processor 8 has demonstrative impact over the output cost that too because of heavy input and processing at the processor 8. On the other hand, quality processing has greater impact over the information processing coefficients. Sensitivity of the output cost with respect to quality processing follows the similar trend as that for information processing coefficients. Specially quality of input from the processor 6 to processor 2, and processor 8 to processor 6 have maximum impact on the output cost.

Cost is sensitive to the constant at the processing unit 8; this shows the level of processing load on the processor. Other units have almost similar values. Slopes with respect to information flow rate show substantial variations in output cost and, again, processors 8 and 6 have the maximum influence on it. These two processors also show greatest impact on the output cost as compared to the other processing points in the system.

The results of sensitivity analysis help in deciding selective improvement in design, setting priorities for control, maximising the quality of information, and reducing the cost of information.

REFLECTIONS

This chapter presents a sound modelling framework for performing a key stage of MIS planning exercise, i.e. analysis of MIS architecture. Architecture has essentially been the basis of all kinds of planning. MIS architecture not only represents a model of the system to be developed but also facilitate further analysis to arrive at the detailed design specifications of individual information linkages. This would lead to certain prior decisions essential for physical designing and implementation of MIS project. MIS architecture has been considered as important planning tool to MIS development and proposed a balanced approach for design and analysis of the same. This chapter specifically delineated a modelling strategy for analysing the MIS architecture to derive detailed design specifications. New approaches to measure information flow rates, quality and cost associated with the flow were suggested to help the modelling process. The model offers a comparative analysis of information linkages in terms of flow, quality, and cost; this helps us to know the effect of various information processing units on quality of output information and identify the source of trouble/ inefficiency about cost and productivity. In other words, the model provides a good measure of information processing capacity of the system, predicts input-output quality and their cost repercussions that help in capital allocation and investment planning on MIS project. A sensitivity analysis can also be carried out to study the impact of variation in these variables. The major limitation of the model is that it requires strict quantification of some design variables that may not be, however, a constraint in some situations. More flexible models need to be developed to overcome these constraints.

Notes

1. Galliers R.D. (1987), Information Systems Planning in the UK and Australia: a Comparison of Current Practices, *Oxford Surveys in Inf. Tech*, 4, 223-255; Kelleher D.(1995), Business Programs and IS

Methodologies, *Inf. Syst. Jl*, 5, 137-157; Kumar P. G. and King W.R. (1994), Organisational Characteristics and IS Planning, *Inf. Syst. Res.*, 5(2), 75-109; King, W.R. (1988), How Effective is Your IS Planning, *Long Range Planning*, 21(5), August, 103-112; King, W.R. (1988), Strategic Planning for Information Resources: Evolution of Concepts and Practices, *Inf. Resources Mgt. Jl.*, Fall, 108; Ward, J.P. Griffiths and P. Whitemore (1990), *Strategic Planning for Information Systems,* John Wiley, New York.

2. Bluementhal S.C. (1973) *MIS: A Framework for Planning and Development,* Prentice Hall, New Jersey; King, W.R. (1978), Strategic Planning for Management, *MIS Qtly.*, March, 26-37; MacFarlan, F.W.(1971), Problems in Planning the IS, *Har. Bus. Rev.*, March-April, 75-89; Mclean E.R. and Soden J.V. (1977), *Strategic Planning for MIS*, John Wiley, New York; Zani W.M. (1970), Blueprint for MIS, *Har. Bus. Rev.*, 48, 6 91-100.
3. Reich B.H. and Benbasat I. (1996), Measuring the Linkage between Business and IT Objective, *MIS Qtly.*, 20(1), 55-77.
4. Earl M.J. (1993), Experience in Strategic Planning, *MIS Qtly,* 17 (1), 1-24; Leaderer A.L. and Gardiner V. (1991), The Success of Strategic Information Planning, *Jl. Strategic IS*, 1 (2); Leaderer A.L. and Sethi V. (1988), The Implementation of Strategic Information System Methodologies, *MIS Qtly.*, Sept., 445-461.
5. Adrains W. and Hoogakkar J.T. (1989), Planning an IS at Netherlands Gas, *Long Range Planning,* 22 (3), 64-74; Brancheau J. Schuster L. and March S.T. (1989), Building and Implementing Information Architecture, *Database*, Summer, 23-45; Galliers, R.D. (1991), Strategic IS Planning: Myths, Reality and Guidelines for Successful Implementation, *Euro. Jl. Inf. Syst.*, 1 (1), 55-64;
6. Gupta, M.P. and Sushil (1993a), Methodology for Design of MIS Architecture, *Systems Research*, 10 (2), 53-82; Gupta, M.P. and Sushil (1993b), Criticality Analysis of MIS Architecture, *Systems Research*, 10 (2), 83-105.
7. Ahituv N. Neumann S. and Zviran M.(1989), Factors Affecting the Policy for Distributing Computing Resources, *MIS Qtly.*, 13 (4), 389-401; Evi Dor P. and Segev E. (1982) Organizational Context and Success and MIS Structure: Some Empirical Evidence, *MIS Qtly.* September, 55-68.
8. Gupta and Sushil (1993a, 1993b) *op cit.*
9. Zadeh L.A. (1965), Fuzzy Sets, *Infor. and Cont.* 338-353.
10. Saaty, T.L. (1980), *The Analytic Hierarchy Process*, McGraw-Hill, New York.
11. Cyert and March (1963), *Behavioural Theory of the Firm*, Prentice Hall, Inglewood Cliff, N.J.
12. Swanson, E.B. (1974), Management Information System: An Appreciation and Involvement, *Mgt. Sci.*, 21, 178-188.
13. Gupta M.P. (1994), Multi Attribute Measure of Information Quality, in G.V.K. Reddy, Nadarajan R. and Venkatsubramanian N.K.(Eds.), *Stochastic Models, Optimisation Techniques and Computer applications*, Wiley Eastern, N. Delhi, 481-490.
14. De Luca, A., and Termini, S. (1972), A Definition of a Probabilistic Entropy in the Setting of Fuzzy Sets Theory, *Inf. & Cont.*, 20, 301-312.
15. Jacoby, J. and Olson, J.C. (Ed.) (1985), *Perceived Quality*, Lexington, Mass.
16. Gupta and Sushil (1993a, 1993b) *op cit.*

APPENDIX 17A

Flow, Quality, Cost and Constraint Equations for Processor 2

Flow Equations	Quality Equation (Backward Analysis)	Quality Equation (Forward Analysis)
$y_{32} = k_{32}.y_{o2}$	$z_{32} = q_{32}.z_{o2}$	$z_{o2} = p_{32}.z_{32}$ +
$y_{62} = k_{62}.y_{o2}$	$z_{62} = q_{62}.z_{o2}$	$p_{62}.z_{62}$ +
$y_{52} = k_{52}.y_{o2}$	$z_{52} = q_{52}.z_{o2}$	$p_{52}.z_{52}$ +
$y_{42} = k_{42}.y_{o2}$	$z_{42} = q_{42}.z_{o2}$	$p_{42}.z_{42}$

Cost Equation

$$\begin{aligned} -x_{o2} = {} & k_{32}.x_{32} + \\ & k_{62}.x_{62} + k_{52}.x_{52} + \\ & k_{42}.x_{42} + f_2(y_{o2}) + \\ & g_2.(z_{o2}) \end{aligned}$$

Structural Constraints

Continuity Equations	Compatibility Quality	Quality Transfer Equation
$y_{21} = y_{o2}$	$x_{21} = -x_{o2}$	$z_{21} = t_{21}.z_{o2}$
$y_{32} = y_{o3}$	$x_{32} = -x_{o3}$	$z_{32} = t_{32}.z_{o6}$
$y_{62} = y_{o6}$	$x_{62} = -x_{o6}$	$z_{62} = t_{62}.z_{o6}$
$y_{52} = y_{o5}$	$x_{52} = -x_{o5}$	$z_{52} = t_{52}.z_{o5}$
$y_{42} = y_{o4}$	$x_{42} = -x_{o4}$	$z_{42} = t_{42}.z_{o4}$

APPENDIX 17B

(a) Process and Transfer Coefficients Associated with the Information Resource Flows

Information Resource	Flow Rate (kbyte)	Information Processing Coefficient (k_{ji})	Quality of the Resource (z_{ji})		Quality Transfer Coefficient (t_{ji})	Quality Processing Coefficient (q_{ji})	Modified Quality Processing Coefficient (p_{ij})
i7/i7'	2734.40	1.6000	0.6211/	0.7116	0.8729	0.8950	0.3371
j7/j7'	2988.30	1.7486	0.6409/	0.7506	0.8539	0.9236	0.3570
k7/k7'	3339.80	1.9542	0.5751/	0.7860	0.7818	0.8287	0.4447
l8/l8'	7812.50	2.1978	0.4076/	0.6722	0.6064	0.6437	1.5535
p9/p9'	3007.80	1.3750	0.6828/	0.7757	0.8802	0.9930	1.0070
q10/q10'	3125.00	1.6326	0.5033/	0.7025	0.7165	0.7310	1.3680
r11/r11'	3515.60	1.5384	0.5970/	0.6969	0.8566	0.8932	1.1196
73/o7	1709.00	1.0938	0.5989/	0.6940	0.8630	1.0115	0.9886
86/o8	3554.70	2.1667	0.4552/	0.6332	0.7188	0.7292	0.5731
m6/m6'	2490.20	1.5179	0.5476/	0.7049	0.6917	0.7812	0.3748
n6/n6'	2460.90	1.5000	0.5302/	0.6911	0.7672	0.8494	0.3406
95/o9	2187.50	1.3576	0.5971/	0.6876	0.8684	0.9516	0.5604
10,5/o10	1914.10	1.1879	0.4926/	0.6886	0.7153	0.7850	0.5944
11,4/o11	2285.16	1.5294	0.5605/	0.6683	0.8386	0.8481	1.1791
32/o3	1562.50	1.1189	0.5426/	0.5921	0.5786	0.5006	0.4948
62/o6	1640.60	1.1748	0.4379/	0.6242	0.7016	0.6399	0.4064
52/o5	1611.30	1.1538	0.4385/	0.6274	0.6988	0.6407	0.3986
42/o4	1494.14	1.0699	0.4781/	0.6608	0.7234	0.6986	0.3390
21/o2	1396.50	-	0.5392/	0.6843	0.7879	-	-

(b) Value of Parameters Associated with Unit Cost of Output Information

Processing Unit (Figure 17.3)	c_i	m_i	m'_i
7	49.25	0.0021	85
8	55.00	0.0011	140
9	46.50	0.0012	95
10	51.50	0.0014	87
11	45.00	0.0011	95

3	59.00	0.0016	112
6	77.50	0.0024	165
5	64.50	0.0025	140
4	56.25	0.0021	120
2	101.50	0.0038	185

APPENDIX 17C

(a) Flow of Input Information for Desired Flow at the Top (Backward Analysis)

	Flow y_{21} (kbyte) 1396.0	4000.0	7000.0	9000.0	12000.0
Information Resource				Flow y_{ji}	(kbyte)
21/o2	1396.50	4000.00	7000.00	9000.00	12000.00
32/o3	1562.00	4475.47	7832.08	10069.82	13426.42
62/o6	1640.60	4699.18	8223.56	10573.15	14097.53
52/o5	1611.30	4615.25	8076.69	10384.32	13845.76
42/o4	1494.14	4279.67	7489.42	9629.26	12839.01
73/o7	1709.00	4895.09	8566.42	11013.96	14685.28
86/o8	3554.70	10181.74	17818.04	22908.91	30545.22
m6/m6'	2490.20	7132.69	12482.21	16048.55	21398.07
n6/n6'	2460.90	7048.76	12335.34	15859.72	21146.29
95/o6	2187.50	6265.66	10964.91	14097.74	18796.99
10,5/o10	1914.10	5482.56	9594.49	12335.77	16447.69
11,4/o11	2285.16	6545.39	11454.44	14727.13	19636.18
i7/i7'	2734.40	7832.15	13706.26	17622.34	23496.45
j7/j7'	2988.30	8559.40	14978.95	19258.65	25678.19
k7/k7'	3339.80	9566.20	16740.85	21523.95	28698.60
l8/l8'	7812.50	22377.37	39160.40	50349.08	67132.11
p9/p9'	3007.80	8615.25	15076.69	19384.32	25845.76
q10/q10'	3125.00	8950.95	15664.16	20139.63	26852.84
r11/r11'	3515.60	10069.75	17622.06	22656.93	30209.24

(b) Quality of Input Information for Desired Quality at the Top (Backward Analysis)

Quality z_{21}	0.1000	0.3000	0.5000	0.7000	1.0000
Information Resource			Quality z_{ji}		
o2	0.1269	0.3808	0.6346	0.8884	1.0000
32	0.0635	0.1906	0.3177	0.4447	0.6353
62	0.0812	0.2437	0.4061	0.5685	0.8122
52	0.0813	0.2440	0.4066	0.5692	0.8132
42	0.0887	0.2660	0.4433	0.6207	0.8867
o3	0.1098	0.3294	0.5491	0.7687	1.0000
o6	0.1158	0.3473	0.5788	0.8103	1.0000
o5	0.1164	0.3491	0.5818	0.8146	1.0000
o4	0.1226	0.3677	0.6128	0.8579	1.0000
73	0.1111	03332	0.5554	0.7776	1.0000
86	0.0844	0.2532	0.4221	0.5909	0.8442
m6	0.0904	0.2713	0.4521	0.6330	0.9043
n6	0.0983	0.2950	0.4916	0.6883	0.9833
95	0.1107	0.3322	0.5537	0.7752	1.0000
10,5	0.0914	0.2741	0.4568	0.6395	0.9135
11,4	0.1039	0.3118	0.5197	0.7276	1.0000
o7	0.1287	0.3861	0.6435	0.9010	1.0000
o8	0.1174	0.3523	0.5872	0.8221	1.0000
m6’	0.1307	0.3922	0.6537	0.9151	1.0000
n6’	0.1282	0.3845	0.6408	0.8972	1.0000
o9	0.1275	0.3826	0.6376	0.8927	1.0000
o10	0.1277	0.3831	0.6385	0.8939	1.0000
o11	0.1240	0.3719	0.6198	0.8677	1.0000
i7	0.1152	0.3456	0.5760	0.8064	1.0000
j7	0.1189	0.3566	0.5944	0.8321	1.0000
k7	0.1067	0.3200	0.5333	0.7467	1.0000
l8	0.0756	0.2268	0.3780	0.5292	0.7560
p9	0.1266	0.3799	0.6331	0.8864	1.0000
q10	0.0933	0.2800	0.4667	0.6534	0.9335
r11	0.1107	0.3321	0.5536	0.7750	1.0000
i7’	0.1320	0.3959	0.6598	0.9238	1.0000
j7’	0.1392	0.4176	0.6960	0.9745	1.0000
k7’	0.1458	0.4373	0.7288	1.0000	1.0000
l8’	0.1247	0.3740	0.6233	0.8726	1.0000
p9’	0.1439	0.4316	0.7193	1.0000	1.0000
q10’	0.1303	0.3909	0.6515	0.9120	1.0000
r11’	0.1292	0.3877	0.6462	0.9047	1.0000

(c) For Varying Demand of Flow and Quality of Information Resource at the Top, Unit Cost of Output Information at the Other Processors

		Flow y_{21} (kbyte) 1396.50	2000.0	7000.0	10000.00
Quality z_{21}	Processing Unit	Unit Cost (₹)			
0.3	7	243.95	242.40	229.55	221.84
	8	196.02	194.33	180.33	171.93
	9	130.71	129.58	120.18	114.54
	10	151.77	151.77	151.77	151.77
	11	290.44	286.17	250.78	229.55
	3	360.23	357.45	334.45	320.64
	6	676.98	671.62	627.19	600.53
	5	467.08	463.80	436.62	420.31
	4	541.43	533.55	468.19	428.98
	2	2436.25	2412.33	2214.10	2095.17
0.6	7	276.77	275.22	262.37	254.66
	8	245.35	243.66	229.66	221.26
	9	166.67	165.54	156.14	150.50
	10	185.10	185.10	185.11	185.10
	11	325.77	321.50	286.11	264.88
	3	433.02	430.25	407.24	393.44
	6	841.16	835.79	791.36	764.70
	5	604.37	601.09	537.91	557.60
	4	639.59	631.70	566.34	527.13
	2	2973.99	2950.06	2751.84	2632.91
0.9	7	296.13	294.58	281.73	274.02
	8	286.69	285.00	271.00	262.60
	9	188.75	187.62	178.22	172.58
	10	205.44	205.44	205.44	205.44
	11	350.11	345.84	310.46	289.23
	3	491.09	488.32	465.31	451.51
	6	981.15	975.78	931.35	904.69
	5	700.76	697.48	670.29	653.98
	4	708.58	700.69	635.34	596.12
	2	3388.45	3364.52	3166.30	3047.37

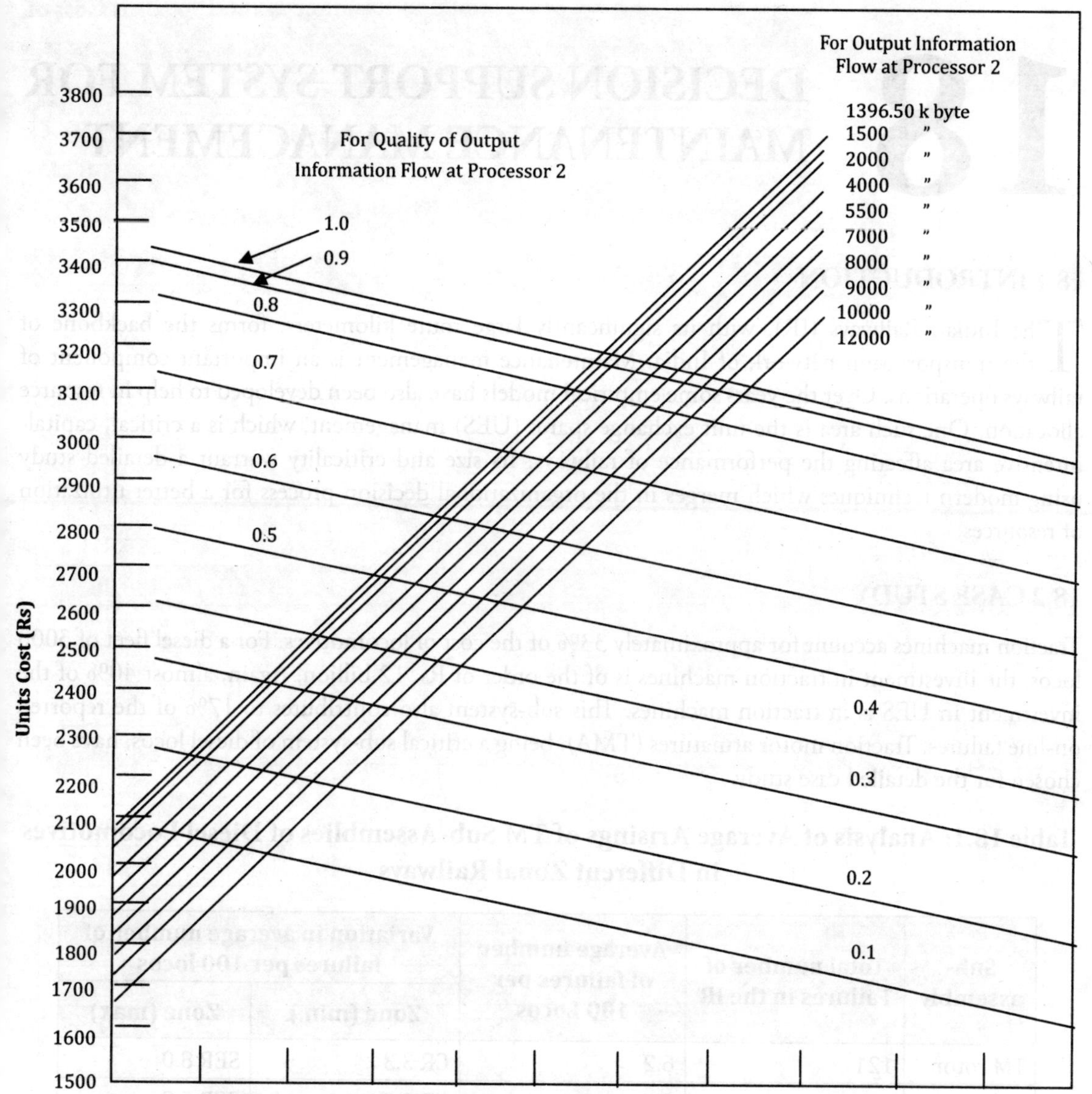

Figure 17.4: Variation in Unit Cost of Output Information at Processor 2 w.r.t. Variation in Information Flows and its Quality

18 DECISION SUPPORT SYSTEM FOR MAINTENANCE MANAGEMENT

18.1 INTRODUCTION

The Indian Railways (IR), with its significantly large route kilometers, forms the backbone of the transportation network of India. Maintenance management is an important component of railways operations. Over the years some empirical models have also been developed to help in resource allocation. One such area is the unit exchange spares (UES) management, which is a critical, capital-intensive area affecting the performance of railways. Its size and criticality warrant a detailed study using modern techniques which merges in the organizational decision process for a better utilization of resources.

18.2 CASE STUDY

Traction machines account for approximately 33% of the cost of locomotives. For a diesel fleet of 3000 locos, the investment in traction machines is of the order of Rs. 12 billion. Again, almost 40% of the investment in UES is in traction machines. This sub-system also contributes to 17% of the reported on-line failures. Traction motor armatures (TMA), being a critical sub-system of diesel locos, have been chosen for the detailed case study.

Table 18.1: Analysis of Average Arisings of TM Sub-Assemblies of Diesel Locomotives in Different Zonal Railways

Sub-assembly	Total number of Failures in the IR	Average number of failures per 100 Locos	Variation in average number of failures per 100 locos	
			Zone (min.)	Zone (max)
TM rotor	121	6.2	CR 3.3	SER 8.0
TM stator	58	2.5	ER 0.8	SER 5.8

ER: Eastern Railway; CR: Central Railway; SER: South-Eastern Railway

Normal scheduled maintenance requires overhauling of the traction motor (TM) during a 3-year schedule. Although the preventive maintenance policy determines the arising for scheduled maintenance, the major work load arising is from non-scheduled maintenance needs. This may be due to line failures or defects identified during periodic inspection in sheds. Because of different operating conditions and maintenance practices followed by different sheds, there is a wide variation in the non-scheduled maintenance requirements. Again, if we look into the indentured items of TM then a lot of variation in non-scheduled arisings is reflected in the failure rates of rotors and stators. A sample of railway-wise traction machines failure data is given in Table 18.1. This table clearly shows that there is a large variation in repair arisings of the sheds and a differential approach in UES provisioning is required. The differential decision needs of multiple levels in UES management, along with the proposed models that can be used and the decision problems that can be addressed by these models are summarized in Table 18.2.

Table 18.2: Decision Levels, Proposed Models and Associated Decisions in UES Repair-Inventory System

Decision Level	Proposed Model	Decision Problems
Base shed and repair shop	Overhauling model	Overhauling capacity for repair arisings and unserviceable inventory Repair cycle and its effect on unserviceable inventory Cost of capital and its effect of overhauled TMA cost
Rebuilding and zonal repair shop managerial level	Rebuilding model	Rebuilding repair capacity Capital cost and its effect on rebuilt TMAs Comparative evaluation of multiple repair facilities Repair and despatch policies
Corporate level strategic decision making	Integrated model	System wide unserviceable UES inventory Logistics policies Capital accumulation in unserviceable inventory Technological selection for rebuilding shops Repair and despatch policies

18.3 THE MODELLING METHODOLOGY

Flow Modelling

Both PST and SD are flow modelling methodologies, where the flow of physical resources through various transformation, transportation and storage processes are captured graphically to give a vivid representation of the system. In PST, the graphic representation includes flow of material and exogenous inputs made at various transformation and other processes along with wastages, if any. It does not include explicit information about policies governing each flow. The PST system graph enables aggregation of processes and provision of storages to reflect process cycle time. The repair-inventory problem under study is a forward flow problem and its representation in simplified PST is quite straightforward. In fact, the planning and costing methodology followed in the rebuilding shop of the case problem under study is same as PST representation, and this gives a one-to-one correlation for decision-makers appreciation.

System Dynamics flow models have an important feature, in that it incorporates information flow as feedback loops and policies governing the flows from storages. This helps in making the system representation explicit but also leads to a more involved flow diagram, making it less tangible to the decision makers.

In PST, the system equations for storage stages are given by differential equations. A typical storage process is represented by component terminal graph in Figure 18.1. The component terminal equations for this storage process are as follows:

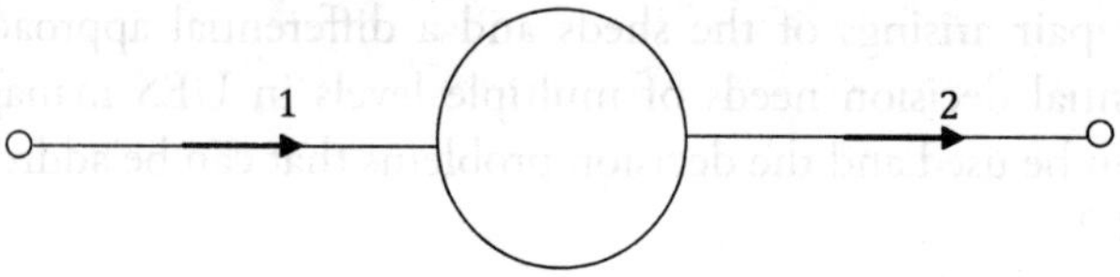

Figure 18.1: PST Component Model of a Storage Process

$$\frac{dM}{dt} = y_1 - y_2$$

$$\frac{dS}{dt} = x_1 y_1 - x_2 y_2 + h$$

$$x_2 = \frac{S}{M}$$

where,

- y_1 and y_2 are flow rates
- x_1 and x_2 are cost per unit
- *M*, *S* are material and capital accumulation respectively

This model simultaneously considers both the material flow and cost variables in an integrated manner in the same model with one set of parameters. The parameters are defined at a single stage independently facilitating modelling of the components.

In case of System Dynamics, the corresponding flow model of a typical storage process will be as given in Figure 18.2.

The first level equations in SD have recursive relationship, which can be simplified by a suitable selection of time interval DT as unity. This further simplifies the equations which then helps in solving the same. For the case given in Figure 18.2, the equations can be written as follows:

$M(t) = M(t-1) + y_1(t-1) - y_2(t-1)$

$S(t) = S(t-1) + x_1(t-1).y_1(t-1) - x_2(t-1).y_2(t-1) + h(t)$

$x_2(t) = S(t)/M(t)$

$y_2(t)$ = Despatch policy

$h(t)$ = Cost of capital

Integrated Scheme of Modelling

A block diagram of the integrated scheme of PST and SD is given in Figure 18.3. Exogenous inputs into the PST models are the material input flow, material costs, process costs and technological coefficients for transformation processes. The PST model gives the flow and cost variables for the system. These are used as inputs in SD formulation to work out different storage levels. Using policies to govern inflows and outflows from different storage points the material accumulation, the capital accumulation, the outflow from the system and the cost of the product can be simulated over a time period. To present the integrated scheme in proper perspective, inputs and outputs for PST and SD as individual systems and of PST and SD as integrated system have been given in Figures 18.4a, 18.4b and 18.4c respectively.

It will be seen that in case of PST alone, as shown in Figure 18.4a, for a given material cost, process cost, input flow and technological coefficient for transformation process, the material outflow and cost can be computed for a particular time. In case of only SD formulation, the dynamic behaviour of the levels for multiple policies governing the flow will be available as shown in Figure 18.4b. In case of an

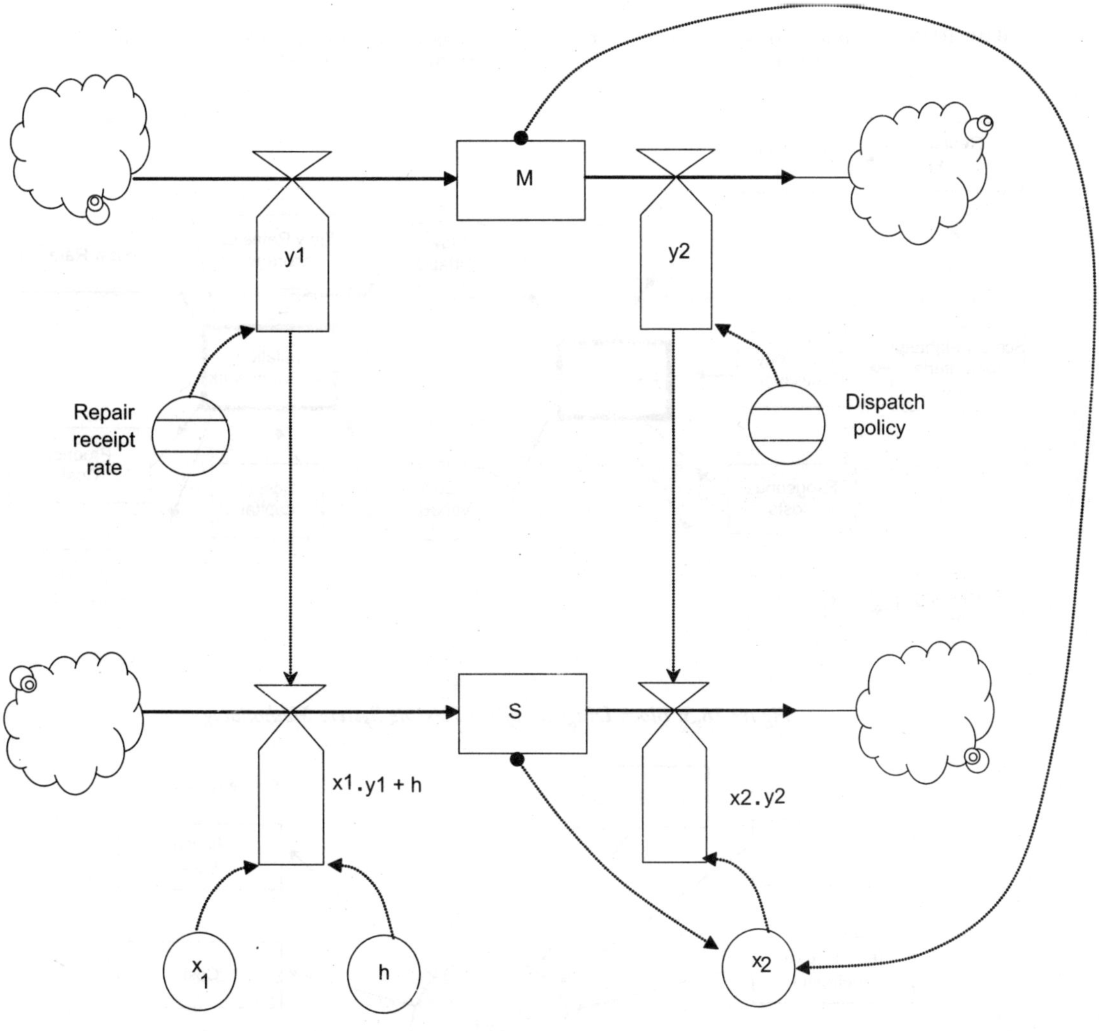

Figure 18.2: System Dynamics Model of a Storage Process

integrated model of PST and SD, the material flow, material accumulation cost of the products and capital accumulation can be studied as a dynamic system and the user can experiment with multiple policies to generate a satisfying solution as depicted in Figure 18.4c.

18.4 UTILITY OF INTEGRATED METHODOLOGY FOR DSS DEVELOPMENT

The proposed integrated PST-SD methodology has strong positive features which makes it suitable as a DSS modelling tool. The flow modelling process required minimum abstraction, has one-to-one correspondence with the real life system, and the problem formulation using mathematical equations all make it amenable for direct end user participation. As a general methodological tool, the integrated methodology supports:

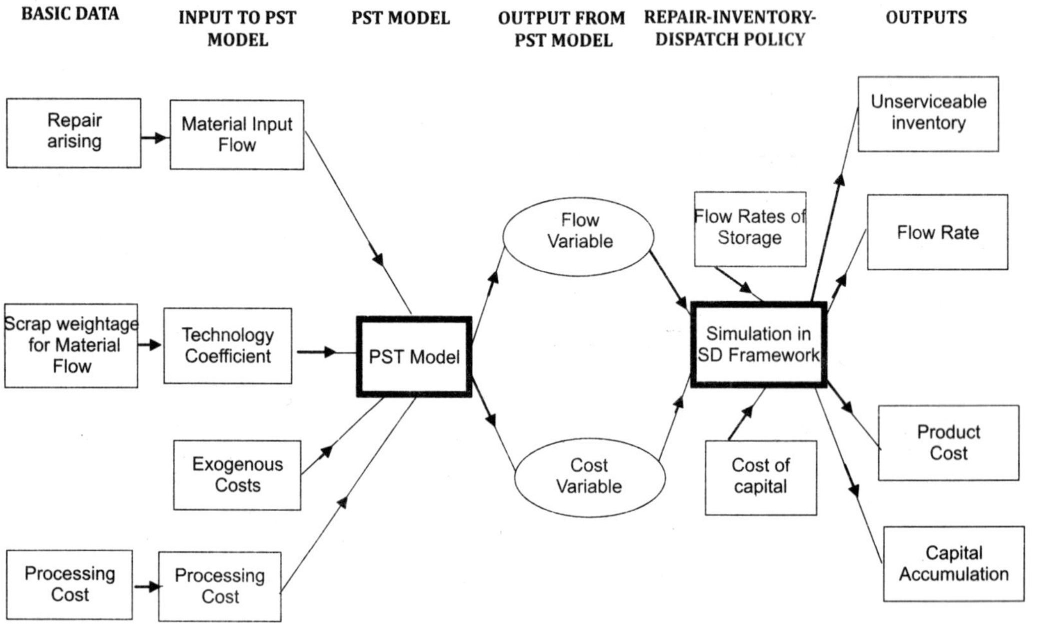

Figure 18.3: Block Diagram of the Flexible System Methodology

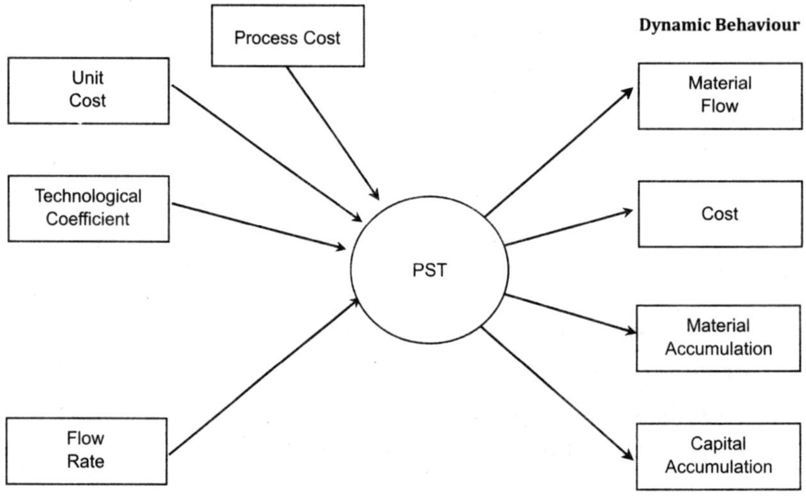

Figure 18.4 a: Input-Output of PST Model

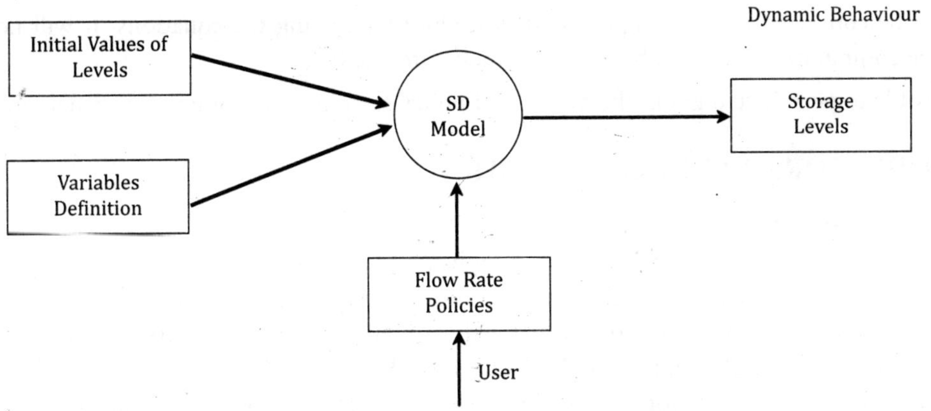

Figure 18.4 b: Input-Output of SD Model

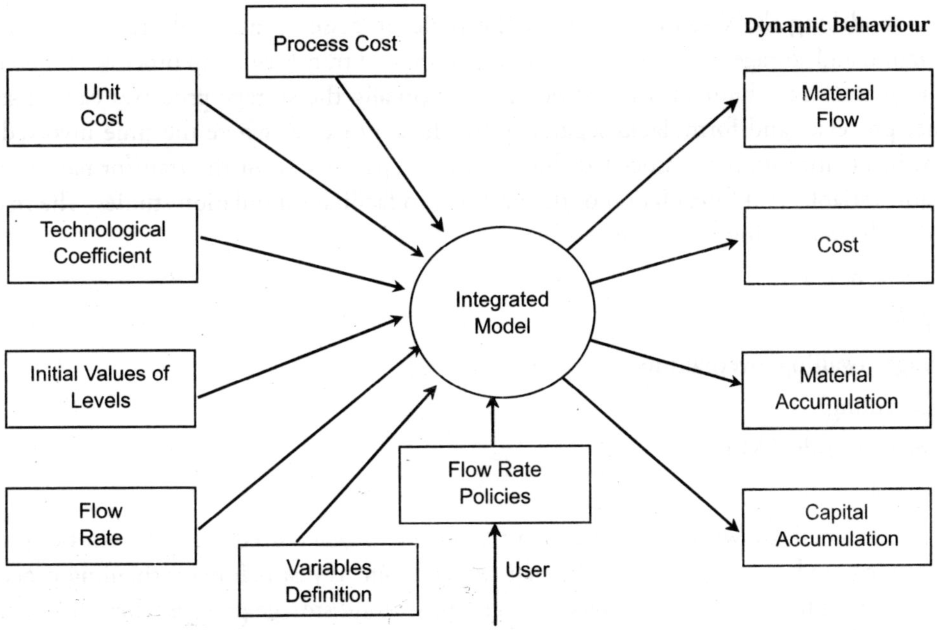

Figure 18.4 c: Input-Output of the PST-SD Integrated Model

i. End user participation in all phases of DSS design, development and analysis.
ii. Its modelling and solution procedures are transparent, inspiring user confidence.
iii. Interactive simulation capabilities.
iv. 'Learning' aspect while modelling and analysis helps in gaining more insight into the problem domain.

v. Can be converted into a computerized procedure for solving the equations. It will not require large computations and can be solved on moderate size PCs.
vi. Flexible to permit changes in the model and policy structure to facilitate experimentation.

18.5 MODEL FOR OVERHAULING UNIT

System Description

The overhaul of TMAs are undertaken in running base sheds during three yearly schedule, and during the periodic overhauls in back shops undertaken during six yearly overhaul of locomotives. All the TM armatures, removed from motors during scheduled maintenance or out of course repairs, are subjected to mechanical and electrical examination. Armatures which qualify for overhaul are cleaned of all carbon and grease and the top insulation is renewed through the application of varnish or insulating dye. Turning of commutators and mica undercutting is also carried out for correction of ovality. Overhauls are generally undertaken on time basis and the work involved is much less compared to rebuilding, and is considered as a part of preventive maintenance activity. The flow model for this system has been developed using integrated methodology of PST and SD.

While modelling the system using PST, the basic processes considered are the transportation, transformation and storage processes. The transportation and transformation processes inherently have some storage element inbuilt into it. However, conceptually the storage processes can be segregated from other processes and formulated separately for those processes where the time involved is large. This helps in identification of associated flow policies, separately from the transformation processes, as a decision variable, and formulation of the problem to facilitate simulation studies. The main stages associated with overhauling process are as follows:

i. Inspection of TMAs
ii. Reshafting
iii. Storage, awaiting, overhauling
iv. Overhauling
v. Storage of ready TMAs
vi. Despatch

The system graph drawn in the subsequent section is a generalized case wherein both reshafting and overhauling are being carried out. The first storage point is just before overhauling process where TMAs have to wait for service facility to be free, and the second storage point is when the armatures are ready awaiting despatch to some other location. Shed repair can be considered as a special case where no reshafting takes place, and the second storage is the end usage point with no despatch process involved as the shed themselves use these armatures for their Traction Motor assemblies.

Component Models

The component terminal equations consist of a cost equation and material flow equations for transformation stages. For storage, stages material and capital accumulation equations and flow equations are developed in SD framework with DT=1. The component terminal graphs for different components are shown in Figure 18.5.

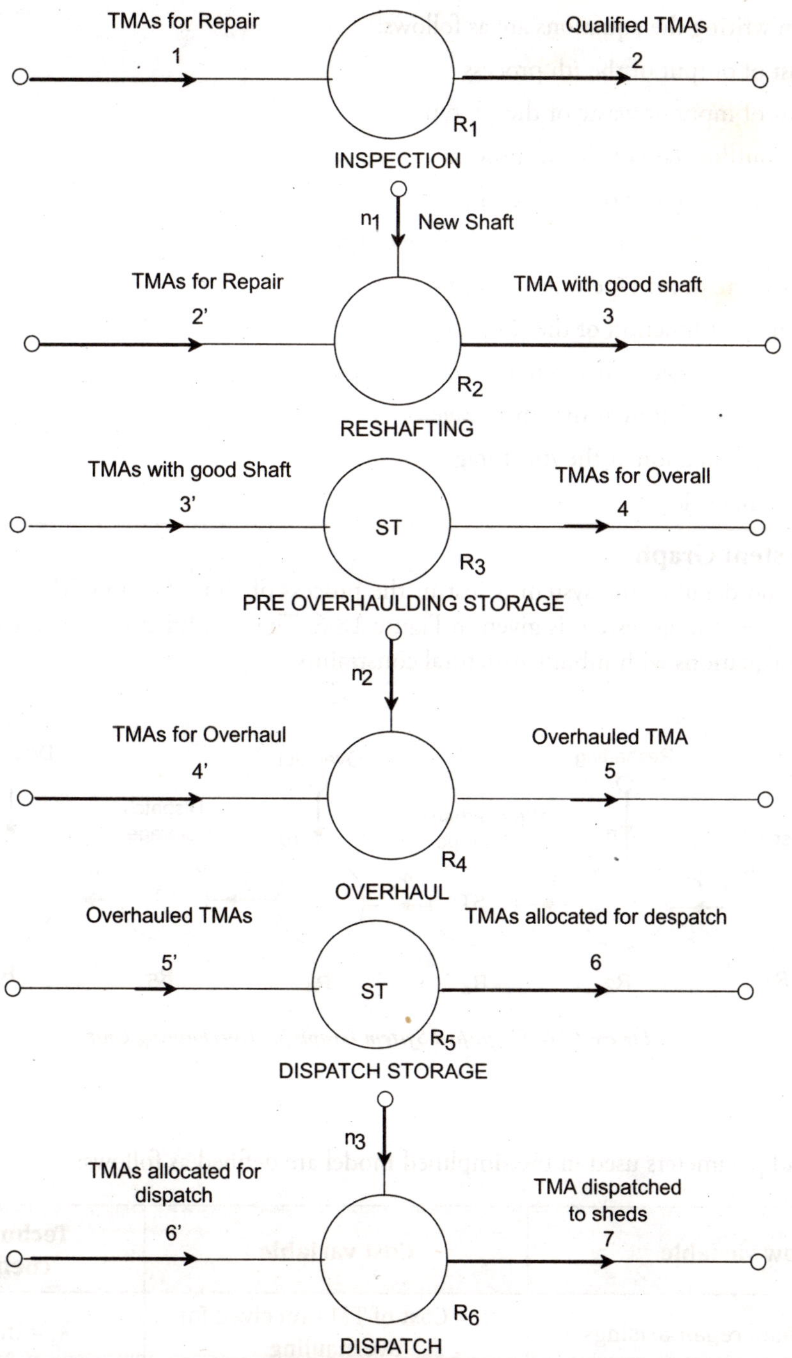

Figure 18.5: Component Terminal Graphs of the Overhauling Model

Symbols used in writing the equations are as follows:

x_i = Unit cost of output of the *i*th process
x_j = Unit cost of input or waste of the *j*th type
y_i = Material outflow rate of the *i*th process
y_j = Flow of input or waste of the *j*th type
k_j = Technological coefficient of input, as waste of the *j*th type
= Input or waste of *j*th type/output of process *i*
f_i = Processing cost function of the *i*th process
h = Capital cost (storage – maintenance costs) per period
M_i = Material accumulation at the *i*th storage
S_i = Capital accumulation at the *i*th storage
DT = Time internal = 1

Simplified System Graph

It is found that no detail of the system is lost in the process of simplification. The simplified system graph of TMA overhauling system is given in Figure 18.6. The simplified model equations have only the component equations with inbuilt structural constraints.

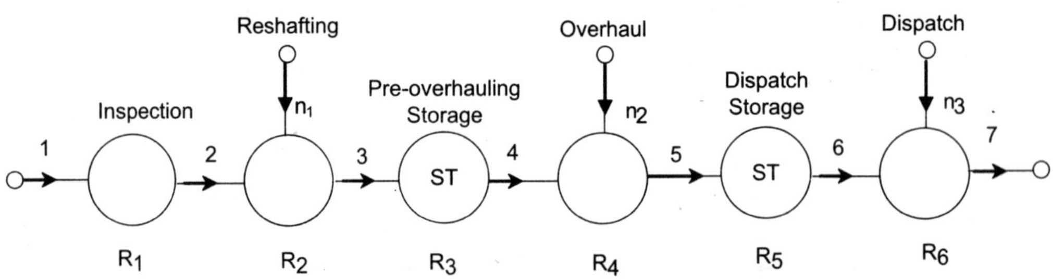

Figure 18.6: Simplified System Graph for Overhauling Unit

i. Legend

The variable and parameters used in the simplified model are defined as follows:

Flow variable	Cost variable	Technological coefficient
y_1 = Overhaul repair arisings	x_1 = Cost of TMA received for overhauling	k_1 = Inspection
y_2 = Armatures with repairs identified	x_2 = Cost of TMA after inspection	k_2 = Reshafting
y_3 = Overhaul repair arisings	x_3 = Cost of TMA after reshafting	k_4 = Overhauling

y_4 = Job from pre overhaul storage	x_4 = Cost of TMA after pre overhaul storage	k_6 = Packaging
y_5 = Output after overhaul process to dispatch store	x_5 = Cost of overhauled TMA	k_{n1} = New shaft usage
y_6 = From dispatch store for dispatch	x_6 = Cost of TMA after despatch store	k_{n2} = Overhaul material usage
y_7 = Dispatches to sheds	x_7 = Cost of TMA despatched	k_{n3} = Packaging material usage
y_{n1} = New armature shafts	x_{n1} = Cost of new shaft	
y_{n2} = Overhaul material flow	x_{n2} = Cost of material used for overhauling	
y_{n3} = Packaging material flow	x_{n3} = Cost of packaging material	
Process cost function	**Material accumulation**	**Capital accumulation**
f_2 = Unit cost for inspection	M_4 = Pre-overhauling store material accumulation	S_4 = Pre-overhauling store capital accumulation
f_3 = Unit cost of re-shafting	M_6 = Dispatch store material accumulation	S_6 = Dispatch store capital accumulation
f_5 = Unit cost of overhauling		
f_7 = Unit cost of packaging		
h = Cost of capital per period		

ii. System equations

System equations of the simplified system graph for the overhauling unit are as follows:

Process Stage	Type of Equations	Equations
Inspection	Flow Cost	$y_1 = k_1 y_2$ $x_2 = k_1 x_1 + f_1$
Re-shafting	Flow Cost	$y_2 = k_2 y_3$ $y_{n1} = k_{n1} y_3$ $x_3 = k_2 x_2 + k_{n1} x_{n1} + f_3$
Pre-overhauling Storage	Material accumulation Capital accumulation Unit cost Repair feed rate	$M_4(t) = M_4(t\text{-}1) + y_3(t\text{-}1) - y_4(t\text{-}1)$ $S_4(t) = S_4(t\text{-}1) + y_3(t\text{-}1).x_3(t\text{-}1) - y_4(t\text{-}1).x_4(t\text{-}1) + S_4(t\text{-}1).h$ $x_4(t) = S_4(t) / M_4(t)$ $y_4(t)$ = Repair Capacity Policy

Overhauling	Flow Cost	$y_4 = k_4.y_5$ $y_{n2} = k_{n2}.y_5$ $x_5 = k_4 . x_4 + k_{n2} . x_{n2} + f_5$
Dispatch Storage	Material accumulation Capital accumulation Unit Cost Dispatch to sheds	$M_6 (t) = M_6 (t\text{-}1) + y_5 (t\text{-}1) - y_6 (t\text{-}1)$ $S_6 (t) = S_6 (t\text{-}1) + y_5 (t\text{-}1). x_5 (t\text{-}1) - y_6 (t\text{-}1). x_6 (t\text{-}1) + S_6 (t\text{-}1).h$ $x_6 (t) = S_6 (t) / M_6 (t)$ y_6 (t) = Despatch Policy
Dispatch	Flow Cost	$y_6 = k_6.y_7$ $y_{n3} = k_{n3}.y_7$ $x_7 = k_6 . x_6 + k_{n3} . x_{n3} + f_7$

The continuity and compatibility of the structural constraints have already been incorporated in the above equations.

iii. Mass balance equations

As there is no process loss of unit exchange spares and they are assumed to be repairable, the mass balance equations for the system can be written as follows:

Repair arising $y_1 = y_2$

Intermediate flow $y_2 = y_3$

Repair capacity dependent flow $y_4 = y_5$

Dispatch policy dependent flow $y_6 = y_7$

Note: All x and y variables are also a function of time but (t) has been omitted in the equations to retain clarity in expressions.

SD Model of Storages

The SD flow diagrams for pre-overhauling storage and dispatch storage stage have been drawn and given in Figures 18.7 and 18.8 respectively. This is to bring out the feedback and policy decision components, affecting the behaviour of the system explicitly, which are otherwise not apparent in the simplified system graph of the repair-inventory system. Also, the level, rate and other associated auxiliary equations of one of the storage stage namely pre-overhauling storage have been written using DYNAMO equations for existing overhaul system structure of the overhauling system as already described. This has been done only as an example to show the nature of equations and further analysis in DYNAMO form alone has not been presented.

Policy Structure Diagram

A simplified integrated system graph of overhauling unit with storage related policy structure is given in Figure 18.9. This novel method of representation has been chosen to bring into focus the policies governing flows to and from the storage levels. This has been done with twin objectives of retaining simplicity in the diagrammatic representation of the system as well as to get the end users explicitly identify flow policies for simulation studies. Some of the alternative policies relevant in this case for the pre-overhauling storage outflow and for dispatch store policies are given in Table 18.3.

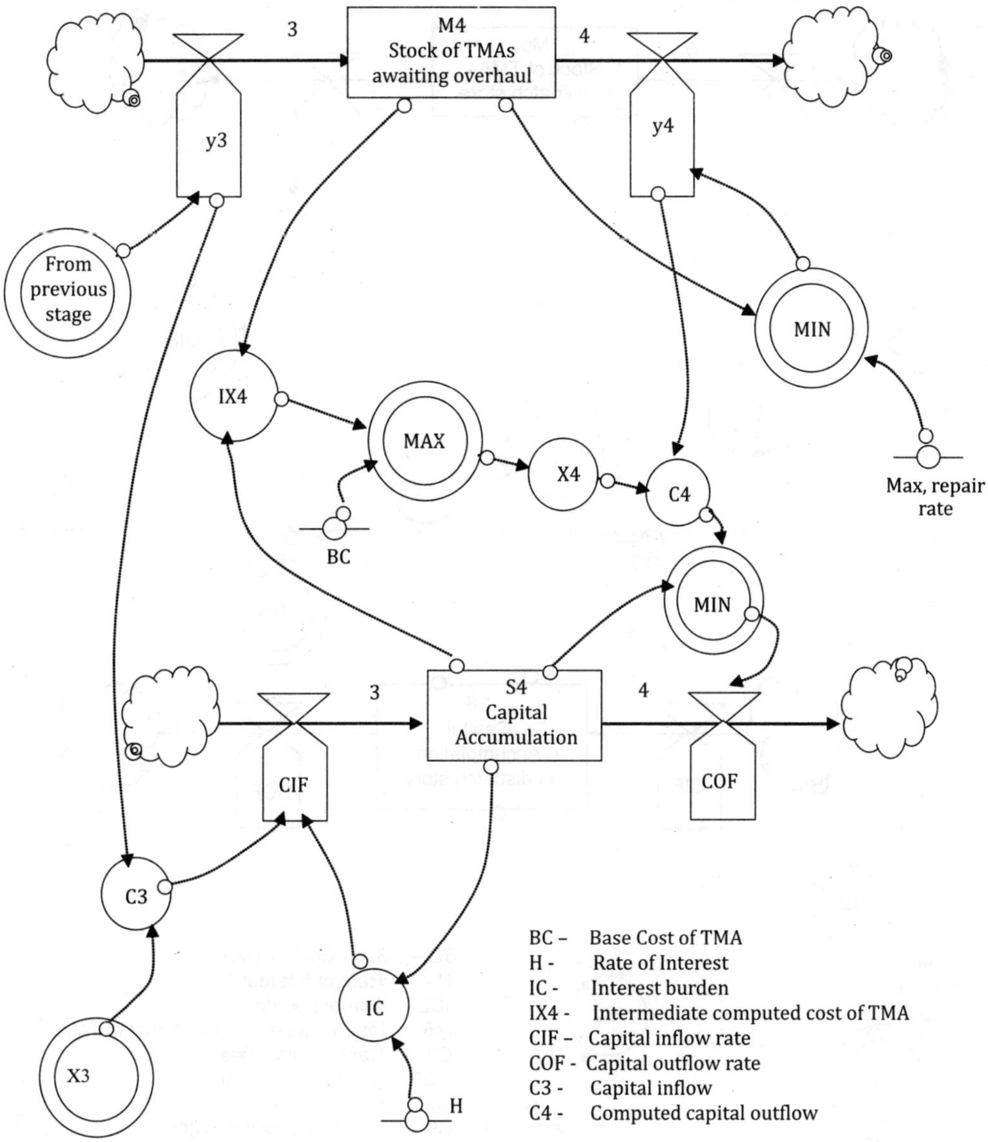

Figure18.7: SD Diagram of Pre-overhauling TMA Storage Stage

DYNAMO equations for pre-overhauling storage are given in Appendix 18 A as an example to show the nature of equations that will be required in SD. The DYNAMO equations for only one storage stage has been given as an illustration and no further analysis have been carried out on these.

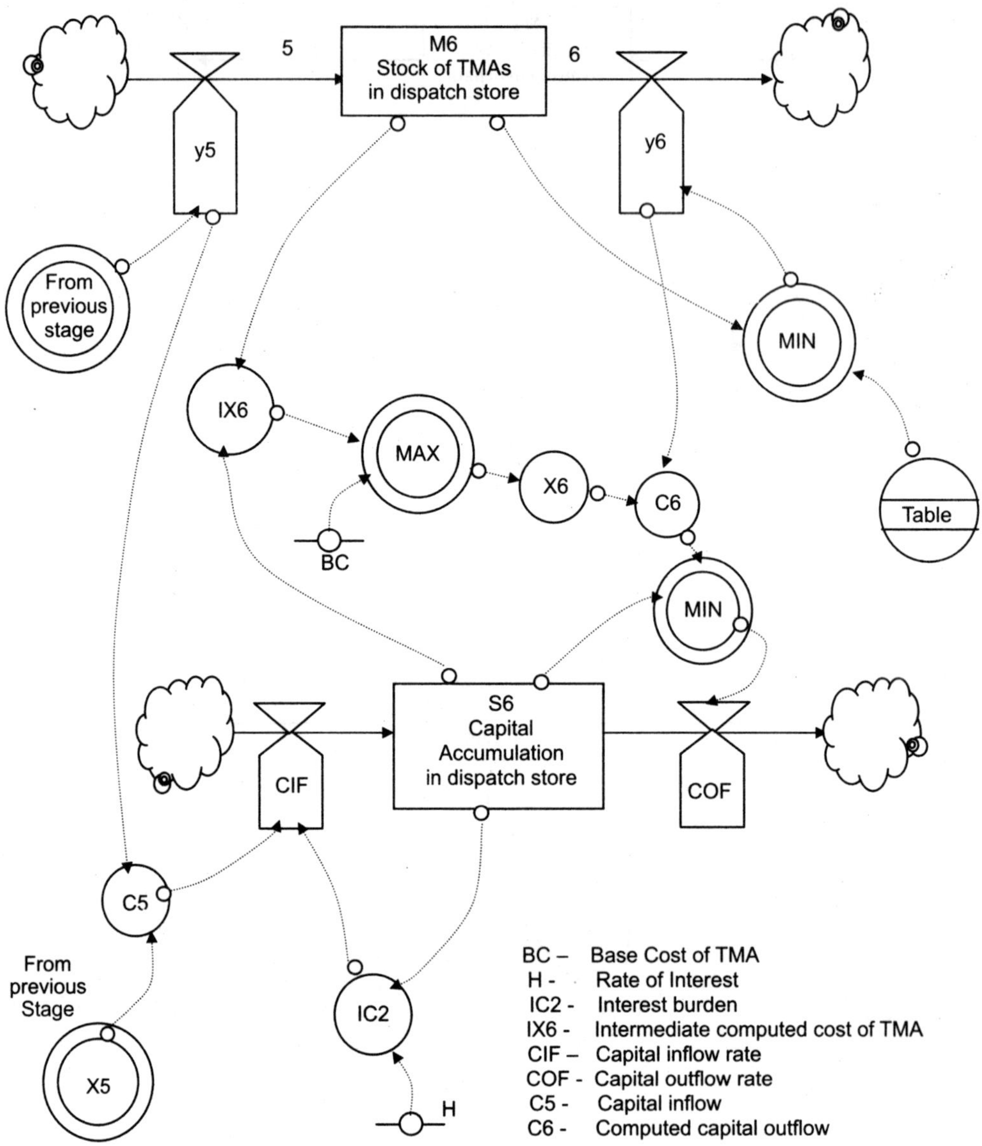

Figure 18.8: SD Diagram of Ready TMA Dispatch Store Stage

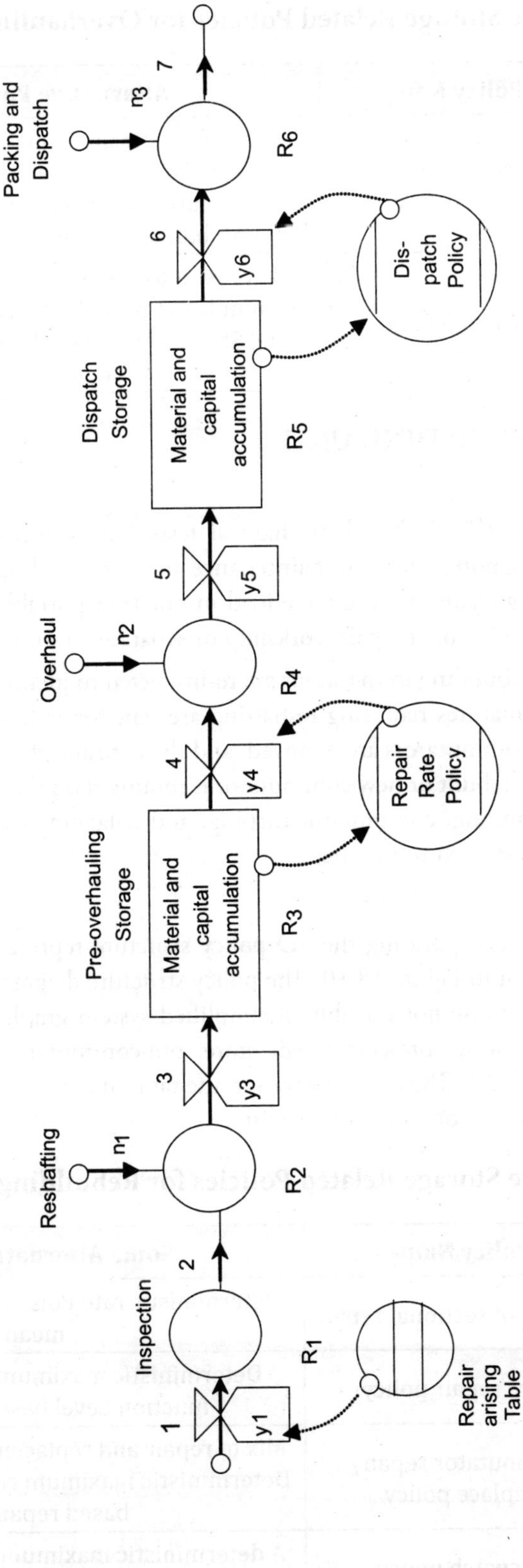

Figure 18.9: Simplified PST-SD Integrated System Graph of Overhauling Unit with Storage Related Policy Structure

Table 18.3: Some Storage Related Policies for Overhauling Model

Stage	Policy Name	Alternative Policies
Pre-overhauling Storage	Overhauling repair rate policy	Deterministic uniform repair rate Table function of cumulative probability distribution of past data, All of previous month balance Storage level based repairs
Dispatch Storage	Dispatch Policy	All of previous month balance Table function of cumulative probability distribution of past data, Storage level based dispatches Multiple of blocking factor

18.6 MODEL FOR TMA REBUILDING UNIT

System Description

TMAs which do not qualify in electrical and mechanical tests during receipt inspection stage of scheduled maintenance or during non-scheduled maintenance are sent for rebuilding. The sheds or the shops where such loads arise, aggregate the repairable load in one transportable lot and send them to the back shop, which may be within zonal repair workshop or separate rebuilding unit.

The armatures received for rebuilding in the shops are re-inspected to identify job content involved in rebuilding the armatures. Armatures requiring reshafting are sent for reshafting first and then the armature coils are stripped, the commutators are removed, and the core and the commutators are taken separately for repairs. A repaired, rebuilt or new commutator is mounted on the repaired core. It is then sent for winding, varnishing, banding, commutator turning, and balancing. The rewound armatures are subsequently packed and dispatched to the sheds.

Policy Structure Diagram

The integrated PST-SD model, incorporating the SD policy structure representation for the storage stages, has been made and is given in Figure 18.10. The policy structure diagram brings out the policy variables explicitly, which is otherwise not possible in simplified system graph alone. Some alternate policies relevant for the receipt store, pre-core repair store, pre-commutator repair store, and the despatch store are given in Table 18.4. The policies given in the table are only indicative of the possible variation in decision problem and is not an exhaustive list.

Table 18.4: Some Storage Related Policies for Rebuilding Model

Process Stage	Policy Name	Some Alternate Policies
Receipt Store	Receipt of sectional repair	Deterministic rate Poisson arising with some mean
Core repair storage	Core repair policy	Deterministic maximum repair rate Table function Level based repair rate
Commutator repair storage	Commutator repair/ replace policy	Mix of repair and replacement of commutator Deterministic maximum repair Inventory level based repair rate
Dispatch store	Dispatch policy	A deterministic maximum rate in multiples of some number All of previous month balance

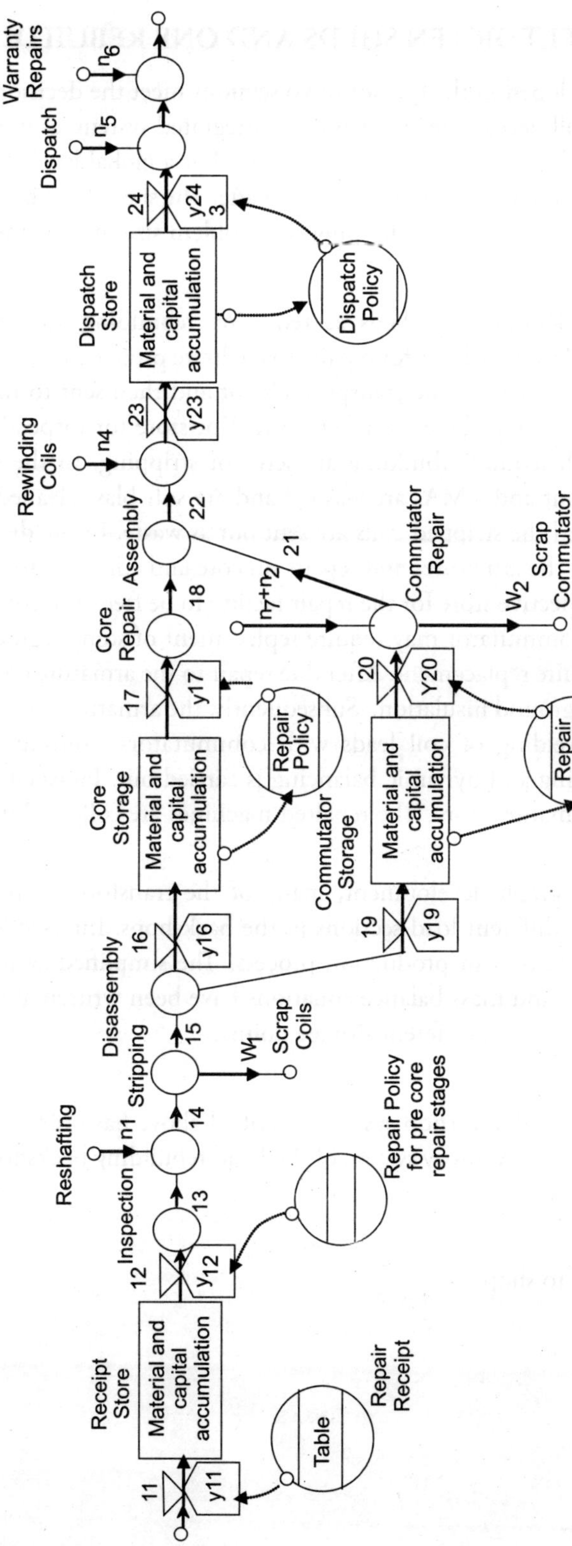

Figure 18.10: Simplified System Graph of a Rebuilding Unit with Storage Related Policy Structure

18.7 INTEGRATED MODEL FOR TEN SHEDS AND ONE REBUILDING UNIT

While the modular models developed in the previous two sections meet the decision needs of a certain decision level, a system study will necessitate an analysis of integrated system, as in real life UES repair inventory problem is a multi-echelon-multi location problem. For a global level analysis, a combined model of feeding base sheds and a rewinding unit has been made. The sheds have been classified in three groups based on their maintenance arisings for keeping the problem size manageable.

System Description

Armatures which have been identified by sheds as requiring rebuilding are collected to form a transportable lot and despatched to backshops for repair. To facilitate proper control over transportation of UES, the failed TMAs are collected in one transportable lot and then sent to the back shop. After the armatures are received in the shop these are taken to receipt store for inspection and assessment of repairs, and armatures which require rebuilding are sent for stripping. In the stripping stage the coils from TMAs are stripped out and TMAs are washed and are soft blast cleaned for removal of all varnish and carbon particles, and the stripped coils are sent out as waste. In the disassembly stage, the commutators are removed from the armatures and repairs on core and commutators assessed. The core and commutator are kept in respective store for the repair facility to be free. The core and commutators are then repaired wherein the commutator may require replacement of some segments and restacking and some commutator may require replacement. After due repair to the armature core and commutator these are reassembled with new ground insulation. Subsequently, the armatures are taken for rewinding process where winding, end welding of coil leads with commutators, commutator turning, mica undercutting, banding, varnishing and dynamic balancing is carried out. Finished armatures are then tested and kept in a repaired armature store. The repaired machines are collected in transportable lots and despatched to the sheds.

For modelling and system graph development, many of the transformation stages and storage stages have been clubbed as per different load sections in the backshops. Intersectional storage points act as buffers to avoid any disruption in production process. The simplified system graph has been drawn and the system equations and mass balance equations have been written along with the capital and material accumulation equations at different storage points.

Simplified System Graph

A flow model of the integrated shed and shop system described above has been made and is given in Figure 18.11. The combined repair-inventory system of shed and rebuilding backshop has the following transformation and storage stages:

(a) Storage of TMAs in shed
(b) Transportation from shed to shop

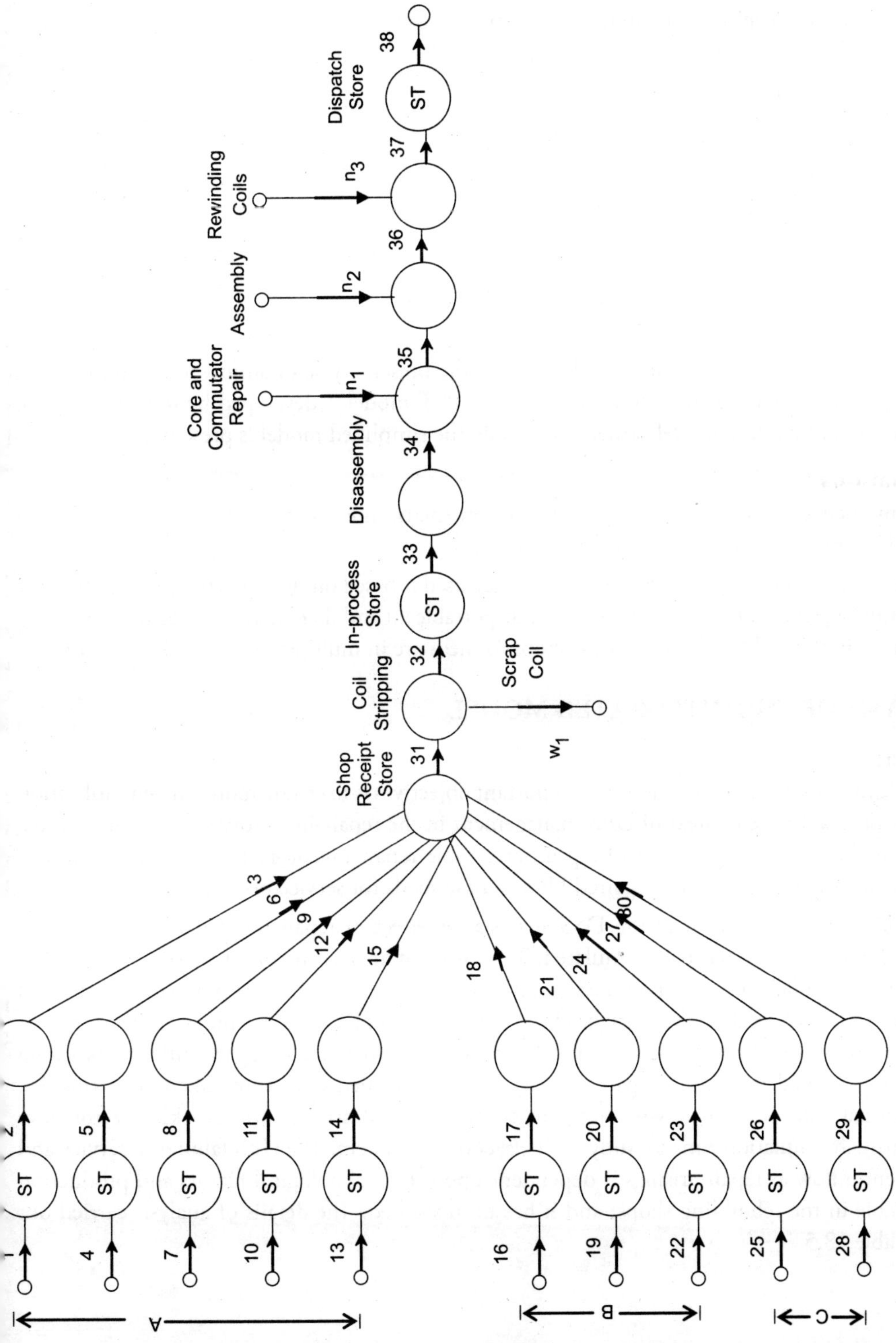

Figure 18.11: Simplified System Graph of Integrated System of Ten Sheds and One Rebuilding Unit

(c) Storage of TMAs received for rebuilding in shop
(d) Stripping of coils
(e) Storage awaiting rebuilding
(f) Disassembly of TMA into core and commutator
(g) Core and commutator repair
(h) Assembly of core and Commutator
(i) Rewinding
(j) Storage of ready TMAs
(k) Despatch

Four storage points have been considered to reflect the present cycle of almost four months from the time of repair arisings to repair and despatches. The PST model is developed by following all the steps given for the overhauling model earlier. Here, only the simplified model is presented.

System Equations

Using the simplified system graph of the integrated ten sheds and one rebuilding shop, the system equations and the mass balance equations are developed.

In the integrated model analysis, the operating policy used is based on the concept of blocking factor where UES requiring repair are collected to form a transportable lot and then despatched to the rebuilding shop for repairs. Similarly, the despatches from shops to sheds are in multiple of the blocking factor.

18.8 ANALYSIS OF THE INTEGRATED MODEL

Analysis Plan

In a large stable functioning system, one of the important objectives is to ensure more efficient utilization of available resources. In the context of UES management in the repair-inventory system under study, this can be stated as evolving policies to keep unserviceable repair-inventory to a minimum level to ensure maximum availability of good repaired UES to base sheds for service utilization.

In view of the above, in most of the DSS analysis the effect of repair-inventory policies on the unserviceable UES inventory has been simulated. The thrust of analysis reported in this chapter is on the effect of logistics policies on the unserviceable inventory. The combined effect of repair capacity and blocking factor on unserviceable inventory has been studied through multiple simulation runs for one set of repair arising values. Scenario analysis for different average repair arisings for different blocking factors and a pre-defined maximum repair capacity has been carried out. Three likely scenarios, i.e. optimistic, normal, and pessimistic with different combination of arisings and blocking factor have been tested through simulation runs to study their effect on the system. The UES failures in service and their subsequent inflow as repair arisings is dependent upon the technological factors and production facilities available in the rebuilding shops, and it has been studied. The details of analysis carried out are given in Table 18.5.

Table 18.5: DSS Analysis Plan for the Integrated Model

Policy Type	Decision Criteria	Test Condition	Decision Problem
Logistics policy	Unserviceable inventory Capital accumulation	Repair arisings of A, B and C sheds with Poisson mean of 3, 5 and 7 Blocking factor 4, 6, 8, 10, 14 and 18 Repair capacity 45 TMA per month Cost of capital at 18 per cent per annum	What is the effect of blocking factor on unserviceable inventory level? What is the effect of blocking factor on capital accumulation in unserviceable inventory?
Repair capacity	Unserviceable inventory	Average repair arising of 43.9 TMAs for 10 sheds. The maximum repair capacity at 41, 45 and 51 TMAs per month Blocking factor of 6 and 8	What is the effect of repair capacity on unserviceable inventory for a given blocking factor? What is the effect on unserviceable inventory for different blocking factors?
Scenario analysis with different repair arisings	Unserviceable inventory	Maximum repair capacity of 45 TMAs per month Average repair arisings of 42.4, 42.6 and 45.4 TMA units per month Blocking factor of 6, 12 and 18	What is the expected inventory for a range of repair arisings with different blocking factor for a maximum repair capacity?
Analysis for three likely scenarios	Unserviceable inventory Capital accumulation	Optimistic scenario Normal scenario Pessimistic scenario	What is expected unserviceable inventory and capital accumulation for the three scenarios?
Technology selection	Unserviceable inventory	Poisson repair arising with mean 2, 4 and 6 for A, B and C group of sheds respectively. Repair capacity of 47 TMA per month Blocking factor 6 Failure rates of 5, 10 and 15 per cent	What is the effect on inventory for different failure rates?

Assumptions made for Simulation Runs

For carrying out simulation studies some general assumptions have been made which are given below. Any specific assumption made for a particular simulation run has been mentioned along with the description of that case. The general assumptions are:

i. For operational control over receipts and dispatches of armatures, the base shed can despatch and receive only one consignment every period. The unserviceable armatures are collected in a transportable lot for despatch and sent to rebuilding shop for repairs.

ii. The return from the back shop has been simulated on the basis of despatches made from the shop and repaired TMAs once despatched from shop has been assumed to be out of the unserviceable inventory system.

iii. The repair capacity of rebuilding shop is deterministic as the shops have production linked incentive scheme. This ensures that normal monthly production targets are achieved.

iv. Cost of capital is 1.5 per cent per month.

v. The cost of transportation per unit remains same for different blocking factors. Similarly, processing cost function does not change when repair capacity lies in a small range.

The dynamic behaviour of the integrated system has been studied through simulation runs of the DSS model for various cases. The total unserviceable UES inventory in the system is a function of repair capacity, repair arisings, repair cycle and logistics factors and these have been studied through simulation experiments.

Data used in simulation study are given in Appendix 18 B. These are only base run values and have been modified to meet the specific test requirements. In Appendix 18 B, sample values of one set of simulation run are given.

Logistics Policy Analysis

Simulation runs to study the effect of logistics policies on the unserviceable inventory have been undertaken. In the simulation experiments, reported in the previous cases, every run was treated as independent run and a new series of random numbers was generated using the cumulative probability table function. In the present set of analysis, one set of values was generated for repair arisings using Poisson distribution cumulative probability table with mean 3, 5 and 7 for A, B and C group of sets respectively. This unique set of repair arising values was then used for one set of simulation run and other parameters like blocking factor of repair capacity were changed to study their effect on unserviceable inventory.

i. Unserviceable inventory for different blocking factors

The first set of runs has been taken for blocking factors 4, 6, 8, 10, 14 and 18. The maximum repair capacity of the system is defined at 45 TMA units per month and the despatches from shop to shed are equal to previous month's production. The expected average inventory for 22 months simulation run, for different blocking factors the system tends to attain a steady state level which is increasing as the blocking factor is increased from six onwards. This appears logical as for higher blocking factors, the shed store inventory will increase till the repairable load in sheds form a transportable lot. Figure 18.12 indicates that when blocking factor is 4, which is less than the mean arising of group B and C sheds, leads to a very high unserviceable inventory level. As the blocking factor is increased to 6 to 8, the unserviceable inventory comes down considerably. The total unserviceable inventory curve takes a trough shape in this region of blocking factors and as it increases to 10, 14 and 18 expected average inventory level increases.

ii. Effect on capital accumulation for different blocking factors

The effect of blocking factor on capital accumulation has also been studied and two sets of simulation runs for blocking factors 6 and 8 were taken when the average repair arisings are at 43.9 TMA units per month and repair capacity is at 51 TMAs per month and the results have been presented in Figure 18.13.

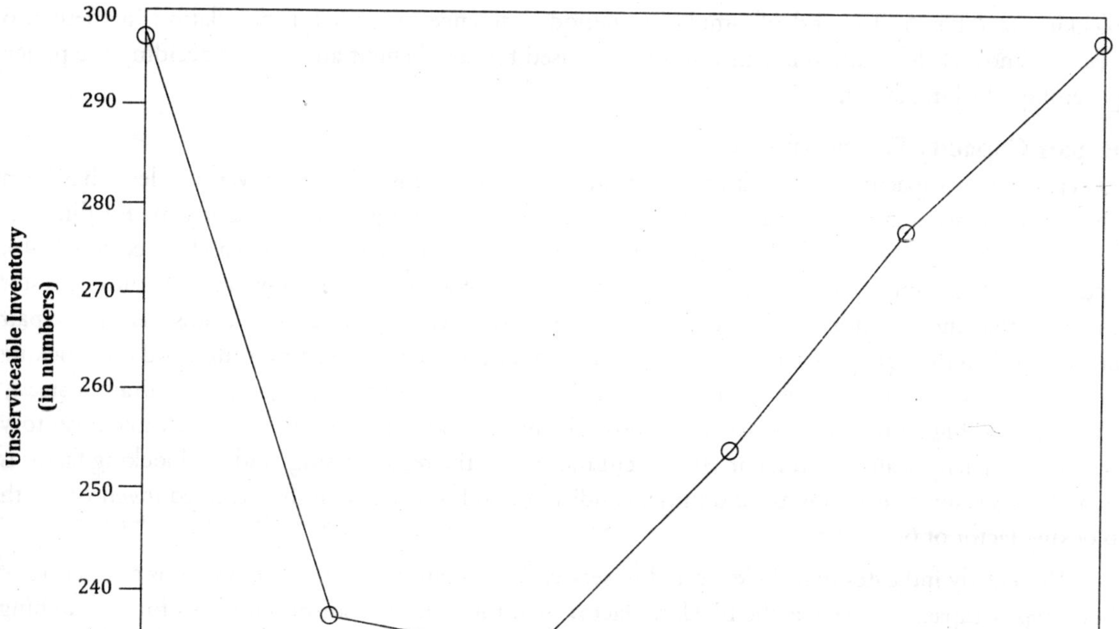

Figure 18.12: Average Unserviceable Inventory for Different Blocking Factors when Repair Capacity is 45, and Arisings 43.9 TMA Per Month

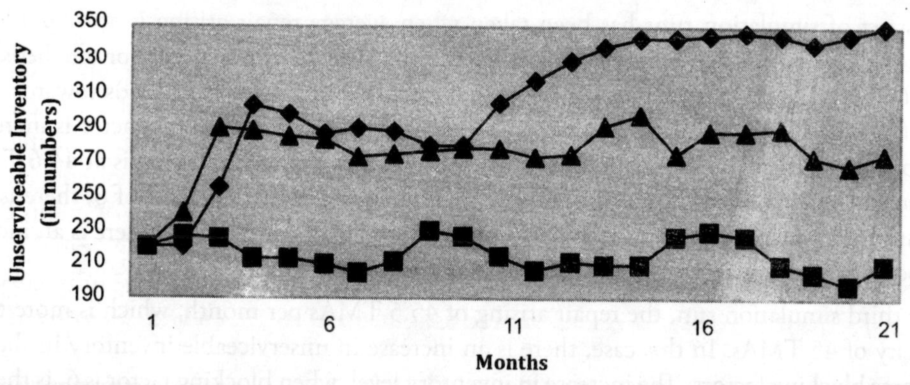

Figure 18.13: Unserviceable Inventory of TMAs for Optimistic, Normal and Pessimistic Scenarios

A study of the results indicate the effect of blocking factor in financial terms and shows that the capital accumulation is lower when blocking factor is 8 vis-à-vis the capital accumulation when the blocking factor is 6. The end of simulation period difference in capital accumulation in these two cases is almost Rs four million. This result can be used for cost-benefit analysis for deciding the policy governing blocking factor.

Repair Capacity Policy Analysis

Effect of repair capacity and blocking factor on the expected unserviceable inventory level has been studied in the second set of simulation runs. Here, the repair arisings has been kept same for different runs with an average of 43.9 TMA units per month. The repair capacity has been changed to 41, 45, and 51 TMAs per month and the experiment repeated for blocking factors of 6 and 8. A study of result indicates that the inventory of TMAs will rise rapidly when the repair capacity is less than the repair arisings. When the repair capacity is equal or 10 per cent more than the repair arisings, and the blocking factor is 6, there is an increase in the unserviceable inventory and the system tries to attain a steady state which is higher than the initial conditions. The results show the expected unserviceable inventory when the capacity is almost equal or 10 per cent more than the repair arisings and the blocking factor is 8, and the steady state level is around initial conditions, and is lower than the expected inventory with blocking factor of 6.

This clearly indicates that the expected unserviceable inventory level in the system is a function of both, repair capacity as well as the blocking factor used for forming transportable lots for despatching repair jobs from sheds to rebuilding shop.

Scenario Analysis for Different Repair Arisings

In the integrated model analyses, the repair arisings of the sheds have been read from Poisson probability distribution table and multiple simulation runs indicated that the average repair arisings will be between 42 and 45.5 TMAs per month. Simulation experiments to carry out scenario analysis for this range of repair arisings with different blocking factors have been carried out. In these experiments, the repair arisings pattern for an average arisings has been kept the same and by changing the blocking factor the simulation runs have been carried out to study the impact of logistics policies on the unserviceable inventory for a maximum repair capacity.

The first set of simulation runs has been taken when average repair arising is at 42.4 TMAs per month and the repair capacity is at 45 TMAs per month. The blocking factor for all sheds is kept uniform at 6, 12, and 18. It can be seen that the unserviceable inventory level tends towards a steady state level of 220 to 250 TMAs. The unserviceable inventory, when the blocking factor is increased to 18, is at about 300 TMAs. In the second set of simulation runs, the repair arising is at 44.6 TMA per month when the repair capacity is at 45 TMAs per month. For a blocking factor of 6, there is a small increase in inventory level to 240 to 260 range. For blocking factors of 14 to 18, there is an increase in initial period and then it stabilizes around 280 to 300 TMAs range.

For the third simulation run, the repair arising of 45.5 TMAs per month, which is more than the repair capacity of 45 TMAs. In this case, there is an increase in unserviceable inventory in the system for all the three blocking factors. The increase in inventory level, when blocking factor is 6, is the lowest, while in the other two cases, the increase is quite sharp in the initial period.

The above scenario analysis covering a range of repair arisings for different blocking factors and with the deterministic repair capacity brings into focus the learning capabilities aspect of DSS. The decision-maker by his intuitive knowledge of the system and understanding of the present status can choose to run the experiments for different repair arisings and logistics policies and generate possible scenarios for his decision making. In the present case, the effect of blocking factor for different repair arisings vis-à-vis repair capacity can be seen. It is found that an increase in repair capacity coupled with blocking factor in the range of 6 to 12 will lead to a steady state inventory level which is lower than the case when the blocking factor is 18.

Analysis for Three Likely Scenarios

For an application of the effect of multiple combination of repair arisings and the blocking factors on unserviceable inventory and resultant capital accumulation, analysis for three likely scenarios viz.– optimistic, normal, and pessimistic for the existing system has been carried out. The test conditions of these test runs are given in Table 18.6.

Table 18.6: Test Condition for Scenario Analysis and Expected Average Unserviceable Inventory and Capital Accumulation

Scenarios	Expected arising (in nos.)	Blocking factor (in nos.)	Repair capacity (in nos.)	Expected average unserviceable inventory	Average capital accumulation (Rs million)
Optimistic	42	8	45	215.3	30.236
Normal	45	14	45	277.7	36.873
Pessimistic	46.5	18	45	307.6	40.092

The simulation runs have been carried out over 22 periods and the expected unserviceable inventory and the capital accumulation for the three scenarios have been computed, and the results presented in graphic form in Figure 18.13 and 18.14 respectively.

A study of unserviceable inventory graph given in Figure 18.13 reveals that for an optimistic scenario when the repair capacity is more than expected arisings, and with optimum blocking factor of 8, the system will be in a steady state and the unserviceable inventory of TMAs will fluctuate around initial conditions as reflected by inventory curve for optimistic conditions.

In the second set of simulation runs, for the normal expected operating conditions, the repair capacity is kept equal to the arisings at 45 TMAs per month. Transport from sheds to backshops and back is by rail vehicles with a blocking factor of 14 to ensure almost full utilization of vehicle capacity. It is found that the unserviceable inventory level, indicated by inventory curve for normal conditions in the Figure 18.13, increases quickly in the initial period and then stabilizes at a higher level.

The pessimistic scenario wherein the repair arisings are more than the repair capacity can occur if there is an increase in repair arising, sudden change in transportation requirement, dislocation maintenance activities due to some emergencies. In such case, it has been assumed that the repair arisings will exceed repair capacity by about 4%. Also, the logistics policy may be to use full capacity of railway vehicle, hence a blocking factor of 18 is used. Under these conditions the unserviceable repair-inventory system shows a sharp rise in the inventory with a rising trend over time as shown by curve for pessimistic conditions in Figure 18.13.

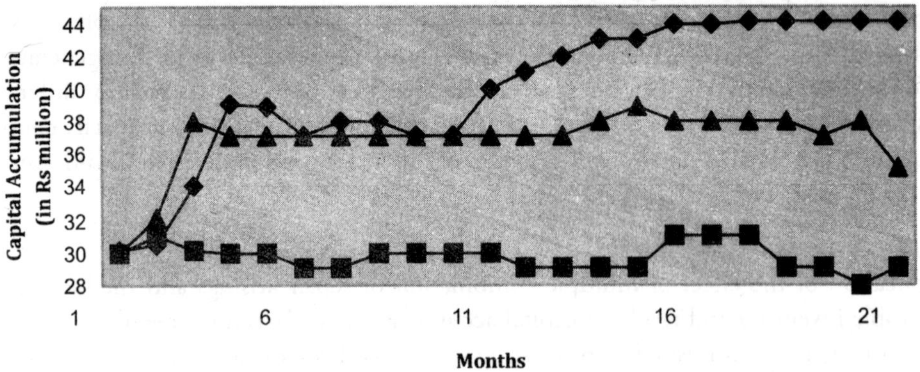

Figure 18.14: Capital Accumulation for Optimistic, Normal and Pessimistic Scenarios

A study of the capital accumulation graph for the above three conditions in Figure 18.14 indicates that in the optimistic scenario the capital accumulation is about Rs 30 million, which increases to Rs 38 million under the normal conditions and will increase to Rs 45 million at the end of 22 period for the pessimistic scenario.

REFLECTIONS

The results of integrated model analysis bring out some salient features of the TMA repair-inventory system. Repair of TMA in backshop will require transportation from sheds to shop and back involving time and money, and will present new control problems. The logistics policies have a significant effect on the unserviceable inventory and capital accumulation. For a given set of repair capacity and repair arisings, the expected average unserviceable inventory curve for different blocking factors assumes a trough shape, giving a range of blocking factors that can be chosen for a repair-inventory system. It is observed that a blocking factor which is less than the mean monthly arisings of all sheds leads to increase in unserviceable inventory. When blocking factor is such that it covers the monthly arisings of all sheds then the inventory level lies in the lower part of the total inventory curve. With an increase in blocking factor beyond this, the inventory level again increases. This shows that there is a need for flexible policy towards blocking factor and a contingent approach, depending upon the repair arisings of the shed, may be adopted. When there is repair capacity constraint the inventory build-up will take place in all cases and the total inventory curves are found to be similar in nature. However, the initial build-up is much more when the blocking factor is on the higher side. When capacity is more than arisings, choice of correct blocking factors can lead to a lower inventory unserviceable TMAs in the system.

Analysis for pessimistic scenario indicates that when arisings exceed the repair capacity and the logistics policy requires use of railway parcel vans only, the capital accumulation is quite large. This shows that while planning repair capacity for costly spares, the pessimistic conditions may be assumed for any cost-benefit analysis. This will ensure that the system generally operates in the optimistic condition range and does not reach the pessimistic conditions.

APPENDIX 18A

DYNAMO Equations of Pre-overhauling Storage

L	M4.K	=	M4.J + DT* (Y3.JK – Y4.JK)
			M4 = TMAs awaiting overhaul inventory
			Y3 = Inflow from previous stage
			Y4 = Overhaul rate
N	M4	=	60
A	Y3.KL	=	TABLE (Y3T, TIME.K, MIN, MAX, INC)
T	Y3T	=	TABLE
R	Y4.KL	=	MIN (M4.K, RC)
			RC = Maximum repair capacity
C	RC	=	25
L	S4, K	=	S4.J +DT* (CIF.JK-COF.JK)
			S4 = Capital accumulation
			CIF = Capital inflow rate
			COF = Capital outflow rate
A	C3.K	=	Y3.JK*X3.K
A	X3.K	=	Table
			Y3 = Inflow rate of armatures
			X3 = Cost of armatures
			C3= Computed capital inflow rate
A	IC.K	=	S4.K*H
C	H	=	.015
			IC = Interest cost
			H = Rate of interest per period
R	CIF.KL	=	C3.K + IC.K
R	COF.KL	=	MIN (C4.K, S4.K)
A	C4.K	=	X4.K*Y4.JK
A	X4.K	=	MAX (IX4.K,BC)
A	IX4.K	=	S4.K/M4.K
C	BC	=	250
			IX4 = Intermediate computed X4
			C4 = Computed Capital Outflow

BC = Base Cost

APPENDIX 18B

DATA USED IN SIMULATION STUDY

Unit Cost of Material Input and Waste

Process Stage	Variable Name	Unit Cost (in Rs Thousand)
Core and commutator repair	x_{n1}	60
Assembly	x_{n2}	20
Rewinding	x_{n3}	160
Scrap conductors	x_{w1}	10

Values of Technological Co-efficients

Process Stage	Variable Name	Value
Technological coefficient for transportation process from shed to shop	k_2; k_5; k_8; k_{11}; k_{14}; k_{17}; k_{20}; k_{23}; k_{26}; k_{29};	1
Stripping	k_{31}	1
Disassembly	k_{33}	1
Core and commutator repair	k_{34}	1
Assembly	k_{35}	1
Re-winding	k_{36}	1
Scrap coils	k_{w1}	1
New commutators	k_{n1}	1
Assembly	k_{n2}	1
New coils	k_{n3}	1

Values of Process Cost Functions

Process Stage	Variable Name	Value (in Rs Thousand)
Transportation from shed 1... shed 10 to shop	f_3; f_6; f_9; f_{12}; f_{15}; f_{18}; f_{21}; f_{24}; f_{27}; f_{30}	5
Stripping	f_{32}	5

Contd...

Disassembly	f_{34}	2
Core and commutator repair	f_{35}	10
Assembly	f_{36}	5
Rewinding	f_{37}	10

Values of Flow Variables

Flow of TMA		Flow values at different period (in numbers)						
Variable description	**Variable name**	**1**	**2**	**6**	**10**	**14**	**18**	**22**
Repair arisings shed 1	y_1	4	1	3	2	3	1	3
From shed store	y_2	8	0	0	0	8	0	8
Transportation to shop	y_3	8	0	0	0	8	0	8
Repair arisings shed 2	y_4	1	4	2	4	5	2	3
From shed store	y_5	8	0	0	8	8	0	0
Transportation to shop	y_6	8	0	0	8	8	0	0
Repair arisings shed 3	y_7	3	4	3	4	2	4	4
From shed store	y_8	8	0	8	0	8	0	0
Transportation to shop	y_9	8	0	8	0	8	0	0
Repair arisings shed 4	y_{10}	4	4	1	4	1	5	1
From shed store	y_{11}	8	0	8	8	0	8	0
Transportation to shop	y_{12}	8	0	8	8	0	8	0
Repair arisings shed 5	y_{13}	4	4	4	7	2	1	0
From shed store	y_{14}	8	0	8	8	8	0	8
Transportation to shop	y_{15}	8	0	8	8	8	0	8
Repair arisings shed 6	y_{16}	3	8	7	8	1	3	2
From shed store	y_{17}	8	0	0	0	8	0	8
Transportation to shop	y_{18}	8	0	0	0	8	0	8
Repair arisings shed 7	y_{19}	8	5	3	7	5	7	2
From shed store	y_{20}	8	8	8	8	8	8	0
Transportation to shop	y_{21}	8	8	8	8	8	8	0

Repair arisings shed 8	y_{22}	4	6	3	5	6	5	9
From shed store	y_{23}	8	0	8	0	0	0	8
Transportation to shop	y_{24}	8	0	8	0	0	0	8
Repair arisings shed 9	y_{25}	6	2	9	11	6	9	10
From shed store	y_{26}	8	0	8	8	8	8	8
Transportation to shop	y_{27}	8	0	8	8	8	8	8
Repair arisings shed 10	y_{28}	3	7	5	6	7	10	4
From shed store	y_{29}	8	0	8	0	8	8	8
Transportation to shop	y_{30}	8	0	8	0	8	8	8
From shop receipt store	y_{31}	40	38	45	45	45	45	38
After stripping	y_{32}	45	45	45	45	45	45	45
After in-process store	y_{33}	45	45	45	45	45	45	45
After disassembly	y_{34}	45	45	45	45	45	45	45
After core repair	y_{35}	45	45	45	45	45	45	45
After assembly	y_{36}	45	45	45	45	45	45	45
After re-winding	y_{37}	45	45	45	45	45	45	45
Outflow from shop dispatch store	y_{38}	45	45	45	45	45	45	45

Values of Cost Variables

Cost of TMA		Cost at different period (in Rs thousand)						
Variable description	**Variable name**	**1**	**2**	**6**	**10**	**14**	**18**	**22**
Received for repairs	x_1	100	100	100	100	100	100	100
After shed store	x_2	100	102.5	104.2	105.9	104.3	104	102.7
With transportation cost	x_3	105	107.5	109.2	110.9	109.3	109	107.7
Received for repairs	x_4	100	100	100	100	100	100	100
After shed store	x_5	100	105	105.3	103.4	103	105.8	106.4
With transportation cost	x_6	105	110	110.3	108.4	108	110.8	111.4
Received for repairs	x_7	100	100	100	100	100	100	100
After shed store	x_8	100	103	103.8	103.5	103.4	103.9	105.2

With transportation cost	x_9	105	108	108.8	108.5	108.4	108.9	110.2
Received for repairs	x_{10}	100	100	100	100	100	100	100
After shed store	x_{11}	100	102.5	103.1	103.1	105.4	104.5	108.9
With transportation cost	x_{12}	105	107.5	108.1	108.1	110.7	109.5	113.9
Received for repairs	x_{13}	100	100	100	100	100	100	100
After shed store	x_{14}	100	102.5	103.7	103.4	102.9	106.3	105.6
With transportation cost	x_{15}	105	107.5	108.7	108.4	107.9	111.3	110.6
Received for repairs	x_{16}	100	100	100	100	100	100	100
After shed store	x_{17}	100	103	104.9	106.9	103.1	104.6	102.8
With transportation cost	x_{18}	105	108	109.9	109.9	108.1	109.6	107.8
Received for repairs	x_{19}	100	100	100	100	100	100	100
After shed store	x_{20}	100	101.5	103	101.7	103.6	102.6	102.1
With transportation cost	x_{21}	105	106.5	108	106.7	108.6	107.6	107.1
Received for repairs	x_{22}	100	100	100	100	100	100	100
After shed store	x_{23}	100	102.5	104.2	102.9	103.6	103.4	103.2
With transportation cost	x_{24}	105	107.5	109.2	107.9	108.6	108.4	108.2
Received for repairs	x_{25}	100	100	100	100	100	100	100
After shed store	x_{26}	100	101.9	101.8	103.2	103.2	102.5	102.3
With transportation cost	x_{27}	105	106.9	106.8	108.2	108.2	107.5	107.3
Received for repairs	x_{28}	100	100	100	100	100	100	100
After shed store	x_{29}	100	103	101.9	103.7	102.3	102.3	103.2
With transportation cost	x_{30}	105	108	106.9	108.7	107.3	107.3	108.2
After shop receipt	x_{31}	105	100.8	103	101.6	102.2	101.3	103.1
After stripping	x_{32}	100	95.8	98	96.6	97.2	96.3	98.1
After-in-process store	x_{33}	100	101.5	98.1	96.6	96.6	97.6	98.4
After disassembly	x_{34}	102	103.5	100.1	98.6	101.6	99.9	100.4
After core repair	x_{35}	118	119.5	116.1	114.6	117.6	115.9	116.4
After assembly	x_{36}	143	144.5	141.1	139.6	142.6	140.9	141.4
After re-winding	x_{37}	308	309.5	306.1	304.6	307.6	305.9	306.4
After shop dispatch store	x_{38}	308	313.5	308.9	312.7	310.2	311.7	308.9

PART V

APPLICATIONS IN MANAGEMENT OF SOCIO-ECONOMIC SYSTEMS

19 INTERSECTORAL MODELLING OF NATIONAL ECONOMY

19.1 INTRODUCTION

Waste flow in the national economy is the aggregate of waste generated in individual subsystems having complex interrelations with each other and the environment. From a systems viewpoint, the term waste can be defined as any unnecessary input to or any undesirable output from the system[1]. It is stressed here that waste parameters deserve explicit consideration in national planning, in view of their important role in various systems.

A review of the literature on national planning indicates that this gives explicit consideration primarily to two dimensions, viz. input to the system, and output from the system. The input-output model is a very widely accepted tool for defining the technological constraints, and for resource balancing for any system. The pioneering work on input-output modelling was done by Leontief[2] in the context of economic planning. Some extensions to the conventional input-output model have been attempted in the past to incorporate environmental interactions. Leontief[3] proposed an elegant theoretical formulation including both residual generation and abatement activities in a single framework.

Probably, Ayres and Kneese[4] were the first to present a generation equilibrium material balance model, in which economic activities were linked to the environment by including waste parameters. Ayres[5] has presented Maler's model, and the formulation by d'Arge and Kogiku in which the optimal growth or extraction models, and environmental externalities have been integrated by treating the environment as a kind of exhaustible resource (neglecting the possibilities of recycling). Isard et. al[6] have also done considerable work on regional development by considering the ecological-economic linkages.

There has been increasing interest in the linkages between environment and economy, the efforts have been made to incorporate environmental parameters in economic planning. The Canadian Environmental Advisory Council (CEAC)[7] has published a review of the literature on this aspect. According to the CEAC review, Leontief, Ford, Den Hartoy, Houweling, Nijkamp, Paelinck, Coupe, Muller and many others have tried to extend the input-output framework to incorporate different types of environmental pollution caused by the production activities of various sectors.

Though many of the formulations in the past have incorporated waste parameters, their efforts were limited to deal with environmental aspects only. Detailed resource balancing for different types of resources, with explicit consideration of waste, is lacking. Further, a survey of waste management (WM) practices in Indian context, has revealed a lack of database on waste flow in the national economy.

Recognizing the importance of waste as an independent parameter, remodelling the economy in the framework of an I-O-W (input-output-waste) model has been proposed, which will be able to define the resource balance more realistically by incorporating the reduction, recycling, abatement, disposal and related functions of WM in the national planning process[8].

In order to incorporate the aspects of WM explicitly and to make national planning goal-oriented, this chapter extends the work of Koenig et. al[9] by integrating a physical system theory model of the national economy, with a multi-objective goal-programming model[10]. The integrated model is used to determine the optimal/satisficing sectoral waste flow and allocation of resources for the desired standard of living and environmental conditions, equitable use of resource base, sustained economic growth and ecological balance of nature.

A graph theoretical system model of the economy is developed conceptualizing the economy as consisting of '*s*' production sectors and '*z*' WM sectors. The system model is dovetailed into a multi-objective decision-making model in a goal-programming framework for developing an integrated national planning model for the WM point of view. The flexible system methodology for integrated national planning is given in Figure 19.1. The proposed model has been illustrated by the example of a two-sector economy having one production sector and one WM sector. The two sector goal-programming model has been run for alternate objective functions and sensitivity analysis has been carried out.

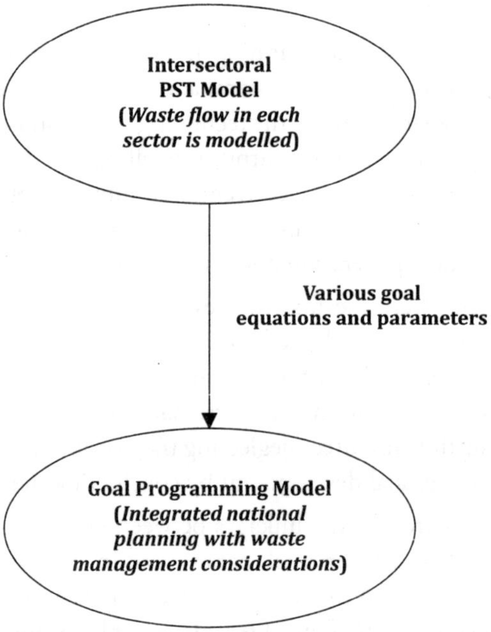

Figure 19.1: Flexible System Methodology for Integrated National Planning

19.2 SYSTEMS MODELLING: INTERSECTORAL RESOURCE AND WASTE FLOWS

Graph Theoretical System Model

Various sectors have complex interrelationships. Output from one sector acts as input to some other sectors, e.g. agricultural equipment manufactured in the industrial sector act as input to the agricultural sector, while various agricultural products serve as input to various industrial processes. In the generalized system model, it is hypothesized that each sector takes input from all other sectors and gives output to all other sectors. All the sectors take direct input from nature and give output to the habitat.

The production sectors have been visualized as transformation or production processes, and are taken as output oriented. The output from each sector will be the planned output. The auxiliary output or waste generated is either to be recycled or directly disposed of to nature.

In practice, recycling of wastes takes place within the sectors but, conceptually, recycling processes have been considered as separate WM sectors to visualize the waste flow clearly. The WM processes have been considered as input dependent. The WM process takes input from all the sectors and from nature, gives output to all the sectors and to the human habitat, and disposes off waste to nature.

The generalized system model of a national economy having *s* production sectors and *z* WM sectors has been developed. The 'across' variable is taken to be the unit cost of the resource and is denoted by *x*, while the flow of the resource is a 'through' variable denoted by *y*. By applying the concepts of physical system theory, the response-stimulus relations for the production and WM sectors are developed as given in Appendix 19 A.

This is a generalized model for economy having $s + z$ sectors. The system graph and structural constraints (continuity, compatibility and mass balance equations) can be written only for a specific example with a definite number of sectors. The notations used in the model are given as follows:

	Parameters and variables		***Subscripts/superscripts***
k	Technological coefficient	*g*	goal
kw	Technological coefficient of waste flow	*h*	Habitat
o	Capacity ratio	*i,j*,1	Production sector
ww	Wastivity	*l,m*,2	Waste management sector
x	Unit cost of resources	*n*	Nature
xw	Unit cost of wastes	*o*	Output
y	Resource flow	*p*	Cost category
yw	Waste flow		

Multi-Objective Model of Waste Management Embedded in the Economic System

The system equations of the graph theoretical model are used to formulate a multi-objective model of the economic system, including WM goals, in a goal-programming (GP) framework, with the following objective goals and constraints. The equations of the goals for an economy with *s* production sectors and *z* WM sectors are given in Appendix 19A.

Objective Goals

(a) Minimize production cost.

(b) Minimize recycling cost

(c) Minimize waste generation in various sectors

(d) Minimize direct disposal of waste to nature

(e) Minimize waste generation from the human habitat

For an economy with *s* production sectors and *z* WM sectors, the total number of objective goals $= 3s + 3z + 2$.

Constraints

(a) Continuity or material balance constraints (total number of continuity and mass balance equations $= 2s + 3z$).

(b) Maximum permissible waste from each sector depending on various factors. The permissible level of wastivity (wastivity = ratio of waste to input, for any given system) can be defined. Also the input should be at least equal to the output. These constraints can be represented as a summation of the technological coefficients.

(c) Resource availability constraint. Maximum total input to various sectors from nature, depending upon the availability of resources to achieve sustained economic growth.

(d) Standard of living/growth rate or demand constraints. Minimum desired output from each sector for a particular planned economic growth rate. Total output to the human habitat from each sector is a policy decision to be taken for a particular standard of living.

(e) Environmental standards constraint. This defines the maximum permissible waste direct to nature, and can be determined from the maximum permissible pollutant level for the desired quality of environment, and the pollution intensiveness of the waste disposed off by various sectors and habitat.

(f) Constraint on waste generation in the habitat. Based on economic status and lifestyles, the norms for the waste generation should be specified.

(g) Waste reduction constraints. The cost of waste control should be less than a fraction of the total cost of production or WM in the respective sector.

(h) Technological constraints. These constraints put lower limits on the intersectoral resource flows.

(i) Non-negativity constraints.

Total number of constraints (except non-negativity constraints) $= 6s + 8z + (s + z)^2 + 3$.

Total variables (technological coefficients, k) =

k's of intermediate inputs

+ k's of direct inputs from nature

+ k's of final output to habitat

+ k's of WM (recycling, abatement, etc.)

+ k's of waste disposal to nature

+ k's of dummy links

$= (s + z)(s + z) + (s + z) + (s + z) + (s + z + 1)z + (s + z + 1) + m$

In practice, some of the variables, like required outputs, cost parameters, etc. can be taken from other planning models, which will simplify the solution procedure.

19.3 DATA REQUIREMENTS FOR THE SYSTEM MODEL

The output from different sectors can be treated as exogenous, and can be determined on the basis of existing demand patterns. Thus, there is a requirement for data on the demand patterns of major sectors.

The data on the cost parameters of the output of different sectors are also required. The total cost of the output of any sector can be determined by aggregating the total cost of various products of the sector. The unit cost is assessed by dividing the total cost by the total output, taken as equal to the demand.

Further, it is necessary to specify various environmental goals. The data relating to pollution standards are accumulated until they yield a specific environmental goal. The waste disposed off to the environment falls under different categories and causes different types of pollution, viz. air, water, land, etc. The air pollution itself is due to different classes of pollutants, e.g. carbon monoxide, nitrogen dioxide, carbon content, sulfur content, etc. In the data accumulation, it is assumed that in an industrialized society almost all types of pollution are increasing.

The priorities assigned to different goals comprise a policy decision depending upon various techno-socio-economic, environmental and political considerations.

19.4 ILLUSTRATIVE EXAMPLE

The proposed graph theoretical GP model has been illustrated for a hypothetical economy consisting of one production sector and one WM (recycling) sector.

The Two Sector Graph Theoretical Model

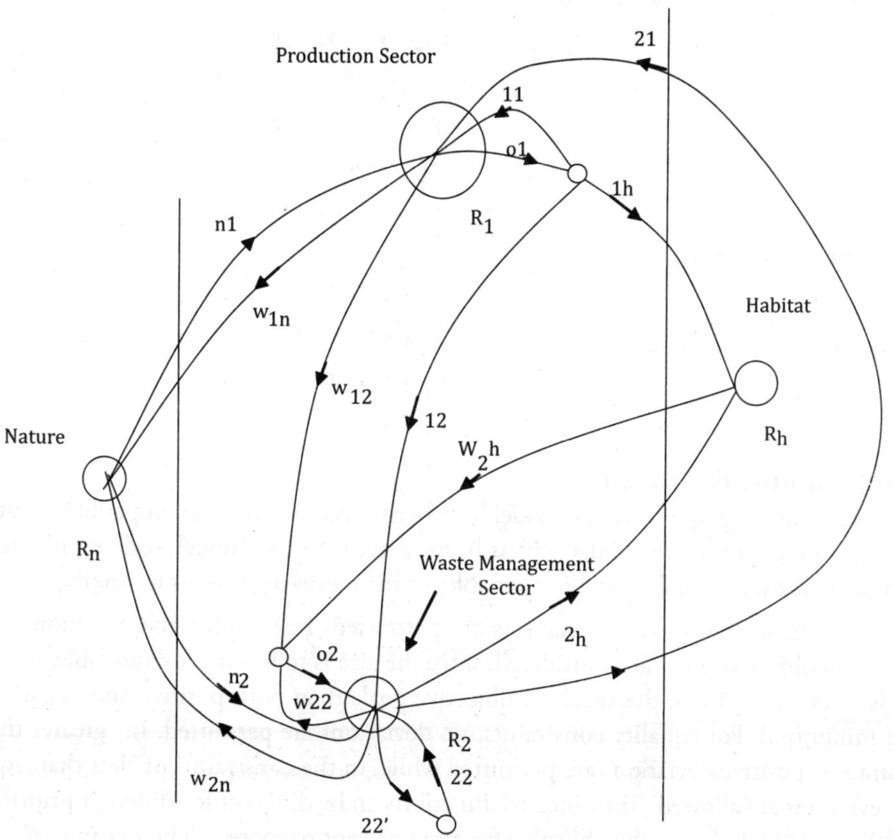

Figure 19.2: Simplified System Graph of the Two Sector National Economy

The simplified system graph of the economy exhibiting the intersectoral resource and waste flow is shown in Figure 19.2. The link 22' in the simplified system graph is a dummy link. For the sake of simplicity in illustration only one cost category, i.e. monetary cost, is taken into consideration in the two-sector model. The design capacity of the production sector, i.e. sector number 1, is assumed to be 100 units. The capacity ratio matrix is shown in Table 19.1. The design capacity of the WM sector, i.e. sector number 2, can be calculated by knowing the capacity ratio.

Table 19.1: Capacity Ratio Matrix for Two Sector Economy

$$\begin{bmatrix} o_{11} & o_{12} \\ o_{21} & o_{22} \end{bmatrix} = \begin{bmatrix} 1.0 & 5.0 \\ 0.2 & 1.0 \end{bmatrix} \tag{19.1}$$

where

$$o_{11} \text{ is } \frac{y_{01}}{y_{01}}, o_{12} \text{ is } \frac{y_{01}}{y_{02}}, o_{21} \text{ is } \frac{y_{02}}{y_{01}}, o_{22} \text{ is } \frac{y_{02}}{y_{02}}$$

The continuity, mass balance and compatibility equations for the simplified system graph are written as follows:

Continuity and mass balance equations

$$y_{o1} + yw_{12} + yw_{1n} - y_{n1} - y_{11} - y_{21} = 0 \tag{19.2}$$

$$y_{o2} + y_{12} + y_{n2} - y_{2h} - y_{21} - yw_{22} - yw_{2n} = 0 \tag{19.3}$$

$$y_{o1} - y_{11} - y_{12} - y_{1h} = 0 \tag{19.4}$$

$$y_{o2} - yw_{12} - yw_{22} - yw_{h2} = 0 \tag{19.5}$$

$$y_{22} - y_{22}' = 0 \tag{19.6}$$

Compatibility equations

$$x_{o1} + x_{11} = 0 \tag{19.7}$$

$$x_{o1} + x_{12} = 0 \tag{19.8}$$

$$x_{o1} + x_{1h} = 0 \tag{19.9}$$

$$x_{22}' + x_{22} = 0 \tag{19.10}$$

Goal-programming Formulation

The cost variables in the graph theoretic model have been specified under a compatibility restriction, and a standard goal-programming formulation has been derived. In this model, scale economies have been ignored in the cost equations. The decision variables of the model are various technological coefficients.

In objective goals, both sided deviations are permitted. In the objective function, only positive deviations from objective goals are considered, since the aim is to minimize those objectives. But, if the objective is to exactly achieve the specified objective goals then both positive and negative deviations should be minimized. For equality constraints, no deviations are permitted. In `greater than equal to' type constraints, positive deviations are permitted while, in the constraints of `less than equal to' type, negative deviations are allowed. The objective function can be different for different priority structures. The deviations not included in the objective function are not supposed to be minimized.

The goal-programming formulation for the two-sector economy with a typical objective function, considering the objective goals on the first priority and the constraint goals on the second priority, is derived using the generalized model given in Appendix 19 A.

Alternative Objective Functions: Sensitivity Analysis

The GP model developed in previous sections was implemented using the goal programming package of Lee[11]. Successive runs of the GP model were obtained by varying the number of goals, priority structure and by changing the values of some of the goals to perform sensitivity analysis. The objective goals were given higher priority than the constraint goals in all the objective functions considered in this study. Some of the constraints were converted into constraint goals, while others were treated as constraints to be satisfied, and the deviations for these constraints were not included in the objective function. In general, the priority structures have been changed on the following lines.

(i) Keep all the objective goals on one priority and all the constraint goals on the other priority.

(ii) Have more than two priority levels as decided by the planners, e.g. the priority levels for production goals, environmental or WM goals; and other goals may be kept separate.

(iii) Keep all the goals on the same priority level.

(iv) Consider exact achievement of some of the goals.

(v) Have fractional weightage for the positive and negative deviations of some of the goals.

The development of the model in terms of adding the goals in stages, and the alternative objective functions considered for sensitivity analysis of a GP model with different numbers of goals may be described as follows using 19 runs of the model. The final model had 29 goals.

Analysis of Results

An economy with one production sector and one WM sector having capacities of 100 units and 20 units, respectively, has been analysed by applying the GP model. Any deviations from the assumed values of the goals have been mentioned at the appropriate places. The sensitivity analysis has been performed by obtaining different runs of the GP model. The output gives the values of different variables, the slack analysis, and the evaluation of the objective function mentioning the values of the under-achievement against each priority.

Analysis of Variables

The intermediate output of the production sector is about 50% of the total output, which may vary from 25% to 75%. The direct output of the habitat is also 50%, against the specified goal of 40% (with a variation from 25% to 75%). The intermediate input to the production sector is about 60% of the total output (with a variation from 30% to 75%), while the direct input is 40% (with a variation from 35% to 70%).

The fraction of intermediate input to waste recycling is very high in comparison with that of the natural inputs; the total inputs required for waste recycling vary from 10% to 30% of the total waste recycled. The direct output of the recycling sector is around 20-40% of the total waste recycled when the specified goal is 25%. The overall constitution of the waste recycled is around 40% from the production sector, 45% from the habitat, and 15% from the recycling sector itself. The total waste disposed off to nature constitutes about 60-65% from the production sector, 25-30% from the habitat and 5-10% from the WM sector.

The minimum and maximum of the ranges of variation of the parameters are 0.0499 and 0.725, respectively, in all the 19 runs, while for the runs of the model with 29 goals these values are 0.00 and 0.35, respectively. Similarly, the minimum and maximum values of the standard deviations of the values of different variables in 19 runs are 0.0139 and 0.1915, respectively, while for the model with 29 goals these values are 0.00 and 0.102, respectively. This indicates that the range of variation as well as the standard deviation of the variables in different runs is reduced in the model with 29 goals compared with all other models, i.e. the sensitivity of different variables gradually decreased with the increase in the number of goals.

The model is almost insensitive to the weightages assigned to the negative deviations. Since these weightages are more or less a subjective estimate, any error in assigning the same will not affect the results significantly. The proposed model is comparatively sensitive to changes in the priorities of positive deviations. This is also sensitive to the consideration of negative deviations in the objective functions, since it changes the aim of the model significantly. If no negative deviations are considered in the objective function, then the goals are free to take values less than assumed.

Depending upon the range, standard deviation and frequency of changes in the values, variables are classified into three categories of sensitivity. The first category (of most sensitive variables) includes k_{11}, k_{n1}, k_{1h}, k_{2h}; the second category (of moderately sensitive variables) includes k_{12}, k_{21}, kw_{12}, kw_{22}, kw_{h2}; and the third category (of less sensitive or even insensitive variables) includes k_{22}, k_{22}', k_{n2}, kw_{1n}, kw_{2n}, kw_{hn}. That is, the inputs to the production sector, and the direct output to the habitat defining the standard of living are more sensitive to the priorities assigned. While the amount of waste to be recycled (from different sectors) is moderately sensitive, indicating that the best possible satisficing level of recycling is much dependent on the priorities assigned to the different goals. If the waste generation goal is given higher priority, the recycling level increases. Moreover, the cost of production/ recycling has a significant impact on the level of recycling. The amount of waste to be disposed off to nature by different sectors and the habitat is somewhat less sensitive to change in the priorities.

The incorporation of negative deviations in the objective function reduces the value of k_{21}, kw_{12}, kw_{h2}, and increases the value of k_{n1}, kw_{22}, kw_{1n}, kw_{hn}. Since by minimizing the negative deviations the goals of waste disposal cannot take values less than assigned, the amount of waste to be disposed off increases, resulting in a subsequent decrease in the fraction of waste to be recycled.

Analysis of Goals

The analysis of the goals has been carried out only for the computer runs of the proposed model with 29 goals. A marked variation in the sensitivity of goals with respect to different priority structures is observed, and the goals are grouped into four sensitivity categories, viz. highly sensitive, moderately sensitive, less sensitive and insensitive goals. The goals of input from nature to production, standard of living, and environment were found to be highly sensitive. The goal of natural inputs is most sensitive, maybe because of the variation in the direct input requirements due to the variation in recycling. The environmental standards will be maintained if higher priorities are given to these goals. However, even if these goals are assigned lower priority, the extent of environmental degradation is not too severe. But, if no priority is assigned, then waste disposal is left unchecked, leading to tremendous pollution. Moderate sensitivity is shown by the goals of WM sectors. The following goals have been identified to be less sensitive: (a) waste generation in the habitat, (b) waste disposal by the habitat, (c) total input to the WM sector, (d) output of the WM sector (total as well as direct), (e) cost of waste control, and (f) constraints on technological coefficients.

The fraction of the output consumed by the habitat and thrown to nature does not vary much. Unless the character of the products is changed there is bound to be a certain fraction left as residue. This can be reduced only by changing the goals for the standard of living, thus reducing the fraction of the output to be consumed, or by tightening the environmental goals for reduced waste disposal; the net option is increased recycling. The continuity constraints are to be satisfied exactly, and are bound to be insensitive.

19.5 APPLICABILITY FOR NATIONAL PLANNING

The model presented in this chapter will prove useful in integrating the waste management aspects of national planning by treating the waste flows explicitly. This model will be able to take the waste management aspects in totality and can be utilized to plan waste control, collection, recycling, treatment, and disposal in all sectors of the economy. The model will be able to portray the intersectoral interaction in terms of resources/wastes.

The allocation of resources to various sectors for production and WM functions can be derived with reference to the priorities of the planning process. The model takes care of the multiple goals of planning such as cost minimization, waste generation, waste disposal, environmental standards, resource availability, waste recycling, standard of living, sustained economic growth, waste reduction and technology, which make the planning process more realistic. In a country like India, where the wastivity level is comparatively high and a resource crisis exists, consideration of WM in national planning is very desirable. This will help in monitoring the productivity of the economy, leading to sustained economic growth.

Sensitivity analysis of the model will help in setting up the priorities for deciding the goals and their weightings in the planning process. This will also aid in understanding the behaviour of different goals/variables with respect to changes in the model structure, which will help in effective control of the economic system. For example, a goal which is more sensitive requires tighter control of the system than does that of a less sensitive goal. The methodology presented in this chapter can be applied to the planning of national/regional economies, subject to data availability. Most of the data are available for the intersectoral resource flow in the input-output tables. The additional data on waste have to be collected. It has been observed that data on WM are rarely maintained and there is hardly any database on waste at the national level.

The major contribution of this approach will be to make national planning more realistic and goal-oriented.

REFLECTIONS

The need to incorporate WM aspects in national planning has been emphasized. Waste has been observed to be a neglected parameter in national planning, and the lack of emphasis on explicit consideration of waste from a systems viewpoint has been pointed out. The principles and concepts of physical system theory are applied in conjunction with goal-programming to determine the optimal sectoral waste flow. The model formulated in this chapter will prove useful for integrated national planning by providing information on intersectoral resource flow, type of technology to be employed, efforts needed for waste reduction and recycling, environmental quality standards, standard of living, sustained economic growth, etc. The model takes care of the temporal priorities in national planning, with explicit consideration of WM.

The use of the proposed model has been illustrated by the example of a two-sector economy. The results of the sensitivity analysis indicate that the goals of national planning show marked variation in sensitivity with respect to change in the priority structure, indicating the need for carefully judged and systematic choices of the priorities in national planning.

The proposed model incorporates the advantages of both physical system theory (that defines the input-output structure of the economy) and the goal programming (that takes care of the priorities) and gives a flexible systems approach to national planning. By applying the techniques of multilevel analysis the same model can be applied from micro to macro levels. A major limitation of the proposed model is that it treats economic as well as environmental processes in a similar framework, whereas the time and space relationships of economic and environmental flows may be, at times, significantly different. This particular limitation may be overcome by employing the proposed framework in the context of multilayered models, as proposed in CEAC[12]. The lower level regional models and higher level economic models can be developed by using the proposed integration of physical system theory and goal programming. In view of the lack of a database on WM, the implementation of the model is restricted. There is a pressing need to develop effective information systems for the purpose of integrated national planning.

Notes

1. Sushil (1989) *Systems Approach to National Planning: A Study in Waste Management,* Anmol Publication, New Delhi.
2. Leontief W. (1966) *Input Output Economics,* Oxford University Press, New York.
3. Leontief W. (1970) *Rev. Economics Statistics*, 52, 262; Leontief W. (1975) *Proc. Inst. Elect. Electron*, Engrs, 63, 345.
4. Ayres R.U. and Kneese A.V. (1969) *Am. Econ. Rev.,* 59, 282.
5. Ayres R.U. (1978) *Resources, Environment and Economics: Applications of Materials / Energy Balance Principle,* Wiley, New York.
6. Isard W., Bassatt K.B., Choguill C.L., Furtado J.G., Izumita R.M., Kissin J., Seyforth R.H. and Tatlock R. (1968) *Ecologic-Economic Analysis for Regional Development,* Harvard University Graduate School of Design for the US Department of Commerce, Environmental Science Services Administration.
7. CEAC (1986) *Examining Environment-Economy Linkages,* Canadian Environmental Advisory Council, Ottawa.
8. Sushil and Vrat P. (1987) *Int. J. Systems Sci.*, 18, 2043.
9. Koenig H.E., Edens T.C. and Cooper W.E. (1975) *Proc. Inst. Elect. Electron. Engrs.,* 63, 501.
10. Sushil (1993), *Inter J. Systems Sci.,* 24 (5), 957-984.
11. Lee S.M. (1972) *Goal Programming for Decision Analysis*, Philadelphia, Auerbach.
12. CEAC (1986) *op. cit.*

APPENDIX 19A

System Model for a Multi-sector National Economy

A.1 Component Models

(i) Production sectors (transformation processes)

$i = 1, \ldots s$	Production sectors
$j = 1, \ldots s$	Production sectors
$j = m = s + 1, \ldots s + z$	WM sectors
$j = h$	Habitat
$i/j = n$	Nature

Resource/waste flow equations

$$y_{ji} = k_{ji} y_{oi} \tag{19.11}$$

$$y_{ih} = k_{ih} y_{oi} \tag{19.12}$$

$$y_{ni} = k_{ni} y_{oi} \tag{19.13}$$

$$yw_{in} = kw_{in} y_{oi} \tag{19.14}$$

Cost equations

$$-x^p_{oi} = \sum_{j=1}^{s+z} k_{ji} x^p_{ji} + k_{ni} x^p_{ni} kw_{ni} xw^p_{in} + \sum_{m=1}^{s+z} o_{mi} kw_{im} xw^p_{im} + f^p_i (y_{oi}) \tag{19.15}$$

where x_{oi} is the unit cost of output of the *i*th production sector, y_{oi} is the output (design capacity) of the *i*th production sector, and *p* is the cost category (e.g. monetary cost, energy cost, etc.).

(ii) Waste management sectors (transformation processes)

$j = l = s+1, \ldots s + z$	WM sectors
$j = 1, \ldots s$	Production sectors
$j = h$	Habitat
$i/j = n$	Nature

Resource/waste flow equations

$$y_{nl} = k_{nl} y_{ol} \tag{19.16}$$

$$y_{jl} = k_{jl} y_{ol} \tag{19.17}$$

$$yw_{jl} = kw_{jl} y_{ol} \tag{19.18}$$

$$yw_{hl} = kw_{hl} y_{ol} \tag{19.19}$$

$$yw_{ln} = kw_{ln} y_{ol} \tag{19.20}$$

Cost equations

$$-x^p_{ol} = \sum_{j=1}^{s+z} kw_{ji} xw^p_{jl} + kw_{hl} xw^p_{hl} + k_{nl} x^p_{nl} + \sum_{j=1}^{s+z} k_{jl} x^p_{jl} + kw_{ln} xw^p_{ln} + f^p_l (y_{ol}) \tag{19.21}$$

where x_{ol} is the unit cost of WM in the WM sector *l*, and y_{ol} is the input of waste to be managed (design capacity) in sector *l*.

A.2 Goals

The equations here correspond to the objectives and constraints defined in the model. There may be more than one equation under a particular objective/constraint.

(i) Objective Goals

(a) Unit cost of production in sector $i = x^g_{oi}$ $i = 1, \ldots, s$ (19.22)

the superscript *g* specifies the goal, i.e., x^g_{oi} is the goal of x_{oi}

(b) Unit cost of WM in sector $l = x^g_{ol}$, $l = s+1, \ldots, s+z$ (19.23)

(c) Waste generated in production sector $i = yw^g_i$, $i = 1, \ldots, s$ (19.24)

Waste generated in WM sector $l = yw^g_l$, $l = s + 1, \ldots, s + z$ (19.25)

(d) Waste disposed to nature from production sector $i = yw^g_{in}$, $i = 1, \ldots, s$ (19.26)

Waste disposed to nature from WM sector $l = yw^g_{ln}$, $l = s+1, \ldots, s+z$ (19.27)

Waste disposed to nature from habitat $= yw^g_{hn}$ (19.28)

(e) Waste generated in habitat $= yw^g_h$ (19.29)

(ii) Constraints

(a) Continuity or mass balance constraints $= 2s + 3z$

(b)
$$\begin{bmatrix}\text{Total resource input to}\\ \text{production sector } i\end{bmatrix} \le \frac{1}{1 - WW_i}\begin{bmatrix}\text{Total output of}\\ \text{sector } i\end{bmatrix}, i = 1, \ldots, s \quad (19.30)$$

$$\begin{bmatrix}\text{Total resource input to}\\ \text{production sector } i\end{bmatrix} \ge \begin{bmatrix}\text{Total output of}\\ \text{sector } i\end{bmatrix}, i = 1, \ldots, s \quad (19.31)$$

$$\begin{bmatrix}\text{Total resource input to}\\ WM \text{ sector } I\end{bmatrix} \le kir^{g1}_l \begin{bmatrix}\text{Total waste managed in}\\ \text{sector } i\end{bmatrix}, i = s + 1, \ldots, s + z \quad (19.32)$$

$$\begin{bmatrix}\text{Total resource input to}\\ WM \text{ sector } I\end{bmatrix} \le kir^{g2}_l \begin{bmatrix}\text{Total waste managed in}\\ \text{sector } i\end{bmatrix}, i = s + 1, \ldots, s + z \quad (19.33)$$

$$\begin{bmatrix}\text{Total output of}\\ WM \text{ sector } I\end{bmatrix} \le f^{g2}(WW_i)\begin{bmatrix}\text{Total waste}\end{bmatrix}, i = S + 1, \ldots, s + z \quad (19.34)$$

(c)
$$\begin{bmatrix}\text{Total input of economy}\\ \text{(s production sectors + z WM sectors)}\end{bmatrix} \le yi^g_n \quad (19.35)$$

(d) $$\left[\begin{array}{c}\text{Total output of } i\text{th production}\\ \text{sector to habitat}\end{array}\right] \geq yi^{g}_{ih,}\ i = 1, \ldots, s \tag{19.36}$$

$$\left[\begin{array}{c}\text{Total output of } i\text{th WM}\\ \text{sector to habitat}\end{array}\right] \geq yi^{g}_{ih,}\ i = s+1, \ldots, s+z \tag{19.37}$$

(e) $$\left[\begin{array}{c}\text{Total waste directly disposed to nature}\\ \text{x pollution intensiveness}\end{array}\right] \leq yi^{g}_{n} \text{ (permissible pollution level)} \tag{19.38}$$

(f) $$\left[\begin{array}{c}\text{Waste generated in habitat}\end{array}\right] \leq yw^{g}_{h} \tag{19.39}$$

(g) $$\left[\begin{array}{c}\text{Total cost of waste control in}\\ \text{production sector } i\end{array}\right] \leq c^{g}_{i}\left[\begin{array}{c}\text{Total cost of production}\\ \text{in sector } i\end{array}\right], i = 1, \ldots, s \tag{19.40}$$

$$\left[\begin{array}{c}\text{Total cost of waste control in}\\ \text{WM sector } i\end{array}\right] \leq c^{g}_{i}\left[\begin{array}{c}\text{Total cost of}\\ \text{WM in sector } I\end{array}\right], I = s+1, \ldots, s+z \tag{19.41}$$

(h) $$\left[\begin{array}{c}\text{The output of } i\text{th production}\\ \text{sector required in } j\text{th}\\ \text{production sector}\end{array}\right] \geq k^{g}_{ij}\left[\begin{array}{c}\text{Total output of } i\text{th}\\ \text{production sector}\end{array}\right], i = 1, \ldots, s, j = 1, \ldots, s, \tag{19.42}$$

$$\left[\begin{array}{c}\text{The output of } i\text{th}\\ \text{production sector used in}\\ m\text{th WM sector}\end{array}\right] \geq k^{g}_{im}\left[\begin{array}{c}\text{Total waste}\\ \text{managed in}\\ \text{sector } I\end{array}\right], i = 1, \ldots, s, I = m = s+1, \ldots s+z, \tag{19.43}$$

$$\left[\begin{array}{c}\text{The output of } i\text{th WM}\\ \text{sector required in } j\text{th}\\ \text{production sector}\end{array}\right] \geq k^{g}_{lj}\left[\begin{array}{c}\text{Total output of}\\ j\text{th production}\\ \text{sector}\end{array}\right], I = s+1, \ldots s+z, m = s+1, \ldots, s+z \tag{19.44}$$

$$\left[\begin{array}{c}\text{The output of } I\text{th WM}\\ \text{sector required in } m\text{th}\\ \text{WM sector}\end{array}\right] \geq k^{g}_{lm}\left[\begin{array}{c}\text{Total waste}\\ \text{managed in}\\ \text{sector } I\end{array}\right], I = s+1, \ldots s+z, m = s+1, \ldots, s+z \tag{19.45}$$

20 MODELLING OF NATIONAL BUDGET

20.1 INTRODUCTION

A government budget is a statement of estimates of expenditure for a specified period of time together with proposals for financing them. It is through the instrumentality of a budget that coordination of resources and expenditures is planned. According to Arthur Smithies:

The process of decision-making that results in a budget contains three necessary ingredients:

i. Determination of the variety of policy objectives the government intends to pursue, such as defence, education, or law enforcement.
ii. Estimation of the cost of pursuing each of these objectives in varying degrees.
iii. An assessment of the willingness and ability of the public to pay for the government's program as a whole.

According to Aaron Wildavsky, a budget may service diverse purposes and can be a political act, a plan of work, a prediction, a source of enlightenment, a means of obfuscation, a mechanism of control, an escape from restrictions, a means to action, a brake on progress, even a prayer that the powers that will deal gently with the best aspirations of fallible men.

Different types of budget classifications have been evolved to serve diverse purposes, such as object or line item classification, economic classification, functional and program classification. No single budget classification has been found to serve equally well all the purposes, for instance, control over inputs, depicting well the state of the economy, and providing an effective tool for the planning and management of government programs and activities. Line item or object wise classification serves the purposes of financial accountability and ensuring legality and regularity of expenditure. Program/activity classification explains the objectives of the organization in terms of the programs and activities. It lays greater emphasis on the economy and efficiency aspects of program planning and management. Performance budgeting lays greater emphasis on the techniques of cost accounting and scientific management, program budgeting is thought to derive its core ideas from economics and systems analysis.

20.2 EXISTING BUDGETING SYSTEM

Process

Budgeting process in our context refers to the pre-budget scrutiny, by the Finance Ministry, of the proposals made by the various administrative ministries and departments. Under this system, the budget estimates are prepared with reference to the basic units of appropriation by the agency concerned. The agency may be either a subordinate office or an attached office. The preparation of the estimates involves an assessment of the next year's actual expenditure and are framed having regard to the spending and administrative capacity of the organization. The estimates so prepared are reviewed by the next superior agency and are finally consolidated at the level of the administrative ministry. The review enables an appreciation of the overall needs, and in the process, certain proposals which may not fit into the priority pattern are eliminated.

At the time of the preparation of budget estimates, the periphery of each agency is confined to its own needs. The needs of a local agency may not, however, be viewed with the same urgency in the context of a broad-based review of the activities. The schemes that are scrutinized by the Ministry of Finance and found acceptable are then finally included in the budget. The procedure in respect of the plan outlays is different in that the extent, magnitude and composition of the planned expenditure are usually determined in consultation with the Planning Commission. The planning process involves the formulation of schemes, setting forth their targets, measuring their costs and identification of other related aspects. It is generally conceded that the process governing the plan outlays is somewhat more elaborate and better articulated than the one governing the non-plan outlays. The fact that there are two agencies responsible for loaning and budgeting has given rise to problems notwithstanding certain organizational aspects that help foster a basic unity of purpose. Among those that have helped strengthening the bonds and facilitated a common budgetary base are:

- The framework of the plans is decided upon by the cabinet as a whole and its decisions form the final policy pronouncements in this regard.
- The Finance Minister is a member of the Planning Commission and plays a constructive role even during the initial stages of formulating the plans.
- The Economic Secretary in the Union Ministry of Finance usually acts as the Chairman of the Working Group on Resources set up by the Planning Commission.
- There is a close coordination between the Ministry of Finance and the Planning Commission in formulating the plans.

Features of the Existing Budgeting System

The features of the system are:

- The budget is prepared and exhibited with a predominant bias for the objects of expenditure.
- It does not indicate the objectives of governmental programs, the economics of the operations, the proposed operations of each of the governmental agencies and the benefits that flow from it or the relationship between the resources and operations.
- It lays more emphasis on the cost aspects without corresponding indication of the results.

20.3 NEED OF NEW BUDGETING MODEL

The increasing public expenditures which brought with them a good deal of complexity, led to two significant questions:

(i) How could an element of built in efficiency be introduced into the expenditures?

(ii) How to control and regulate the increasing public expenditure?

These questions, which to some extent are synchronized with the emergence of developmental planning, led to a critical search for new techniques and systems that would enable better determination of the aggregates and detailed allocations, render clarity to the purposes and direction of expenditures and generally serve as a guideline for developmental purposes.

In order to answer these questions, several new budgeting models are prepared. Here, in this chapter an attempt has been made to systematically analyze the national budget and giving outlays to different sectors on the basis of chosen priorities. Attempt has also been made to access the performance of the budget and the sensitivity of each outlays on the performance.

This Budget model resembles the planning programming budgeting in its overall assessment. In the next part, we will discuss about planning program budgeting.

20.4 PLANNING PROGRAMMING BUDGETING

A major purpose of the system of planning programming budgeting is considered to be improving the basis for major program decisions, by providing the decision-maker with more choices and better analytical data to support his decisions. It remains, however, important to spell out clearly what the decisions are and what is sought to be accomplished thereby.

The need for allocation of resources arises because the amount of resources available to a government at a time is limited. This scarcity of resources lends importance to the need for cost analysis. In this context, laying of priorities in the selection of programs and projects assumes significance.

The planning programming budgeting system provides a methodology for the allocation of funds in the budget. This is based on analytical studies directed to an assessment of alternative means for the achievement of the specified objectives. The system involves defining various programs, identifying inputs required with a calculation of the related outputs expected, and trying to obtain the lowest possible input-output ratio. The presumption is that by incorporating a scheme of planning in budgetary process the new technique called PPBS (Planning Programming Budgeting System) – it would help employ public resources in the most efficient and economical manner. The allocation of budget resources according to cost-benefit/effectiveness analysis is considered to achieve the best possible input-output relationship. Despite the various difficulties in defining the scope of cost-benefit/ effectiveness analysis and the problems of measuring conveys that a theory of budgeting can be built by incorporating various stages of policy analysis and planning into the planning programming budgeting system. PPBS will enable us to:

- Identify our national goals with precision and on a continuing basis.
- Choose among those goals the ones that are most urgent.
- Search for alternative means of reaching those goals most effectively at the least cost.
- Inform ourselves not merely for next year's costs–but on the second, and third, and subsequent years' costs–of our budgets.
- Measure the performance of our programs to insure a rupees's worth if served for each rupee spent.

The flexible system methodology for modelling of overall structure of the central budget is shown in Figure 20.1. Overall structure of the central budget has two major receipt and expenditure sections, as shown in Figure 20.2.

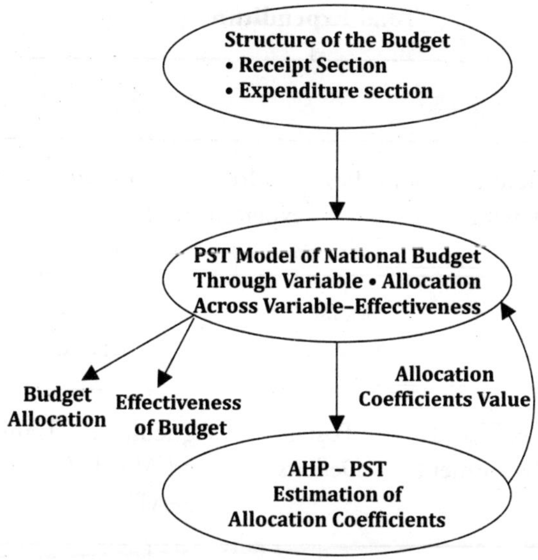

Figure 20.1: Flexible System Methodology for Modelling of National Budget

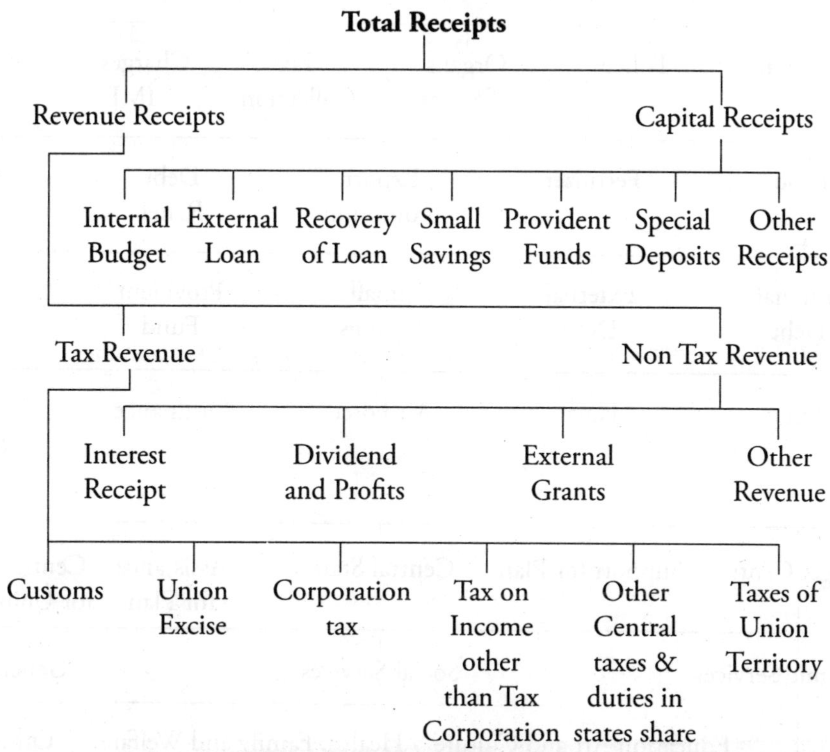

Figure 20.2 (a): Structure of Central Budget – Receipt Section

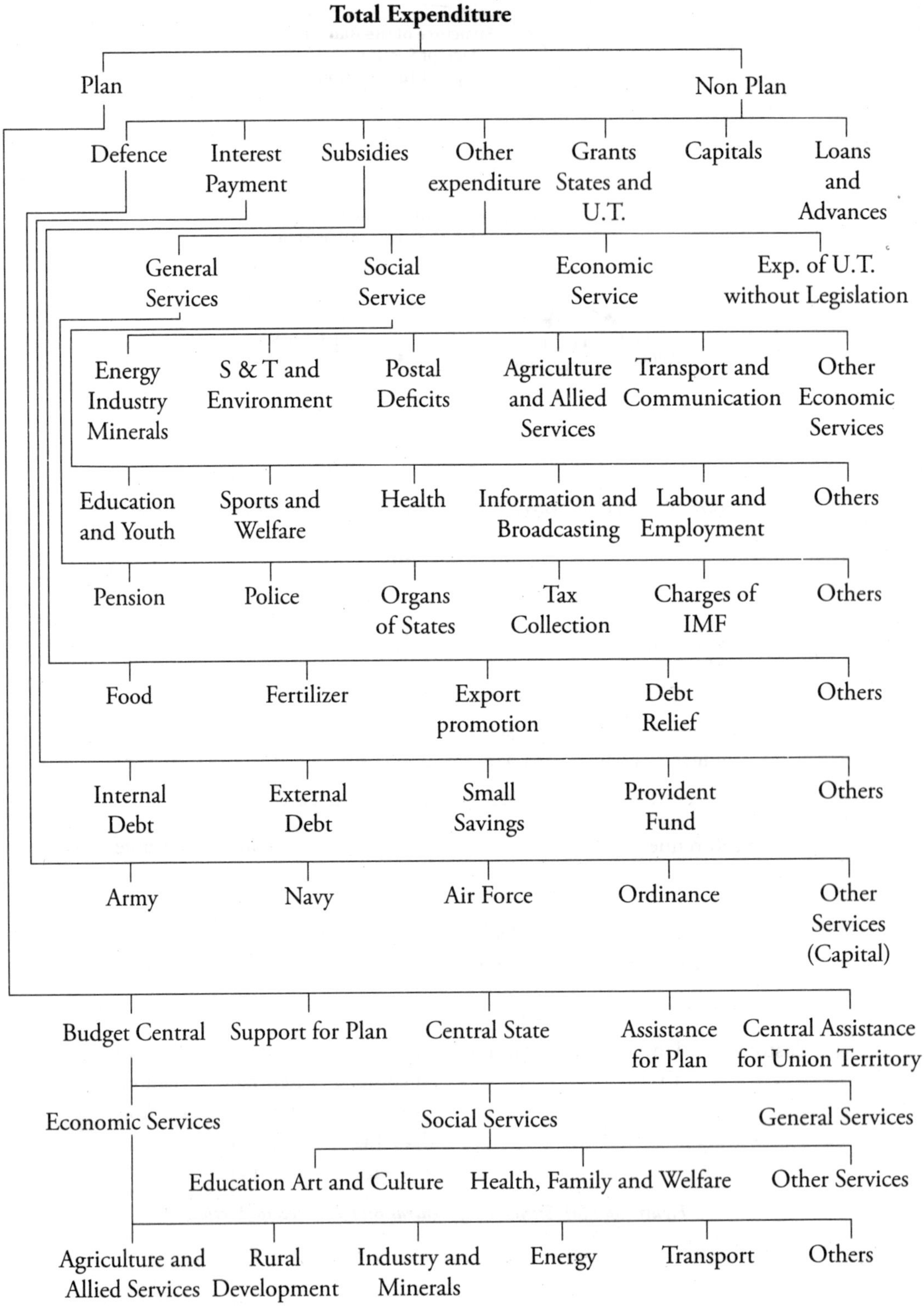

Figure 20.2 (b): Structure of Central Budget – Expenditure Section

20.5 DEVELOPMENT OF PHYSICAL SYSTEM MODEL

Component Models

Each component of the budgeting system is taken and its component graph is plotted. Naming is done for all the nodes and branches in the following fashion.

We select nodes from the left corner of the data hierarchy and start giving it a number (Starting from 1 and increasing so forth). The output of each node is named by node number preceded by 'o'.

For each component of the node, performance equation, flow equation and funds balance equation are determined for both the sections. Two variables are chosen for analysis.

(*a*) *x* variable (across variable) –Performance Index

(*b*) *y* variable (through variable)– Outlays value

Sample component models are shown in Figure 20.3.

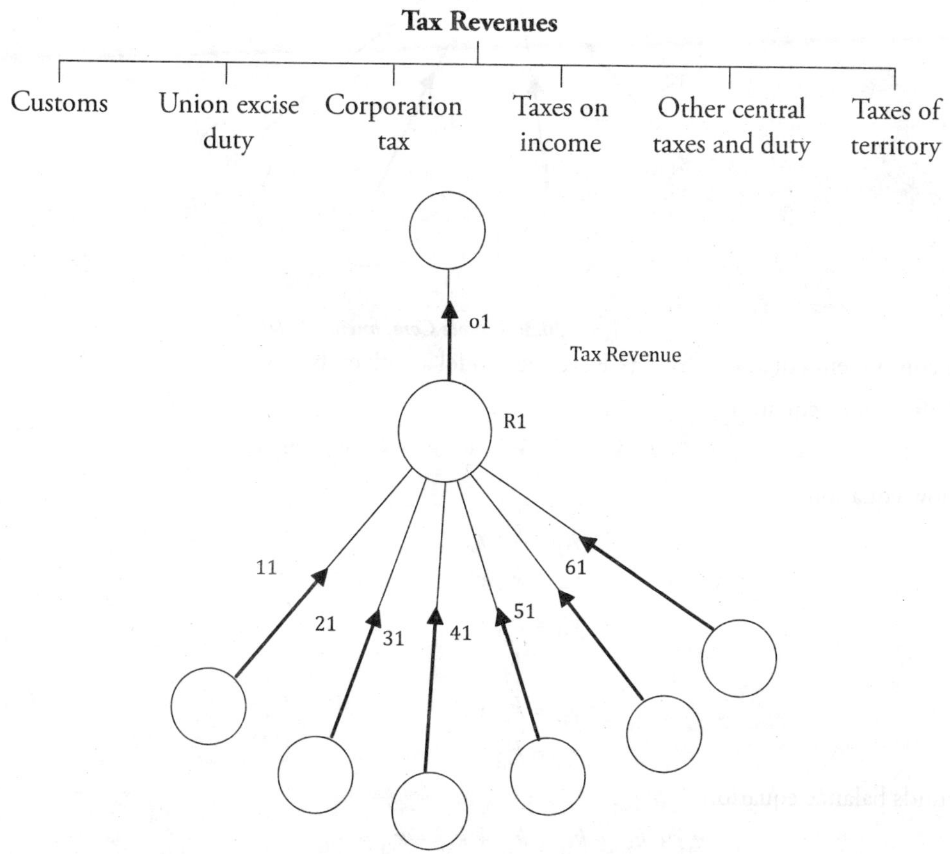

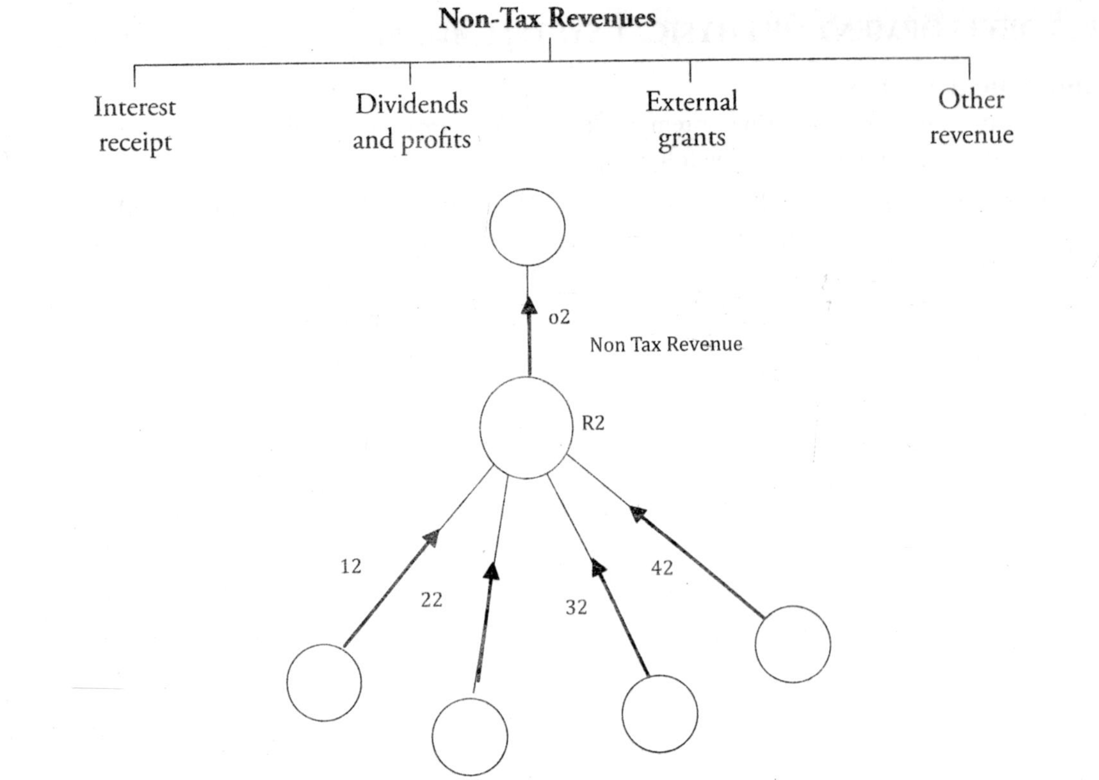

Figure 20.3: Sample Component Models

Sample component equations for Tax Revenue model are given below:

(i) Performance equation

$$x_{o1} = k_{11}\, x_{11} + k_{21}\, x_{21} + k_{31}\, x_{31} + k_{41}\, x_{41} + k_{51}\, x_{51} + k_{61}\, x_{61}$$

(ii) Flow equations

$$y_{11} = k_{11}\, y_{o1}$$
$$y_{21} = k_{21}\, y_{o1}$$
$$y_{31} = k_{31}\, y_{o1}$$
$$y_{41} = k_{41}\, y_{o1}$$
$$y_{51} = k_{51}\, y_{o1}$$
$$y_{61} = k_{61}\, y_{o1}$$

(iii) Funds balance equation

$$k_{11} + k_{21} + k_{31} + k_{41} + k_{51} + k_{61} = 1$$

Simplified System Graph

This model has been simplified as it satisfies all the three necessary requirements given in chapter three. As there is one-to-one correspondence, the links can be replaced by a single link. The continuity and compatibility equations can be absorbed in the component terminal equations.

The simplified system graphs are shown for both the receipt and expenditure sections in Figures 20.4 and 20.5 respectively. The above simplification will result in the number of edges in the system graph half. The number of variables and the number of system equations also reduce to half of the original one.

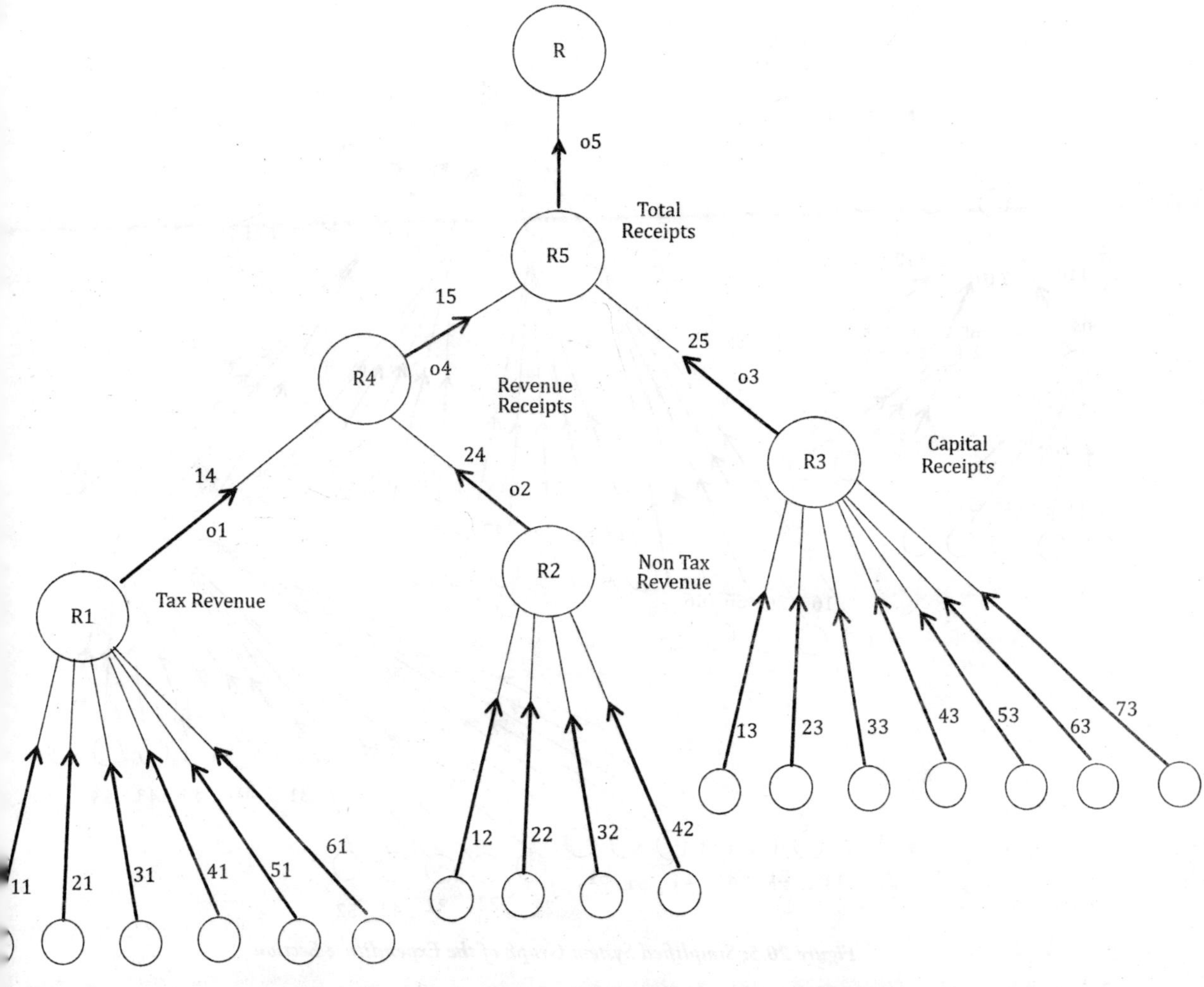

Figure 20.4: Simplified System Graph of the Receipt Section

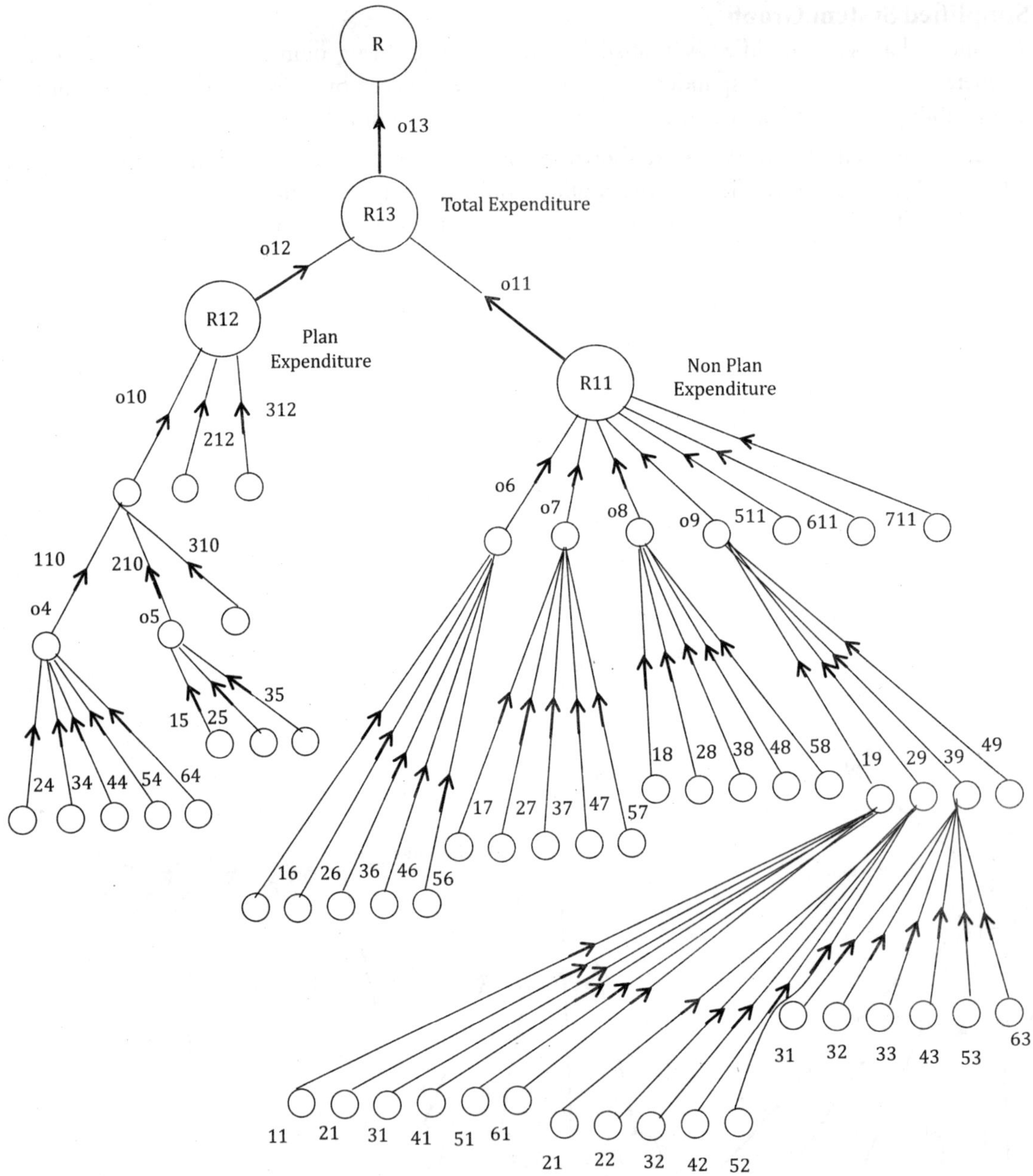

Figure 20.5: Simplified System Graph of the Expenditure Section

20.6 FLOW ANALYSIS

Allocation coefficient, for each branch is calculated by using flow equations. At each node, one allocation coefficient value is determined by funds balance equation.

All data with heads name and code with all flow equations and funds balance equations are put in the spreadsheet. In the spreadsheet, it is seen that the calculated value for flow variable is nearly equal to the actual value of the flow variable. So in this way the given model is validated for the flow analysis.

The model is not validated for the performance variable because of non-availability of the data about the performance of each head.

Generating Own Allocation Coefficient

For generating own allocation coefficients, we have made use of AHP technique.

Flow Analysis for Generated Allocation Coefficients

Flow analysis is done for the generated value of the allocation coefficient and the new values of the outlays of each head are determined on spreadsheet.

Graphical Presentation

A comparison has been made between old value of allocation and new value of allocation on bar graphs. The relative importance of the various heads at the same node for the new value of allocation is shown by Pie graph.

We obtained 13 bar graphs and 13 pie graphs in the expenditure section. We obtained five bar graphs and five pie graphs in the receipt section. These graphs can also be generated on the screen by using macros.

Results from the Analysis of Bar and Pie Graphs

Receipt Section

i. Tax revenue should be decreased from the existing value of outlays and corresponding amount of increase should take place in non-tax revenue.
ii. At capital receipt node, the outlays for internal debt and external loan and saving should be increased as compared to existing outlay.
iii. Under non-tax revenue head, interest receipt should cover 54.7% of the outlay for non-tax revenue.

Expenditure Section

i. Non plan expenditure should be reduced.
ii. There should be cut in defence expenditure and in interest payments.
iii. There should be increase in subsidy for export and debt relief with cut in subsidy for fertilizer and food.
iv. There should be cut in internal debt and increase in savings.
v. Savings should cover 38.9% of the total interest payments.

In fact, from each bar graph, we can estimate what is the change in the outlay for each head from the existing value. From each pie graph, we can estimate the relative percentage of all the heads under one node for the generated values of allocation coefficients.

20.7 GENERATING OWN ALLOCATION COEFFICIENTS USING AHP

The reason for using AHP technique for generating allocation coefficient for each head of both the sections is that there are multiple criteria used in the decision-making of how much to allocate and many of the criteria are of qualitative nature. Here, an attempt has been made to select the criteria and converting the qualitative nature of the criteria in quantitative nature by using AHP technique.

Allocation coefficients are developed in the following steps. Three criteria are selected in the expenditure section.

(a) **Economic Service** means to become self reliant (by export promotion and important substitution), technology upgradation, infrastructure development (communication, transport, power, water), and profits from different heads.

(b) **Social Service** means employment generation, pollution control, security, human resource development (health, education), and infrastructure development (food, housing, hospitals, clothing)

(c) **General Service** means for police, for interest payments, subsidies, grants, etc.

Four criteria are selected for the receipt section.

(a) To make product internationally cost effective

(b) To have equitable distribution of money

(c) To create habit of saving

(d) To promote and protect certain industries

Feeding Data for AHP Modelling

For AHP modelling, the heads at the same hierarchical level from the node are considered as alternatives. Now pairwise comparisons are made between each criteria and then for each criteria, pairwise comparison are made between various alternatives. The values are fed in the software package and the value of global priority vector obtained for each alternative are the value of allocation coefficient for the respective head.

Different weightages are also determined for different criteria using AHP model. These weightage will be useful in performance analysis.

20.8 FINDING VALUE OF ACROSS VARIABLE

A scale is chosen (ranging from 0.1 to 0.9) with the following meaning

Performance	Value
Very high	0.9
High	0.7
Medium	0.5
Low	0.3
Very Low	0.1

According to the expected performance of each head for the each criterion, a value ranging between the 0.1 and 0.9 is placed on the basis of above scale for the exogenous (lowermost nodes) variables only. The scale values are multiplied by the respective criteria weightage and are summed up to get out the value of performance index of exogenous variables.

The value of performance index for the higher order exogenous variables is determined by using performance equations of the PST model.

20.9 SENSITIVITY ANALYSIS

Sensitivity analysis means finding the value of ∂_x/∂_k. So, for doing sensitivity analysis, each value of allocation coefficient is increased by a small amount (here by 0.1) one by one and each time the value of performance index is noted down.

$$\text{Sensitivity Value} = \frac{(\text{New performance index} - \text{old performance index})}{(\text{New allocation coefficient} - \text{old allocation coefficient})}$$

Results from Existing Allocation Coefficients

The performance of the whole budget using existing allocation coefficients for both the sections are

	Performance
(i) For receipt section	0.53685
(ii) For expenditure section	0.40525

Results from Generated Allocation Coefficient

The performance of the whole budget using generated allocation coefficients using AHP for both the sections are

	Performance
(i) For receipt section	0.55786
(ii) For expenditure section	0.44231

Receipt Section Sensitivity Analysis

The most sensitive head in the receipt section is Internal Department and its sensitivity value is 0.246613. If the allocation coefficient of this section is increased by value of 0.1 (increase of 37%), the performance of the overall receipt section is increased from 0.53685 to 0.58253 (increase of 9%).

Expenditure Section Sensitivity Analysis

The most sensitive head in the expenditure section is States Plan and its sensitivity value is 0.1353375. If the allocation coefficient of this head is increased by value of 0.1 (increase of 30%), the performance of the overall expenditure section is increased from 0.44231 to 0.45585 (increase of 3%).

REFLECTIONS

This model is useful in

(i) Comparing the performance of the future budget with the existing budget and showing its comparative performance both analytically and graphically.

(ii) Studying the trends in the changing of allocation of resources to different heads over the years.

(iii) Studying the variations in the sensitivity of the heads and quickly observing the effect of changing one particular outlay on the overall budget.

APPENDIX 20A

Values of Allocation Coefficients

Receipt

(1) $k_{11} = 0.354$
$k_{21} = 0.247$
$k_{31} = 0.065$
$k_{41} = 0.162$
$k_{51} = 0.032$
$k_{61} = 0.200$

(2) $k_{12} = 0.351$
$k_{22} = 0.128$
$k_{32} = 0.164$
$k_{42} = 0.143$
$k_{52} = 0.214$

(3) $k_{13} = 0.269$
$k_{23} = 0.204$
$k_{33} = 0.177$
$k_{43} = 0.154$
$k_{53} = 0.103$
$k_{63} = 0.092$

(4) $k_{14} = 0.116$
$k_{24} = 0.201$
$k_{34} = 0.334$
$k_{44} = 0.162$
$k_{54} = 0.171$
$k_{64} = 0.049$

(5) $k_{15} = 0.278$
$k_{25} = 0.264$
$k_{35} = 0.459$

Expenditure

(1) $k_{11} = 0.474$
$k_{21} = 0.326$
$k_{31} = 0.134$

$k_{41} = 0.030$
$k_{51} = 0.010$
$k_{61} = 0.025$
(2) $k_{12} = 0.353$
$k_{22} = 0.057$
$k_{32} = 0.064$
$k_{42} = 0.127$
(3) $k_{13} = 0.272$
$k_{23} = 0.120$
$k_{33} = 0.156$
$k_{43} = 0.017$
$k_{53} = 0.054$
$k_{63} = 0.247$
$k_{73} = 0.134$
(4) $k_{14} = 0.782$
$k_{24} = 0.218$
(5) $k_{15} = 0.663$
$k_{25} = 0.336$
(6) $k_{16} = 0.502$
$k_{26} = 0.055$
$k_{36} = 0.128$
$k_{46} = 0.009$
$k_{56} = 0.305$
(7) $k_{17} = 0.443$
$k_{27} = 0.078$
$k_{37} = 0.195$
$k_{47} = 0.267$
$k_{57} = 0.018$
(8) $k_{18} = 0.207$
$k_{28} = 0.377$
$k_{38} = 0.218$
$k_{48} = 0.094$
$k_{58} = 0.104$
(9) $k_{19} = 0.502$

$k_{29} = 0.168$
$k_{39} = 0.247$
$k_{49} = 0.082$
(10) $k_{110} = 0.791$
$k_{210} = 0.208$
$k_{310} = 0.001$
(11) $k_{111} = 0.245$
$k_{211} = 0.324$
$k_{311} = 0.165$
$k_{411} = 0.175$
$k_{511} = 0.049$
$k_{611} = 0.018$
$k_{711} = 0.024$
(12) $k_{112} = 0.574$
$k_{212} = 0.390$
$k_{312} = 0.035$
(13) $k_{113} = 0.319$
$k_{213} = 0.681$

21 ENERGY PLANNING IN INTERSECTORAL FRAMEWORK

21.1 INTRODUCTION

Energy has become a critical resource in the present world due to the rapidly increasing energy requirements related to economic growth–depleting fossil fuel reserves like coal and petroleum–and the environmental impact of waste generated through various energy production and consumption processes.

It is being increasingly realized that rational management and conservation of available energy resources is of the utmost importance, until we are able to develop alternative, renewable sources of energy, and bring these to the level of commercial exploitation.

Energy conservation efforts have two kinds of effects:

- The efficiency of energy conservation and utilization processes affect the extent of direct energy waste.
- As energy is one of the inputs for any production process, any reduction or recycling of waste from the process indirectly affects the energy waste.

This chapter gives an energy model of the national economy, dealing with the following inter-related issues.

- The efficiency of various energy conservation and utilization processes in terms of the macro-level technological coefficients.
- The direct and indirect energy waste as calculated on the basis of technological coefficients.
- The impact of energy waste on the unit cost of energy available to the consumers.
- The quantum of environmental pollution due to energy wastes disposed off to nature.
- The consumption of natural resources based on the direct and indirect energy demand.

The flexible system methodology for energy planning is given in Figure 21.1.

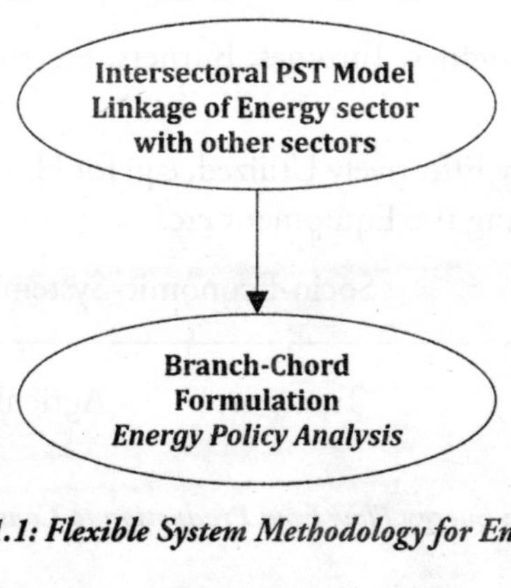

Figure 21.1: Flexible System Methodology for Energy Planning

The principles of physical system theory have been applied, by considering the national economy to consist of n aggregated sectors broadly categorized as energy producers and energy consumers. It is hypothesized that each sector takes inputs from nature and all other sectors, and gives output to all other sectors and the human population. Leontief's concept of input-output economics has been used with one modification[1].

$$\Sigma\text{Input} = \Sigma\text{Output} + \Sigma\text{Waste}$$

In terms of the modified concept, the total energy demand consists of the direct energy demand of the human population, the intermediate demand of other sectors which require energy as input for their production, and the energy wastes which cannot be recycled and have to be disposed off to nature.

The model has been solved using the physical system theory and the generalized solution gives a set of simultaneous equations in the form:

$$[K]\,[V] = [K_1]\,[V_1] + [R]$$

where V is a matrix of unknown variables, K is a matrix containing the value of technological coefficients, V_1 is a matrix of specified variables, K_1 is another matrix of technological coefficients, and R is a matrix of constants indicating recycling costs. These can be solved through the standard matrix inversion techniques to get the value of all unknown variables.

21.2 ENERGY FLOW AND ECOSYSTEM CONCEPT

Figure 21.2 shows the energy flow from the production to the consumption stage through energy conservation or transformation processes, transmission and distribution processes, and energy utilization processes[2].

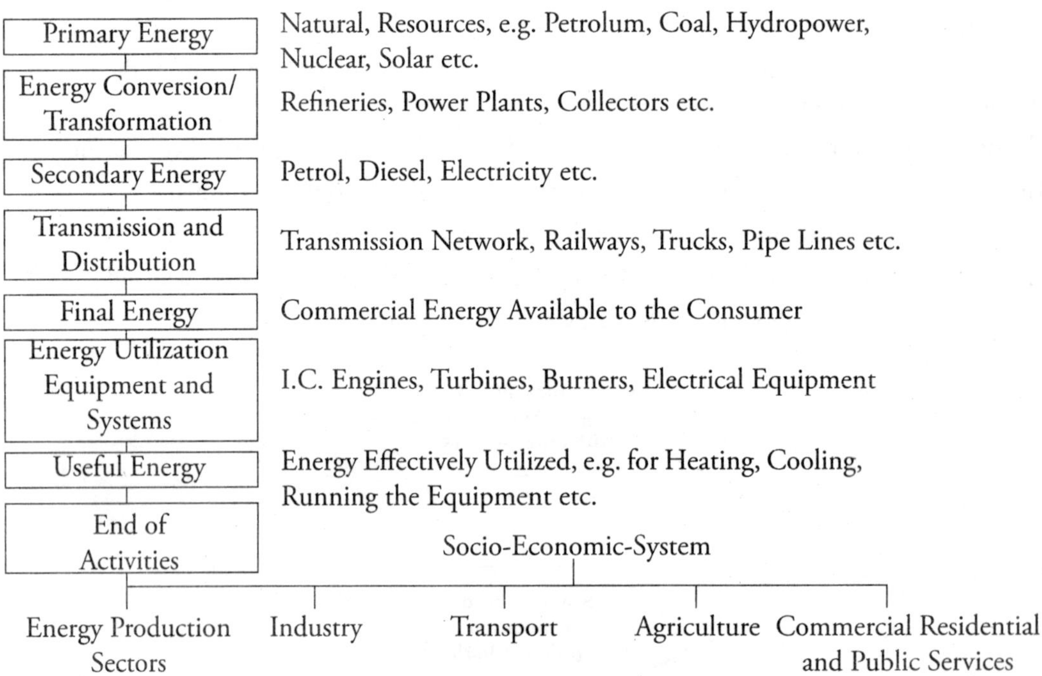

Figure 21.2: Energy Flow from Production to Consumption Stage

The energy flow can be conceptualized as an ecosystem. The ecosystem of an industrialized society is visualized as a system of natural environmental components, and man-made transformation, transportation and storage processes as shown in the energy flow diagram. The components of the natural environment have limited processing capacities which represent the ecological constraints. Systems analysis employing an ecosystem has been used in developing a generalized energy model for the national economy covering aspects of direct and indirect energy wastes and related issues.

21.3 SYSTEMS MODELLING: TWO-SECTOR ECONOMY MODEL

The present model starts with the simplified case of a two-sector economy, say the energy sector i and the non-energy sector j. The notation used in the model are given below:

Variables and Parameters

- y resource flow rate
- x unit cost of resource
- k technological coefficient
- n number of sectors
- R/r recycling cost per unit output per waste recycled
- c constant related to recycling cost
- t tree branches
- t' co-tree chords
- v number of vertices
- e number of edges

Subscripts and Superscripts

- i,j transformation/recycling process
- oi main output of ith process
- i/j from ith to jth process
- w_{in} waste of ith transformation process disposed off to nature
- w_ir waste of ith transformation process recycled
- n nature
- h human population

Transformation Processes

The free-body diagram of the ith sector is shown in Figure 21.3. The ith sector takes the input from nature, from the sector j and partially from its own output, and gives the output to all sectors and the population. It also generates the auxiliary output or the waste to be recycled or to be disposed off to nature.

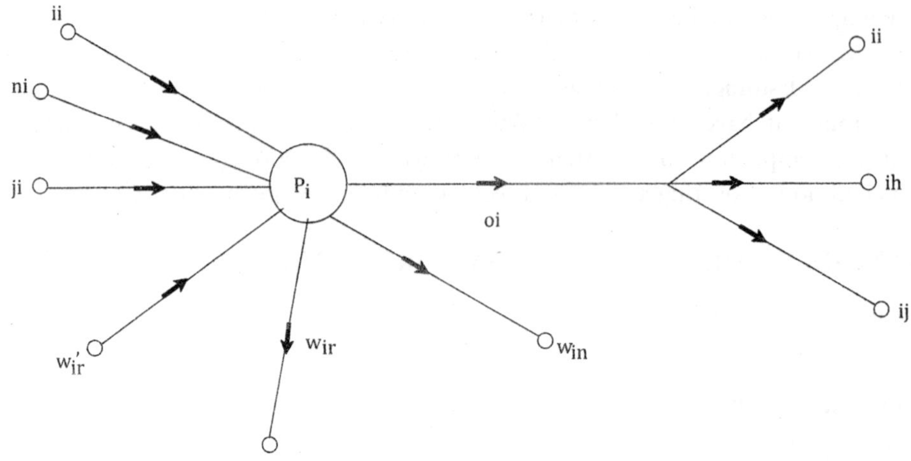

Figure 21.3: Free Body Diagram for ith Sector in Two-sector Economy Model

Waste Recycling Processes

In the present model, waste is defined as any unnecessary input or undesirable output. It is assumed that each sector is generating waste, some of which can be recycled after processing at an additional cost. The fraction of waste which can be recycled will depend on the conservation efforts and the cost of recycling. Both these parameters, i.e. recycling ratio and recycling cost, are determined exogeneously.

For the sake of simplicity, the recycling process has been treated as a transfer function, showing the recycling cost as a function of waste which can be recycled, or as a function of the output of the sector.

Waste that cannot be recycled is disposed off to nature. It includes the waste from the transformation processes as well as waste recycling processes. Waste disposal to nature also takes place at a notional cost. For example, industrial effluents may be disposed off in the sea or in the river after some effluent treatment processes. The waste disposal cost should include the cost of treatment processes and the notional cost of polluting the water. The notional environmental cost may be estimated either on the basis of damage likely to occur through the disposal of waste to nature, or in terms of the cost of restoring the environment back to its proper state (ecological balance).

Transportation and Storage Processes

All the transportation costs are assumed to be centralized by taking transport as a sector within the economy. Storage processes become relevant only in the dynamic situation. The present model is static, hence the storage processes have been ignored.

System Graph

The system graph of the inter-sectoral flows taking nature as well as the human population as part of the eco-system is shown in Figure 21.4. The linkages depict the flow of exogenous variables and waste flow, as well as other input and output variables for each process.

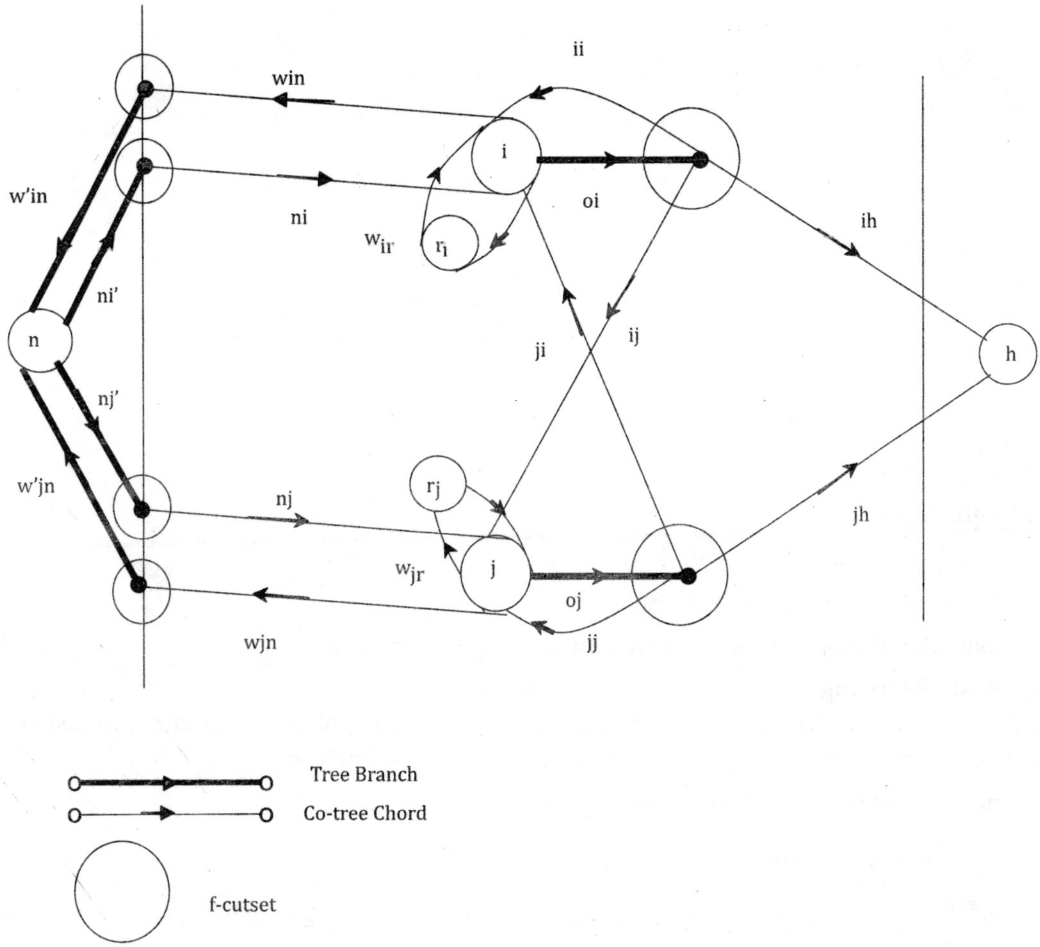

Figure 21.4: System Graph for Two-sector Economy Model

Mathematical Formulation

Applying the concepts of physical theory, response-stimulus relations for each transformation process are written as follows:

(a) Sector (Transformation Process)

For the *i*th process, y_{ji} is the material flow rate from the *j*th sector to the *i*th sector, y_{ii} is the material flow rate of the *i*th sector output back to the *i*th sector, y_{oi} is the output capacity of the *i*th sector, y_{win} is the waste disposed of to nature, x_{oi} is the cost per unit output of the *i*th process, k_{ji} is the *j*th to *i*th technological processing coefficient, and k_{win} is the disposable waste technological coefficient.

Similar notations may be used for the *j*th process.

Resource Flow Equations

For the *i*th sector

$$y_{ji} = k_{ji}\, y_{oi}$$
$$y_{ii} = k_{ii}\, y_{oi}$$
$$y_{ni} = k_{ni}\, y_{oi}$$
$$y_{ih} = k_{ih}\, y_{oi}$$
$$y_{win} = k_{win}\, y_{oi}$$

Similarly for the jth sector

$$y_{ij} = k_{ij}\, y_{oj}$$
$$y_{jj} = k_{jj}\, y_{oj}$$
$$y_{nj} = k_{nj}\, y_{oj}$$
$$y_{jh} = k_{jh}\, y_{oj}$$
$$y_{wjn} = k_{wjn}\, y_{oj}$$

Cost Equations

$$x_{oi} = -\,k_{ji}\, x_{ji} - k_{ii}\, x_{ii} - k_{ni}\, x_{ni} - k_{win}\, x_{win}$$
$$x_{oj} = -\,k_{ij}\, x_{ij} - k_{jj}\, x_{jj} - k_{nj}\, x_{nj} - k_{wjn}\, x_{wjn}$$

Economies of scale have been ignored for the sake of simplicity.

(b) Waste Recycling

For the *i*th process y_{wir} is the waste of the *i*th sector recycled back to the sector after processing, k_{wjr} is the recycled waste technological coefficient, and x_{wjr} is the recycling cost.

Similar notation may be used for the *i*th process.

Resource Flow Equations

$$y_{wir} = k_{wjr}\, y_{oi}$$
$$y_{wjr} = k_{wjr}\, y_{oj}$$

Cost Equation

Waste recycling has been assumed to be a transfer function, showing the cost of recycling as a function of recycled waste

$$\begin{aligned} x_{wir} &= r_i f(y_{wir}) \\ &= r_i f(k_{wir}\, y_{oi}) \\ &= r_I k_{wir.} f(y_{oi}) = R_i f(y_{oi}) \end{aligned}$$

where $R_i = r_i k_{wir}$. Similarly, $x_{wjr} = R_j f(y_{oj})$. For a linear relationship, this may be expressed as:

$$y_{wir} = R_i\, y_{oi} + c_i$$
$$y_{wjr} = R_j\, y_{oj} + c_j$$

By adding the cost of waste recycling, the cost equation for the *i*th sector is

$$x_{oi} = -\,k_{ji}\, x_{ji} - k_{ii}\, x_{ii} - k_{ni}\, x_{ni} - k_{win}\, x_{win} - R_i\, y_{oi} - c_i$$
$$x_{oj} = -k_{ij}\, x_{ij} - k_{jj}\, x_{jj} - k_{nj}\, x_{nj} - k_{wjn}\, x_{wjn} - R_j\, y_{oj} - c_j$$

Unit cost of output = from *i*th sector	Cost attributable to intermediate inputs (from all sectors including *i*th sector)	+ Cost attributable to direct inputs (from nature)
	+ Cost of wastes disposed off (to nature)	+ Cost of waste recycling

The system graph is divided into tree (T) and co-tree (T'). Tree (T) is a connected subgraph, containing all the nodes or vertices, and has no closed path.

The edges of the system graph contained in the tree are called branches. The tree branches must include all the specified across-variables or across-drivers. In the present case, the across-drivers are the cost of natural inputs and the cost of waste disposal to nature.

All the edges of the system graph other than branches are called chords and form the co-tree T'. Specified through-variables or through-drivers are included in the co-tree. In the present case, the through-drivers are flows to the human population or the final demands of each sector.

In the system graph shown in Figure 21.4, the number of vertices (v) is seven, and vertices n, i, j, h, r_i and r_j are considered to be at the same datum, i.e. these are taken as one. The number of edges (e) is 16, the specified across-variables are x'_{ni}, x'_{nj}, x'_{win}, x'_{wjn}, and the specified through-variables are y_{ih} and y_{jh}. Finally, the number of tree branches is six, and the number of chords is 16–6=10.

Structural Constraints

Continuity Conditions

The continuity conditions are a generalization of Kirchhoff's current law (KCL) and these are formed by taking each branch in turn separately. In the matrix form, the continuity conditions (fundamental cutset equations) are:

$$\left[\begin{array}{cc|cc} U & O & C_{11} & C_{12} \\ O & U & C_{21} & C_{22} \end{array}\right] \left[\begin{array}{c} y_{bd} \\ y_{b2} \\ \hline y_{c1} \\ y_{cd} \end{array}\right] = 0$$

where bd represents the branches corresponding to the specified across-variables, $b2$ represents the branches corresponding to the other across-variables, cd represents the chords corresponding to the specified through-variables, and $c1$ represents the chords corresponding to the other through variables.

The continuity equations (fundamental cutset equations) are shown in Appendix 21A.

The number of continuity conditions is $v - 1 = 6$

Compatibility Conditions

The compatibility conditions are a generalization of Kirchhoff's voltage law (KVL) and these are formed by taking each chord, in turn separately. The compatibility equations (fundamental circuit equations) are

$$\begin{bmatrix} B_{11} & B_{12} & U & O \\ B_{21} & B_{22} & O & U \end{bmatrix} \begin{bmatrix} X_{bd} \\ X_{b2} \\ \hline X_{c1} \\ X_{cd} \end{bmatrix} = 0$$

So we have, the f-circuit equations as shown in Appendix 21A.

The number of compatibility conditions is $e - v + 1 = 10$

Branch-Chord Formulation

The response-stimulus equations are in a hybrid form, i.e. it is a mixed formulation, so it can be solved as a set of simultaneous equations by the branch-chord formulation to get the values of unknown variables in terms of exogenous variables.

The response-stimulus equations may be expressed in the hybrid form as follows:

$$\begin{bmatrix} W_{11} & W_{12} \\ W_{21} & W_{22} \end{bmatrix} \begin{bmatrix} X_{b2} \\ X_{c1} \end{bmatrix} = \begin{bmatrix} Z_{11} & Z_{12} \\ Z_{21} & Z_{22} \end{bmatrix} \begin{bmatrix} Y_{b2} \\ Y_{c1} \end{bmatrix} + \begin{bmatrix} F_1 \\ F_2 \end{bmatrix}$$

The branch-chord formulation gives a solution of the form,

$$\begin{bmatrix} Z_{11}C_{11} - Z_{12,} & W_{11} - W_{12}B_{12} \\ Z_{21}C_{22} - Z_{22,} & W_{21} - W_{22}B_{12} \end{bmatrix} \begin{bmatrix} Y_{c1} \\ X_{b2} \end{bmatrix} = \begin{bmatrix} W_{12}B_{11,} & -Z_{11}C_{22} \\ W_{22}B_{11,} & -Z_{21}C_{22} \end{bmatrix} \begin{bmatrix} X_{bd} \\ Y_{cd} \end{bmatrix} + \begin{bmatrix} F_1 \\ F_2 \end{bmatrix}$$

Hence the intended solution is given in Appendix 21B.

On the right-hand side, all the variables are specified. Hence the above set of simultaneous equations may be solved by the standard matrix inversion methods to get the values of all unknown variables on the left-hand side.

21.4 MULTISECTOR ECONOMY MODEL

For the purpose of analysis, the national economy may be considered to consist of n aggregated sectors, broadly categorized as energy producers and energy consumers. The above categorization is only for logical understanding, because mathematically the generalized model assumes that each sector takes the input from nature and all the sectors, and gives the output to all the sectors and the human population.

For the present work, the national economy has been divided into seven aggregated sectors:

(a) petroleum sector,

(b) non-petroleum energy sector,

(c) non-conventional energy sector,

(d) industry,

(e) transport,

(f) agriculture,

(g) commercial, residential and public services.

The first three sectors may be considered to be energy producers and the remaining four as energy consumers. The reason for grouping all the non-petroleum energy sectors together, and considering petroleum sectors and non-conventional energy sectors separately, is that we propose to study the petroleum sector and non-conventional energy sector in more detail.

The model is based on the assumption of a self-sufficient economy, interacting only with nature and the human population across the boundaries. Foreign trade components can be considered by taking imports as a source of input in addition to nature, and exports as another consumer, in addition to human population.

Free-body Diagram

The free-body diagram of the *i*th production sector, in a multi-sector model is shown in Figure 21.5. The assumptions of waste to be recycled or to be disposed off to nature, are the same as for the two-sector model.

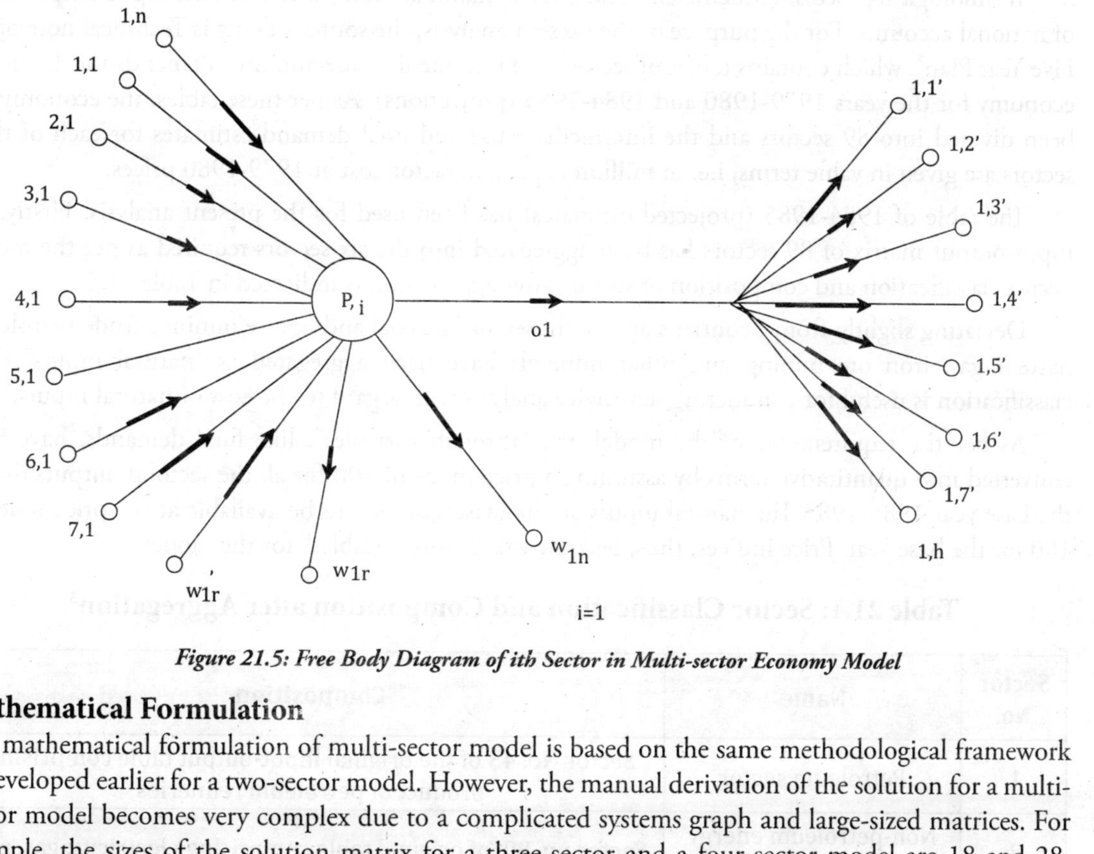

Figure 21.5: Free Body Diagram of ith Sector in Multi-sector Economy Model

Mathematical Formulation

The mathematical formulation of multi-sector model is based on the same methodological framework as developed earlier for a two-sector model. However, the manual derivation of the solution for a multi-sector model becomes very complex due to a complicated systems graph and large-sized matrices. For example, the sizes of the solution matrix for a three-sector and a four-sector model are 18 and 28, respectively, which makes it very difficult to solve the model for more than four-sectors manually.

This constraint has been overcome by generalizing the two-sector model, and considering i and j as subscripts having values 1 to n. This generalization is a step forward in the solution methodology for the multi-sector economy model. It provides substitution. It also renders the methodology amenable to programming logic so that a matrix generation and solution are possible with the computer.

Illustration with a Three-sector Economy Case

For n=3, the mathematical formulation is shown in Appendix 21C. The solution is derived by substituting the subscripts i and j by the numerals in the generalized model. However, the same result is achieved if a three-sector model is solved from first principles.

21.5 INPUT DATA

The data required in the model are "parameters', such as technological processing coefficients, disposable waste technological coefficients, and waste recycling ratios, the "exogenous through variables", such as final demands of all the sectors, and the "exogenous across variables", such as cost of natural inputs, cost of waste disposal to nature and cost of waste recycling.

Technological processing coefficients and final demands are computed from the input-output tables of national accounts. For the purpose of the present analysis, the source of data is Technical note on 6th Five Year Plan[3], which contains tables of sector-wise intermediate use and final demands for the Indian economy for the years 1979-1980 and 1984-1985 (projections). As per these tables, the economy has been divided into 89 sectors and the intermediate use and final demand estimates for each of these sectors are given in value terms, i.e. in million rupees, at factor cost at 1979-1980 prices.

The table of 1984-1985 (projected estimates) has been used for the present analysis. Firstly, the input-output matrix of 89 sectors has been aggregated into the six sectors required as per the model. Sector classification and composition of sectors after aggregation is indicated in Table 21.1.

Deviating slightly from Leontief's approach, sector like coal and lignite mining, crude petroleum, natural gas, iron ore mining and other minerals have been aggregated as "natural inputs". This classification is useful for conducting sensitivity analysis with regard to the cost of natural inputs.

As per the requirements of the model, the "through variables", like final demands, have been converted into quantitative terms by assuming a price index of 100 for all the sectoral outputs during the base year 1984-1985 The natural inputs are likewise assumed to be available at the price index of 100 for the base year. Price indices, thus, become the "across variables" for the model.

Table 21.1: Sector Classification and Composition after Aggregation[3]

Sector No.	Name	Composition
1.	Petroleum sector	Sector No. 43 of the original input-output table comprising product of petroleum refineries
2.	Non-petroleum energy sector	Sector no. 80 of original table comprising electricity, gas, etc.

3.	Industry sector	Sector no. 20-42, 44, 79 of original table comprising sugar, textiles, paper, rubber, edible oils, leather, chemicals, cement, iron and steel, machine tolls, electrical equipment, transport equipment, construction, etc.
4.	Transport sector	Sector no. 81 and 92 of original table, comprising railways, and other surface, air and water transport systems.
5.	Agriculture sector	Sector no. 1-15 of original table, comprising rice, wheat and other major crops, plantations, milk and milk products, forestry and logging, fishing etc.
6.	Commercial, residential and public services	Sector no. 80 of original table comprising communications, trade, storage and warehouses, banking and insurance, education, medical health, real estate and other services.

Note: Sectors 16-19 of the original table, comprised of coal, petroleum, natural gas, iron ore and other minerals, are treated as natural inputs in the model.

The conventional input-output tables have no estimates of sector-wise waste generation or waste recycling in the economy. In the present analysis, these variables are assumed to be exogenous to the model, so that sensitivity analysis can be conducted by changing these variables externally. Waste generation and waste recycling are considered in terms of incremental values over the base run, because it is difficult to estimate their absolute values.

Data on sector-wise waste disposal costs and waste recycling costs are also not available. As a rough estimate, the cost of waste disposal for all the sectors is assumed to be 20% excess over the cost of output, i.e. the waste disposal is priced at an index of 120. Similarly, waste recycling is also priced at an index of 120. As the present analysis is aimed at understanding the general behaviour of the economy through changes in the policy variables, quantitative errors in the above cost estimates will not lead to any serious distortion in the results. Table 21.2 shows the input data used in the present analysis.

21.6 MODEL CALIBRATION

The model has been calibrated on the basis of the input -output of the Indian economy shown in Table 21.2. The results indicate that the following relationship between aggregate inputs, aggregate outputs and aggregate waste disposed is maintained:

$$\Sigma \text{ natural inputs} + \Sigma\text{primary inputs} = \Sigma\text{final flows} + \Sigma\text{waste disposed.} \quad (21.1)$$

If

$$\Sigma\text{waste disposed} = 0,$$

then

$$\Sigma \text{ natural inputs} + \Sigma\text{primary inputs} = \Sigma\text{final flows} \quad (21.2)$$

which is close to the conventional input-output concept. Base run results are derived on the same assumption, and these tally reasonably with the aggregate outputs results of the Planning Commission table[3].

21.7 RESULTS OF SENSITIVITY ANALYSIS

Sensitivity runs of the model have been conducted to observe the effect of changes in the policy variables, such as waste management, final demand and natural resource pricing, on the aggregate output, natural resource requirements, and output price index of the petroleum sector. The impact of such changes on the total economy in terms of the aggregate natural resource requirements, aggregate primary inputs, aggregate output, and aggregate quantum of waste disposal is also analyzed.

Table 21.2: Input Data on Technological Coefficients and Final Demands of Six Aggregated Sectors (Units of Final Demand Equivalent to Million Rupees at Factor Cost at 1979-1980 Prices)

From \ To	Petroleum sector	Non-petroleum energy sector	Industry sector	Transport sector	Agricultural sector	Services sector	Final demand
Petroleum sector	0.00091	0.036315	0.010623	0.250059	0.017732	0.001038	-5031.1
Non-petroleum energy sector	0.004983	0.229680	0.019654	0.007885	0.004469	0.019663	10850.9
Industry sector	0.000743	0.051156	0.313114	0.156901	0.097113	0.112847	587314.3
Transport sector	0.000188	0.043890	0.036215	0.017864	0.004939	0.048854	57760.6
Agricultural sector	0.000110	0.000000	0.097474	0.014235	0.119942	0.012417	406223.0
Services sector	0.036006	0.084387	0.158927	0.128120	0.037716	0.122970	344915.0
Natural inputs	0.802109	0.110029	0.025124	0.006688	0.000019	0.000323	
Primary inputs	0.155770	0.444345	0.338869	0.543248	0.718071	0.681888	

Note: Negative sign indicates import of petroleum

Impact of Policy Variable Changes on Petroleum Sector

The specifications of the different sensitivity runs that have been taken are shown in Table 21.3. The following observations can be made on the basis of these runs:

(1) All the results related to waste generation; waste recycling, and increase in the final demand justify the assumption that waste generation or increase in the final demand in any sector leads to increase in the requirement of natural inputs, aggregate output and/or price index of the petroleum sector. Similarly, increases in the waste recycling leads to a reduction in the above variables.

Table 21.3: Specifications of Sensitivity Runs

Sector No.	Specifications	Results Reference
1.	Base run Waste generation nil; Waste recycling nil; Price index of natural inputs 100; Price index of waste disposal 120; Price index of waste recycling 120.	Same as initial states for all the runs with Nos. 2, 4, 6, 8, 9
2.	Petroleum sector Waste generation increasing up to 25% of aggregate sectoral output; Waste recycling nil; Price indices same as in the base run	Figure 21.6
3.	Petroleum sector Waste generation is 25% of aggregate sectoral output; Waste recycling increasing up to 15% of aggregate sectoral output; Price indices same as in the base run	Figure 21.7
4.	Industry sector Waste generation is 25% of aggregate sectoral output; Waste recycling increasing up to 15% of aggregate sectoral output; Price indices same as in the base run	Figure 21.8
5.	All sectors Waste generation increasing up to 25% of aggregate sectoral output; Waste recycling nil; Price indices same as in the base run	Figure 21.9
6.	All sectors Waste generation is 25% of aggregate sectoral output; Waste recycling increasing up to 15% of aggregate sectoral output; Price indices same as in the base run	Figure 21.10
7.	All sectors Waste generation nil; Waste recycling nip; Price indices same as in the base run	Figure 21.11

(2) The waste generation run for the petroleum sector (Figure 21.6) shows that aggregate output is almost unchanged, but the price index of output increases with the increase in its waste level. Conversely, the waste generation run for the industry sector shows that the aggregate output of the petroleum sector increases with the increase in waste level of the industry, but the price index of its output is almost constant.

The explanation for this behaviour is that waste generation in the petroleum sector itself necessitates changes in its technology in the form of higher technological coefficients which results in a higher price index. But there is negligible change in its aggregate output, because there are no changes in its final demand or the intermediate requirements of the other sectors. Some minor increase takes place due to a higher intermediate requirement of the petroleum sector itself.

On the other hand, waste generation in the industry causes an increase in the requirement of intermediate flow from the petroleum sector to the industry, thereby necessitating higher aggregate output from the petroleum sector. But there is no change in the output price index of the petroleum sector, because its technological coefficients remain the same.

In short, both direct and indirect wastes necessitate higher inputs to the petroleum sector, but the effect of direct waste is reflected in the form of higher output price index of the petroleum sector, whereas indirect waste affects the quantum of its aggregate output.

(3) A comparative study of waste generation of the petroleum and the industry sector shows that the natural resource requirements of the petroleum sector are more sensitive to waste generation in the industry (indirect waste) than waste generation in the petroleum sector itself (direct waste). At a waste generation level of 25% in the petroleum sector, its natural resource requirements go up by 30%, whereas for the same waste generation level in the industry sector, the natural resource requirements of the petroleum sector increase by 43%. Hence, for energy conservation, it is as important to control the indirect waste as the direct waste.

(4) The waste recycling runs of the petroleum sector and the industry sector (Figure 21.7 and 21.8, respectively) show similar results. However, it is to be emphasized that in spite of the additional cost of waste recycling, the output price index of the petroleum sector declines due to a reduction in the inputs required per unit output.

(5) A sensitivity run based on an increase of waste levels simultaneously in all the sectors (Figure 21.9) shows the alarming dimension of wastivity in the economy. The curve of natural resource requirements for the petroleum sector is highly pronounced for simultaneous increase of waste levels in all the sectors compared to the similar curve for direct waste in the petroleum sector only. When the waste generation in the economy goes up to 25% in all the sectors, the natural resource requirements of the petroleum sector increase above 400% of the base level, i.e. an average increase of over 12% for every 1% increase in the base level. This average increase will be higher if the waste generation level increases further because the growth curve is of nonlinear nature. The aggregate output of the petroleum sector also increases above 300% of its initial level to compensate for the 25% waste in the economy. The output price index of the petroleum sector increases to 163 as compared to 100 in the base run.

This accentuated effect of wastivity is the result of the multiplier effect of sectorial inter-linkages in the economy.

O(FLOP)	**Aggregate**	**Output**	**Petroleum**	**Sector**	
C(PSTOP)	**Output**	**Price Index**	**Petroleum**	**Sector**	
N(FNLP)	**Natural**	**Resource**	**Requirement**	**Petroleum**	**Sector**
		O(FLOP)	**C(PSTOP)**	**N(FNLP)**	
	Maximum Values	**3.699E+04**	**1.5507E+02**	**3.8456E+04**	
	Maximum Values	**3.679E+04**	**1.0000E+02**	**2.9662E+04**	

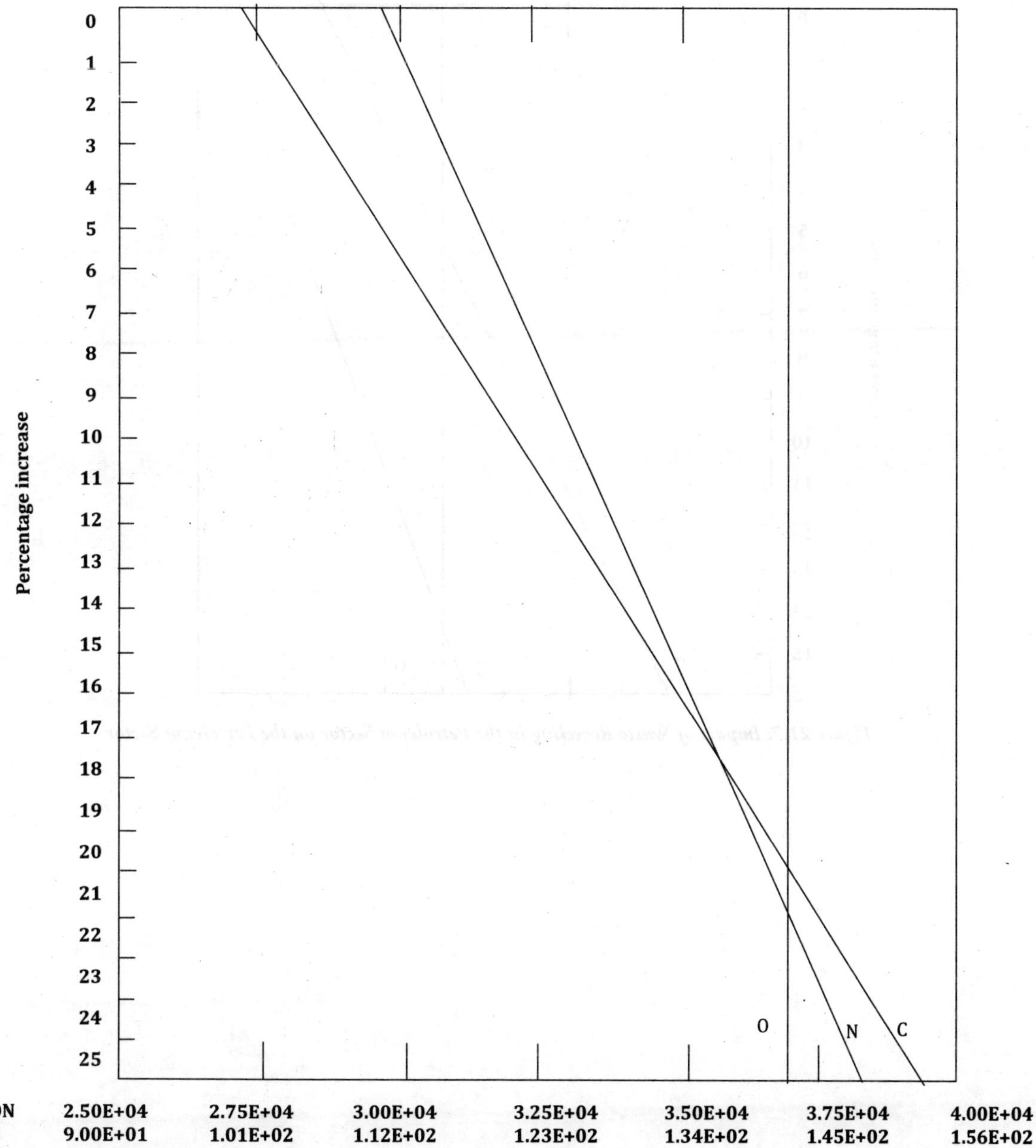

Figure 21.6: Impact in Waste Generation in the Petroleum Sector on the Petroleum Sector

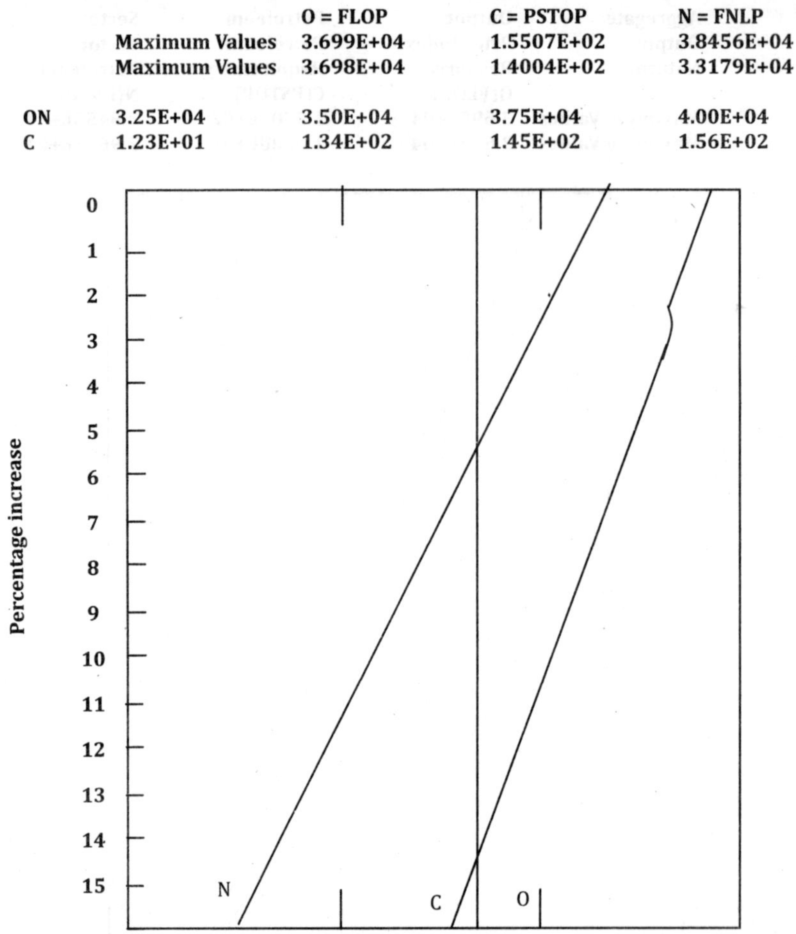

Figure 21.7: Impact of Waste Recycling in the Petroleum Sector on the Petroleum Sector

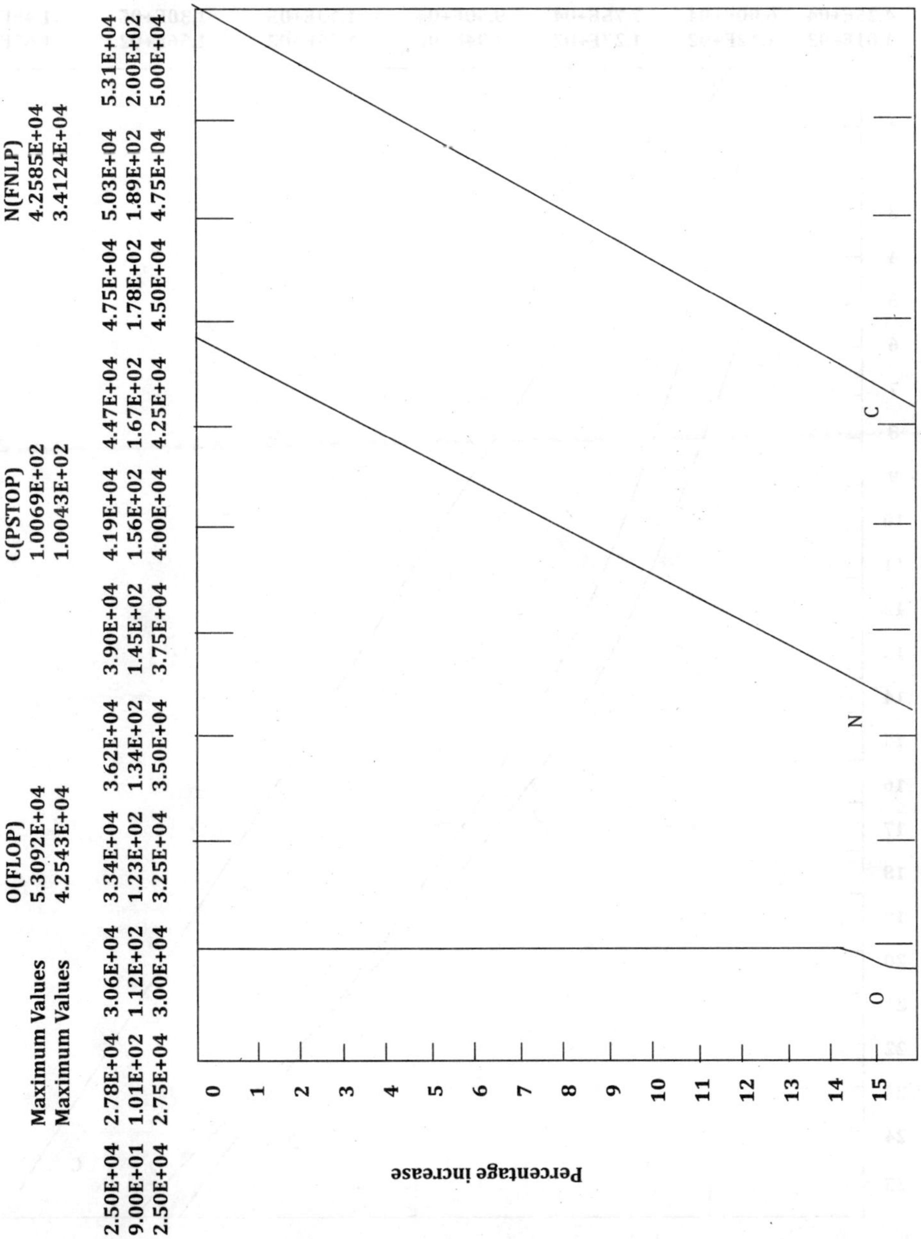

Figure 21.8: Impact of Waste Recycling in the Industry Sector on the Petroleum Sector

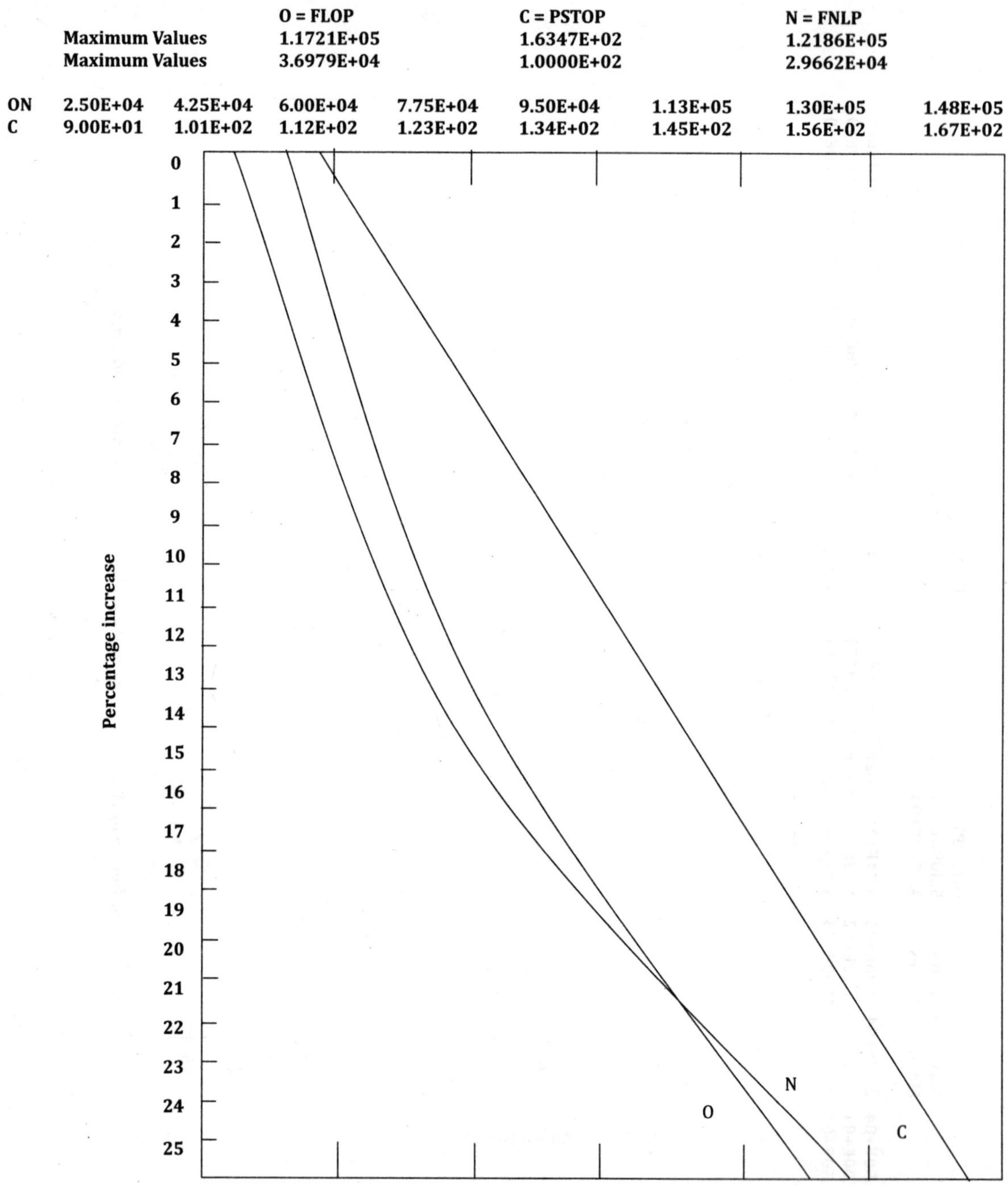

Figure 21.9: Impact of Increase in Waste Generation in All the Sectors on the Petroleum Sector

(6) A sensitivity run of waste recycling simultaneously in all the sectors of the economy (Figure 21.10) also shows a similar impact on the petroleum sector. By simultaneous waste recycling to the extent of 15%, the natural resource requirements of the petroleum sector can be reduced by 43% or to a level only 78% higher than the base run. The aggregate output requirements of the petroleum sector decline by almost 50%, and the output price index also comes down to 144 as compared to 163.

(7) A sensitivity run for an increase in the final demands of all thc sectors (Figure 21.11) shows that the increase in the requirement of inputs, aggregate output and price index of the petroleum sector are not as pronounced as with waste generation. A simultaneous increase of 25% in the final demands of all the sectors results in a 25% increase in the natural resource requirements, a 25% increase in the aggregate output and no impact on the output price index of the petroleum sector.

(8) Finally, the sensitivity run for the increase in the cost of the natural inputs for the petroleum sector shows that a 25% increase in the cost of natural inputs raises the output price index of petroleum to 120, i.e. a 20% increase. By comparing it with the effect of waste generation in the petroleum sector, this level of increase in the output price index will occur at nearly a 9% waste in the petroleum sector, which shows that the crude petroleum prices will have less significant effects on the output price of the petroleum sector as compared to the waste management policies.

Impact of Policy Variable Changes on Total Economy

The results of the sensitivity runs specified in Table 21.3 (except run 9) on the total economy are shown in Table 21.4. The impacts of such changes on the total economy are shown in the form of aggregate natural inputs, aggregate primary inputs (value added), aggregate output and aggregate quantum of waste disposal in the economy.

The following observations can be made on the basis of this analysis:

(1) These runs show the input-output-waste relationship, which confirms the model calibration.

(2) Aggregate natural inputs to the economy are more sensitive to the waste generation and waste recycling in the industry sector as compared to the petroleum sector. When the waste generated in the industry sector increases to 25% the aggregate natural resource requirements of the economy rise to 155% of the base level, whereas for a similar increase of waste generation in the petroleum sector, aggregate natural resource requirements go up to 113% of the base level.

The effect of simultaneous waste generation in all the sectors of the economy is phenomenal. When the waste generation level throughout the economy increases by 25%, aggregate natural resource requirements reach up to 335% of the base level.

Similar effects are observed for waste recycling in the economy.

(3) Aggregate primary inputs (or value added) in the economy are less sensitive to changes in the policy variables, as compared to the aggregate natural inputs. An increase in waste generation by 25% throughout the economy raises the aggregate requirement of the primary inputs to 172%. However, the relative impact of policy changes on the primary inputs reflects a pattern similar to that of the natural inputs, though at a lower scale.

(4) The quantum of aggregate waste disposal shows a similar pattern of sensitivity to the waste generation and waste recycling parameters. An increase of waste generation to 25% in the

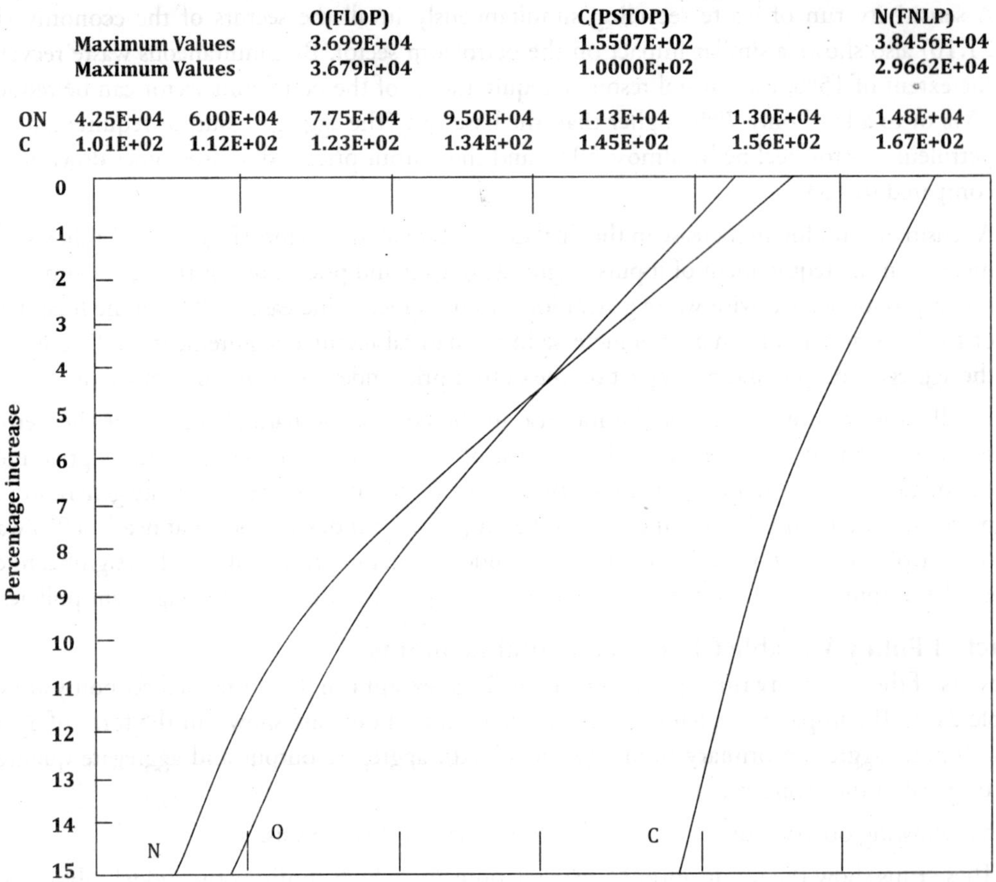

Figure 21.10: Impact of Waste Recycling in all the Sectors on the Petroleum Sector

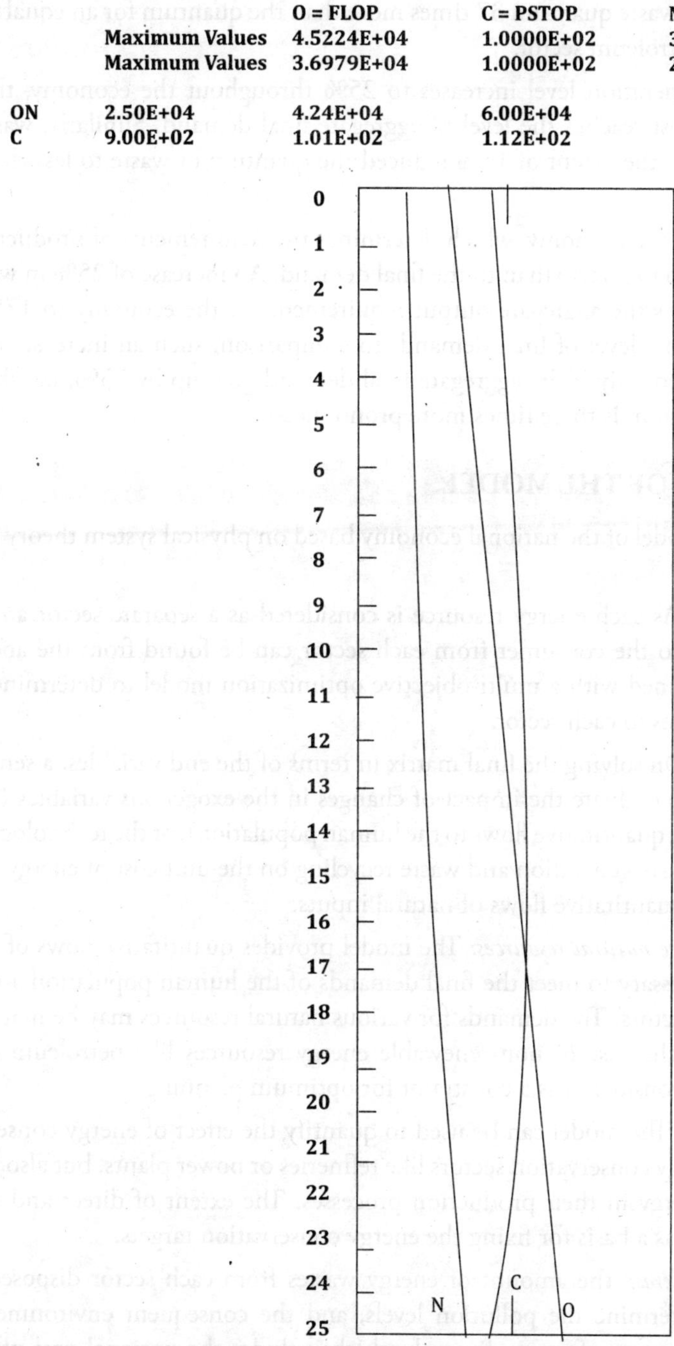

Figure 21.11: Impact of Increase in the Final Demand of all the Sectors on the Petroleum Sector

industry results in a waste quantum 37 times more than the quantum for an equal degree of waste generation in the petroleum sector.

When the waste generation level increases to 25% throughout the economy, the quantum of waste disposed almost reaches the level of aggregate final demand. Similarly, waste recycling in the total economy to the extent of 15% reduced the quantum of waste to less than one-third of its original level.

(5) Aggregate output of the economy, which determines the requirements of production capacity, is also more sensitive to the waste than to the final demand. An increase of 25% in waste generation in the economy raises the aggregate output requirements of the economy to 175% of the base level to meet the same level of final demand. In comparison, such an increase in the aggregate output will be needed only if the aggregate final demand goes up by 75%, i.e. the effect, in the case of waste generation, is three times more pronounced.

21.8 APPLICATIONS OF THE MODEL

The generalized energy model of the national economy based on physical system theory can be used for the following purposes.

(a) *Resource allocation:* As each energy resource is considered as a separate sector and the unit cost of energy available to the consumer from each sector can be found from the above model, the model can be combined with a multi-objective optimization model to determine the optimum allocation of resources to each sector.

(b) *Sensitivity analysis:* On solving the final matrix in terms of the end variables, a sensitivity analysis can be performed to evaluate the impact of changes in the exogenous variables (like the cost of natural inputs, or the quantitative flows to the human population), or the technological coefficients of various inputs, waste generation and waste recycling on the unit cost of energy available to the consumer, and the quantitative flows of natural inputs.

(c) *Consumption of scarce natural resources:* The model provides quantitative flows of various natural resource inputs necessary to meet the final demands of the human population and intermediate demands of other sectors. The demands for various natural resources may be matched with their availability, and in the case of non-renewable energy resources like petroleum and coal, their availability may be considered as a constraint for optimum planning.

(d) *Energy conservation:* The model can be used to quantify the effect of energy conservation efforts, not only in the energy conservation sectors like refineries or power plants, but also in other sectors which consume energy in their production processes. The extent of direct and indirect energy wastes may be used as a basis for fixing the energy conservation targets.

(e) *Environmental problems:* The amount of energy wastes from each sector disposed off to nature may be used to determine the pollution levels, and the consequent environmental problems. By incorporating the cost of waste disposal, which includes the notional cost of environmental pollution, a basis is established for total cost which the consumer has to bear in terms of the environmental costs for each unit of energy. This may help in determining the taxation and subsidy levels for each energy sector to cover the gap between the real energy cost and its market price.

Table 21.4: Impact of Policy Variable Change on the Total Economy (Units Equivalent to Million rupees at Factor Cost at 1979-1980 Prices)

	Base run	Petroleum sector waste generation =25% waste recycling = 0%	Petroleum sector waste generation=25% waste recycling = 15%	Industry sector waste generation=25% waste recycling = 0%	Industry sector waste generation=25% waste recycling = 15%	All sectors waste generation=25% waste recycling = 0%	All sectors waste generation=25% waste recycling = 15%	All sectors final demand increases by =25%
Aggregate natural inputs	.64,874.058	73,681.129	68,396.286	101,050.450	77,366.523	217,377.950	104,114.970	81,092.580
Aggregate primary inputs	1,337,156.700	1,337,596.900	1,337,332.500	1,642,287.700	1,442,524.700	2,304,551.000	1,606,654.900	1,671,446.000
Aggregate intermediate	1,149,108.700	1,149,853.000	1,149,405.700	1,764,377.200	1,361,572.800	3,077,572.600	1,685,372.600	1,436,385.700
Aggregate inputs	2,551,139.458	2,561,131.029	2,555,134.486	3,507,715.060	2,881,464.023	5,599,501.550	3,396,142.470	3,188,924.280
Aggregate final demands	1,402,032.700	1,402,032.700	1,402,032.700	1,402,032.700	1,402,032.700	1,402,032.700	1,402,032.700	1,752,540.900
Aggregate intermediate	1,149,108.700	1,149,853.000	1,149,405.700	1,764,377.200	1,361,572.800	3,077,572.600	1,685,372.600	1,436,395.700
Aggregate outputs	2,551,141.400	2,551,885.700	2,551,438.400	3,166,409.900	2,763,605.500	4,479,605.300	3,087,405.300	3,188,926.600
Aggregate waste disposed	nil	9247.477	3698.368	341,308.100	117,860.900	1,119,900.600	308,740.320	nil
Aggregate outputs plus waste	2,551,141.400	2,561,113.177	2,555,136.768	3,507,718.000	2,881,466.400	5,599,505.900	3,396,145.620	3,188,926.600

21.9 INTEGRATION WITH OTHER MODELS

As indicated earlier, the model may be integrated with an optimization model to determine the resource allocations for different energy sectors, with the objectives of minimizing the average unit cost of energy and the direct disposal of waste to nature, the constraints being the minimum desired output from each sector for a particular planned economic growth rate, the availability of natural resources, and the maximum permissible direct waste to nature based on the maximum permissible pollutant levels for the desired quality of environment.

Similarly, the model may be integrated with a simulation model like `system dynamics' to project the effect of various causal variables, on the variables which are exogenous to the present model like final demands of various sectors, the cost of natural inputs, and the technological processing coefficients. Such integration will overcome the limitations of a physical system model, such as its static character, without compromising its present simplicity.

The model can also be integrated with the technological forecasting models such as Delphi to incorporate the opinions of experts on factors such as the desired level of economic growth, technological innovations and conventional and non-conventional energy sectors, energy conservation and waste recycling efforts, and their likely impact on the environmental costs, etc.

21.10 POLICY IMPLICATIONS

Energy conservation

The sensitivity analysis of the model clearly indicates that energy conservation should be the focal point of energy policy, especially for developing countries where the paucity of resources with regard to crude oil, minerals and other natural resources severely constraints economic growth. There also exists the need to define energy conservation in a broader perspective, covering the direct as well as indirect energy wastes. The model also emphasizes that, even with the additional cost incurred, waste recycling and waste reduction is a viable policy because the savings in terms of input requirements more than offset the additional cost involved.

Aggregate energy demand

The model indicates that aggregate demand of the economy depends on the final demands of different sectors, technology-mix and waste management policies. The model also shows that the aggregate energy demand is more sensitive to the technology-mix and waste management policies than to the final demand. Decisions regarding expansion of capacity in the energy sector are based on the projections of the aggregate energy demand. Hence, the technology-mix and waste management will have a significant bearing on capacity decisions in the energy sector.

Energy pricing

The sensitivity analysis indicates the impact of policies with regard to natural resource pricing and waste management on the unit cost of energy. It has been demonstrated that the waste generation and waste recycling influence the output price index of energy more than the natural resource prices. These observations may help in evolving more effective energy pricing policies.

Ecological and environmental issues

The model demonstrates the impact of percentage waste generation and waste disposal in each sector on the quantum of waste disposal to nature. The notional cost of waste disposal to nature influences the output price index of energy. This should help in fixing appropriate policies on taxation and subsidies to reflect the environmental costs in the energy output prices.

The estimates of natural resource requirements, based on appropriate policy assumptions regarding final demands, technology-mix and waste management, should help in more effective planning with respect to non-replenishable resources.

National accounting incorporating waste management

In the conventional input-output modelling, proper attention has not been given to the waste parameter. Though some extensions of the input-output model are attempted by Leontief, Ayres and Kneese, and others to incorporate the waste parameter, their efforts are limited to dealing with environmental aspects only. The sensitivity analysis demonstrates the detrimental effects of waste in the economy. There is a need to develop a new paradigm of national consistency modelling to incorporate waste management aspects in national planning in an input-output-waste framework.

REFLECTIONS

In this chapter, the ecosystem concept and the principles of physical system theory have been used to model the energy-related issues of the national economy. The model uses the Leontief concept of input-output economics with the modification:

$$\Sigma \text{ Input} = \Sigma\text{Output} + \Sigma\text{Waste}$$

The advantages of structural information in input-output models are combined with waste management aspects, which make the model useful in dealing with many interrelated energy policy issues, such as: resource allocation to different sectors; constraints on the availability of natural resources; the impact of changes in the cost of natural inputs or in the technological coefficients; the impact of direct and indirect energy waste on the unit cost of energy; environmental problems due to the wastes disposed off to nature; the impact of waste recycling on the unit cost of energy and pollution; etc.

The systems analysis methodology has been simplified by deriving a generalized mathematical model and getting the solution matrix for *n* sectors directly by substitution. The model can be integrated using any of the optimization models, e.g. goal programming, or with a simulation model like `system dynamics', or with the technological forecasting models like Delphi for system planning at various levels of aggregation.

Notes

1. Sushil (1984) System Modeling of Waste Management in National Planning, Ph.D. Thesis, *Indian Institute of Technology, Delhi*; Ivars G. (1975) *Proc. Inst. Elect. Electron, Engrs,* 63-431; Leontief W. (1970), *Rev. Econ. Statist.,* 52, 262.
2. Hoffman K.C. (1973) A Unified Framework for Energy System Planning, *Energy Modelling-Special Energy Policy Publication*, England, IPC Science and Technology Press.
3. Technical Note on 6th Five Year Plan of India (1980-85), Perspective Planning Division, Planning Commission, Government of India.

APPENDIX 21A

Structural Constraints

(a) f-cutset equations

$$
\left[\begin{array}{cccccc:cccccccc:cc}
1 & & & & & & -1 & & & & & & & & & \\
 & 1 & & 0 & & & & -1 & & 0 & & & & & & \\
 & & 1 & & & & 0 & & -1 & & & & & & 0 & \\
 & & & 1 & & & & & & -1 & & & & & & \\
\cdashline{7-16}
 & 0 & & & 1 & & & & & & -1 & 0 & -1 & 0 & -1 & 0 \\
 & & & & & 1 & & & 0 & & 0 & -1 & 0 & -1 & 0 & -1
\end{array}\right]
\mathrm{X}
\left[\begin{array}{c}
Y'_{ni} \\ Y'_{nj} \\ Y'_{win} \\ Y'_{wjn} \\ \hdashline
Y_{oi} \\ Y_{oj} \\ \hdashline
Y_{ni} \\ Y_{nj} \\ Y_{win} \\ Y_{ii} \\ Y_{jj} \\ Y_{ij} \\ Y_{ji} \\ \hdashline
Y_{ih} \\ Y_{jh}
\end{array}\right] = 0
$$

(b) f-circuit equations

$$
\left[\begin{array}{cccc:cc:cccccccccc}
1 & 0 & 0 & 0 & 0 & 0 & 1 & & & & & & & & & \\
1 & 0 & 0 & 0 & 0 & 0 & & 1 & & & & & & & & \\
1 & 0 & 0 & 0 & 0 & 0 & & & 1 & & & & & & & \\
1 & 0 & 0 & 0 & 0 & 0 & & & & 1 & & 0 & & & & \\
1 & 0 & 0 & 0 & 0 & 0 & & & & & 1 & & & & & \\
1 & 0 & 0 & 0 & 0 & 0 & & & & & & 1 & & & & \\
1 & 0 & 0 & 0 & 0 & 0 & & & & & & & 1 & & & \\
1 & 0 & 0 & 0 & 0 & 0 & & 0 & & & & & & 1 & & \\
\cdashline{1-6}
1 & 0 & 0 & 0 & 0 & 0 & & & & & & & & & 1 & \\
1 & 0 & 0 & 0 & 0 & 0 & & & & & & & & & & 1
\end{array}\right]
\left[\begin{array}{c}
X'_{ni} \\ X'_{nj} \\ X'_{win} \\ X'_{wjn} \\ \hdashline
X_{oi} \\ X_{oj} \\ \hdashline
X_{ni} \\ X_{nj} \\ X_{win} \\ X_{wjn} \\ X_{ii} \\ X_{jj} \\ X_{ij} \\ X_{ji} \\ \hdashline
X_{ih} \\ X_{jh}
\end{array}\right] = 0
$$

APPENDIX 21B

Branch-Chord Formulation

$$
\left[\begin{array}{cc:cccccccc}
0 & & & & 0 & & & & & \\
\hdashline
1 & 0 & k_{ni} & 0 & k_{win} & 0 & k_{ii} & 0 & 0 & k_{ji} \\
0 & 1 & 0 & k_{ni} & 0 & k_{win} & 0 & k_{ii} & k_{ii} & 0
\end{array}\right]
\left[\begin{array}{c}
X_{oi} \\ X_{oj} \\ \hdashline X_{ni} \\ X_{nj} \\ X_{win} \\ X_{wjn} \\ X_{ii} \\ X_{jj} \\ X_{ij} \\ X_{ji}
\end{array}\right] =
$$

$$
\left[\begin{array}{cc:cccccccc}
-k_{ni} & U & 1 & & & & & & & \\
0 & -k_{nj} & & 1 & & & & & & \\
-k_{wjn} & 0 & & & 1 & & 0 & & & \\
0 & -k_{wjn} & & & & 1 & & & & \\
-k_{ii} & 0 & & & & & 1 & & & \\
0 & -k_{jj} & & & 0 & & & 1 & & \\
0 & -k_{ij} & & & & & & & 1 & \\
-k_{ji} & 0 & & & & & & & & 1 \\
\hdashline
-r_i & 0 & 0 & 0 & 0 & 0 & 0 & 0 & 0 & 0 \\
0 & -r_j & 0 & 0 & 0 & 0 & 0 & 0 & 0 & 0
\end{array}\right]
\left[\begin{array}{c}
Y_{oi} \\ Y_{oj} \\ \hdashline Y_{ni} \\ Y_{nj} \\ Y_{win} \\ Y_{wjn} \\ Y_{ii} \\ Y_{jj} \\ Y_{ij} \\ Y_{ji}
\end{array}\right] +
\left[\begin{array}{c}
0 \\ 0 \\ 0 \\ 0 \\ 0 \\ 0 \\ 0 \\ -C_i \\ -C_j
\end{array}\right]
$$

$$\text{L.H.S.} = \begin{bmatrix} -1 & 0 & 0 & 0 & k_{ni} & 0 & k_{ni} & 0 & 0 & 0 \\ 0 & -1 & 0 & 0 & 0 & k_{nj} & 0 & k_{nj} & 0 & 0 \\ 0 & 0 & -1 & 0 & k_{win} & 0 & k_{win} & 0 & 0 & 0 \\ 0 & 0 & 0 & -1 & 0 & k_{wjn} & 0 & k_{wjn} & 0 & 0 \\ 0 & 0 & 0 & 0 & (k_{ii}-1) & 0 & k_{ii} & 0 & 0 & 0 \\ 0 & 0 & 0 & 0 & 0 & (k_{jj}-1) & 0 & k_{jj} & 0 & 0 \\ 0 & 0 & 0 & 0 & 0 & k_{ij} & -1 & k_{ji} & 0 & 0 \\ 0 & 0 & 0 & 0 & k_{ji} & 0 & k_{ji} & -1 & 0 & 0 \\ 0 & 0 & 0 & 0 & r_i & 0 & r_i & 0 & (1-k_{ii}) & -k_{iji} \\ 0 & 0 & 0 & 0 & 0 & r_j & 0 & r_j & -k_{uhi} & (1-k_{hh}) \end{bmatrix} \begin{bmatrix} Y_{ni} \\ Y_{nj} \\ Y_{win} \\ Y_{wjn} \\ Y_{ii} \\ Y_{jj} \\ Y_{ij} \\ Y_{ji} \\ X_{oi} \\ X_{oi} \end{bmatrix}$$

$$\text{R.H.S.} = \begin{bmatrix} & & & & -k_{ni} & 0 \\ & & & & 0 & -k_{ni} \\ & & & & -k_{win} & 0 \\ & & & & 0 & -k_{wjn} \\ & & 0 & & k_{ii} & 0 \\ & & & & 0 & -k_{jj} \\ & & & & 0 & -k_{jj} \\ & & & & -k_{ji} & 0 \\ k_{ni} & 0 & k_{win} & 0 & -r_i & 0 \\ 0 & k_{nj} & 0 & k_{wjn} & 0 & -r_j \end{bmatrix} \begin{bmatrix} X'_{ni} \\ X'_{nj} \\ X'_{win} \\ X'_{wjn} \\ Y_{ih} \\ Y_{jh} \end{bmatrix} + \begin{bmatrix} 0 \\ 0 \\ 0 \\ 0 \\ 0 \\ 0 \\ 0 \\ -C_i \\ -C_j \end{bmatrix}$$

APPENDIX 21C

Three Sector Model

$$
\text{L.H.S.} = \left[\begin{array}{cccccccccccccccccc}
-1 & 0 & 0 & 0 & 0 & 0 & kn_1 & 0 & 0 & kn_1 & kn_1 & 0 & 0 & 0 & 0 & & & \\
 & -1 & 0 & 0 & 0 & 0 & 0 & kn_2 & 0 & 0 & 0 & kn_2 & kn_2 & 0 & 0 & & & \\
 & & -1 & 0 & 0 & 0 & 0 & 0 & kn_3 & 0 & 0 & 0 & 0 & kn_3 & kn_3 & & & \\
 & & & -1 & 0 & 0 & k_{w1n} & 0 & 0 & k_{w1n} & k_{w1n} & 0 & 0 & 0 & 0 & & & \\
 & & & & -1 & 0 & 0 & k_{w2n} & 0 & 0 & 0 & k_{w2n} & k_{w2n} & 0 & 0 & & & \\
 & & & & & -1 & 0 & 0 & k_{w3n} & 0 & 0 & 0 & 0 & k_{w3n} & k_{w3n} & & & \\
 & & & & & & (k_{11}-1) & 0 & 0 & k_{11} & k_{11} & 0 & 0 & 0 & 0 & & & \\
 & & & & & & 0 & (k_{22}-1) & 0 & 0 & 0 & k_{22} & k_{22} & 0 & 0 & & 0 & \\
 & & & & & & 0 & 0 & (k_{33}-1) & 0 & 0 & 0 & 0 & k_{33} & k_{33} & & & \\
 & & & & & & 0 & k_{12} & 0 & -1 & 0 & k_{12} & k_{12} & 0 & 0 & & & \\
 & & 0 & & & & 0 & 0 & k_{13} & 0 & -1 & 0 & 0 & k_{13} & k_{13} & & & \\
 & & & & & & k_{21} & 0 & 0 & k_{21} & k_{21} & -1 & 0 & 0 & 0 & & & \\
 & & & & & & 0 & 0 & k_{23} & 0 & 0 & 0 & -1 & k_{23} & k_{23} & & & \\
 & & & & & & k_{31} & 0 & 0 & k_{31} & k_{31} & 0 & 0 & -1 & 0 & & & \\
 & & & & & & 0 & k_{32} & 0 & 0 & 0 & k_{32} & k_{32} & 0 & -1 & & & \\
 & & & & & & r_1 & 0 & 0 & r_1 & r_1 & 0 & 0 & 0 & 0 & (1-k_{11}) & -k_{21} & -k_{31} \\
 & & & & & & 0 & r_2 & 0 & 0 & 0 & r_2 & r_2 & 0 & 0 & -k_{12} & (1-k_{22}) & -k_{32} \\
 & & & & & & 0 & 0 & 0 & r_3 & 0 & 0 & 0 & r_3 & r_3 & -k_{13} & -k_{23} & (1-k_{33})
\end{array}\right]
\left[\begin{array}{c}
Y_{n1} \\ Y_{n2} \\ Y_{n3} \\ Y_{w1n} \\ Y_{w2n} \\ Y_{w3n} \\ Y_{11} \\ Y_{22} \\ Y_{33} \\ Y_{12} \\ Y_{13} \\ Y_{21} \\ Y_{23} \\ Y_{31} \\ Y_{32} \\ X_{02} \\ X_{03}
\end{array}\right]
$$

$$
\text{R.H.S.} = \begin{bmatrix}
 & & & & & & -k_{n1} & 0 & 0 \\
 & & & & & & 0 & -k_{n2} & 0 \\
 & & & & & & 0 & 0 & -k_{n3} \\
 & & & & & & -k_{w1n} & 0 & 0 \\
 & & & & & & 0 & -k_{w2n} & 0 \\
 & & & & & & 0 & 0 & -k_{w3n} \\
 & & & & & & -k_{11} & 0 & 0 \\
 & & & & & & 0 & -k_{22} & 0 \\
 & & & & & & 0 & 0 & -k_{33} \\
 & & & 0 & & & 0 & -k_{12} & 0 \\
 & & & & & & 0 & 0 & -k_{13} \\
 & & & & & & -k_{21} & 0 & 0 \\
 & & & & & & 0 & 0 & -k_{23} \\
 & & & & & & -k_{31} & 0 & 0 \\
 & & & & & & 0 & 0 & -k_{23} \\
 & & & & & & -k_{31} & 0 & 0 \\
 & & & & & & 0 & -k_{32} & 0 \\
k_{n1} & 0 & 0 & k_{w1n} & 0 & 0 & -r_j & 0 & 0 \\
0 & k_{n2} & 0 & 0 & k_{w2n} & 0 & 0 & -r_2 & 0 \\
0 & 0 & k_{n3} & 0 & 0 & k_{w3n} & 0 & 0 & -r_3
\end{bmatrix}
\begin{bmatrix} X'_{n1} \\ X'_{n2} \\ X'_{n3} \\ X'_{w1n} \\ X'_{w2n} \\ X'_{w3n} \\ Y_{2h} \\ Y_{2h} \\ Y_{3h} \end{bmatrix}
+
\begin{bmatrix} 0 \\ 0 \\ 0 \\ 0 \\ 0 \\ 0 \\ 0 \\ 0 \\ 0 \\ 0 \\ 0 \\ 0 \\ 0 \\ 0 \\ 0 \\ -C_1 \\ -C_2 \\ -C_3 \end{bmatrix}
$$

22 STEEL PLANNING IN INTERSECTORAL FRAMEWORK

22.1 INTRODUCTION

Iron and steel constitute one of the most critical inputs to the infrastructure growth of an economy. Steel production needs large investments and has a long gestation period. Further, appropriate technology and manpower is also required. The depleting reserves of quality iron ore, rising costs of inputs, lowering of import duties on steel, growing concern for the environmental impact of steel production, and emerging importance of waste management are some of the issues confronting the Indian iron and steel industry today.

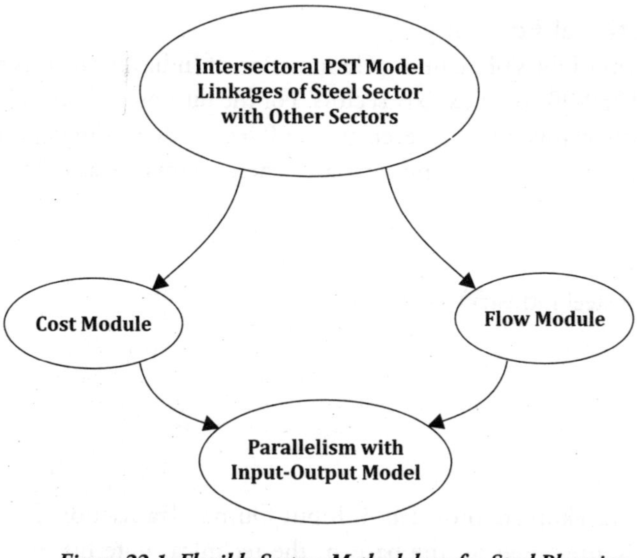

Figure 22.1: Flexible System Methodology for Steel Planning

Physical System Theory (PST) has certain characteristics, which make structural modelling and analysis efficient and flexible. It models simultaneously both the flow and cost variables in an integrated manner with one set of parameters. Thus, PST model development provides an effective quantitative management planning tool. These models can be subjected to sensitivity analysis with respect to changes in the technology[1].

In this chapter, structural model of the iron and steel sector of Indian economy is being developed using physical system theory framework. This model is required to represent interlinked sectoral structure of the national economy. Sectoral classification of national economy has been suitably evolved for the purposes of modelling. The developed model is required to deal with the following interrelated issues; sectoral demands specifically steel demand, trade account with exports and imports at sectoral output and flow. The model is required to appropriately represent the structure of the economy.

In the chapter, firstly national economy modelling would be discussed along with sectoral classification. PST would be discussed and enriched by modelling of seven-sector structure of the Indian economy. Simplification of PST modelling is covered along with discussion of national economy model in simplified PST framework. This structural model is ultimately given the shape of a software. The flexible system methodology for steel planning is given in Figure 22.1.

22.2 NATIONAL ECONOMY MODELLING

All the entries in the Input-Output Transaction Table (IOTT)[2] are at factor cost, i.e. excluding trade and transport charges and net indirect taxes (i.e. indirect taxes less subsidies). The IOTT, to begin with, is prepared at original purchasing price, i.e. at the price, in which the actual transaction takes place. The entries at factor cost are arrived after removing the components of trade and transport margins and net indirect taxes. These have been shown in separate rows in the table. The row of net indirect taxes thus depicts the taxes paid by the industries on intermediate inputs used in the process of production of industrial output.

Sectorization of National Economy

Input-Output Transaction Tables of Planning Commission of India are taken as the primary document. The table for the year 1989-90 contains sixty sectors. For the purpose of convenience, these sixty sectors are aggregated into seven sectors. One can even go for 60 sectors depending upon the requirements; this flexibility is in-built in this PST model and software. These sectors are as follows:

- Agriculture
- Steel
- Industry (excluding steel industry)
- Construction
- Transport
- Energy
- Services

Details of sectoral breakup are provided in Input Output Transaction Tables (CSO, 1986, 1990, 1996). This breakup is attempted on the basis of the technical note on sixth plan of India. In this grouping, steel sector consists of iron ore, steel industry, and steel alloy. Agriculture sector consists of all farm produces, fishing, animal husbandry, forestry etc. Industry sector consists of all industries. Energy sector includes petroleum, coal and lignite, and electricity. Transport sector includes rail, road, and other transports. Service sector includes trade, communication, and other services.

22.3 SECTORAL MODEL OF NATIONAL ECONOMY

Multisector Economy

For the purpose of analysis, the national economy may be considered to consist of n aggregated sectors, broadly classified as steel producing and steel consuming sectors. This categorization is only for logical understanding, because mathematically the generalized model assumes that each sector takes the input from all other sectors including itself and gives the output to all sectors and to itself. Waste is being disposed off to nature and some waste is being recycled. It is assumed that some cost is involved in waste disposal and waste recycling. The sectoral outputs are being consumed by the habitat. The import and export is taking place with each sector at output level. Natural inputs are taken as part of respective sectors. Most of the steel exports are iron ore exports, because of the underdeveloped nature of the industry. However, provision for natural resource sector has been made in the model to incorporate flexibility in modelling and analysis. For the present work, the national economy has been divided into seven aggregated sectors that are agriculture, steel, industry, construction, transport, energy, and services.

Unlike earlier works, export and import are also shown as part of the economy. This trade is taking place at output level. Here natural inputs for steel sector such as iron ore etc. are part of that sector.

Transformation Process

A free body diagram of *i*th sector of two-sector economy is shown in Figure 22.2. Here *i*th sector takes the input from the *j*th sector and partially from its own output. It provides output to all sectors and the population. It also generates the auxiliary output (waste). This waste is either to be recycled or to be disposed off to nature. Free body diagram for *i*th 'Short Circuit Connection Element' is shown in Figure 22.3. This explains the connection between v^{oi} and v^{ii} (sector also consumes its output partially). Further a system graph for seven sectors in a multi sector national economy has been developed. The *i*th sector takes input from different sectors and partially from its own output, and gives output to different sectors and population. Some of the output is exported and some of the finished product is imported. It also generates the auxiliary output or the waste to be recycled or to be disposed off to nature.

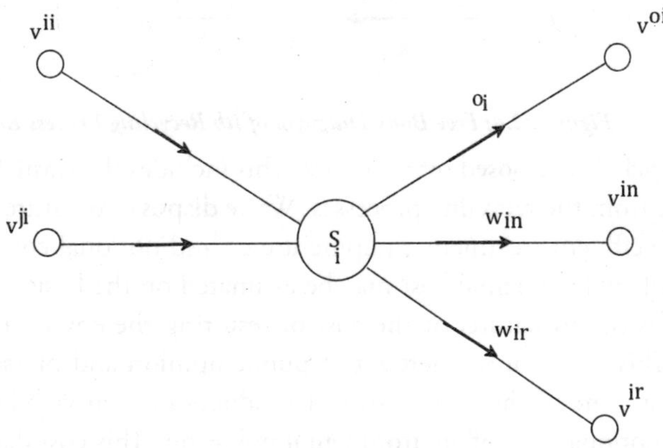

$$\begin{bmatrix} Y_{ji} \\ Y_{ii} \\ Y_{wir} \\ Y_{win} \\ \hline X_{oi} \end{bmatrix} = \left[\begin{array}{cccc|c} & & & & K_{ji} \\ & & & & K_{ii} \\ & & 0 & & K_{wir} \\ & & & & K_{win} \\ \hline -K_{ji} & -K_{ii} & -k_{wir} & -k_{win} & 0 \end{array}\right] \begin{bmatrix} X_{ji} \\ X_{ii} \\ X_{wir} \\ X_{win} \\ \hline Y_{oi} \end{bmatrix} + \begin{bmatrix} 0 \\ 0 \\ 0 \\ 0 \\ \hline V_i \end{bmatrix}$$

Figure 22.2: Free Body Diagram for ith Transformation Process in a Two Sector Economy

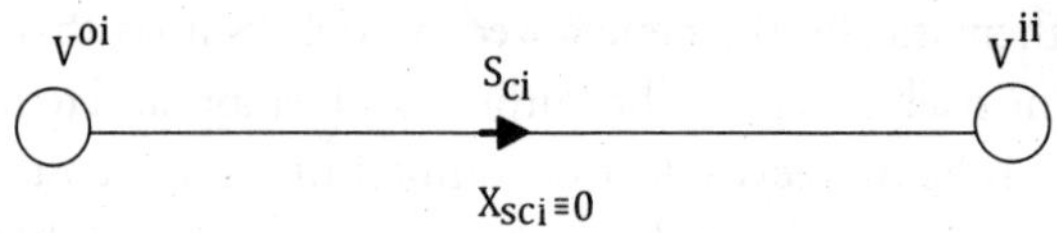

Figure 22.3: Free Body Diagram for ith "Short Circuit" Connection Element

Mass Balancing

In input-output–waste model, waste generation and waste recycling becomes part of the system. In considering waste management, the technological coefficients need changes accordingly. This change is effected through mass balancing of inflows and outflows for each transformation process. It is assumed that quantum of change in inputs due to waste generation and/or waste recycling is uniformly distributed over all the inputs mass balance. This will revise the technological coefficients.

Waste Recycling Process

Waste is defined as an unnecessary input or undesirable output[4]. Assuming each sector generates some waste. Some portion of this waste is being recycled at an additional cost. Rest of the waste is being disposed off to nature (not shown). The fraction of the waste that can be recycled will depend on the conservation efforts and recycling cost. Both these parameters, i.e. recycling ratio and recycling cost, are to be determined exogenously.

The free body diagram of the *i*th recycling process r_i is shown in Figure 22.4. For the sake of simplicity, the recycling process has been treated as a transfer function, shown the recycling cost as a function of waste that can be recycled, or as a function of the output of the sector.

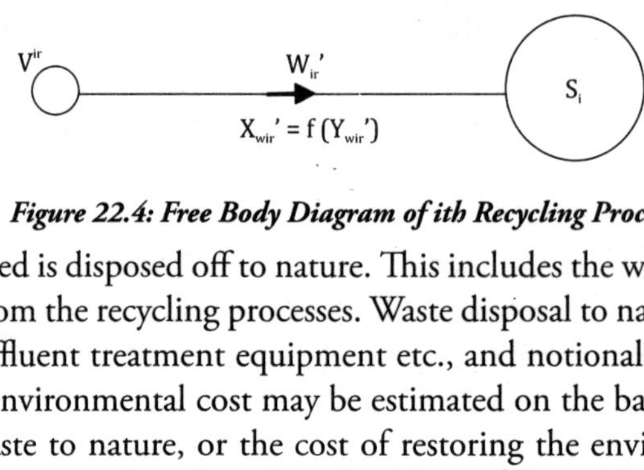

Figure 22.4: Free Body Diagram of ith Recycling Process Ri

Waste that is not recycled is disposed off to nature. This includes the waste from the transformation processes as well as waste from the recycling processes. Waste disposal to nature takes place at a notional cost. There may be some effluent treatment equipment etc., and notional cost of pollution (air, water, noise, etc.). This notional environmental cost may be estimated on the basis of damage likely to occur through the disposal of waste to nature, or the cost of restoring the environment to its proper state, i.e. ecological balance. This is one area where lot of public opinion and pressure is coming up and it is assumed to develop into one of the major costs of production, even viability of old and new units would depend upon its notional cost of environmental pollution. This cost data is not available and is assumed in percentage terms.

Assume x_{wir} represents cost of waste recycled by sector *i*.

y_{wir} represents the amount of waste being recycled.

Waste recycling is assumed to be a transfer function, and the cost of recycling is treated as a function of recycled waste.

Thus, $x_{wir} = r_i f(y_{wir})$

$$= r_i f(k_{wir} \cdot y_{oi})$$

$$= r_i\, k_{wir} \cdot f(y_{oi})$$
$$= R_i\, f(y_{oi})$$

where $Ri = ri \,.\, k_{wir}$

For linear relationships

$$x_{wir} = Ri\, y_{oi} + c_i$$

Thus, the cost of waste recycling is equivalent to the summation of output recycled and value addition fraction to total output. Here k_{wir} is the recycled waste technological coefficient and r_i denotes the percentage of recycling being undertaken. c_i is the cost index and it denotes a physical quantity.

Transportation and Storage Processes

All the transportation costs are assumed to be part of the model as transport is assumed as a sector of the economy. Thus, transportation process is not considered in developing a two-sector model.

Storage processes become relevant only in a dynamic situation. The model being developed is static in nature; hence, the storage processes have been ignored.

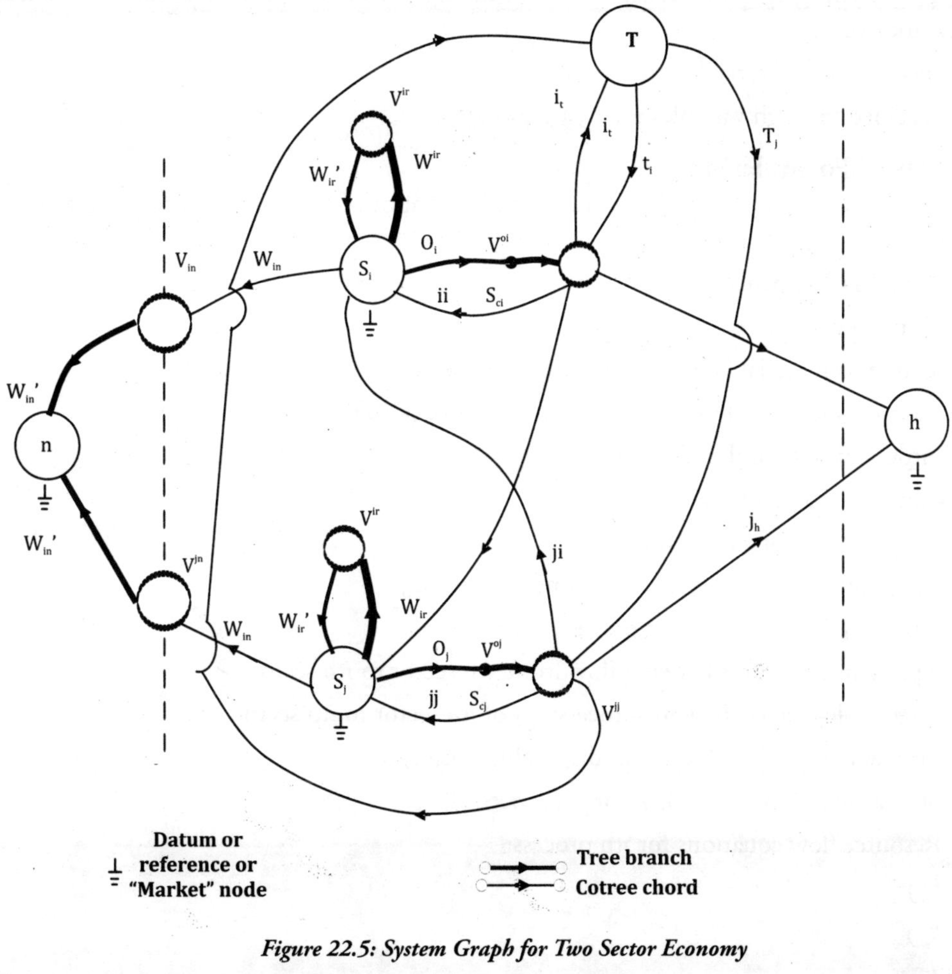

Figure 22.5: System Graph for Two Sector Economy

$$Y_{oi} = Y_{ii} + Y_{ij} + Y_{it} - Y_{ti} + Y_{ih}$$
$$Y_{oj} = Y_{ji} + Y_{jj} + Y_{jt} - Y_{tj} + Y_{jh}$$

$$\begin{bmatrix} Y_{oi} \\ Y_{oj} \end{bmatrix} = \begin{bmatrix} K_{ii} & K_{ij} \\ K_{ji} & K_{jj} \end{bmatrix} \begin{bmatrix} Y_{oi} \\ Y_{oj} \end{bmatrix} + \begin{bmatrix} Y_{it} \\ Y_{jt} \end{bmatrix} - \begin{bmatrix} Y_{ti} \\ Y_{tj} \end{bmatrix} + \begin{bmatrix} Y_{ih} \\ Y_{jh} \end{bmatrix}$$

$$= [K][Y_o] + [E] - [I] + [Y_h]$$

System Graph for Two Sector Economy

The system graph of the inter-sectoral flow as discussed is shown in Figure 22.5. The linkage depict the flow of exogenous variables and waste flow, as well as other input and output variables for each process. The system graph has vertices (v), considering the vertices n, i, j, h, r_i, r_j and t. The number of edges becomes 16.

The specified across variables are x'_{win}, x'_{wjn}.

The specified through variables y_{ih}, y_{jh}, y_{jt} and y_{it}.

Mathematical Formulation

On applying the concepts of PST, response–stimulus relations for various transformation processes are as follows:

(a) Transformation processes

For *i*th process;

y_{ji} – represents material flow rate from *j*th to *i*th sector.

y_{ii} – represents material flow rate of the *i*th sector back to to *i*th sector.

y_{oi} –output capacity of the *i*th sector.

y_{win} – waste disposed off to nature.

y_{ti} – represents imports to sector *i*.

y_{ij} – represents output of sector *i* consumed by sector *j*.

x_{oi} – represents cost/unit output of the *i*th process.

x_{ji} – represents costs/unit material flow from, *j*th sector to *i*th sector.

k_{ji} – is technological coefficient (processing) of *j*th sector to *i*th sector.

k_{win} – is technological coefficient of disposable waste.

Similar notation to be taken for *j*th process.

Resource flow equations for *i*th process:

$$y_{ji} = k_{ji}\, y_{oi}$$
$$y_{ii} = k_{ii}\, y_{oi}$$
$$y_{ih} = k_{ih}\, y_{oi}$$

$y_{win} = k_{win} y_{oi}$

Similar equations can be written for the *j*th process.

After ignoring economies of scale, cost equations become

$x_{oi} = - k_{ii} x_{ii} - k_{ji} x_{ji} - k_{win} x_{win} - R_i y_{oi} - c_i$

and

$x_{oj} = - k_{ij} x_{ij} - k_{jj} x_{jj} - k_{wjn} x_{wjn} - R_j y_{oj} - c_j$

This implies that

Unit cost of output from the *i*th sector = cost attributable to intermediate inputs and input from the sector itself + cost of waste disposal to nature + cost of waste recycling.

22.4 MATHEMATICAL FORMULATION OF SEVEN SECTOR MODEL

Applying the concepts of physical system theory, response-stimulus relations for each transformation processes are written. The mathematical formulation of a multi-sector model is based on the simplified PST methodological framework as discussed earlier in the book. Leontief's concept of input-output economics has been used with one modification.

Thus, Input = Output + Waste

Using the notations as given earlier, free body diagram for *i*th sector in a multi-sector economy is shown in Figure 22.6.

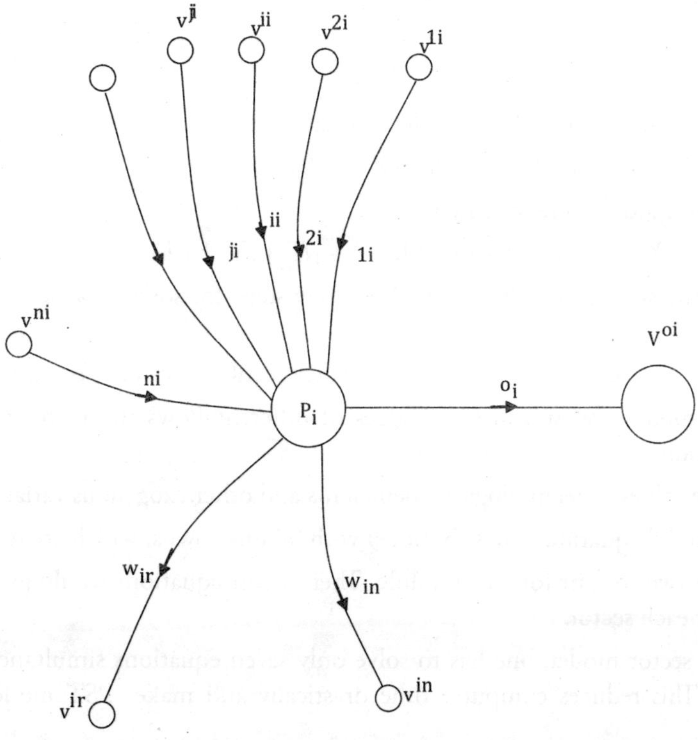

Figure 22.6: Free Body Diagram for ith Sector in a Multi-sector Economy

The flow and cost equations for the seven sector model are given in Appendix 22A.

The flow equations represent flow module of seven-sector economy. On the basis of these flow equations a generalized flow module can be represented because data of total sectoral consumption by the population and sectoral exports and imports are available with IOTT.

For cost equations, the cost of output from sector i = (sum of cost of input from sector 1 to 7) x (Percent of respective output from different sector) + cost of waste disposal to nature + cost of waste recycling + value addition.

The cost equations given in Appendix 22A represent cost module. After generalizing they can be represented in matrix form as:

$$[X_{oi}] = [K_{ij}]\ [X_{ij}] + [K_{win}]\ [X_{win}] + [K_{wir}]\ [X_{wir}] + [V_i] \tag{22.1}$$

$$[X] = [K]\ [X] + [KW]\ [XW] + [KR]\ [XR'] + [V] \tag{22.2}$$

Output cost equals cost of inputs, cost of waste disposal, cost of waste recycling and value addition.

22.5 MODEL SOLUTION

In order to make the framework of physical system theory more practicable, it is necessary to resort to the process of simplification of the model. The purpose of simplification is to create a balance between the variety of the system to be handled by the analyst and the analyst resources. A generalized format of equations has been evolved and a solution methodology is developed. The simplified PST model consists of two modules.

i. Flow Module
ii. Cost Module

In this national economy active system flow module is represented as in generalized form.

$$[Y_{oi}] = [K_{ij}]\ [Y_{oi}] + [K_{it}]\ [Y_{oi}] + [K_{ih}]\ [Y_{oi}] - [Y_{oi}] \tag{22.3}$$

Generalized cost module is represented as

$$[X_{oi}] = [K_{ij}]\ [X_{ij}] + [K_{win}]\ [X_{win}] + [K_{wir}]\ [X_{wir}] + [V_i] \tag{22.4}$$

Following are the simplified solution methodology steps for solving 'n' sector national economy model.

(i) Observe that nodes are simplified and redundancy rectified from the equations obtained.

(ii) Obtain flow module by substituting values of different flows from flow equations into mass balance equations.

(iii) Substitute the values of technological coefficients and other exogenous variables.

(iv) Obtain a set of 'n' equations (mass balance) with 'n' unknowns, which are to be determined.

(v) Observe steps two to four for cost module. Their n cost equations would give solution values of cost index for each sector.

To solve seven sector model, one has to solve only seven equations simultaneously each for flow and cost module. This reduces computer time drastically and makes PST modelling user friendly.

The solution of simultaneous equations is based on Gauss Jordan algorithm. The details of computer software are described in the subsequent section. An input-output structure of the simplified PST model is shown in Figure 22.7.

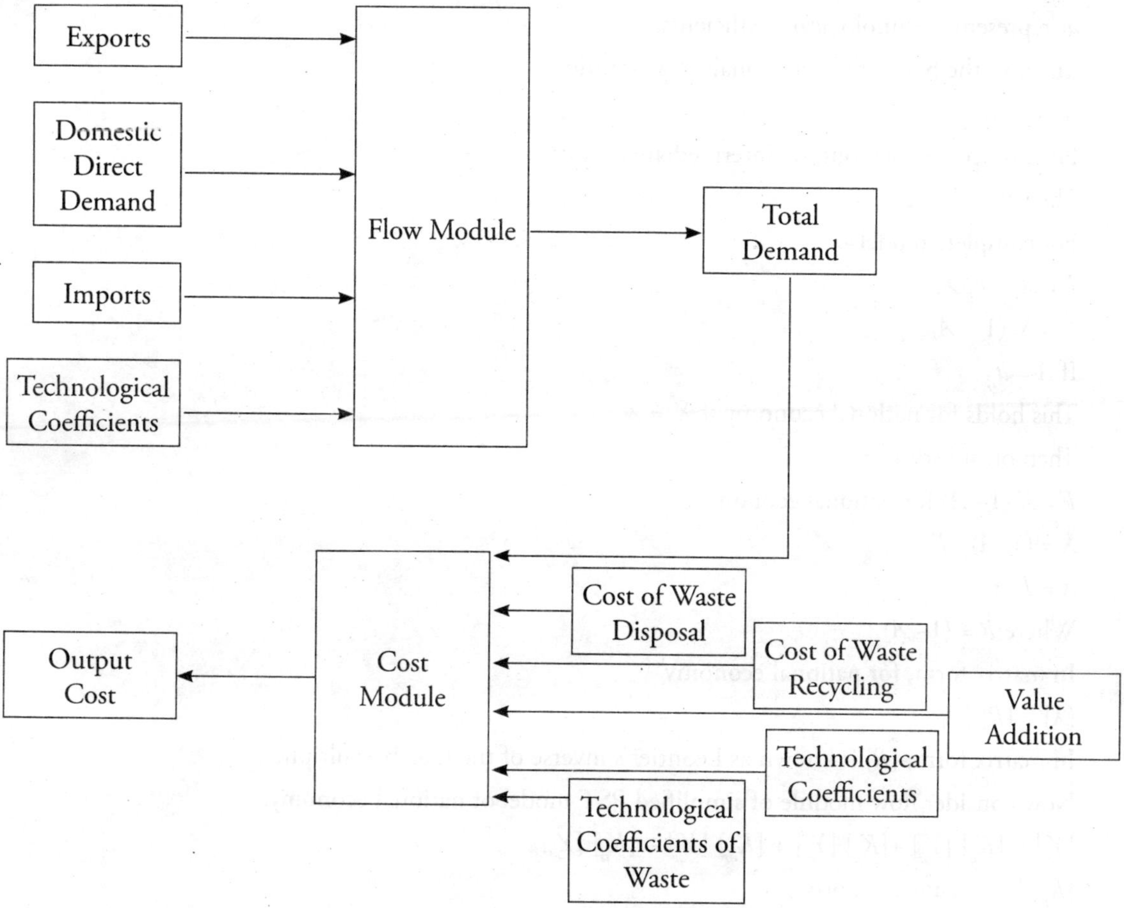

Figure 22.7: Simplified Structural Model

22.6 PARALLELISM WITH INPUT-OUTPUT MODEL

The input-output table provides a systemic description of the interdependence among different sectors in the economy. This table may also be regarded as a desegregation of the production accounts in a national accounting system. Here the part of the total gross output used as input for further production of goods and services is termed as intermediate use. Consider an input-output structure of national economy based model. The formal structure of the model can be described as follows.

The economy is assumed to consist of '*n*' sectors. Its model is developed after assuming that there is a fixed proportion of Leontief's type production function. This means that the quantity of each input used by any other sector is constant proportion only of the level of output of that sector. This implies that the amount of each kind of input utilized by a sector varies in direct proportion to its output.

If X_{ij} represents intermediate output

X_j represents total output

F represents final output

a_{ij} represents technological coefficients.

Then on the basis of proportionality assumption.

$X_{ij} = a_{ij} X_j$

Final output = total output-intermediate output

Then $F = X_j - X_{ij}$

For complete model

$F = X_j - a_{ij} X_j$

$X = X_j (1 - A)$

If $A = a_{ij}$

This holds for national economy if $X_j = X$

Then on substituting

$F = X(1 - A)$ for national economy

$X = (1-A)^{-1} F$

$X = R.F.$

Where $R = (1 - A)^{-1}$

In matrix form, for national economy

$[X] = [R]\ [F]$

In matrix form, $[R]$ is known as Leontief's inverse of the matrix multiplier.

Now consider flow module of simplified PST model of national economy.

$[Y_{oi}] = [K_{ij}]\ [Y_{oi}] + [K_{it}]\ [Y_{oi}] + [K_{ih}]\ [Y_{oi}] - [K_{ti}]\ [Y_{oi}]$

$[K_{ti}]\ [Y_{oi}]$ denotes exports

$[K_{it}]\ [Y_{oi}]$ denotes imports

Intermediate output equals

$[K_{it}]\ [Y_{oi}] + [K_{ih}]\ [Y_{oi}] - [K_{ti}]\ [Y_{oi}] = [Y_h]$

Flow module becomes,

$[Y] = [Y]\ [K] + [E] - [I] + [Y_h]$

$[E]$ denotes exports

$[I]$ denotes imports

$[Y_h]$ denotes habitat demand

This implies that for the simplified PST model;

i. Flow module is nothing but Leontief's inverse model of Input-Output Table.

ii. Cost module is also on similar steps to Leontief's equation.

It is observed that, the developed and simplified model is an extension of Leontief's Input-Output model, which is also being considered for cost as well as flow modelling.

22.7 SOFTWARE

The Structural model of the national economy is developed with the help of physical system theory. This micro model consists of two modules: the cost and flow modules. Utmost care has been taken for making these modules independently. The flow module provides the values of total output, while cost module calculates the cost index of various sectors of the national economy.

This software utilizes Gauss Jordan method of solving simultaneous equations. It can be conveniently run on a personal computer or PC compatible. Following are some of the outcomes of simplification of PST modelling and the associated software.

The sectoral expansion is very easy and this doesn't change the accuracy and time taken for solution accomplishment.

Cost and flow modules are independent and cost and flow values are calculated independently.

Data entry is flexible; one can open a file of one's choice and enter data or make changes in the file.

The amount of waste disposal and waste recycling of the percentage of total production is entered for different sectors and the model provides them proper placement.

The cost of waste disposal and waste recycling is also entered in the order of sectors and in percentage terms over and above the unit cost of each sector.

The amounts of sectoral exports, sectoral imports and direct demand are fed in terms of million rupees and that too in separate files.

Value addition constants of each sector are fed to the system in a separate file.

Basically, the software is developed for input-output-waste modelling but natural resources consumption for each sector can also be taken into consideration, by substituting the values sector wise. In case one doesn't consider natural resources in the model, one is at liberty to substitute zero values in this file.

Flow values are calculated in millions of rupees and they denote the total output of the sector. Flow module takes input from technological coefficient, export, habitat demand, and natural resources waste disposal to nature and waste being recycled. After completing data entries in these files one has to run the main program to obtain total output of the respective sectors.

The cost module needs data for value addition, cost of waste disposal, and cost of waste recycling only. In this program, cost is executed to obtain relative cost of each sector. This module assumes the same values of technological coefficients, natural resources, waste disposal, and recycling amounts as specified in the flow module unless otherwise changed.

The program has been used extensively for validation runs, and sensitivity analysis of the structural model of national economy. Lastly, the model is to become a part of the integration steel model for steel policy analysis, projections, and scenario, generation.

REFLECTIONS

The structure of the national economy has been modelled on the basis of input-output structure using physical system theory framework. This model can incorporate socio-economic issues, viz. exports, imports, value addition, waste generation-disposal-recycling and sector interlinkages, and their effects over the sectoral costs. Developed model has been validated and compared successfully with Leontief's Input-Output Model. This has been further generalized and used to model the seven sectors of Indian economy. This model can further be expanded to sixty sectors, which is the normal sectorization plan of the Indian economy.

Notes

1. Sushil (1991) Simplified Physical System Theory Model of a Closed Chain Production System, *Int. J. System Science,* 22(6), 987-999.
2. Central Statistical Organization (1990) *Input-Output Transactions Table 83-84,* Department of Statistics, Ministry of Planning, Government of India; Central Statistical Organization (1995) *Input-Output Transactions Table 89-90,* Department of Statistics, Ministry of Planning, Government of India.
3. Central Statistical Organization (1989) *National Accounts Statistics-Sources and Methods,* Department of Statistics, Ministry of Planning, Government of India.
4. Sushil (1984) *System Modelling of Waste Management in National Planning,* Ph.D. Thesis, IIT Delhi.

APPENDIX 22A

Writing the flow equations as shown in Figure 22.6 and extending it for seven sectors.

$y_{11} = k_{11} y_{o1}$

$y_{21} = k_{21} y_{o1}$

$y_{31} = k_{31} y_{o1}$

$y_{41} = k_{41} y_{o1}$

$y_{51} = k_{51} y_{o1}$

$y_{61} = k_{61} y_{o1}$

$y_{71} = k_{71} y_{o1}$

$y_{w1n} = k_{w1n} y_{o1}$

$y_{w1r} = k_{w1r} y_{o1}$

$y_{t1} = k_{t1} y_{o1}$

$y_{1t} = k_{1t} y_{o1}$

Similar equations for other sectors on the same lines can be written down. Writing mass balance equations for seven sectors as evident from Figure 22.6.

$y_{o1} = y_{11} + y_{12} + y_{13} + y_{14} + y_{15} + y_{16} + y_{17} \, y_{1h} + y_{1t} - y_{t1}$

$y_{o2} = y_{21} + y_{22} + y_{23} + y_{24} + y_{25} + y_{26} + y_{27} \, y_{2h} + y_{2t} - y_{t2}$

$y_{o3} = y_{31} + y_{32} + y_{33} + y_{34} + y_{35} + y_{36} + y_{37} \, y_{3h} + y_{3t} - y_{t3}$

$y_{o4} = y_{41} + y_{42} + y_{43} + y_{44} + y_{45} + y_{46} + y_{47} \, y_{4h} + y_{4t} - y_{t4}$

$y_{o5} = y_{51} + y_{52} + y_{53} + y_{54} + y_{55} + y_{56} + y_{57} \, y_{5h} + y_{5t} - y_{t5}$

$y_{o6} = y_{61} + y_{62} + y_{63} + y_{64} + y_{65} + y_{66} + y_{67} \, y_{6h} + y_{6t} - y_{t6}$

$y_{o7} = y_{71} + y_{72} + y_{73} + y_{74} + y_{75} + y_{76} + y_{77} \, y_{7h} + y_{7t} - y_{t7}$

Substitution from flow equations to mass balance equations and bringing all variables to the left-hand side:

$y_{o1} (k_{11}) + k_{12} y_{o2} + k_{13} y_{o3} + k_{14} y_{o4} + k_{15} y_{o5} + k_{16} y_{o6} + k_{17} y_{o7} + y_{1h} + y_{1t} - y_{t1} - y_{o1} = 0$

on rearranging one obtains flow module of the seven-sector model.

$(k_{11}-1) \, y_{o1} + k_{12} y_{o2} + k_{13} y_{o3} + k_{14} y_{o4} + k_{15} y_{o5} + k_{16} y_{o6} + k_{17} y_{o7} + y_{1h} + y_{1t} - y_{t1} = 0$

$k_{21} y_{o1} + (k_2-1) \, y_{o2} + k_{23} y_{o3} + k_{24} y_{o4} + k_{25} y_{o5} + k_{26} y_{o6} + k_{27} y_{o7} + y_{2h} + y_{2t} - y_{t2} = 0$

$k_{31} y_{o1} + k_{32} y_{o2} \, (k_{33}-1) \, y_{o3} + k_{34} y_{o4} + k_{35} y_{o5} + k_{36} y_{o6} + k_{37} y_{o7} + y_{3h} + y_{3t} - y_{t3} = 0$

$k_{41} y_{o1} + k_{42} y_{o2} + k_{43} y_{o3} + (k_{43}-1) \, y_{o4} + k_{45} y_{o5} + k_{46} y_{o6} + k_{47} y_{o7} + y_{4h} + y_{4t} - y_{t4} = 0$

$k_{51} y_{o1} + k_{52} y_{o2} + k_{53} y_{o3} + k_{54} y_{o4} + (k_{55}-1) \, y_{o5} + k_{56} y_{o6} + k_{57} y_{o7} + y_{5h} + y_{5t} - y_{t5} = 0$

$k_{61} y_{o1} + k_{62} y_{o2} + k_{63} y_{o3} + k_{64} y_{o4} + + k_{65} y_{o5} \, (k_{66}-1) \, y_{o6} + k_{67} y_{o7} + y_{6h} + y_{6t} - y_{t6} = 0$

$k_{71} y_{o1} + k_{72} y_{o2} + k_{73} y_{o3} + k_{74} y_{o4} + k_{75} y_{o5} \, k_{76} y_{o6} \, (k_{77}-1) \, y_{o7} + y_{7h} + y_{7t} - y_{t7} = 0$

For sector 1,

$$x_{o1} = -\ k_{12}\ x_{11} + k_{21}\ x_{21} + k_{31}\ x_{31} + k_{41}\ x_{41} + k_{51}\ x_{51} + k_{61}\ x_{61} + k_{71}\ x_{71} + k_{w1n}\ x_{w1n} + k_{w1r}\ x_{w1r}) + v_1$$

Similar equations can be obtained for rest six sectors.

$$x_{o2} = -\ (k_{12}\ x_{11} + k_{22}\ x_{22} + k_{32}\ x_{32} + k_{42}\ x_{42} + k_{52}\ x_{52} + k_{62}\ x_{62} + k_{72}\ x_{72} + k_{w2n}\ x_{w2n} + k_{w2r}\ x_{w2r}) + v_2$$

$$x_{o3} = -\ (k_{13}\ x_{13} + k_{23}\ x_{23} + k_{33}\ x_{33} + k_{43}\ x_{43} + k_{53}\ x_{53} + k_{63}\ x_{63} + k_{73}\ x_{73} + k_{w3n}\ x_{w3n} + k_{w3r}\ x_{w3r}) + v_3$$

$$x_{o4} = -\ (k_{14}\ x_{14} + k_{24}\ x_{24} + k_{34}\ x_{34} + k_{44}\ x_{44} + k_{54}\ x_{54} + k_{64}\ x_{64} + k_{74}\ x_{74} + k_{w4n}\ x_{w4n} + k_{w4r}\ x_{w4r}) + v_4$$

$$x_{o5} = -\ (k_{15}\ x_{15} + k_{25}\ x_{25} + k_{35}\ x_{35} + k_{45}\ x_{45} + k_{55}\ x_{55} + k_{65}\ x_{65} + k_{75}\ x_{75} + k_{w5n}\ x_{w5n} + k_{w5r}\ x_{w5r}) + v_5$$

$$x_{o6} = -\ (k_{16}\ x_{16} + k_{26}\ x_{26} + k_{36}\ x_{36} + k_{46}\ x_{46} + k_{56}\ x_{56} + k_{66}\ x_{66} + k_{76}\ x_{76} + k_{w6n}\ x_{w6n} + k_{w6r}\ x_{w6r}) + v_6$$

$$x_{o7} = -\ (k_{17}\ x_{17} + k_{27}\ x_{27} + k_{37}\ x_{37} + k_{47}\ x_{47} + k_{57}\ x_{57} + k_{67}\ x_{67} + k_{77}\ x_{77} + k_{w7n}\ x_{w7n} + k_{w7r}\ x_{w7r}) + v_7$$

Substituting values of different costs from compatibility equations, one obtains for sector one

$$x_{o1} = -\ k_{11}\ x_{o1} + k_{21}\ x_{o2} + k_{31}\ x_{o3} + k_{41}\ x_{o4} + k_{51}\ x_{o5} + k_{61}\ x_{o6} + k_{71}\ x_{o7} + k_{w1n}\ x_{w1n} + k_{w1r}\ x_{w1r}) + v_1$$

In this equation values $r_1\, y_{o1} + c_1$ is equated for $k_{wir}\, x_{wir}$

Thus, cost module becomes

$$(k_{11}-1)\ x_{o1} + k_{21}x_{o2} + k_{31}x_{o3} + k_{41}x_{o4} + k_{51}x_{o5} + k_{61}x_{o6} + k_{71}x_{o7} + k_{w1n}x_{w1n} + r_1\, y_{o1} + c_1 + v_1 = 0$$

$$k_{12}x_{o1} + (k_{21}-1)\ x_{o2} + k_{32}x_{o3} + k_{42}x_{o4} + k_{52}x_{o5} + k_{62}x_{o6} + k_{72}x_{o7} + k_{w2n}x_{w2n} + r_2\, y_{o2} + c_2 + v_2 = 0$$

$$k_{13}x_{o1} + k_{23}x_{o2} + (k_{33}-1)\ x_{o3} + k_{43}x_{o4} + k_{53}x_{o5} + k_{63}x_{o6} + k_{73}x_{o7} + k_{w3n}x_{w3n} + r_3\, y_{o3} + c_3 + v_3 = 0$$

$$k_{14}x_{o1} + k_{24}x_{o2} + k_{34}x_{o3} + (k_{44}-1)\ x_{o4} + k_{54}x_{o5} + k_{64}x_{o6} + k_{74}x_{o7} + k_{w4n}x_{w4n} + r_4\, y_{o4} + c_4 + v_4 = 0$$

$$k_{15}x_{o1} + k_{25}x_{o2} + k_{35}x_{o3} + k_{45}x_{o4}\ (k_{55}-1)\ x_{o5} + k_{65}x_{o6} + k_{75}x_{o7} + k_{w5n}x_{w5n} + r_5\, y_{o5} + c_5 + v_5 = 0$$

$$k_{16}x_{o1} + k_{26}x_{o2} + k_{36}x_{o3} + k_{46}x_{o4} + k_{56}x_{o5}\ (k_{66}-1)\ x_{o6} + k_{76}x_{o7} + k_{w6n}x_{w6n} + r_6\, y_{o6} + c_6 + v_6 = 0$$

$$k_{17}x_{o1} + k_{27}x_{o2} + k_{37}x_{o3} + k_{47}x_{o4} + k_{57}x_{o5} + k_{67}x_{o6}\ (k_{77}-1)\ x_{o7} + k_{w7n}x_{w7n} + r_7\, y_{o7} + c_7 + v_7 = 0$$

23 MODELLING OF A TRANSPORT SYSTEM

23.1 INTRODUCTION

Various studies have been carried out to optimize railway yard operations, train sequencing and scheduling, empty wagon distribution, power (locomotive) distribution, crew scheduling, and facility (yard, maintenance depot) location. A variety of techniques including simulation, analytical models, transportation models, gravity models, network optimization models, and time series analysis have been applied to various facets of rail transportation. These techniques, however, do not consider explicitly the relationship between the flow and the cost of carrying of commodities (passengers, goods) in a rail transport network. Keeping in mind the complexity of the interrelationships and the effects of internal as well as external factors, the principles and concepts of physical system theory are applied to assess the unit cost of carrying passengers for a broad gauge Northern Railway network, given the level of flows through the network. The flexible system methodology used for the analysis of transport network is shown in Figure 23.1.

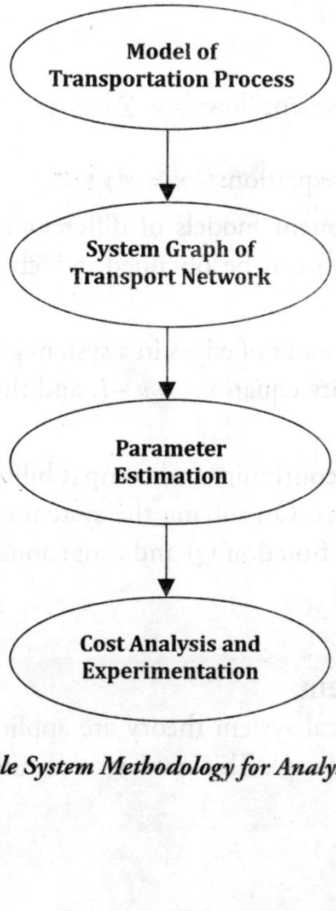

Figure 23.1: Flexible System Methodology for Analysis of Transport Network

23.2 MODELLING A RAILWAY TRANSPORT NETWORK

Only the transportation processes taking place in a railway transport network are modelled; transformation and storage processes have not been considered. In a rail transport network, the links (or sections-in railway parlance) are treated as components and are modelled as transportation processes. The network has interaction with environment at some nodes. The environment as a whole can be treated as one node. It is assumed that there is no waste in the system.

The "across" variable is taken to be the unit cost of flow and is denoted by x, while the flow of passengers is a "through" variable denoted by y. The free body diagram of the ith transportation process (link) is shown in Figure 23.2.

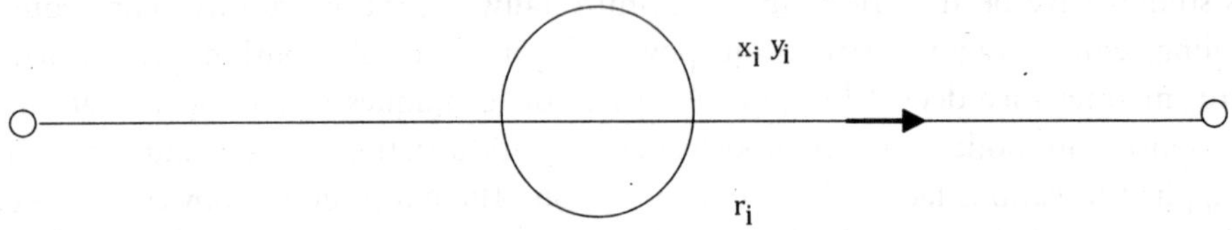

Figure 23.2: Free Body Diagram of the ith Link

The associated parameters and variables are as follows:

y^j_i : flow of the jth type, through link i

y_i : equivalent flow of material through link i

c_j : equivalence factor for jth type of flow

x_i : Unit cost of flow through link i

g : transportation function

$$\text{then equivalent flow: } y_i = \sum_j c_j \times y_i^j \tag{23.1}$$

$$\text{cost equation: } -x_i = g(y_i) \tag{23.2}$$

On bringing together the component models of different transportation processes in a one-to-one correspondence, the system graph can be obtained, which will be governed by continuity and compatibility constraints.

If v = number of nodes and e = number of edges in a system graph, then the number of component equations = e, the number of continuity equations = $v-1$, and the number of compatibility equations = $e-v+1$.

The total number of component, continuity and compatibility equations defines the system model and are referred to as system equations. On solving the system equations, the cost of haulage can be derived in terms of the transportation function (g) and exogenous variables.

23.3 A CASE STUDY

Description of the Physical System

The concepts and principles of physical system theory are applied to the transportation process in a broadgauge network of Northern Railway[1]. This network consists of New Delhi – Delhi – Panipat –

Kurukshetra – Ambala – Rajpura – Dhuri –Jhakhal – Narwana – Jind – Rohtak – New Delhi sections. It also includes main line links of New Delhi –Panipat and Delhi - Rohtak and branch line sections of Panipat-Rohtak, Panipat-Jind and Kurukshetra-Narwana, which fall within the network. For simplicity, Hazarat Nizamuddin station has been clubbed with New Delhi station as a single node. This network has links with the environment at seven nodes, viz. New Delhi, Delhi, Ambala, Rajpura, Dhuri, Jhakhal, and Rohtak. The environmental flows are aggregated to reduce network complexity.

Assumptions

The following major assumptions have been made in the study:

(i) Only passenger trains have been modelled ignoring the freight trains, treating the network as incapacitated.

(ii) The transportation function (*g*) is assumed to be linear. Data for the Northern Railway as a whole is found to fit this assumption well.

(iii) As linkage unit cost data is not available, the gross data for the system as a whole (Northern Railway) has been used for arriving at the link wise data.

ARCS	No.
AB	21
BA	22
AC	23
CA	24
AK	25
KA	26
BK	29
KB	30
CK	35
KC	36

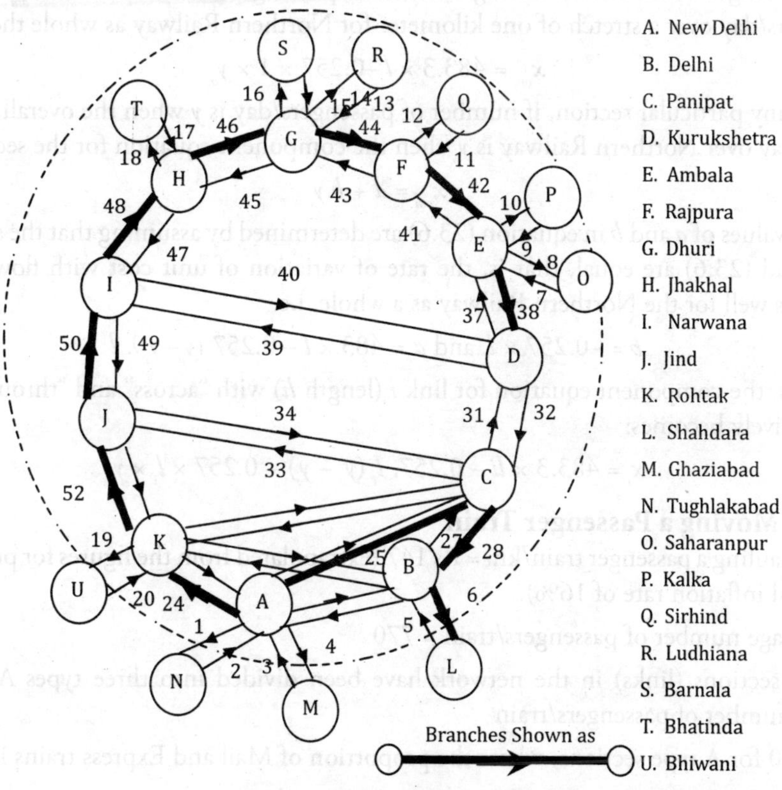

Figure 23.3: System Graph

System Model

The system graph, shown in Figure 23.3, consists of twelve nodes (including one environment node) and fifty-two arcs. The nodes labelled L,M,N,O,P,Q,R,S,T, and U constitute the environmental node and serve as reference node. To form a tree, the eleven branches 6, 24, 25, 28, 38, 42, 44, 46, 48, 50 and 52 are chosen. The remaining forty one links are chords. There are 52 component equations in all.

$$-x_i = a_i + b_i y_i \,\forall_i \tag{23.3}$$

Here a_i and b_i are constants. If some variables are specified, the corresponding component equations are deleted from the system. There are 11 fundamental cutset equations, and 41 fundamental circuit equations. Once the linkwise transportation functions are known and some exogenous variables are specified, the values of remaining variables can be calculated.

Development of Transportation Function

As data about linkwise unit costs is not available, a linear transportation function for the Northern Railway as a whole is developed. The corresponding regression equation is:

$$x_o = 483.3 - 0.257\, y_o \tag{23.4}$$

The relationship between cost of hauling 1000 passengers/km (x_o) over Northern Railway and number of passengers (in thousands) hauled per day (y_o) is given by the equation 23.4.

If l = length of the stretch through which the passengers are hauled, and x_{om} = cost of hauling 1000 passengers/day over a stretch of one kilometer for Northern Railway as whole then

$$x_{om} = 483.3 \times l - 0.257 \times 1 \times y_o \tag{23.5}$$

For any particular section, if number of passengers/day is y_s when the overall number of passengers hauled/day over Northern Railway is y_o then the component equation for the section becomes

$$x_{om} = a + b.y_s \tag{23.6}$$

The values of a and b in equation (23.6) are determined by assuming that the slopes of the equations (23.5) and (23.6) are equal, that is, the rate of variation of unit cost with flow is the same for each section as well for the Northern Railway as a whole, i.e.

$$b = -0.257 \times l, \text{ and } a = 483 \times l - 0.257\,(y_o - y_s).\, l$$

Thus, the component equation for link i (length li) with "across" and "through" variables as x_i and y_i respectively becomes:

$$x_i = 483.3 \times li - 0.257.\, l_i\,(y_o - y_s) - 0.257 \times l_i \times y_i \tag{23.7}$$

Cost of Moving a Passenger Train

Cost of hauling a passenger train/km = Rs 147 (extrapolated from the figures for previous years assuming an annual inflation rate of 16%).

Average number of passengers/train = 770

The sections (links) in the network have been divided into three types A, B and C, assuming average number of passengers/train

= 820 for A type sections, where the proportion of Mail and Express trains is more,

= 770 for B type sections with an average mix, and

= 720 for C types (branch line) sections.

Thus, section-wise cost of hauling 1000 passengers/km is Rs 179 for 'A' sections, Rs 190 for 'B' sections, and Rs 204 for 'C' sections. This is in effect the marginal cost of hauling passenger trains in these sections.

Parameter Estimation and System Analysis

Based on the data published by the Indian Railways[2], the values of *x,y, a,* and *b* for all the links (sections) have been calculated as shown in Table 23.1.

Table 23.1: Parameter Estimation

Sr. No.	Arcs	Trips/Week	Section Type	Pass/ Train	Pass/ Day 000's (y)	Length (l) kms	Slope (b)	Constant (a)	$x=x_o l$	$x^1=a+by$
1.	1,2	238	A	820	27.88	21	-5.39	4092.6	3753.4	3942.4
2.	3,4	106	A	820	12.42	20	-5.13	3818.4	3574.7	3754.7
3.	5,6	252	A	820	29.52	3	-0.77	585.9	536.2	563.2
4.	7,8	119	B	770	13.09	81	-20.78	15478.5	15417.7	15206.4
5.	9,10	54	B	770	5.94	74	-18.99	14005.1	14085.3	13892.3
6.	11,12	161	A	820	18.86	25	-6.42	4814.3	4468.4	4693.3
7.	13,14	52	C	720	5.35	62	-15.91	11724.6	12620.7	11639.5
8.	15,16	35	C	720	3.60	28	-7.18	5282.4	5699.7	5256.5
9.	17,18	63	C	720	6.48	99	-25.40	18750.2	20152.4	18585.6
10.	19,20	42	C	720	4.32	50	-12.83	9442.1	10178.0	9386.7
11.	21,22	63	A	820	7.38	4	-1.03	758.5	714.9	750.9
12.	23,24	70	A	820	8.20	89	-22.84	16895.5	15907.4	16708.3
13.	25,26	27	C	720	2.78	70	-17.96	13191.2	14249.2	13141.3
14.	27,28	49	A	820	5.74	89	-22.84	16839.4	15907.4	16708.3
15.	29,30	77	C	720	7.92	70	-17.96	13283.4	14249.2	13141.3
16.	31,32	103	A	820	12.07	67	-17.19	12785.6	11975.3	12578.1
17.	33,34	21	C	720	2.16	71	-18.22	13368.4	14452.7	13329.1
18.	35,36	28	C	720	2.88	72	-18.48	13570.0	14656.3	13516.8
19.	37,38	103	A	820	2.07	42	-10.78	8014.8	7506.9	7884.8
20.	39,40	28	C	720	2.88	87	-22.32	16397.1	17709.7	16332.8
21.	41,42	217	A	820	12.07	28	-7.18	5439.2	5004.6	5256.5
22.	43,44	49	C	720	5.04	78	-20.15	14744.1	15877.7	14643.2
23.	45,46	45	C	720	4.63	66	-16.94	12468.8	13434.9	12390.4
24.	47,48	49	C	720	5.04	39	-10.01	7372.0	7938.8	7321.6
25.	49,50	63	C	720	6.48	33	-8.47	6250.1	6717.5	6195.2
26.	51,52	63	C	720	6.48	58	-14.88	10985.0	11806.5	10888.5

Three experiments were carried out by specifying different exogenous variables as follows:

Case 1: One across variable x 24 (cost of hauling 1000 passengers along Panipat-New Delhi, Section) and fifteen through variables y1, y3, y5, y9, y11, y13, y15, y17, y19, y21, y23, y35, y39, and y41 (inflows from environmental node, flows along branch lines and flows along the important New Delhi – Delhi and Ambala–Rajpura sections) are specified.

Case 2: Only fifteen through variables (as in Case 1) are specified.

Case 3: Only twelve through variables y1, y3, y5, y7, y9, y11, y13, y15, y17, y19, y21, and y41 (Inflows from environmental node and flows along the important New Delhi – Delhi and Ambala-Rajpura sections) are specified.

The values for across variables are calculated in all the three cases and the results are enclosed in Table 23.2.

Table 23.2: Unit Cost Results

Arc No.	Case 1	Case 2	Case 3	Arc No.	Case 1	Case 2	Case 3	Arc No.	Case 1	Case 2	Case 3
1.	3856	2457	2485	18.	-8817	-10592	-7782	35.	16635	9043	1590
2.	-3856	-2457	-2485	19.	4582	6109	3714	36.	-16635	-9043	-1590
3.	3856	2457	2485	20.	-4582	-6109	-3714	37.	6085	3778	1459
4.	-3856	-2457	-2485	21.	-1965	-726	669	38.	-6085	-3778	-1459
5.	1891	1731	1816	22.	1965	726	-669	39.	11788	9836	741
6.	-1891	-1731	-1816	23.	-15907	-5391	-334	40.	-11788	-9836	-741
7.	1993	4372	6508	24.	15907	5391	334	41.	4301	2948	1401
8.	-1993	-4372	-6508	25.	727	3652	1257	42.	-4301	-2948	-1401
9.	1993	4372	6508	26.	-727	-3652	-1257	43.	1165	833	356
10.	-1993	-4372	-6508	27.	-13942	-4665	335	44.	-1165	-833	-356
11.	6294	7320	7908	28.	13942	4665	-335	45.	1358	2389	230
12.	-6294	-7320	-7908	29.	2691	4378	1926	46.	-1358	-2389	-230
13.	7459	8203	7552	30.	-2691	-4378	-1926	47.	-1121	-162	-1992
14.	-7459	-8203	-7552	31.	7959	3528	2898	48.	1121	162	1992
15.	7459	8203	7552	32.	-7959	-3528	-2898	49.	-1811	-1369	-1405
16.	-7459	-8203	-7552	33.	17936	11975	2234	50.	1811	1369	1405
17.	8817	10592	7782	34.	-17936	-11975	-2234	51.	-1303	-2951	-644
								52.	1303	2951	644

23.4 Model Validation

As linkwise unit cost data is not available, direct validation of the model is not possible. It can however, be validated indirectly in the following manner:

x_i can be calculated in two ways. Firstly from system unit cost of flow (equation (23.4)) and secondly from the component equation for that link (equations (23.7)), a_i, b_i and y_s having been specified. The values for x_i in both the ways have been calculated for all the links (Table 23.1) and they are found to be not statistically different at a significance level of $\alpha = 0.05$.

23.5 APPLICABILITY FOR RAILWAY TRANSPORT PLANNING

This analysis can be extended to include freight trains flows also, provided the total train flow (both passenger and freight) in any link of the network is much below the capacity of the link. Under such circumstances, the capacity of each link can be assumed to be partitioned into two portions, one for passenger traffic and the other for freight traffic and both can be modelled and analysed by decomposition. In the case of freight trains, the through variable (y) can be the gross load in thousands of tonnes and the across variable (x) the cost of haulage per thousand gross tonnes between the two nodes.

This model can be used in transport planning and performance appraisal. The budgeted cost figures for a certain planned level of traffic can be arrived at. Similarly, based on the actual traffic flow during a particular time period, the actual expenditure can be compared to the expenditure predicted by using the unit costs derived from the model.

Sensitivity studies of the transportation function can be carried out and the effect of addition or deletion of links on the network capacity can be estimated. Thus, it can help predict the effect of sudden closure (due to accident or natural calamity) of any link on the redistribution of flows in other links or the effects of constructing new links on the workload of remaining links etc.

REFLECTIONS

The passenger traffic flows in a rail transport network have been modelled. At present, the model suffers from the lack of link wise detailed data. This model assumes traffic flows in all links much below their capacities. The model need to be suitably modified to apply it to sections approaching saturation limits. This model predicts linkwise costs. Thus, the costs at the nodes (which could be a major chunk of the total cost) have to be apportioned to the respective links in an empirical way. Further studies are required to make this distribution more scientific. The model can be extended to study the effects of variation in link capacities, network structure, and technology changes in train operation systems.

Notes

1. Sibal V.K., Sushil and Kanda A. (1988) Application of Physical System Theory to Modelling and Analysis of Rail Transport Network, in *Proceedings of National Conference on Transportation System Studies,* IIT Bombay, December 15-17, Tata McGraw Hill, New Delhi.
2. Indian Railways Annual Statistics (1980-81 to – 1986-87); Northern Railway Passenger Trains Time Table (April, 1988).

24 PROCESS MODELLING OF WASTE MANAGEMENT SYSTEMS

24.1 INTRODUCTION

From time immemorial, human beings have utilized the resources of the earth to support themselves. In early times, the needs were less, resources were available in plenty as the population was much less and consequently wastes were not a significant problem. Waste is a result of human activities from the dawn of civilization which has become more prominent during and after the industrial revolution, and today the accumulation of wastes has become a consequence of life in an industrialized society[1].

The inefficient and improper methods of waste management, particularly in developing countries like India are creating pollution problems in the air, water resources and land which are interfering with community life and development[2]. An important feature of waste is its multiplicity characteristic, i.e. uncontrolled or ill-managed waste leads to the creation of more waste, and the failure to salvage and reuse the waste results in avoidable depletion of natural resources.

From the total resources available within a fixed time frame, a fraction is consumed, a fraction is wasted and the remainder is a reserve. Consequently, the major objective of waste management is to minimize the fraction that is wasted.

Naturally, social costs are also incurred for effective waste management and therefore it is essential to carry out a social cost-benefit analysis prior to the implementation of waste management systems. From the above discussion, it is clear that waste management is associated with the identification, reduction, storage, collection, transfer and transport, reuse and recycling, and processing and disposal of waste keeping in view other environmental conditions involved, in the complete spectrum of the solution to the problem of waste. However, the present study on waste management is restricted to management of municipal solid waste.

24.2 SECTORS IN WASTE MANAGEMENT SYSTEMS

In most developed countries, and also in some developing countries there are basically six sectors to which municipal solid waste can be recycled/disposed off. These are

i. Biodegradable recycling
ii. Non-biodegradable recycling
iii. Composting
iv. Incineration
v. Pyrolysis
vi. Landfills

The sectors of waste management, waste generation and consumption of outputs of waste management are shown in the schematic diagram in Figure 24.1.

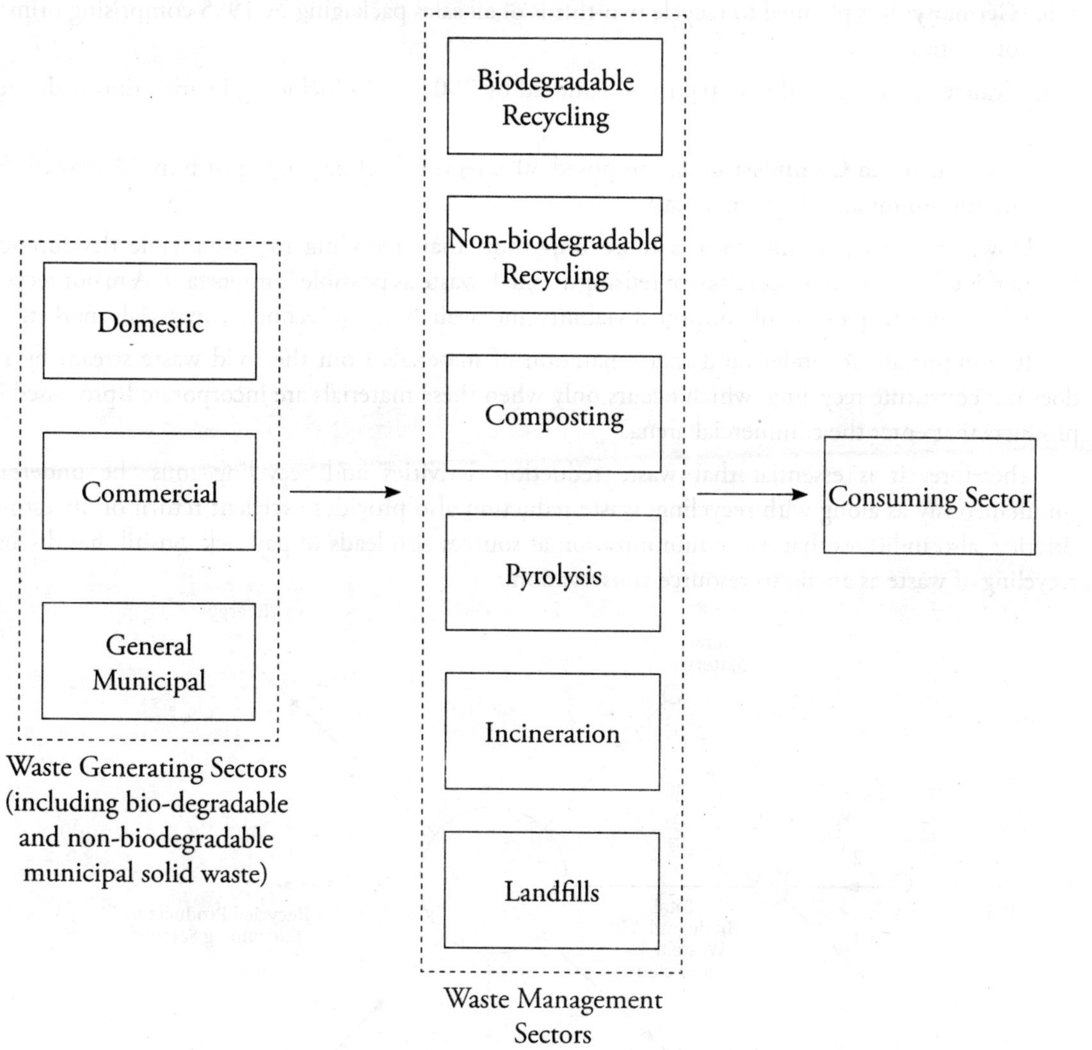

Figure 24.1: Schematic Diagram of Overall Waste Management System

Recycling of Waste

Recycling of waste involves both biodegradable and non-biodegradable recycling of salvable waste[3]. Some broad guidelines that could be followed to dispose off salvable waste include

- feasibility of recycling to be analysed.
- use waste to produce byproducts
- evaluate feasibility of selling waste/scrap as raw material to other factories/agencies /dealers.
- identify proper processing techniques to generate useful products from the waste.

These days plastic recycling is becoming important as various European countries plan to recycle large quantity of plastics primarily used in packaging. For example,

a. **Germany**: It is planned to recycle two-thirds of all sales packaging by 1995 comprising primarily of plastics.
b. **France**: 75% recycled material to be achieved by 2003 A.D., including incineration with energy recovery.
c. **The European Commission** has proposed 60% recycling of packaging with five years with 30% minimum for any single material[4].

However, waste minimization is more important than recycling for sustainable development[5]. Skinner[6] is of the view that recycling or reusing as much waste as possible is important. A major recycling impediment is the question of continued viability and availability of secondary materials market.

It is important to understand that separation of materials from the solid waste stream by itself does not constitute recycling, which occurs only when these materials are incorporated/processed into products that enter the commercial arena.

Therefore, it is essential that waste reduction activities and recycling must be undertaken simultaneously as along with recycling, waste reduction also provides excellent return on investment. Brierley[7] also indicates that waste minimization at sources also leads to payback. Sushil[8] has discussed recycling of waste as an aid to resource conservation.

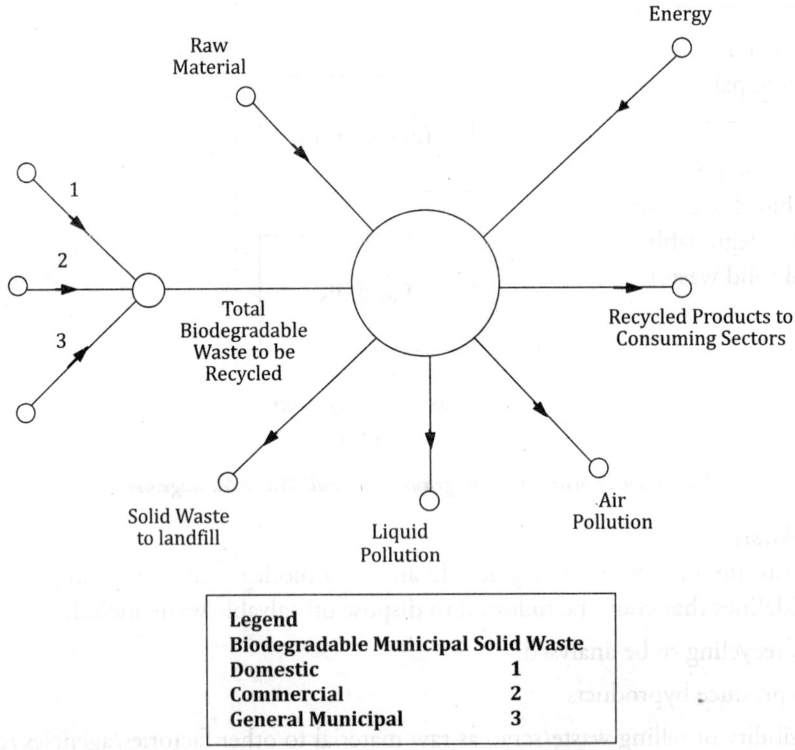

Figure 24.2: Material–Energy Flow Model-Biodegradable Recycling

Based on the above consideration, the material-energy flow modules of both biodegradable recycling and non-biodegradable recycling are shown in Figure 24.2 and Figure 24.3 respectively, indicating flow of materials and the waste generation and useful product streams. Recycling has also been analysed in a Systems Dynamics framework.

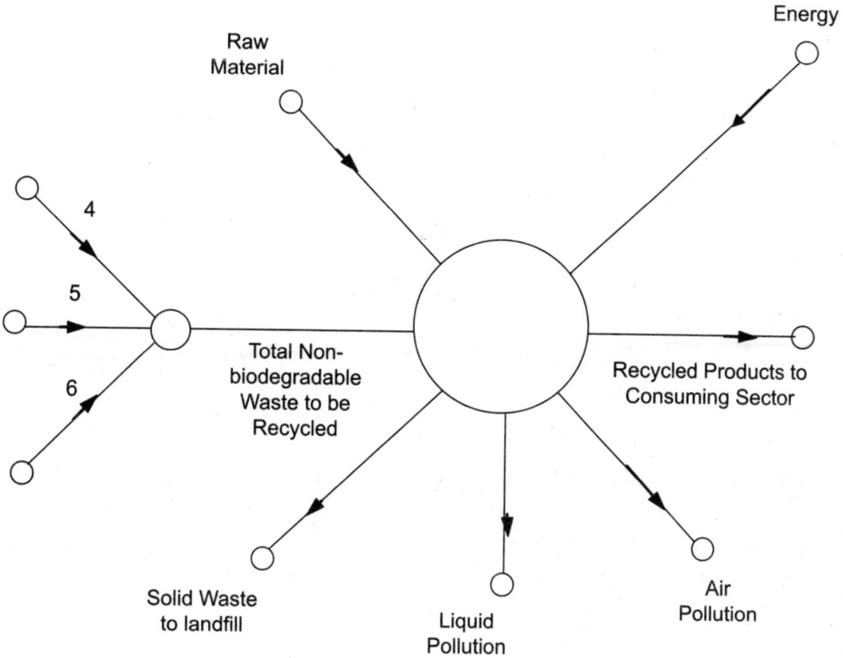

Legend	
Non-Biodegradable Municipal Solid Waste	
Domestic	**4**
Commercial	**5**
General Municipal	**6**

Figure 24.3: Material – Energy Flow Model-Non-biodegradable Recycling

Composting

The flow model of composting process is given in Figure 24.4. Composting could also be considered as a waste disposal recycling technique.

Incineration

It is an important method for waste disposal as incineration involves thermal processing of waste which results in reduction in volume and generation of energy and byproducts.

According to Mashyekhi[9], the incineration sector processes solid waste to produce energy. Incineration produces ash which must be dumped in the landfills. According to Porteous[10], incineration complements other forms of treatment, rather than being a complete alternative. Incineration is a long established waste management option and schematic incineration of household waste was pioneered in UK in late 19th Century[11]. However, it is less cost-effective than recycling and other waste minimization alternatives and it is more expensive than even other waste disposal options, including landfilling. Incineration as a waste management option has been analysed. Skinner integrates solid waste management alongwith incineration.

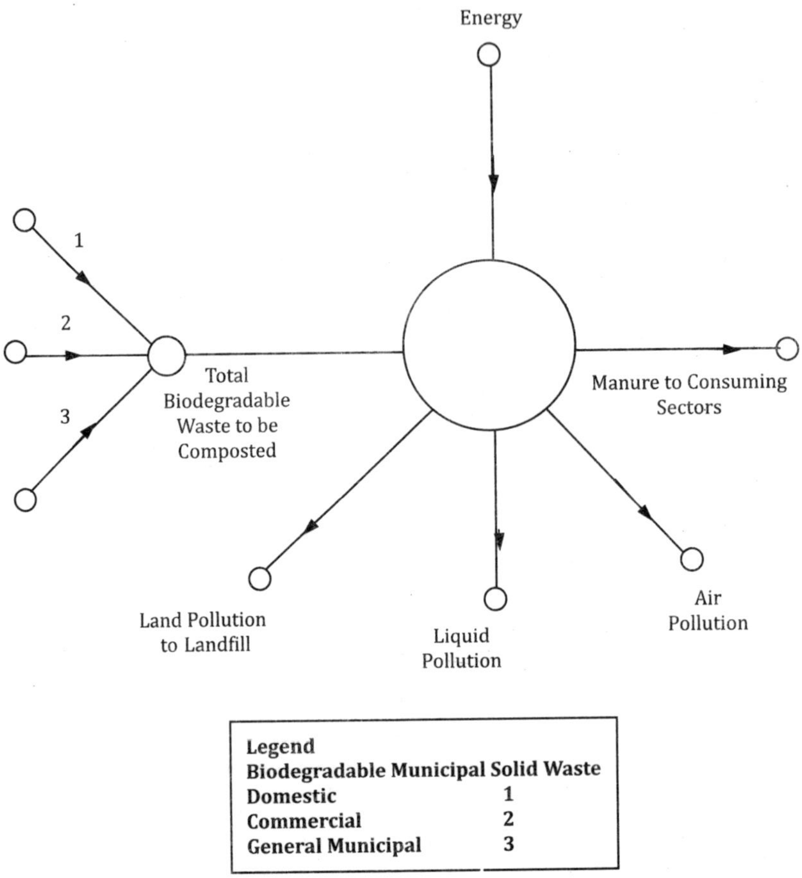

Figure 24.4: Material –Energy Flow Model –Composting

A material-energy flow model of incineration is shown in Figure 24.5 depicting flow of the material wastes involved.

Pyrolysis

According to Srivastava et al.[12], the various components produced in pyrolysis are

Components	% by Weight (Approx.)
- Gaseous products	25%
- Char	17–25%
- Liquid phase	20–80%

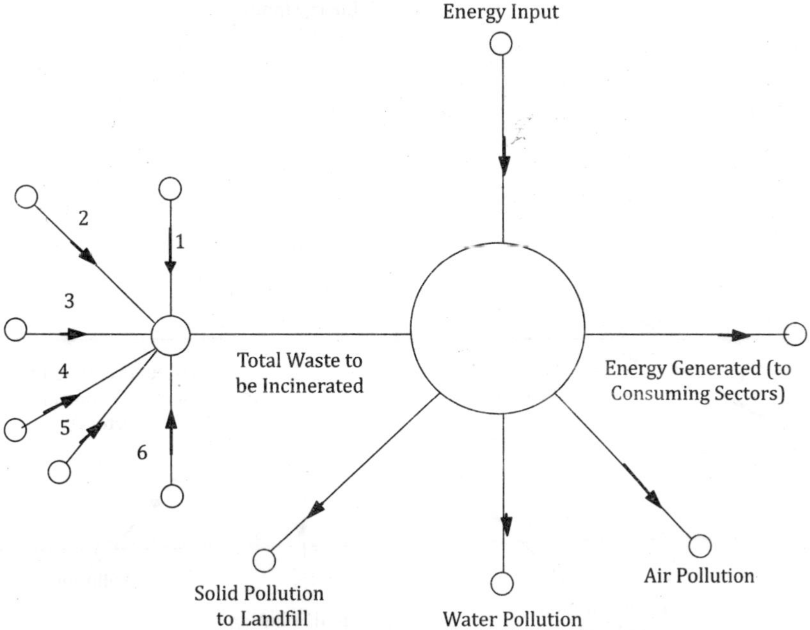

Legend	
Biodegradable Municipal Solid Waste	
Domestic	**1**
Commercial	**2**
General Municipal	**3**
Legend	
Non-biodegradable Municipal Solid Waste	
Domestic	**4**
Commercial	**5**
General Municipal	**6**

Figure 24.5: Material – Energy Flow Model -Incineration

Pyrolysis is considered to be an industrial process rather than a waste management process. The process of pyrolysis has been studied by Bradbury[13], Brunner et al.[14], Jegers et al.[15], Koufpanous[16], etc. However, very little work has been done to integrate pyrolysis along with waste management in order to analyze its importance in its entirety.

A material–energy flow model of pyrolysis as a waste management sector is given in Figure 24.6.

Landfills

Besides acting as a disposal site for various types of waste, landfills are also considered as an ultimate disposal waste management sector for those solid wastes that are of no use and remain after recycling, processing, etc.

Sanitary landfills have been proclaimed for some time as a low cost, safe way of disposing off municipal solid waste. Wilson[17] has discussed the development of landfills in a historical perspective in Europe and highlighted the policy and practice with a simple landfill stepladder.

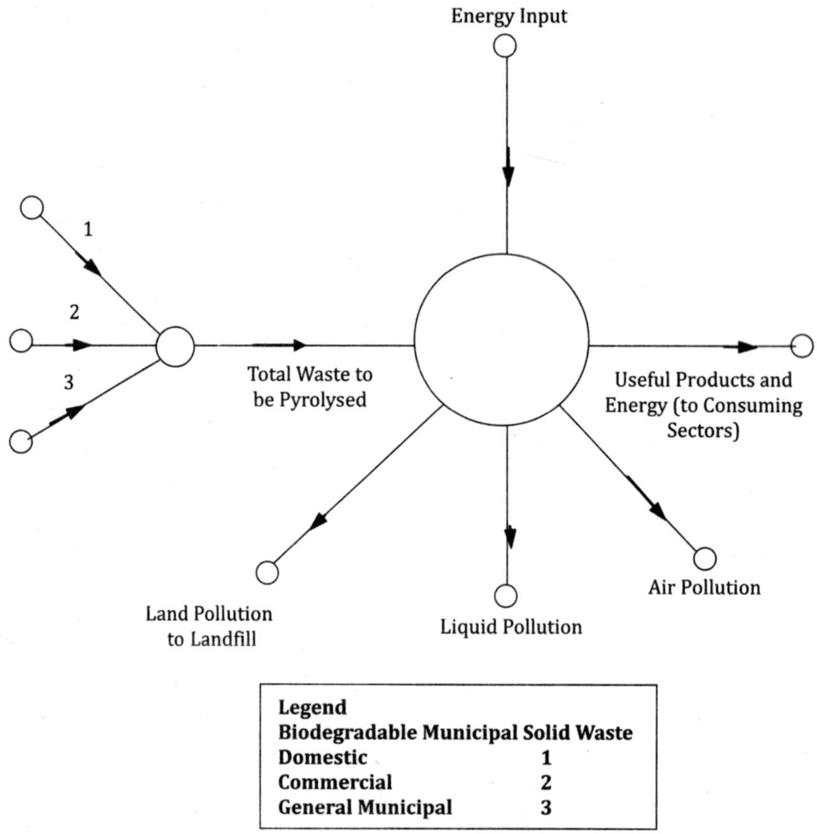

Figure 24.6: Material – Energy Flow Model -Pyrolysis

Accumulated solid waste in the landfills generate pollution in the environment. However, according to Skinner, landfill technology has advanced rapidly over the past decades.

Crawford and Smith[18] have dealt in detail with all aspects of landfill technology, and Harris et al.[19] have developed a strategy for the development of sustainable landfill design including, leachate treatment by development of highly efficient bio-reactor landfills for mixed wastes, with high leaching rates.

A material-energy flow model of a landfill is shown in Figure 24.7.

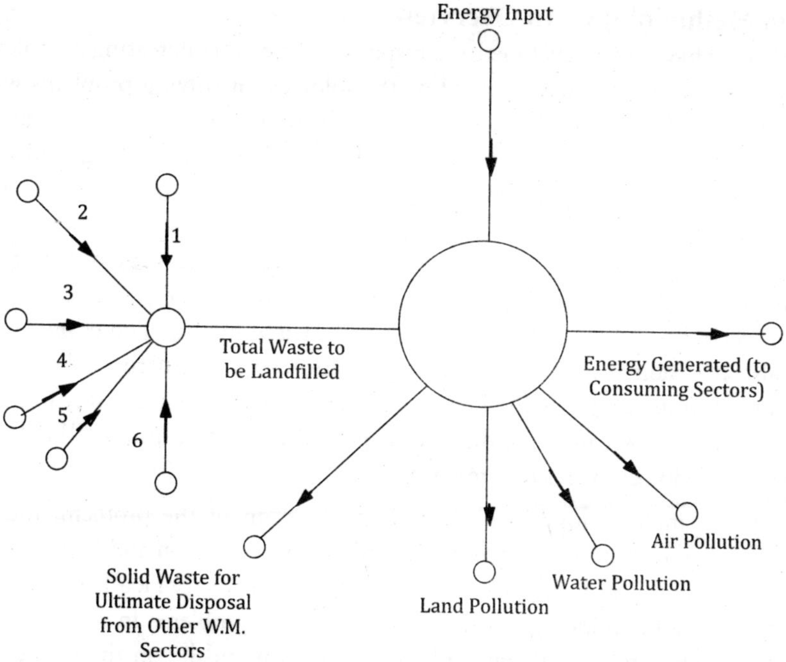

Legend
Biodegradable Municipal Solid Waste

Domestic	**1**
Commercial	**2**
General Municipal	**3**

Legend
Non-biodegradable Municipal Solid Waste

Domestic	**4**
Commercial	**5**
General Municipal	**6**

Figure 24.7: Material – Energy Flow Model -Landfills

24.3 DEVELOPMENT OF MODELLING METHODOLOGY

The modelling methodology is selected based on the conceptualization of the problem being studied and matching the attributes of the problem with the system based techniques. A Physical System Theory (PST) framework integrated with systems dynamics (SD) is proposed as an appropriate modelling methodology, for the waste management problem for supporting multiple levels of decision-making. PST provides a modelling tool for discreet physical flow systems with finite interfaces, which are prevalent in waste management. Systems dynamics, however, provides a framework for studying the system from a dynamic standpoint and with its thrust on the learning paradigm. System dynamics would be able to provide more insight into operation of the system. An integration mechanism of Physical System Theory (PST) and System Dynamics (SD) is presented and the advantages obtained by the synergistic approach are highlighted.

Flexible System Methodology: An Overview

Systems methodologies have been developed as a response to the ever-increasing complexity of systems, and the increasing need to adopt system based methodologies for solving problems with a thrust on implementation in an end user environment. The task of designing or selecting methodology for a particular problem situation is becoming more difficult with an ever-increasing choice of techniques. The integration of qualitati e and quantitative tools is emerging very rapidly to cater to the diverse requirements of management related problems in general, and Municipal Solid Waste Management in particular. Keeping the above in view Sushil[20] suggested a flexible system methodology, which is built on a spectral, integrative and innovative paradigm. It tries to resolve the end of continuum paradoxes as it is based on a spectral paradigm, treating all the system based methodologies on a continuum from hard to soft and also all problem situations from well structured to unstructured. For example, Physical System Theory is hard in its design, while Delphi, Scenario Building etc. are soft in design. However, Systems Dynamics is somewhere in the middle of the hard and soft continuum. The flexibility in using systems dynamics methodology has been presented [21].

The flexible system methodology involves conceptualization of the problem, fuzzy clustering of problem and system based techniques, matching of attributes and selection and integration of techniques which is most appropriate for both the problem and solution. This approach has been followed for the selection of the modelling methodology in the present study as problems in real life being considered in this chapter are not clustered on the ends of the continuum and lie on the whole continuum and more realistically on the middle part than the ends, having some parts structured and some parts unstructured.

Problem Conceptualization

The problem area has been conceptualized in terms of various attributes and its characteristics as presented in Table 24.1.

Table 24.1: Conceptualization of Municipal Solid Waste Management Problem

S.No.	Attributes	Characteristics
1.	Nature	Physical flow of waste through various transportation-transformation processes
2.	Structure	Multiple location, flow structure
3.	Situation Specific Characteristics	Mix of Stochastic (type of waste) and deterministic processes, reasonable steady state
4.	Nature of outcome desired	To support decision-making, learning about the waste management systems facilitate policy analysis
5.	People involved	Not well-versed with quantitative models, good in judgement, wide experience
6.	Decision Environment	Bureaucratic, stress on empirical models, rules of thumb.

From the above conceptualization, it is clear that the waste management system has physical flows of items through discreet components with well-defined finite interfaces. It also involves some managerial planning for high level tactical decisions in a multi-location system. In real life, the system of waste management is dynamic, interacting with internal as well as external factors.

A fuzzy clustering of the attributes identified above is obtained utilizing the flexible systems approach by considering the various attributes of problem of waste management.

In the system proposed in this study, the problem consists of a mix of operational to strategic problem areas and the decision process involves unitary to pluralistic decision-making.

Further, the problem situation indicates a mixed character of operating system and policy decision system indicating that the problem would be somewhere in between the well-structured and unstructured problem as shown below:

Well Structured o————————o————————o Un-Structured

The waste management problem being analyzed

The problem conceptualization indicates that the requirement is for a generalized flow modelling methodology which is capable of covering entire spectrum of decisions as well as meets the requirements of multiple levels of decisions. Further, the model must be flexible so that variables could be incorporated in order to facilitate analysis. In view of the requirements of the problems as mapped on various attributes, two flow modelling methodologies, viz. Physical System Theory and System Dynamics, have been identified to have high possibility of application for analyzing waste management systems.

Matching Attributes with System Techniques

In order to match the problem attributes two system based flow methodologies were selected.

- Physical System Theory (PST)
- System Dynamics (SD)

The details of matching of attributes of PST and SD methodologies are shown in Table 24.2.

Table 24.2: Matching of Problem and Attributes of Techniques

Sr. No.	WM Problem Attributes	PST Attributes	SD Attributes
1.	Physical Flow of Waste	Physical flow model with material and cost	Flow model of physical flows and feedback of information in levels and storage
2.	Multiple levels involving structured and semi-structured decision -making	Structured with focus on deterministic decisions	Unstructured and semi-structured areas with focus on policy analysis
3.	Dynamic system	Analytical solution	Dynamic study through simulation
4.	Transparent modelling and graphical representation scheme	Transparent modelling methodology with minimum ambiguity and abstraction	Relatively involved graphical representation
5.	Solution Procedure	Algebraic linear equations can be solved manually and easily	Special software package like "DYNAMO" or "STELLA" required

From the above it is clear, that as a flow modelling technique, PST facilitates modelling and analysis of physical flow systems with discrete interfaces having a well-defined structure. PST models two variables, viz. the physical flow, and the associated cost or energy inputs. The system equations are written as a set of algebraic linear equations and analytical solution is possible. Application of this methodology requires a well-structured system and is very well suited for operational decisions.

However, in SD, the flow of information is modelled along with the physical flow, the emphasis being on storages and levels. It is, therefore, used for modelling semi-structured and unstructured problems with focus on policy analysis. SD is based on the principle that the behaviour of a system is due to its structure and policies. This method permits experimentation with multiple policies and analysis of the effect of these on the system.

Selection and Integration

Matching of the PST and SD attributes with the problem attributes shows that neither of the methodologies by themselves are able to meet the entire set of problem attributes. It is found that PST and SD exhibit complimentary attributes in many of the areas and an integration of the PST methodology under a SD framework would meet the problem requirements. Therefore, in order to understand the waste management system, it is essential to integrate the PST and SD methodologies and evolve a suitable approach for the WM system. In this system, SD is amalgamated with the PST framework as shown in Figure 24.8.

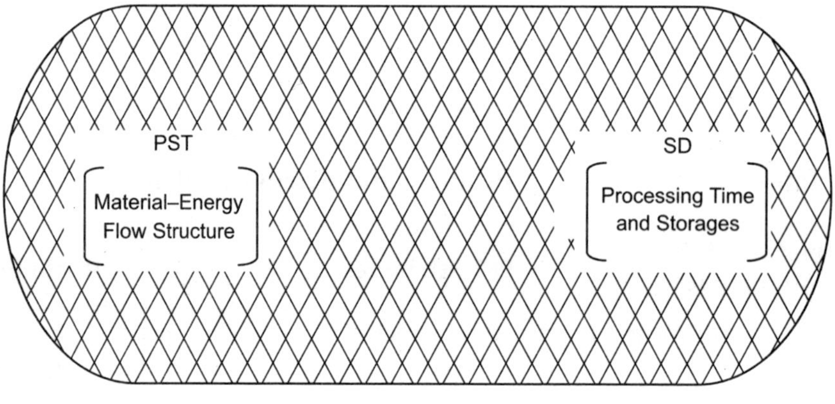

Figure 24.8: Amalgamation of Physical System Theory and System Dynamics

As the waste management problem is analyzed using the above methodologies, there is a dynamic shift in the attributes of the problem. The application of the proposed methodology using both PST and SD can be done flexibility depending upon the cognitive burden and expertise available.

The flexibility being proposed in this study is systemic flexibility having major attributes of spectral, integrative, interactive, innovative and fuzziness characteristics indicating a trend towards flexibility in waste management[22].

The continuum approach to waste management could also involve treating the various waste management sectors ranging from recycling to the ultimate disposal in landfills along a continuum with other sectors falling in the line of the continuum as shown below:

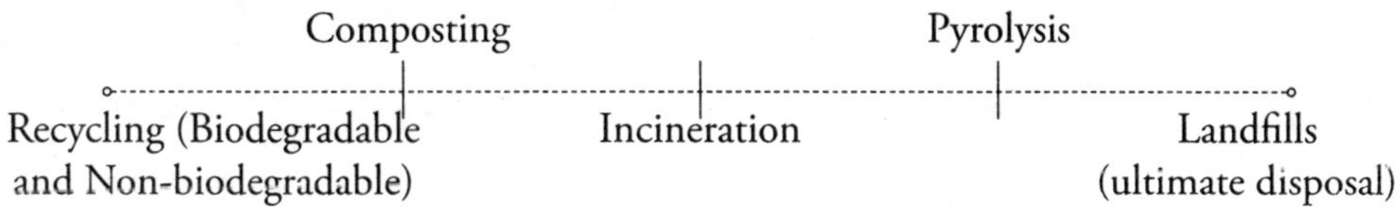

Here the real situation is considered to be dynamic as the composition of waste changes regularly, ambiguously and therefore it is proposed to consider a spectrum of it and attempt to integrate, interact, interplay or innovate capturing its fuzziness. Therefore, for managing waste, the flexible systems management is a natural process.

24.4 MODEL DEVELOPMENT USING PHYSICAL SYSTEM THEORY

Physical System Theory (PST) originated in the field of electrical engineering for the analysis of complex electrical networks and circuits. The physical system theory is based on the graph theoretic approach. The graph theoretic approach has flow variables (quantity input/output to the system) and the costs involved/incurred in the waste management system.

Keeping the above in view it is possible to utilize physical system theory approach to waste management. Sushil[23] has utilized this type of analysis for Waste Management alongwith Goal Programming for national planning. Recent work threw some insight into the applicability of PST to various areas. Consequently, this work has been taken to utilize physical system theory to many critical aspects of waste management and particularly to municipal solid waste management which is, a very important problem requiring immediate solution on a global basis in order to ensure the survival of the planet earth.

As previously discussed, the three major waste generating sectors as far as municipal solid waste is concerned and that are being analyzed in the present work are:

- Domestic Waste
- Commercial Waste
- Municipal Waste

Domestic waste is primarily those generated in residential houses. Commercial wastes include those from industrial and office complexes, and municipal waste is that generated on the roads, sidewalks, pavements, parks etc. and is generally collected by the local authority of the area.

Each type of waste discussed above would ultimately be recycled/ disposed off in one or more of the six waste management sectors, as mentioned earlier.

Each waste management sector has basically three elements, i.e. waste input, output of a useful product/energy and processing of waste leading to further pollution of land, air, and water as the case may be, as indicated in Figures 24.2 to 24.7.

The above concept can be shown schematically in a generalized flow model as given in Figure 24.9 for any waste management sector.

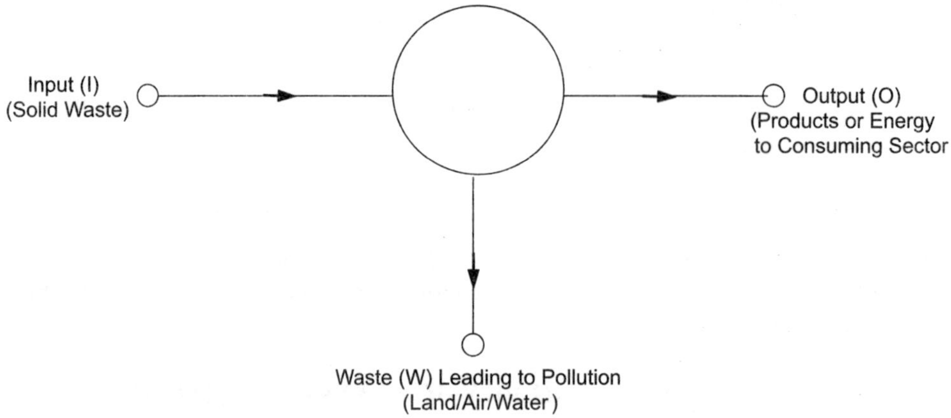

Figure 24.9: Generalized Flow Model of a Waste Management Sector

In the present study, for all waste management sectors we are considering only one consumption sector.

Based on the overall schematic diagram (Figure 24.1) of the waste management systems and the material –energy flow models (Figure 24.2 to 24.7), the cost and flow equations of each type of waste management sector is obtained. The parameters and variables, utilized to develop the programme in `DYNAMO' language, are given below.

Parameters

(a) Technological coefficients are ratios of pollution generated per unit of waste input to the specific waste management sector. The technological coefficients of water, land and air pollution for landfilling waste management sector are (in the computer simulation model sector-6 is the waste management sector allotted for landfilling) K6LP for land pollution, K6WP for water pollution, K6AP for air pollution. Consequently the technological coefficient for gas generation/production is KP6 = 1 – K6LP– K6AP – K6WP.

(b) The technological coefficients for biodegradable and non-biodegradable waste for domestic, commercial and municipal waste are defined as the quantity of waste of a particular type divided by the total quantity of waste fed to the particular waste management sector. The technological coefficients for landfilling in this category are KDB6, KCB6, KMB6, KDNB6, KCNB6, KMNB6 where the first three coefficients are for biodegradable components and the later three are for non-biodegradable components.

Variables

Exogenous Variables

The major variables in this category are

(a) Cost of processing the waste, which for landfilling is referred to as F6.

(b) Cost of handling the wastes, which for landfilling is XDB1, XCB6, XMB6, XDNB6, XCNB6, XMNB6, where the first three variables are for biodegradable waste and the latter three for non-biodegradable waste.

(c) Cost of water, land and air pollution which for landfilling is X6WP, X6LP and X6AP respectively.

Endogenous Variables

These are the variables generated while executing the model and the major ones include:

(a) Allocation coefficients for resource allocation to various waste management sectors. For landfilling the allocation coefficients are CDB6, CCB6, CMB6, CDNB6, CCNB6, CMNB6.

(b) The quantity of product produced alongwith the quantity of solid, liquid and air pollution generated, which for landfilling would be YP6, Y6LP, Y6AP, Y6WP, etc.

(c) The total cost per unit for producing the product, which for landfilling is XO6. Values of other endogenous variables are also computed during the simulation keeping in view the requirements of the problem.

The above equations are based on physical system theory which is subsequently integrated with System Dynamics by incorporating level and rate equations in the PST model.

A typical set of equations for landfilling utilized for simulation in DYNAMO are shown in Appendix 24A. Similar equations have been obtained for other waste management sectors.

24.5 RESULTS AND DISCUSSIONS

The results based on simulation runs utilizing the DYNAMO simulation package for basic level policy are shown in Figure 24.10 to Figure 24.13. Figure 24.10 indicates that the maximum allocation for biodegradable waste at average collection cost is for biodegradable recycling and the minimum allocations for biodegradable waste as shown in Figure 24.10 is for incineration; pyrolysis, composting, and landfilling falling between in the same order.

Figure 24.11 indicates that the maximum allocation at base level policies and average collection cost is for landfilling followed by incineration and the minimum allocation is for non - biodegradable recycling. Figures 24.10 and 24.11 indicate an initial transient response in the allocation for all waste management sectors as initially the allocation for each sector has been assumed to be equal and in about 100 days study-state is achieved for all the waste management sectors and from then on the actual allocation is achieved and maintained till the end of the simulation, i.e. 650 days.

Figure 24.12 indicates that the total cost involved in utilization of the waste management sector is maximum for non- biodegradable recycling and minimum for biodegradable recycling. Incineration, landfilling, composting, and pyrolysis fall in between in the same order. The above results are as expected as the allocation obtained in Figures 24.10 and 24.11 are reverse of the total cost shown in Figure 24.12. Figure 24.13 shows total amount of flow in tonnes per day for each of the waste management sector. From the figure it is clear that the maximum flow is for landfilling followed by incineration, biodegradable recycling, pyrolysis, composting, and non-biodegradable recycling. Figure 24.13 shows the maximum quantity of flow for landfilling as both biodegradable and non-biodegradable waste is fed to the system in a similar fashion to incineration. Further, the final disposal sector for other waste sectors is also landfilling and therefore, the flow is bound to be the highest.

When the policy is changed keeping the collection cost at the average level, it is observed that when the allocation is increased to 0.5 for landfilling, the maximum allocation for biodegradable waste is now to landfilling followed by biodegradable recycling, pyrolysis, composting, and incineration as shown in Figure 24.14. When compared to Figure 24.10, it is observed that the allocation of landfilling has increased from the fourth position to the first position based on the policy constraint imposed. Therefore, the policies can have a significant impact on the actual allocation of resources for the various waste management sectors.

Figure 24.15 appears to be identical to Figure 24.11. However, the overall allocation for landfilling is significantly higher when compared to Figure 24.11. Figure 24.16 shows the total quantity of flows in tonnes per day after incorporation of the policy change and the trend is similar to Figure 24.13. However, the total quantity of flow for landfilling is much higher than the other waste management sector due to change in policy invoked in the simulation model.

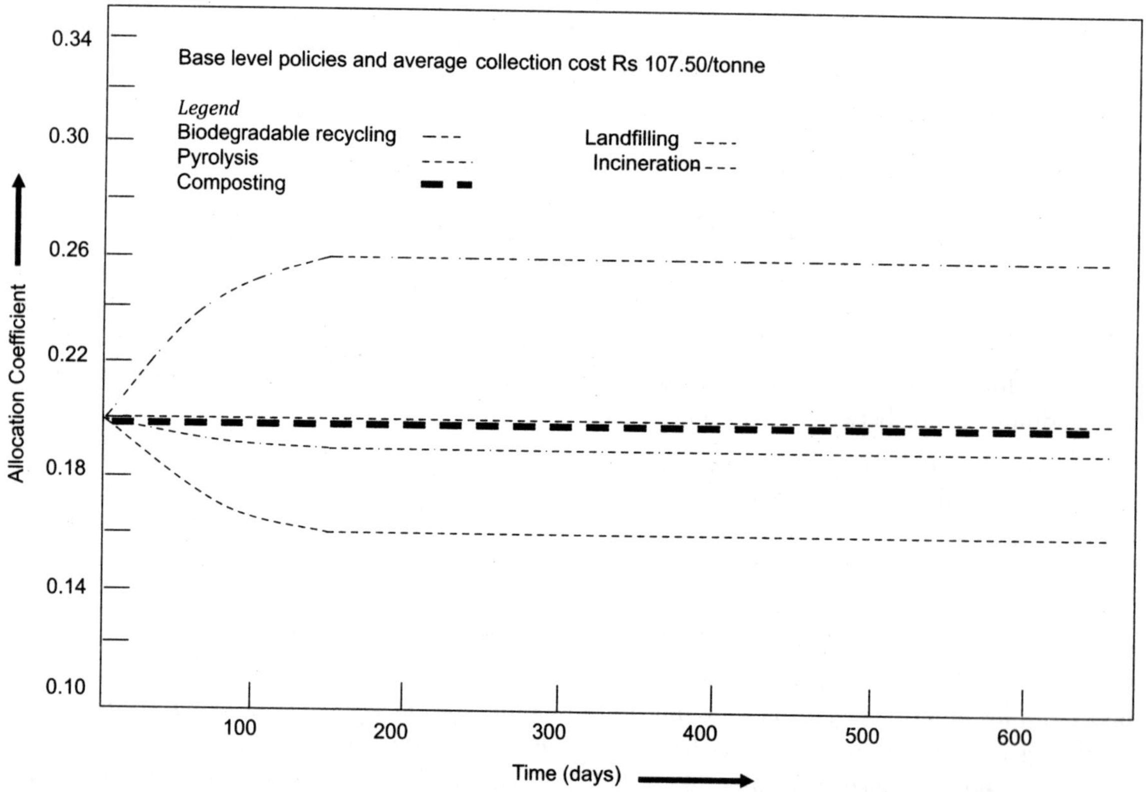

Figure 24.10: Allocation Coefficient (Biodegradable) Vs. Time

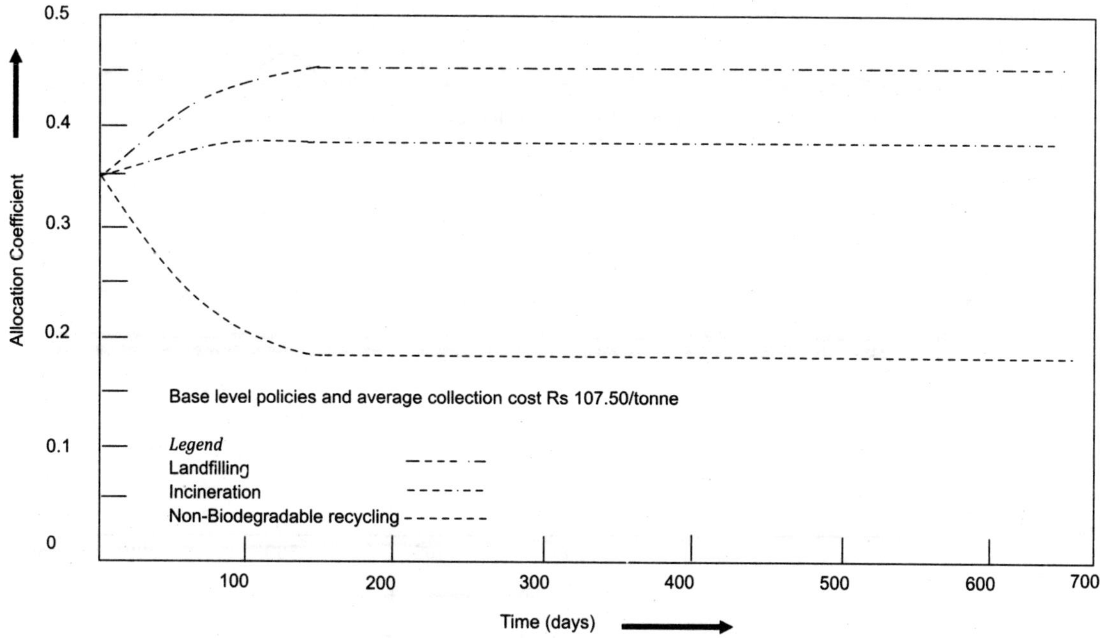

Figure 24.11: Allocation Coefficient (Non-Biodegradable) Vs. Time

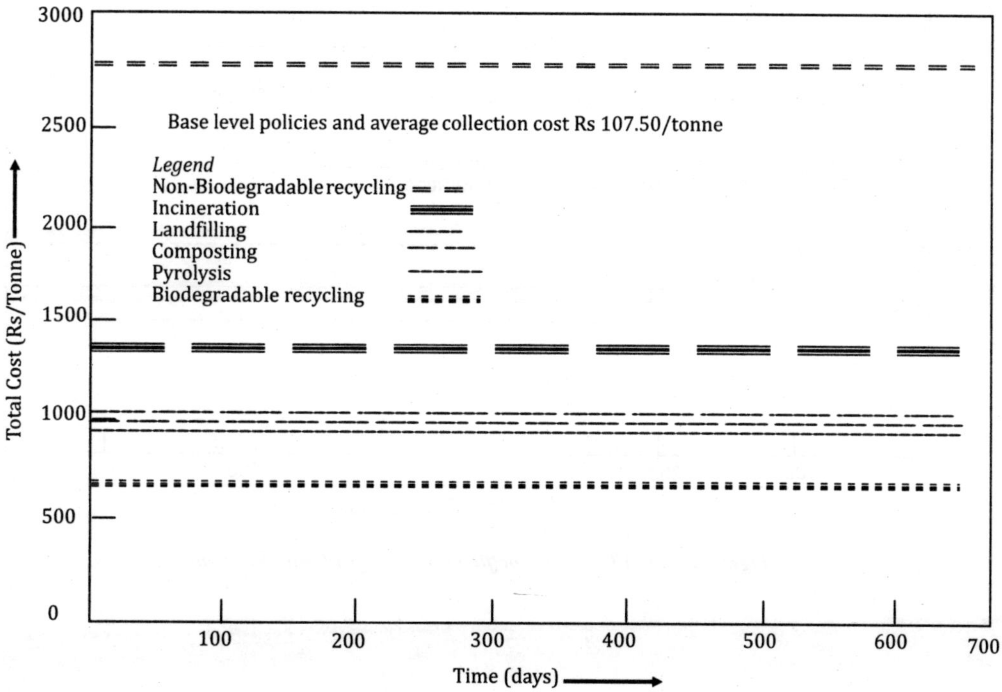

Figure 24.12: Total Cost Vs. Time

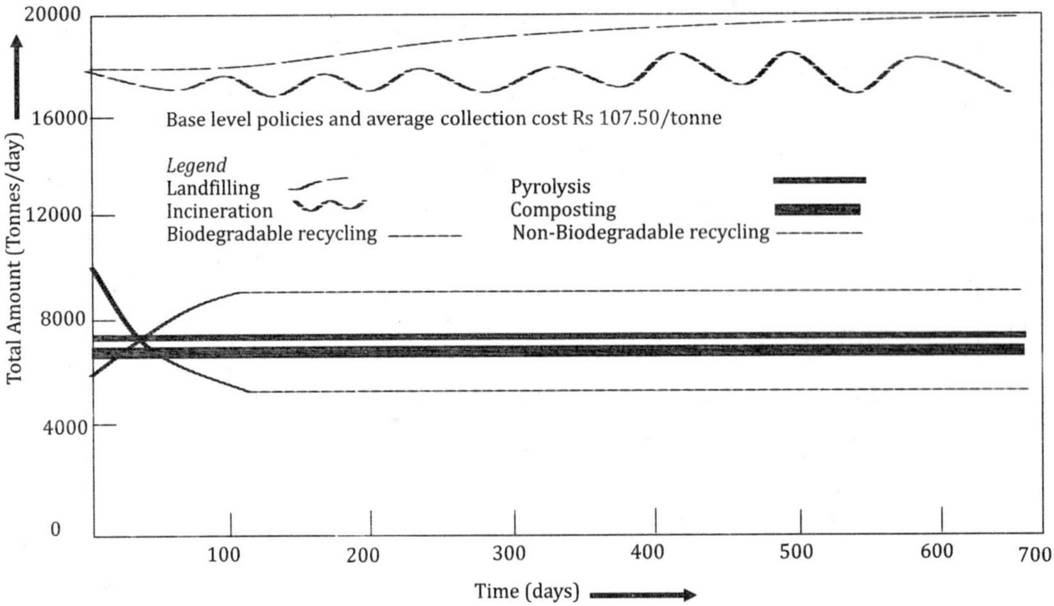

Figure 24.13: Total Amount Vs. Time

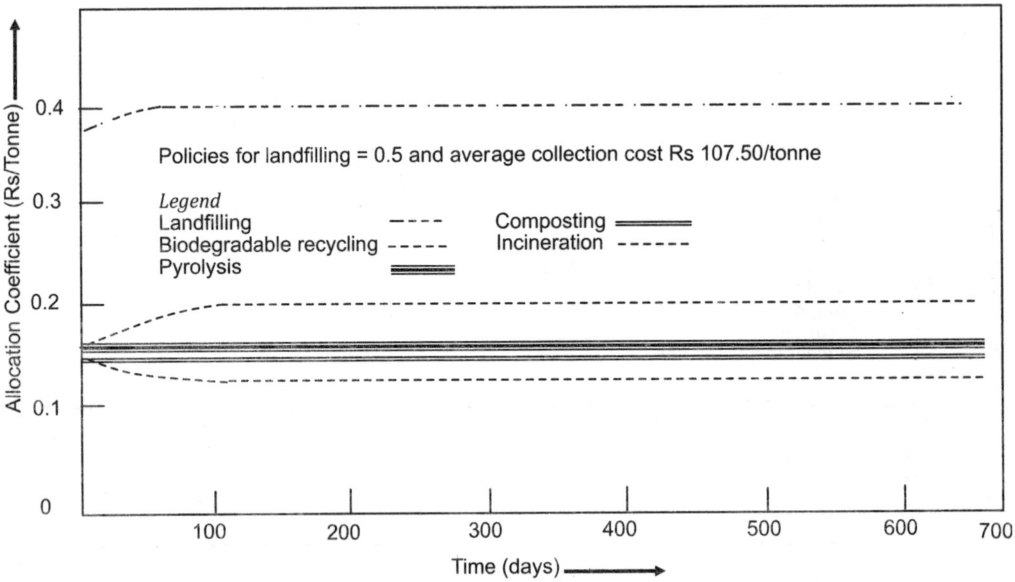

Figure 24.14: Allocation Coefficient (Biodegradable) Vs. Time

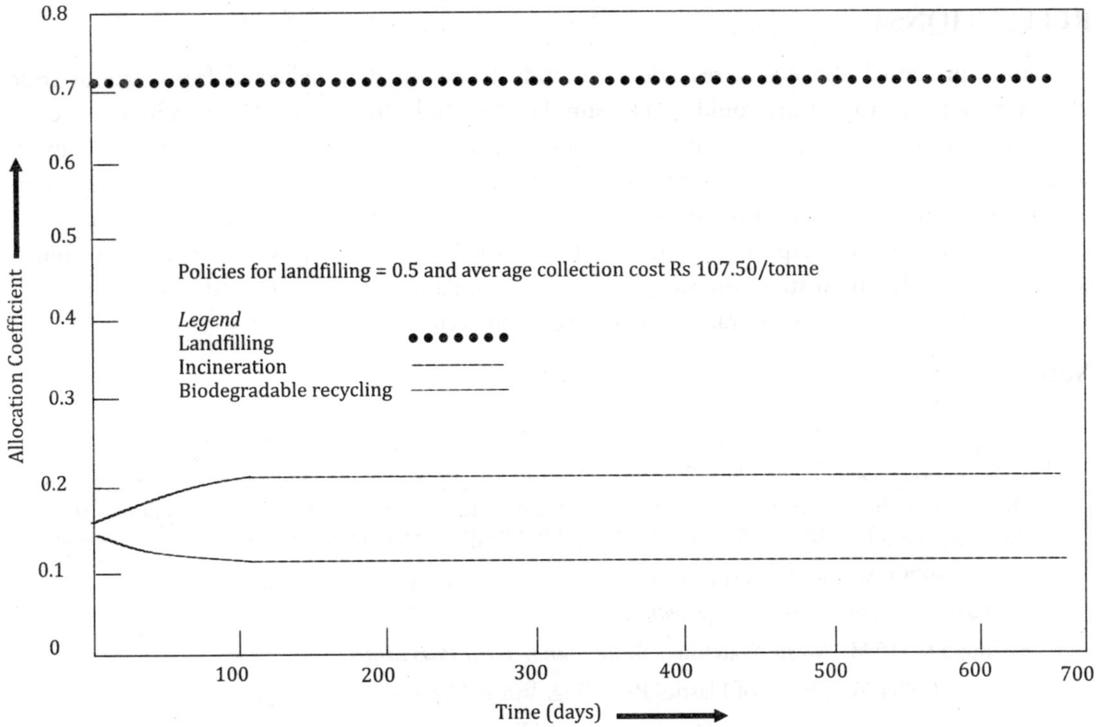

Figure 24.15: Allocation Coefficient (Non-Biodegradable) Vs. Time

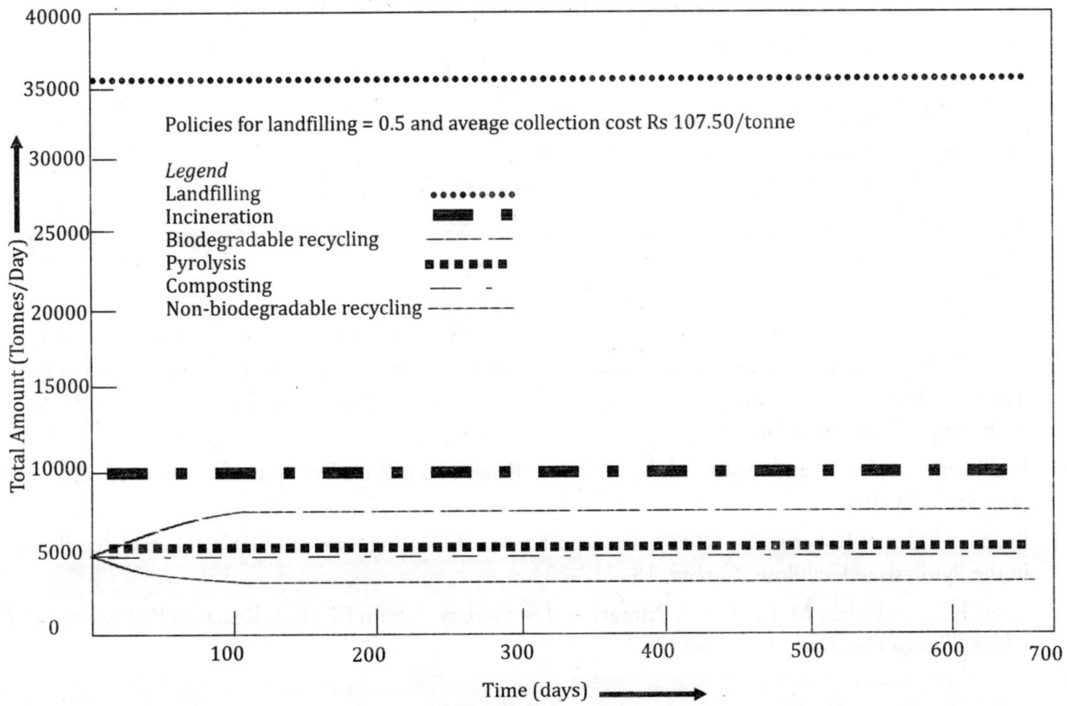

Figure 24.16: Total Amount Vs. Time

REFLECTIONS

The study presented above is a realistic case study based on data collected from various sources. However, an actual case study could also be simulated by collecting actual data required in the model for a particular city/town/area/country. The above model has attempted to show the applicability of simple algebraic equations to simulate relatively complex municipal solid waste management system so that, decisions regarding allocation of resource to various waste management sectors could be achieved. In the present study, the capacity constraint of the various waste management sectors have not been considered in order to illustrate the simplicity of the simulation model and its strength to predict the allocation of resources in the overall waste management scenario.

Notes

1. Sushil, (1980), *Systems Approach to Waste Management in India,* M. Tech. Dissertation, Indian Institute of Technology, Delhi, India; Sushil, (1984), *Systems Modeling of Waste Management in National Planning,* Ph.D. Thesis, Indian Institute of Technology, Delhi, India; Sushil, (1989), *Systems Approach in National Planning: A Study in Waste Management,* Anmol Publication, New Delhi, India.
2. Vrat P., (1979), Waste Management in India: The Need for Organized Systems Approach, Country Paper, *APO Symposium on Waste Management,* Japan.
3. Bellamy D., (1993), Waste Watching, *Waste Management Proceedings*, 3-5.
4. Snow B., (1994), Overview of Plastics Recycling, *Wastes Management*, 39.
5. Kersey J.D. (1994), Sustainable Development and Wastes Management, *Wastes Management,* 48-52.
6. Skinner J., (1994), International Progress in Solid Waste Management, *IWM Proceedings,* 3-5.
7. Brierely M.J., (1993), Environmental Objectives: Going for Gain not Pain, *Wastes Management,* 52-53.
8. Sushil, (1990), Waste Management: A Systems Perspective, *Special Issue Industrial Management and Data Systems,*90 (5), 1 to 67.
9. Mashayekhi A.N., (1993), Transition in the New York State Solid Waste System: A Dynamic Analysis, *System Dynamics Review,* 9(1), 23-47.
10. Porteous A., (1993), The Incineration of Waste, *Wastes Management,* 34.
11. Santen, A.V., (1993), Incineration: Its Role in the U.K. Waste Strategy, *Waste Management Proceedings,* 18-23.
12. Srivastava V.K., and Jalan R.K., (1994), Prediction of Concentration in the Pyrolysis of Biomass Materials-I, *Journal Energy Conversion and Management,* 35 (12), 1031-1040; Srivastava V.K., Sushil and Jalan R.K., (1994), Development of Mathematical Models for Prediction of Concentration in the Pyrolysis of Biomass Materials, Presented at the *Symposium on Advances in Chemical Engineering* at Indian Institute of Chemical Technology, Hyderabad, India.
13. Bradbury A., Sakai Y., and Shafizadeh F., (1979), A Kinetic Model for Pyrolysis of Cellulose, *J. Appl. Polym. Sci.,* 23, 3271-3280.
14. Brunner P.H. and Roberts P.V., (1980), The Significance of Heating Rate on Char Yield and Char Properties in the Pyrolysis of Cellulose, *Carbon*, 18, 217-224.
15. Jegers H.F., and Klein M.T., (1985), Primary and Secondary Lignin Pyrolysis Reaction Pathways, *Ind. Eng. Chem., Process Des. Dev., 24, 173-183.*

16. Koufopanos C.A., Papayannakos N., Maschio G., and Lucchesi A., (1991), Modeling of the Pyrolysis of Biomass Particles: Studies on Kinetics, Thermal and Heat Transfer Effects, *Can. J. Chem. Engg.,* 69, 907-915.

17. Wilson D., (1993), The Landfill Stepladder-Landfill Policy and Practice in an Historical Perspective, *Wastes Management,* 24-28.

18. Crawford J.F., and Smith P.G., (1985), *Landfill Technology,* Butterworths.

19. Harris R.C., Knox K. and Walker N., (1994), Strategy for the Development of Sustainable Landfill Design, *IWM Proceedings,* 26-29.

20. Sushil, (1994), Flexible Systems Methodology, *Systems Practice,* 7(6), 633-651.

21. Sushil, (1993), *System Dynamics-A Practical Approach for Managerial Problems,* Wiley Eastern Limited, New Delhi.

22. Sushil, (1997), Flexible System Management: An Evolving Paradigm, *Systems Research and Behavioral Science,* 14(4), 259-275.

23. Sushil, (1993), Application of Physical Systems Theory and Goal Programming to Modeling and Analysis of Waste Management in National Planning, *Int. J. Systems Sci.,* 24, (5), 957-984.

APPENDIX 24A

DYNAMO EQUATIONS

C K6LP = .965

K6LP is the coefficient of land pollution for waste management sector 6 viz. landfilling.

C K6WP = .01

K6WP is the coefficient of water pollution for waste management sector 6 viz. landfilling

C K6AP = .02

K6AP is the coefficient of air pollution for waste management sector 6 viz. landfilling

A K1PLP.K = K1LP/(Y06.K)

K1PLP.K is the coefficient of land pollution from composting fed to landfilling

A K2PLP.K = K2LP/(Y06.K*10)

K2PLP.K is the coefficient of land pollution from composting fed to incineration

A K3PLP.K = K3LP/(Y06.K*10)

K3PLP.K is the coefficient of land pollution from composting fed to pyrolysis

A K4PLP.K = K4LP/(Y06.K)

K4PLP.K is the coefficient of land pollution from composting fed to biorecycling

A K5PLP.K = K5LP/Y06.K

K5PLP.K is the coefficient of land pollution from composting fed to non-biorecycling

A KP6.K =1 –K6LP-K6WP – K6AP

KP6.K is the coefficient of product formation waste management sector 5 viz. landfilling

C F6 = 500

F6 is the processing cost for landfilling/MT of waste

A XDB6.K = 107.50

XDB6.K is the cost of transport biodegradable domestic waste to waste management sector 6 viz. landfilling

A XCB6.K = 107.50

XCB6.K is the cost of transport biodegradable commercial waste to waste management sector 6 viz. landfilling

A XMB6.K = 107.50

XMB6.K is the cost of transport biodegradable municipal waste to waste management sector 6 viz. landfilling

A XDNB6.K = 107.50

XDNB6.K is the cost of transport biodegradable non-domestic waste to waste management sector 6 viz. landfilling

A XCNB6.K = 107.50

XCNB6.K is the cost of transport biudegradable non-commercial waste to waste management sector 6 viz. landfilling

A XMNB6.K = 107.50

XMNB6.K is the cost of transport biodegradable non-municipal waste to waste management sector 6 viz. landfilling

A KDB6.K = YDB6.K/Y06.K

KDB6.K is the coefficient of flow of domestic biodegradable waste for waste management sector 6 viz. landfilling

A KCB6.K = YCB6.K/Y06.K

KCB6.K is the coefficient of flow of commercial biodegradable waste for waste management sector 6 viz. landfilling

A YDB6.K = CDB6.K*YDB.K

YDB6.K is the actual quantity flow of domestic biodegradable waste for waste management sector 6 viz. landfilling

A YCB6.K = CCB6.K*YCB.K

YCB6.K is the actual quantity flow of commercial biodegradable waste for waste management sector 6 viz. landfilling

A YOD6.K = YDB6.K +YCB6.K+YMB6.K+YDNB6.K+YDNB6.K+YMNB6.K+YCNB6.K

YOD6.K is the total flow of waste for waste management sector 6 viz. landfilling

A YO6.K = DELAY3(YO6D.K,300)

YO6.K is the total flow of waste after considering delay of one year.

A KMB6.K = YMB6.K/YO6.K

KMB6.K. is the coefficient of flow of waste municipal biodegradable waste for waste management sector 6 viz. landfilling

A YMB6.K = CMB6.K*YMB.K

YMB6.K is the actual quantity of flow of municipal biodegradable waste for waste management sector 6 viz. landfilling

A KDNB6.K = YDNB6.K/YO6.K

KDNB6.K is the coefficient of flow of domestic non-biodegradable waste for waste management sector 6 viz. landfilling

A YDNB6.K = CDNB6.K*YDNB.K

YDNB6.K is the actual quantity of flow of domestic non-biodegradable waste for waste management sector 6 viz. landfilling

A KCNB6.K = YCNB6.K/YO6.K

KCNB6.K is the coefficient of flow of commercial biodegradable waste for waste management sector 6 viz. landfilling

A YCNB6.K = CCNB6.K*YCNB.K

YCNB6.K is the actual quantity of flow of commercial biodegradable waste for waste management sector 6 viz. landfilling

A KMNB6.K = YMNB6.K/YO6.K

KMNB6.K is the coefficient of flow of municipal non-biodegradable waste for waste management sector 6 viz. landfilling

A YMNB6.K = CMNB6.K*YMNB.K

YMNB6.K is the actual quantity of flow of municipal non-biodegradable waste for waste management sector 6 viz. landfilling

A YP6.K = KP6.K*YO6.K

YP6.K is the actual quantity of product produced in waste management sector 6 viz. landfilling

A Y6LP.K = K6LP*YO6.K

Y6LP.K is the actual quantity of land pollution from waste management sector 6 viz. landfilling

A Y6AP.K = K6AP*YO6.K

Y6AP.K is the actual quantity of air pollution from waste management sector 6 viz. landfilling

A Y6WP.K = K6WP*YO6.K

Y6WP.K is the actual quantity of water pollution from waste management sector 6 viz. landfilling

A X6LP.K =500

X6LP.K is the cost of treating one tonne of waste from land pollution in waste management sector 6 viz. landfilling

A X6AP.K = 500

X6AP.K is the cost of treating one tonne of waste from air pollution in waste management sector 6 viz. landfilling

A X6WP.K = 100

X6WP.K is the cost of treating one tonne of waste from water pollution in waste management sector 6 viz. landfilling

A XO6.K = KDB6.K*XDB6.K+KCB6.K*XCB6.K+KMB6.K*XMB6.K+KDNB6.K* XDNB6.K +MO6.K

XO6.K is the total cost estimation

A MO6.K = KCNB6.K*XCNB6.K+KMNB6.K*XMNB6.K+K2PLP.K*X6LP.K+K3PLP.K* X6LP.K+M7.K

MO6.K is the total cost estimation

A M7.K = K4PLP.K*XLP.K+K5PLP.K*X6LP.K+K6LP*XLP.K+K6WP*X6WP.K+K6AP* X6AP.K+M8.K

M7.K is the total cost estimation

A M8.K = F6+K1PLP.K*X6LP.K

M8.K is the total cost involved in waste management sector 6 viz. landfilling operation

SPEC DT=1, LENGTH=400, PRTPER=1, PLTPER=10, SAVPER=10

PART VI

LOOKING BEYOND

25 LOOKING BEYOND

25.1 INTRODUCTION

In the previous parts of the book, a range of applications of modelling and analysis of physical systems has been provided using the concepts of physical system theory in a flexible system methodological framework. The graph theoretic physical system theory (PST), which is taken as the base methodology in his book, has been simplified, generalized and integrated with other modelling methodologies to evolve flexible models of physical systems.

In this process, various developments have been reported which helped in overcoming the rigidness and limitations of PST as a linear and deterministic framework to a more flexible framework. The applications have brought out the issues of handling non-linearity, dynamics, stochasticity and ambiguity, on the one hand, and optimization of parameters, on the other. Attempts have also been made to generalize the PST framework to include general variables in the modelling framework that are not fulfilling the rigid definitions of across and through variables and are not governed by the principle of conservation. However, this journey from a theory of physical systems to a theory of flexible systems is very long and the book provides, at best, a good start point only.

In this chapter, on looking beyond, we would like to sketch the pointers for future research in this area, both in terms of new modelling requirements and new application areas. Further, an attempt is made here to not only apply the theory in an analytic mode but to move out of the box and apply it in a synthetic paradigm also. In the end, a gaming and learning mode of modelling is advocated to blend the quantitative as well as qualitative variables linked with it.

25.2 FEATURES FOR FLEXIBILITY IN MODELLING

Though the applications reported in the previous parts of the book indicate different types of flexibility in modelling of physical systems, many more research avenues are to be explored. Here a synoptic view of select proposals in specific areas such as non-linearity, dynamics, stochasticity, fuzziness, and integration in modelling is provided, to lead to more flexible and applicable models. Flexibility in modelling would provide wider options and change possibilities as per the changing situation and complexity of the real systems.

Non-linearity in Modelling

In the applications reported in this book, most of the models developed consist of a set of linear equations. The generalized models of basic processes given in chapter two are basically linear models.

However, it can be seen in chapter ten on technology selection models that the derivation of the cost equation for a multi-stage manufacturing system becomes non-linear. A similar kind of non-linear model has been obtained in chapter twelve in the context of inspection policy for a multi-stage system.

Thus, it can be seen that though the basic models of the components are linear, the derivation in a multi-stage system becomes non-linear in nature.

But if the basic characteristics of managerial systems are examined, it can be easily found that these systems are inherently non-linear. In an analytical manner, it was difficult to build non-linear physical systems models. However, in the simplified form, when the system equations can be solved in a step-by-step manner using simulation, we can handle the non-linearity in systems in a much better manner as is done in system dynamics modelling[1].

The models of the basic processes, given in chapter two, are assumed to be linear. Let us examine the nature of a typical model of a transformation process. There are two types of parameters, viz. technological coefficients, and processing cost coefficients.

The technological coefficient defines the input required or waste generated per unit of output. In the basic model it is assumed to be linear, given by the following equation (Figure 25.1)

$$k_{ji} = y_{ji}/y_{oi}$$

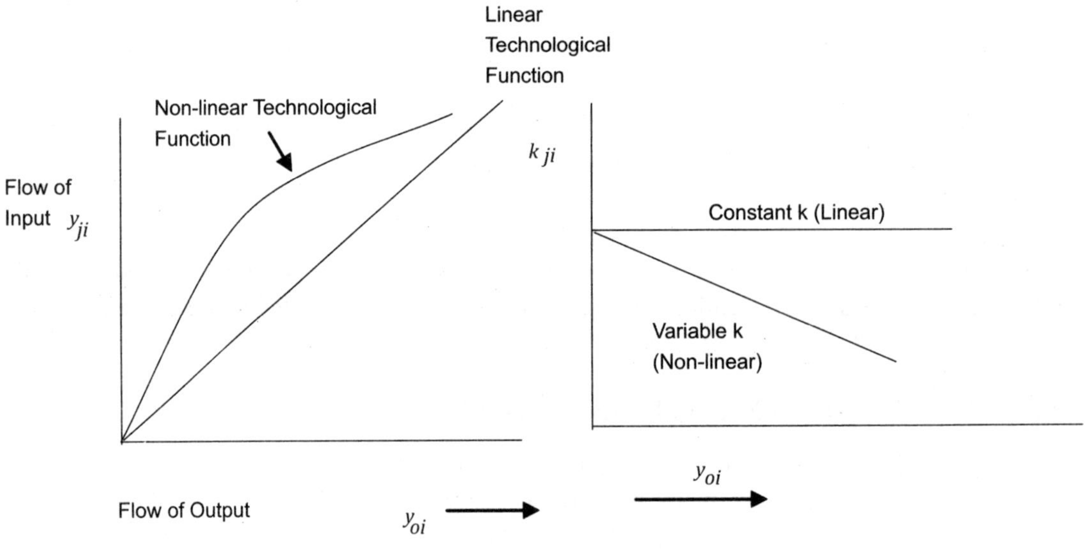

Figure 25.1: Technological Coefficient Relationship

That is, the basic assumption that the relationship between the input and output is linear and the slope of the curve is constant for all values of output, may not be the case in reality. The input requirements may change non-linearity with the output as shown in Figure 25.1. In this case, the technological coefficient is varying with respect to the output. Thus, we have a feedback between the technological coefficient and output of a transformation process, as shown in Figure 25.2.

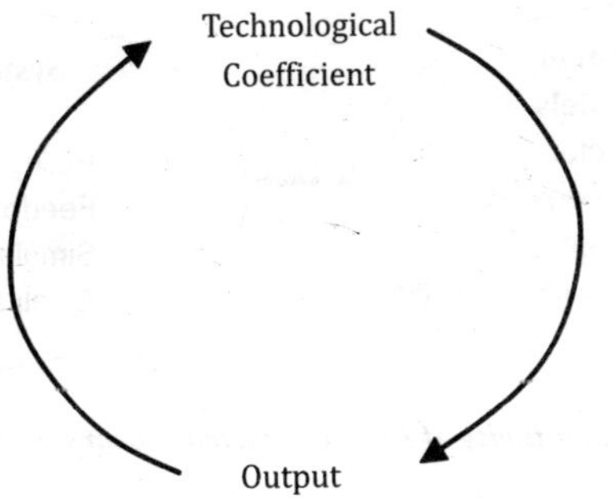

Figure 25.2: Feedback Loop in a Transformation Process

Let us examine the change in the equations of the general model of the transformation process

Cost Equation

$$x_{oi} = k_{ji}\, x_{ji} + f_i\,(y_{oi}) \tag{25.1}$$

where k_{ji} is a function of y_{oi}, such as

$$k_{ji} = k_{ji} \pm n_{ji}\, y_{oi} \tag{25.2}$$

Substituting equations (25.2) in (25.1), we have

$$x_{oi} = (k_{ji} \pm n_{ji}\, y_{oi})\; x_{ji} + f_i\,(y_{oi}) \tag{25.3}$$

The equation (25.3) is a non-linear equation. We can see that one feedback in the system has generated a non-linearity in the basic model. The real world management system may have many feedbacks which can be modelled in a system dynamics (SD) framework. Thus, an integrated PST-SD model will be able to handle different types of non-linearities. A typical example of such a model could be seen in chapter 18 in the form of a DSS for maintenance management. It is envisaged that future works in physical systems modelling would generate innovate ways of handling non-linearity in systems.

Dynamic Modelling

Many of the models presented in the applications considered are static. A few dynamic models are presented in chapters 9, 13, 18, 21, 22, and 24. In traditional PST modelling, the most widely applied dynamic model was 'state-space model' as outlined in chapter three. This model has a rigid framework and may have rate to rate feedbacks which may, at times, give a false dynamics of the managerial systems.

Thus, in order to capture the dynamics effectively, the simplified PST models may be suitably integrated and amalgamated with SD models to capture both the input-output structure and the policy structure of the system.

A critical comparison of PST and SD is given in chapter two. Both the approaches are based on flow modelling and can complement each other as shown in Figure 25.3.

These two approaches can be suitably integrated in various ways as discussed in chapter two (Figure 2.7). With this kind of integration of PST and SD modelling, the systems from operational to strategic levels may be modelled in a synergistic manner. This will provide a variety of options in modelling framework to evolve highly flexible models that can be used as a basis for gaming and learning in an interactive environment.

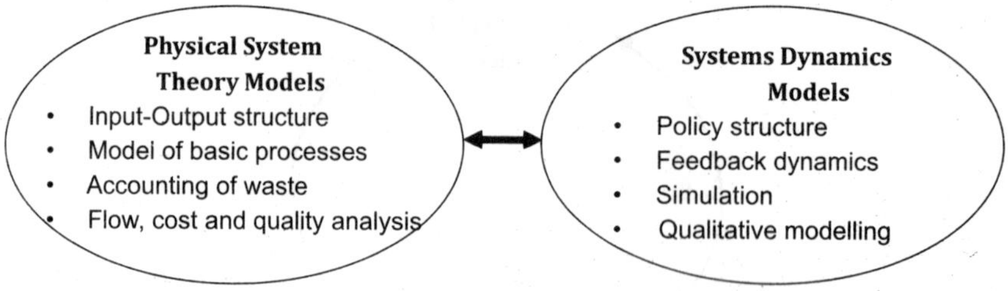

Figure 25.3: Complementarity of Physical System Theory and System Dynamics Models

Modelling Stochasticity in Systems

Uncertainties of different kinds exist in the working of various physical systems, and thus, from a practical stand point the variables and parameters may be treated to be stochastic rather than deterministic. Traditionally, PST modelling had been following a deterministic route; primarily because of the analytic formulation adopted for analysis. With the step-by-step solution procedure, simulation can be effectively applied and various parameters and variables can be treated to be stochastic.

A specific application of stochastic analysis can be seen in chapter 11 for cost and quality analysis. In this case, the quality parameters and indices are estimated by simulation and then the quality of output is derived by using a simulation model.

Other applications of simulation of PST models to treat stochasticity are in chapter 12 on inspection policy and chapter 13 on material handling systems.

These few models provide only an indication about the relevance and applicability of stochastic analysis of PST models. The range of issues of uncertainty in managerial systems is very wide and there is significant research potential on this front.

If in a transformation process model, the probability distribution of technological, processing cost and quality parameters is known and the distribution of the input variables is made available, the distribution of cost/quality of output can be derived. We can also assess the standard deviation of the cost and quality of output rather than only estimating their mean values. This will help a great deal in risk analysis of various systems.

The integrated PST-SD models can involve stochastic parameters and variables to handle stochasticity dynamically.

Developing Models of Ambiguous Systems

The level of uncertainty and ambiguity goes high as we move from the real physical system to conceptual physical systems. The ambiguity in systems can be effectively captured by using fuzzy set theory[2]. The parameters and variables of the physical system models can be fuzzified, which will help a great deal in modelling linguistic variables.

For example, the authority and responsibility as `very low', `low', or `high' can be measured using fuzzy set theory. The component terminal equations in PST models can be used as fuzzy relations and the whole PST model can be simulated using fuzzy arithmetic. A similar experiment done in the context of system dynamics models has proved to be useful, and interesting results were generated[3].

In this book, only one chapter on meta analysis of management information systems (chapter 17) uses fuzzy set theory in an elementary manner. It would be a worthwhile modelling experience, particularly in the context of soft systems such as organizational systems, marketing systems, innovation systems, and knowledge management systems to convert the whole PST model as a set of fuzzy relations.

Intelligent Modelling

The physical system models may be made intelligent models by integrating them with neural nets. The simplest possible modelling could be to utilize a PSD-SD model of a physical system for training a neural net. The trained neural net model will be an intelligent model that can be used to generate system behaviour under changed circumstances.

Intelligent models are highly adaptive in view of continuous learning and thereby lead to effective decision-making. Intelligence in modelling is a rich source of creating flexibility as these models are more organic in nature. By applying such models, the mechanistic view of physical systems graduate to the organic and learning view.

Integrated Modelling using Flexible Systems Methodology

Most of the applications presented in the book have used flexible systems methodology[4] as explained in chapters one and three. First of all, the problem under consideration is conceptualized and a set of systems techniques are selected and integrated to suit the problem attributes most closely. Multiple techniques of modelling complement limitations of each other and thus the underlying assumptions become more realistic. This results in more realistic and valid models that bridge the gap between the problem requirements and systems methodology.

Integration of systems based techniques results in handling competing requirements of the problem such as part structured–part unstructured, data partly quantitative–partly qualitative, some issues are static–some are dynamic, and so on. Dealing with such opposing features of the problem requires flexibility in modelling which can be provided by integrative models. Since these models have more options and can have many change mechanisms in terms of parameters as well as structures, these are inherently flexible and adaptive.

Such integrative and flexible models can be used for better experimentation and building richer scenarios as compared to isolationist models developed by using only one technique. Higher the level of integration higher is the level of flexibility. The easiest way to integrate two models is to use them in succession, which is extensively used in this book. Amalgamation of two different techniques leading to a new one is difficult and can be experimented rarely. One such application has been made in chapter 18 which proves to be very successful in addressing the problem and overcoming the limitations of techniques used.

25.3 NEW APPLICATION AREAS

It is evident from the fundamentals of PST, that it can be easily applied to those physical systems which have a transparent structure. However, by its simplification and generalization, and integration with other system techniques it can be applied to ill-defined and partly ambiguous systems as well. Apart from the applications made in this book and past research, many new areas are opening up that can be resorted to such flexible modelling. A few of the potential areas of application are: value chain analysis, supply chain management, enterprise modelling, knowledge management, innovation management, business process re-engineering, e-business, and so on.

Out of these, indicative research suggestions are provided for the first three areas, i.e.

- Value Chain Analysis
- Supply Chain Management
- Enterprise Modelling

Value Chain Analysis

Value chain analysis provides insight about the high value adding and low value adding areas in an enterprise. As Porter[5] has proposed, there are two types of value chains: (i) industry value chain; and (ii) company value chain. The industry value chain links up all the backward and forward components of an industry as shown in Figure 25.4. The value chain for a company consists of both direct and indirect value adding activities as shown in Figure 25.5.

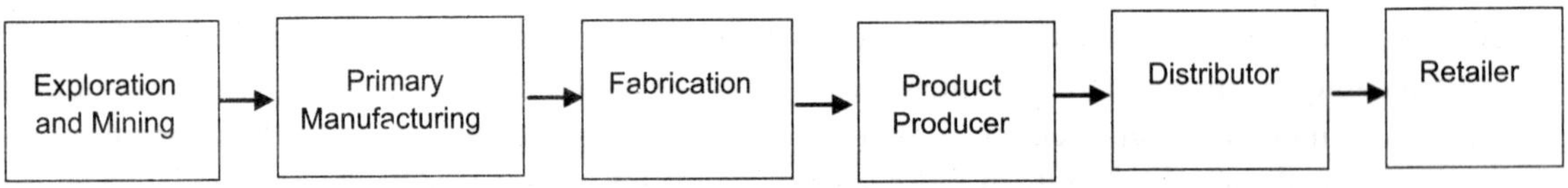

Figure 25.4: A Typical Industry Value Chain for Engineering Products

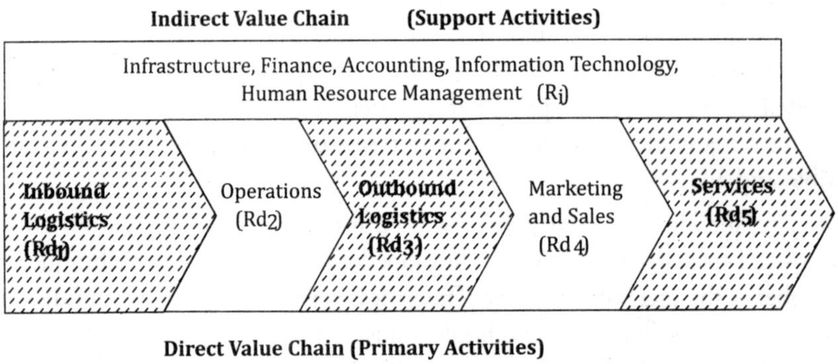

Figure 25.5: A Typical Company Value Chain

In both the cases, physical systems modelling concepts can be applied. Each component of the value chain can be modelled as a transformation process, where the across variable will be 'unit value' and through variable will be flow of 'products' and 'services'. Here, we are taking up the case of a typical company value chain and propose an outline physical systems model for the same. The simplified system graph for a typical value chain is shown in Figure 25.6.

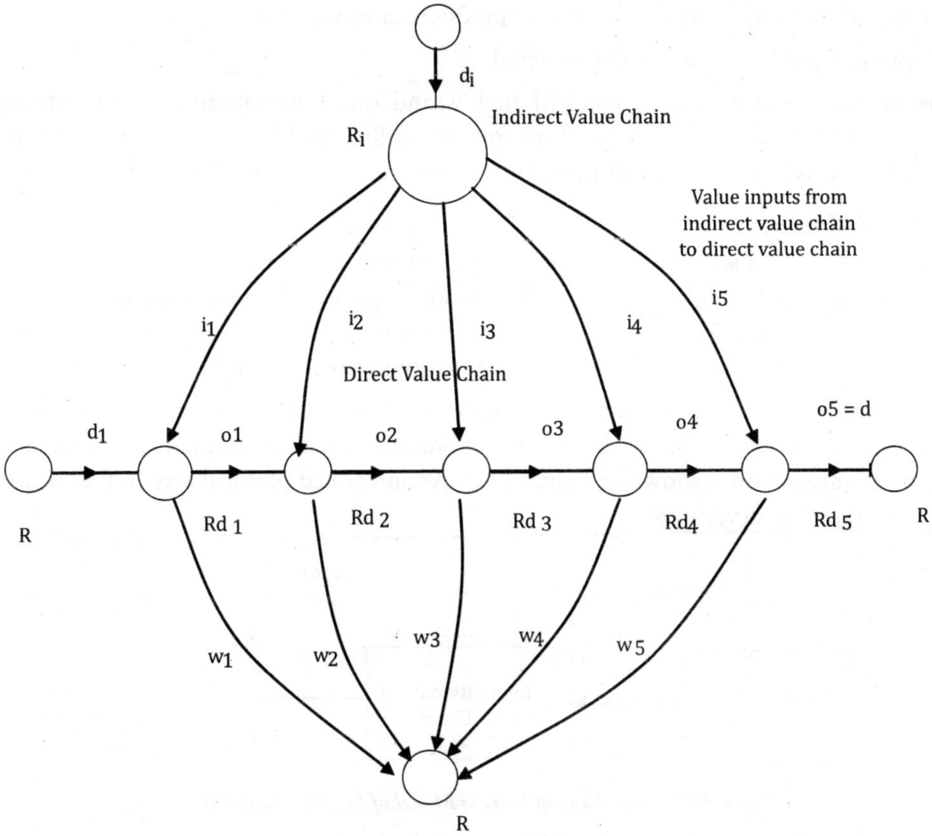

Figure 25.6: Simplified System Graph of a Company Value Chain

The direct value chain stages are treated as transformation processes that take input from outside in terms of raw material and process it through various value chain stages represented as R_{d1}, R_{d2},...R_{d5}. The last stage of the direct value chain, i.e. R_{d5} gives the final output to the market. At each stage some waste is generated, which is represented as w_1, w_2,...w_5. The indirect value chain, i.e. finance, human resources, R&D, MIS, etc. has been conceptualized as one component R_i. It can also be conceptualized as multiple components. The indirect value chain R_i takes input d_i in the form of various resources, and gives output in the form of various services that add value at various stages of the direct value chain.

The value addition at any stage can be computed as:

Value addition (*va*) = Value Output – Direct Value Input

For example in case of stage 1:

$$va_1 = x_{o1} \cdot y_{o1} - x_{d1} \cdot y_{d1} \qquad (25.4)$$

where

va_1 = Value addition at stage 1

x_{o1} = Unit value of output of stage 1

y_{o1} = Output of stage 1 (in terms of products)

x_{d1} = Unit value of direct input (in terms of products or services) to stage 1.

y_{d1} = Input to stage 1 in terms of raw material

In this way, we can estimate the stages with high *va* and low *va*, which can aid in identifying areas of improvement, strengths, and weaknesses. This is only an outline model; many more facets of value chain analysis will be opened once a practical model for a case situation is developed and implemented.

Supply Chain Management

Supply chain management is a useful concept that synchronizes the customer service requirements with the flow of materials from suppliers. This is a flow-modelling concept to integrate various components of the supply chain to minimize the intermediate storage requirements. It helps in balancing the competiting goals of high customer service, low inventory investment, and low operating costs.

Supply chain management is a physical system and can be modelled at multiple levels. At the macro level, we can develop a model based on the Stevens model of supply chain integration[6]. The stage four of the integration is shown in Figure 25.7. An integrated physical system theory and system dynamics model may be developed.

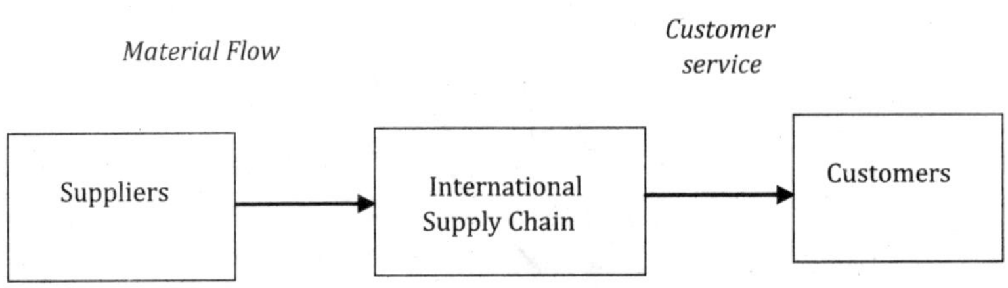

Figure 25.7: Stage Four of Stevens Model of Supply Chain Integration

A flow model can be developed with PST giving the basic structure in terms of transformation processes, and SD modelling dealing with storage and other policies.

At the micro level, similar dynamic models of the various cycles of supply chain process can be developed, e.g.

- Customer order cycle
- Replenishment cycle
- Manufacturing cycle
- Procurement cycle

The sub-models of these cycles can be integrated to give a model of supply chain management.

Enterprise Modelling

Enterprise modelling is an upcoming area to help develop enterprise systems. It helps in integrating different models in an enterprise such as manufacturing, logistics, finance, human resources, and so on. Physical systems modelling can be usefully extended for modelling the enterprise for the purposes of developing effective enterprise systems/enterprise resource planning (ERP).

Some important models that can be developed to analyze the data for decision-making in the prescribed framework would be:

- Manufacturing
- Procurement and Logistics
- Sales and distribution
- Finance
- Manpower planning

These models can be integrated using integrated PST-SD models that will provide the base for Decision Support Systems or Enterprise Support Systems.

25.4 ENHANCING PRACTICAL APPLICABILITY

Though modelling of physical systems is a powerful approach, there are serious limitations in applying the same in practical situations. Lack of a user friendly software platform for integrated modelling hampers the use of flexible systems methodology. Though software packages are available for individual modelling techniques such as system dynamics, optimization, etc. there is no integrated and interactive software platform that can facilitate the flexible modelling using integration of multiple modelling methodologies.

Further, traditionally most of the applications of modelling are done in an analytic mode, i.e. dividing the system under consideration into components and resynthesizing them to get a holistic view. For practical implementations, a synthetic paradigm of inside-out type is also required.

Moreover, the traditional applications of systems modelling followed a hard systems approach focussed on optimization of systems. The soft systems approach or learning orientation of modelling had been very limited.

In order to enhance the practical applicability of ideas discussed in this book, following three issues are highlighted here:

i. Developing integrated software
ii. Complementing modelling with synthetic paradigm
iii. Using in a gaming and learning environment.

Developing Integrated Software

One major limitation in applying flexible systems modelling is the non-availability of user friendly software packages to support the development of integrated models based on a combination of modelling methodologies. At present, integrated models are developed using some programming language, which is an expert mode and can be used for research purposes only. If the modelling approach is to be applied to practical problems in a wider sense, the software packages like DSS generators encompassing various modelling methodologies are to be developed.

Some important modelling methodologies that should be incorporated in the package, *inter alia* are physical system theory, system dynamics, simulation, mathematical programming, fuzzy set theory, neural nets, and so on.

The package should have an intelligent modelling interface with graphic modelling support.

Complementing Modelling with Synthetic Paradigm

The synthetic paradigm is an out of the box approach[8]. It tries to answer the questions 'why' rather than 'how', which is addressed by the analytic paradigm. In the synthetic paradigm, the system under consideration is not considered as the whole rather it is treated as a part of a larger whole. Thus, the cycle of analysis and synthesis is reversed, i.e. first the synthesis is done and then the analysis is done as shown in Figure 25.8.

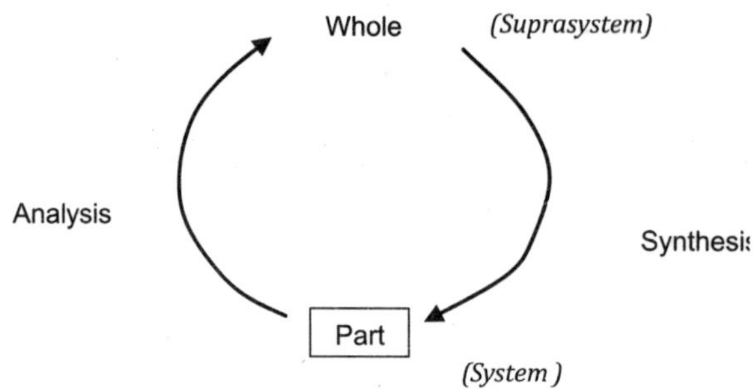

Figure 25.8: Synthetic Paradigm

Thus, the system under consideration is first fitted to a larger whole, i.e. suprasystem. The suprasystem is examined for its characteristics. Then the analysis is performed, i.e. examining the role that the concerned system is going to play in the suprasystem. This is an inside-out approach to look beyond the system boundaries and is more creative and imaginative.

The synthetic paradigm of inquiry should be used as a complement to the analytic modelling of the physical systems.

For example, if we want to develop the model of a multistage manufacturing system to analyze cost and flow (as is done in chapters three, four and six) we should not only divide it into component parts, but also examine the suprasystem in which this manufacturing system is working. It may be the integrated supply chain in which internal manufacturing is a part, linking the supplies and customers. Thus, we can examine the requirements of customers and the limitations of supplies that would give a better picture of the role of manufacturing system in context. In the light of these requirements, the cost and flow analysis should be carried out and interpreted. May be the system model can include some constraints or policy guidelines for studying the dynamics of the system based on the synthetic inquiry in the context of whole supply chain.

Using in a Gaming and Learning Environment

Traditionally, physical system models had been applied with hard systems thinking in a normative or optimization model. The role of models is dual; they can be used as a tool for decision-making on the one hand, and as an object of learning about the problem domain on the other. Decision-making skills of practicing managers can be enhanced if they develop models as a learning exercise. The behaviour of cost, flow, and quality with varying parameters and exogenous variables coupled with sensitivity analysis can generate valuable insight for the managers. They may convert the models in the form of a game, where the model is blackboxed and policy parameters can be decided by the managers. The model can give the feedback in the form of key performance variables such as cost and quality of output.

The gaming environment will be more effective if the models are dynamic. The integrated software platform suggested in the previous sections should have a gaming interface for designing interactive simulation games.

These games, based on flexible models, can be played in a multi-user environment for interactive learning in a micro-world setting. Thus, apart from using models as decision aids these can be used as effective learning aids.

25.5 EVOLVING A THEORY OF FLEXIBLE SYSTEMS

This book deals with a physical view of the management systems that deal with physical quantities, have a tangible structure and are governed by the principle of conservation. It has been observed that physical systems are both real and conceptual. Most of the managerial and socio-economic systems are conceptual physical systems.

During the journey of modelling conceptual physical systems, it has been observed that these systems have uncertainties, ambiguities, non-linearities, and dynamics of different orders. Moreover, many of the variables in these systems do not follow the principle of conservation strictly. Thereby, the need was felt to develop a flexible modelling approach to meet the changing requirements of these systems.

Apart from the above-mentioned complexities, these systems exhibit a chaotic behaviour and are faced with conflicts of different kinds.

Thus, it was observed that only a physical system theory rooted in the concept of a linear graph and rigid constraints may not be sufficient to gain deep insight and develop valid models of the dynamically changing world of management systems. A need for innovation and integration with other modelling methodologies was felt. Flexible systems methodology was applied to evolve flexible models of physical systems by suitably integrating various techniques and generalizing the framework of PST.

It was expressed in chapter two, that there is a movement towards a 'flexible system theory'. This journey from 'physical system theory' to 'flexible system theory' is long and evolutionary. Many new ideas have been developed, yet it's a long way to have a sound theory of flexible systems.

Developing a theory of flexible systems would require a definition of 'what' do we mean by 'flexible systems'; and what are their characteristics. It would also be important to answer 'why' do we need flexibility in systems. And finally, the most important question – 'how' these flexible systems should be designed, operated and evolved should be addressed.

We are at a primitive level in this journey towards a theory of flexible systems. Here, we would briefly attempt to address the first two questions, i.e. 'why' and 'what' of flexible systems. The third and most crucial question of 'how' can only be touched upon and not investigated in depth at this fag end of the book. This is to be attempted by ongoing research efforts as an open question to the system theorists and practitioners.

Why should there be Flexibility in Systems?

Systems are aggregation and assemblage of interacting and interdependent components so as to form an organized whole with a common purpose. These organized wholes of different kinds follow the principles of hierarchy and emergence, and communication and control. They are embedded hierarchically in a dynamically changing and turbulent environment.

Systems for their existence and sustainable growth have to evolve in harmony with the environment. They cannot remain rigid and static when the environment is fast changing and exhibiting complexity, chaos, and contradictions.

In order to cope with the challenge, the systems should be open, adaptive, responsive, and agile. Flexibility is the governing characteristic of living systems. The pressures of the environment force the systems to be more lively thereby demanding higher flexibility on different planes.

A review of trends in management has been given by Sushil[9] mapped on five different planes, i.e. spectral, integrative, interactive, innovative, and fuzziness. The review sums up the trend in management as: 'toward greater flexibility'.

The need for flexibility in systems can hardly be overemphasized in this fast changing world which is full of paradoxes. The manufacturing systems should be agile to meet the diverse customer requirements having flexibility in products, process, volume, tooling, etc. The organizational system should have flexibility at the level of structure, process, culture, and work practices. The strategic system should have flexibility to balance the present and future by multiplicity, emergence, and innovation. Similarly, the need for flexibility in other systems such as, information system, financial system, marketing system, and so on can be examined.

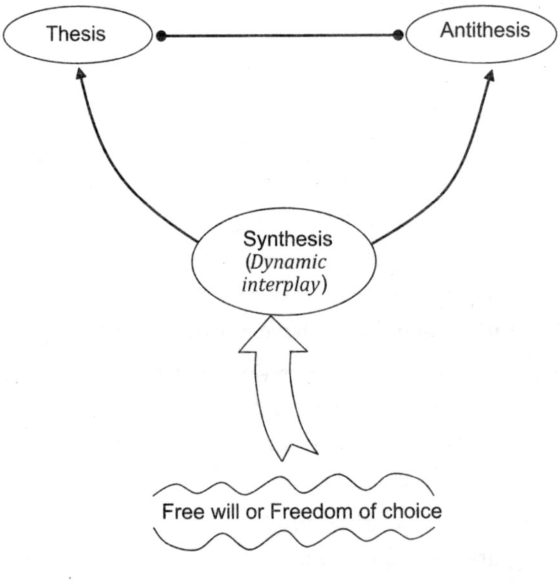

Figure 25.9: Concept of Systemic Flexibility

What is a Flexible System?

It is extremely difficult to define a flexible system in view of a variety of connotations and interpretations about the underlying concept of flexibility. Flexibility is a multi-dimensional concept such as quality and is interpreted as adaptiveness, responsiveness, openness, agility, multiplicity, resilience, versatility, customization, freedom, elasticity, mobility, and so on[10].

In order to develop a cohesive interpretation, the concept of systemic flexibility is defined using the concept of continuum as follows[11] (Figure 25.9).

"*Flexibility is the exercise of free will or freedom of choice on the continuum to synthesize the dynamic interplay of thesis and antithesis in an interactive and innovative manner, capturing the ambiguity in systems and expanding the continuum with minimum time and efforts*".

This means that systemic flexibility is defined by three key words, viz.:

- Options
- Change mechanisms
- Freedom of choice

Keeping this definition of flexibility in view, a very simple definition of a flexible system can be given as:

"A flexible system is system with multiple options, able to implement multiple options in different combinations including conflicting options, can change the options dynamically in view of changed situation and has freedom of choice to make that change happen actively rather being changed passively by the external forces, and above all, does all that with minimum time and efforts."

Thus, a flexible system is supposed to have following capabilities:

- Working with high speed
- Renewing itself
- Managing change
- Generating options
- Carrying opposites simultaneously
- Handling multiple things
- Managing chaos
- Managing uncertainty
- Dealing with ambiguity
- Managing complexity.

In summary, a flexible systems at any point of time is partially defined and continuously evolving to dynamically interplay the seemingly conflicting options.

How is a Flexible System to be Designed, Operated and Evolved?

This is the central question of the theory of flexible systems. Keeping in view the definition of systemic flexibility, we can only express at this stage that in the design, operation and evolution of flexible systems we have to keep in view various paradoxes, a few of them are shown in Figure 25.10.

Thesis	Antithesis
Quantitative	Qualitative
Hard System	Soft System
Analytic	Synthetic
Closed	Open
Static	Dynamic
Deterministic	Stochastic
Crisp	Fuzzy
Order	Chaos
Differentiation	Integration

Figure 25.10: Select Continua for a Flexible System

It can be seen that physical system theory, in the traditional framework, lies on the left-hand side of the continua shown in Figure 25.10. By generalizing its framework and integrating it with other modelling methodologies, a range of options have been attempted across the continua and many more possibilities are explored. However, this is only the beginning of the journey towards a flexible system theory and this is a challenge before the system theorists and practitioners to evolve a sound theory of flexible systems.

REFLECTIONS

It is evident from the developments in systems thinking and practice, and its application in management that there is a gradual evolution towards flexible systems all around. Flexibility is a much-talked about concept in practice, but a theory of flexible systems is still in its infancy. This book can at best be seen as a preparation to move on the path of flexible systems modelling which can be coupled with other works, in the area of flexible manufacturing systems to prepare a base for the development of a theory of flexible systems. It is advocated that the next stepping stone in this journey is `flexibility assessment' at the systemic level.

Notes

1. Forrester J.W. (1961) *Industrial Dynamics*, MIT Press, Cambridge; Sushil (1993) *System Dynamics: A Practical Approach for Managerial Problems,* Wiley Eastern Ltd., New Delhi.
2. Zadeh L.A.(1965) Fuzzy Sets, *Information and Control,* 8, 338-353.
3. Pankaj, Sushil and Seth K. (1994) A Fuzzy Set Theoretic Approach to Qualitative Analysis of Causal Loops in System Dynamics, *European Journal of Operational Research,* 78, 380-393.
4. Sushil (1994) Flexible Systems Methodology, *Systems Practice*, 7 (6).
5. Porter M.E. (1985) *Competitive Advantage: Creating and Sustaining Superior Performance,* The Free Press, New York.

6. Stevens G. (1989) Integrating the Supply Chain, *International Journal of Physical Distribution and Materials Management,* 19 (1), 3-8.
7. Chopra S. and Meindl P. (2001) *Supply Chain Management: Strategy, Planning and Operation,* Pearson Education Asia, Singapore.
8. Sushil (1999) *Flexibility in Management,* Global Institute of Flexible Systems Management, Vikas Publishing House, New Delhi; Ackoff R.L. (1981) *Creating the Corporate Future,* Wiley, New York.
9. Sushil (1999), *op. cit.*
10. Sushil (2000) Concept of Systemic Flexibility, *Global Journal of Flexible System Management,* 1 (1), 77-80.
11. *Ibid.*

INDEX

T

U

V

W